MATERIALS SCIENCE AND TECHNOLOGIES

COMPREHENSIVE GUIDE FOR MESOPOROUS MATERIALS

VOLUME 1

SYNTHESIS AND CHARACTERIZATION

MATERIALS SCIENCE AND TECHNOLOGIES

Additional books in this series can be found on Nova's website
under the Series tab.

Additional e-books in this series can be found on Nova's website
under the e-book tab.

MATERIALS SCIENCE AND TECHNOLOGIES

COMPREHENSIVE GUIDE FOR MESOPOROUS MATERIALS

VOLUME 1

SYNTHESIS AND CHARACTERIZATION

MAHMOOD ALIOFKHAZRAEI
EDITOR

nova publishers
New York

NOTICE TO THE READER

Library of Congress Cataloging-in-Publication Data

ISBN: 978-1-63463-958-3

Published by Nova Science Publishers, Inc. † New York

Contents

PREFACE

This is the first volume of the four volume set of Comprehensive Guide for Mesoporous Materials. The discovery and application of mesoporous materials in the different industries, including the catalyst industry, is one of the major advances over the past two decades. One of the important mesoporous materials is zeolite. Over the past decade, different types of zeolites have been produced in the nanometer range dimensions. Porosities can also be regular or irregular. Due to more research done on mesoporous materials, knowledge of this science has increased. Researchers are now able to make mesoporous materials with dimensions equal to the predetermined requirements to use in engineering, such as with molecular separation membranes.

This volume mainly discusses the synthesis and characterization of mesoporous materials. Some of the mesoporous materials discussed in this volume are alumina, carbon, silica and also polymeric materials. There are 15 chapters in this volume; each one includes examples of these interesting materials, supported by appropriate figures for better clarification.

Chapter 1 - The synthesis of mesoporous alumina has a very interesting area of application, particularly as catalyst or catalyst support. This chapter focuses on the individual synthesis routes to produce mesoporous aluminas with large surface area s and narrow pore size distributions. The utilization of various structure -directing agents (cationic, anionic and neutral) during the preparation of aluminas was studied. The syntheses carried out with aqueous ionic surfactants produced aluminas with poorer textural properties than the aluminas prepared with nonionic methods in alcoholic solutions. It was found that textural properties of the aluminas differ depending on the preparation method. Some catalytic applications of mesoporous aluminas were cited. It was found that the stages of precipitation, aging and elimination of the surfactant, required more detailed experimental information.

Chapter 2 - The process of cationic layered minerals intercalation with inorganic pillars leading to PILC and PCH materials was presented and discussed. PILCs are relatively well known group of clay-based materials, thus authors concentrated mainly on presentation of PCHs (Porous Clay Heterostructures), which are a relatively new group of porous materials that can be obtained from layered clay minerals by a surfactant directed method. PCHs are characterized by the combined micro-mesoporous structure, relatively high surface area, high thermal and hydrothermal stability, surface acidity and ion-exchange properties and therefore, these materials are very promising for application in various processes, including catalysis and adsorption . The synthesis of PCHs and their application in various catalytic and adsorption processes is presented and discussed.

Chapter 3 - This chapter describes the preparation and surface modification of mesoporous films and membranes based on silica colloidal spheres. Silica spheres form close-packed fcc lattice with ordered arrays of interconnected mesopores upon vertical deposition from colloidal solutions. Silica colloidal films can be prepared on various solid supports and suspended in glass or silicon support. Free-standing silica colloidal membranes can be obtained by sintering above 1000 °C. Silica colloidal membranes have a highly ordered geometry ; they are capable of size-selective transport and can be surface-modified in a well-defined and controlled manner with small molecules and polymers . The surface modification can be achieved using silanol chemistry, and polymer brushes with narrow mass distribution and controlled length can be formed on the mesopore surface using atom transfer radical polymerization and ring-opening polymerization . The transport in the resulting mesoporous membranes can be controlled by pH and ionic strength, temperature , light , and small molecule binding. Proton conducting silica colloidal membranes can be prepared by filling the mesopores with sulfonated polymer brushes.

Chapter 4 - The development of new materials , such as those based on mesoporous structures involved in green processes, is a current and hot issue with special relevance on the design of novel, more efficient and environmental-friendly catalysts. In fact, in the last decade, the number of reported novel mesoporous materials with application in catalysis has been notably increased.

In this sense, this chapter is intended as an update of catalytic processes of interest involving mesoporous materials such as functionalized mesoporous silicates and metallosilicates , metal-organic-frameworks and mesoporous carbons. This overview covers the most recent achievements in the application of mesoporous structures as catalysts and even as catalyst supports in acid -base reactions related to Fine Chemistry.

Chapter 5 - Owing to the unique physicochemical and electric properties the ordered mesoporous carbons have been a subject of intense research for many years. They have a well-developed surface area and great pore volume and have been found suitable for a number of applications including catalysis , energy storage and transformation , nanotechnology and as adsorbents in processes of air, water and industrial gas purification .

This chapter presents a survey of methods for the synthesis of mesoporous carbons with particular emphasis on the hard- and soft-template methods. The factors influencing the ordering as well as textural and structural parameters of ordered mesoporous carbons are identified and discussed such as the physicochemical properties of these materials , the so far applied methods of their modification and their potential use as adsorbents of various compounds from the liquid phase , including amino acid s , vitamins and dyes .

Chapter 6 - Nickel ion-loaded mesoporous silica (Ni-M41) was found to show good activity for one-step conversion of ethene or ethanol to propene by us. The conversion of ethene was 68% and the propene selectivity was 48% in a continuous gas-flow reaction at 673 K. The following reaction scheme was suggested based on the partial pressure dependences and the reactions of the intermediate compounds : dimerization of ethene to 1-butene, isomerization of 1-butene to 2-butene, and metathesis of 2-butene and ethene to propene. Various physicochemical techniques were applied to study the nickel species loaded on M41 and elucidated the formation of 2:1 nickel phyllosilicate-like species and the existence of three-coordinated Ni^{2+} ions on five-membered Si-O rings , which were suggested as active site s for the dimerization reaction of ethene. The conversion of ethanol on Ni-M41 was finally introduced and the reaction mechanism s were discussed. Two routes from ethanol to propene were suggested; the dehydration route and the dehydrogenation route. All of the results indicate that the formation of propene from ethene or ethanol could be achieved without using the shape selectivity well known in zeolite catalysis .

Chapter 7 - Over the last decades, nanoporous polymeric materials have been the subject of widespread interest and intense research as they can arguably be considered for nanotechnological applications , e.g., advanced nanofiltration membranes , interlayer dielectrics , high surface area catalytic supports, as well as size/shape -selective nanoreactors . Organic porous materials have unique properties that distinguish them from their inorganic nanostructured analogues, such as tunable mechanical properties , ease to be functionalized, and especially lower production cost . Designing nanoporous polymeric materials with controlled morphology and functionality thus constitutes an active research area. In this context, original approaches using suitable porogen templates have emerged: such templates are able to induce specific structural pores within the residual frameworks. These template-oriented routes are very interesting as a wide array of nanoporous polymers with a well-defined porosity can be derived from miscellaneous macromolecular architectures. Smart methodologies include molecular imprinting , selective removal of a porogen homogeneously blended within a thermostable matrix , selective removal of one partner from (semi-)interpenetrating polymer networks , selective destruction of one block in self-organized block copolymers , as well as removal of self-assembled molecules from supramolecular architectures. This chapter affords a critically selected appraisal of these synthetic strategies developed to create functional nanoporous polymers and investigates the scope and limitations associated with the utilization of such routes.

Chapter 8 - In the synthesis of mesoporous materials , there is a high interest to find simple, efficient and rapid processing methods. Over the last decade, UV irradiation has made a valuable contribution in this field especially for mesoporous silica film s . Initially implemented to eliminate the organic template by photodegradation , this technology has proven its utility for patterning, driving sol-gel polymerization and controlling self-assembly . Most photo-mediated processes rely on in situ photogeneration of Brönsted photoacids to promote hydrolysis and condensation of siloxane species . Many advantages can result from a light -induced polymerization as regard to efficiency (yields, reactivity), environmental impact (solvent-free, low energy consumption) and improved control (temporal and spatial control). This chapter provides for the first time a state -of-the-art and comprehensive review of domains in which light has enabled technical innovations in the synthesis of mesoporous inorganic films.

Chapter 9 - Mesoporous undoped and metal (Fe or Cu) doped carbon aerogels obtained using a sol-gel process, followed by low temperature supercritical drying and various pyrolysis conditions were synthesized. The resulting materials were characterized by means of elemental analysis, nitrogen adsorption -desorption , TEM , SEM, AFM and XRD methods. The presence of graphene-like structures was evidenced in the case of Fe-doped carbon aerogels . Due to their special morpho-structural properties: high surface area and pore volume, narrow mesopore size distribution , low density, etc., these materials were tested in different applications such as: phenol chemical wet air oxidation and phenol and pesticide adsorption from aqueous solutions .

Chapter 10 - Organic porous materials are a class of advanced materials , which have shown promising potential in many technological applications. Polyaniline (a conducting polymer) is a novel organic semiconducting material , which has attracted considerable attention because of its remarkable optical, electrochemical, and conducting properties . Nanostructured polyaniline combines the properties of low-dimensional organic conductors with high surface area materials. Polyaniline has been investigated in a number of catalytic and bio-technological applications because of the unique flexibility with respect to morphological controlled synthesis , tunable pore size , and ion-exchange properties. This chapter covers the state -of-the-art in the synthesis, chemical structure and properties, and electrocatalytic activity of polyaniline for sensing application. Single and simultaneous detection of clinically and pharmaceutically important molecules using polyaniline based electrodes is discussed in detail.

Chapter 11 - Self-assembly of colloidal particles at liquid/liquid interfaces leads to the formation of mesoporous shells: Pickering emulsions result from assembly at interface between immiscible hydrophilic/hydrophobic phases, while shells separating an aqueous core from an aqueous continuous phase give rise to 'colloidosomes'. In both cases, the unique properties of the mesoporous shells provide selective and/or controlled release with unique transport properties. This chapter reviews the characteristics and properties of colloidal assembly at the interfaces of Pickering emulsions and colloidosomes, and the factors controlling transport through the colloidal shells.

Chapter 12 - The development of materials with potential application in CO_2 capture is a topic of scientific interest. In this respect it was established that by adding nitrogen groups is achieved increasing adsorption of this gas. Chemical modification was performed in gas phase with ammonia and ammonium hydroxide solution at 30%, of various activated carbon monoliths obtained by activation of African palm stone with H_3PO_4, $ZnCl_2$ and $CaCl_2$ solutions. The original and modified monoliths were characterized by N_2 adsorption at 77K, infrared spectroscopy , Boehm titration and immersion calorimetry in benzene and water . Was evaluated the ability of CO_2 adsorption in a volumetric equipment at a temperature of 273 K and until 1 bar, was obtained values between 120-334 $mgCO_2g^{-1}$. The experimental results indicated that trough both methods of chemical modification used in this work, is achieved the increasing the content of nitrogen groups in the monolith and thus increases the CO_2 adsorption capacity.

Chapter 13 - X-ray scattering is a very important tool for the investigation of mesoporous materials . In this chapter, the authors gave an overview of their use for the study of periodic porous networks. First, the authors introduced the three types of textures: monoliths, powders and thin films . Then the authors reviewed all the known possible structures encountered for mesoporous materials . The basic notions of X-ray scattering are explained: diffraction , reciprocal space and space-group determination. They are detailed for the different types of structures: 2D or 3D networks, monolith, powders and thin films . More advanced analysis methods based on the measurement of the Bragg peaks intensities in relationship with the form factor contribution are introduced, allowing to a refined structural description of the porosity . Lastly, some examples of in-situ studies under controlled atmosphere or during material formation are given. These examples illustrate the wide range of experimental possibilities offered by X-ray scattering analysis.

Chapter 14 - The reaction of resorcinol and formaldehyde , catalyzed by base , produce a porous gel formed by chain-like aggregates of resin nanoparticles . Conventional drying caused aggregation of the constituent nanoparticles, with collapse of the pores, producing a resin without open pores (xerogel). The pyrolysis of the resin produces a compact carbon . The usual way to avoid pore collapse is to avoid water evaporation and instead drying using supercritical fluids (aerogels), lyophilization (cryogels) or using low surface tension solvents (ambigels). In all cases, the procedures are complex and expensive, including solvent exchange and/or long curing steps. In this chapter the authors describe the use of different species (micelles , polyelectrolytes , fibers or nanoparticles) to stabilize the resin nanoparticles against aggregation during drying. In that way, porous dry resin can be produced by air drying without complex experimental steps. Pyrolysis of the dry porous resins produces porous carbons which are suitable for applications in energy storage (supercapacitors), water purification (deionizers), energy conversion (fuel cell s) or catalysis . Moreover, the size and distribution of pores can be controlled by the concentration of soluble stabilizers . On the other hand, hierarchical porous carbons can be produced by hard templating the resin with silica nanoparticles or cellulose fibers. The materials are characterized and properties relevant to technological applications are measured.

Chapter 15 - Oxygen reduction reaction (ORR) is one of the most interesting research issues in the academia and industries due to its importance in proton exchange membrane fuel cell s (PEMFCs). Despite the advances in all kinds of catalysts, Pt-based materials are still the most efficient electrocatalysts for ORR. In particular, Pt-based nanomaterials with a porous or network-shaped structure possess intriguing physical and chemical properties to generate enormously promising potentials for ORR catalysis . This chapter attempts to summarize the recent advances in the design and synthesis of porous Pt-based nanostructured materials and their application as electrocatalysts for ORR in PEMFCs, including improved understanding of the reaction mechanism s and the development of synthetic strategies for producing catalysts with high activity and stability . Key scientific issues and prospective directions of research in this field are also discussed.

In: Comprehensive Guide for Mesoporous Materials. Volume 1 ISBN: 978-1-63463-958-3
Editor: Mahmood Aliofkhazraei

Chapter 1

SYNTHESIS OF MESOPOROUS ALUMINA FOR CATALYTIC APPLICATIONS

José Luis Contreras[1], Guadalupe Gómez[2], Beatriz Zeifert[3] and José Salmones[3]
[1]Universidad Autónoma Metropolitana-Azcapotzalco
México D.F., México
[2]Síntesis y Aplicaciones Industriales, Tultitlán, México
[3]Instituto Politécnico Nacional-ESIQIE, UPALM Zacatenco,
México, D.F., México

ABSTRACT

The synthesis of mesoporous alumina has a very interesting area of application, particularly as catalyst or catalyst support. This chapter focuses on the individual synthesis routes to produce mesoporous aluminas with large surface area s and narrow pore size distributions. The utilization of various structure -directing agents (cationic, anionic and neutral) during the preparation of aluminas was studied. The syntheses carried out with aqueous ionic surfactants produced aluminas with poorer textural properties than the aluminas prepared with nonionic methods in alcoholic solutions. It was found that textural properties of the aluminas differ depending on the preparation method. Some catalytic applications of mesoporous aluminas were cited. It was found that the stages of precipitation , aging and elimination of the surfactant , required more detailed experimental information.

1. INTRODUCTION

The scientific advance in the design and preparation of heterogeneous catalysts has benefited from the studies of precipitated mesoporous materials using the cationic, anionic or nonionic procedures either silicon oxide (SiO_2), alumina (Al_2O_3) or other oxides as titanium oxide (TiO_2) and zirconium oxide (ZrO_2) [1].

Table 1. Template and synthesis conditions of the micro and mesoporous materials

Example	Template	Synthesis conditions	Surface Area (m^2/g)	Pore volume (cm^3/g)	Pore diameter (nm)	Ref.
MCM-41	Ionic surfactant	basic	1128(550°C), 403(850°C)	0.95	ND	5
MCM-48	Ionic surfactant	basic	1433(550°C), 108(850°C)	1.14	ND	5
SBA-15	Ionic surfactant	acidic	632(550°C), 446(850°C)	0.63	ND	5
HMS	No ionic	basic	1021(550°C), 213(850°C)	0.81	ND	5
Alumina	Octylamine	basic	452(850°C)2. 52:14452:20: 21.6 12 51	0.81	5	10
Alumina	Dodecylamine	basic	418(450°C)	0.83	5	10
Alumina	Hexadecyalmine	basic	434(450°C)	0.64	3.2	10
Alumina	$C_{16}H_{33}NH_2$	basic	410(450°C)	0.60	3.0	10
Alumina	$C_{16}H_{33}NH(C_2H_4NH)_2H$*NH(*4NH*)2*	basic	462(450°C)	0.63	3.1	10
Alumina	$Me(CH_2)_{11}0S0_3Na$	basic	ND (600°C)	ND	ND	13

ND = Non determined.

The design of supports have implications in the mesoporous phenomenon, in the internal mass transfer by diffusion of reactants and in the catalytic reaction, but especially in the controlled adsorption -desorption of reactants and products, for which it is important to control the pore size distribution [2-5], the surface area of materials and achieving high thermal stability (Table 1).

It has been known that alumina used as support or catalyst is most frequently applied in the chemical industry , and any improvements on their properties or in its manufacturing process is desirable [6].

The synthesis of alumina is more complex than the synthesis of mesoporous silicates, and these materials have been documented success in controlling the pore size and the formation of different types of active site s on their surface, For example, the synthesis of MCM-41 mesoporous silicates and aluminosilicates [7].

The successful synthesis of ordered mesoporous aluminas, reviewed by Marquez et al. [8], Cejka et al. [9], Pinnavaia, [10] and Li et al. [11], have resulted in some significant improvements in the textural properties of such mesoporous material, particularly as a catalyst .

The conventional aluminas have surfaces between 50-300 m^2/g, and are prepared by precipitation [12]. These aluminas usually exhibit lower surfaces areas of less than 400 m^2/g and the main disadvantage were their large pore size distribution , sometimes they show even more than one maximum in the range of 3-15 nm. For that reason, the successful synthesis of mesoporous alumina , with areas larger than 500 m^2/g with a narrow pore size distribution is interesting [1].

These materials are characterized by having a high specific surface area and pore size distribution in the region of mesopores (2-10 nm) [13]. Some studies on the synthesis of mesoporous aluminas using surfactants that have been cited where surfaces and pore size distribution were improved. However, these early forms of mesoporous aluminas were amorphous [14].

In accordance with Márquez et al. [8], the first successes in the synthesis of mesoporous alumina were reported by Vaudry et al. [15] using long chain carboxylic acid s , and some attempts of the group of Stucky made lamellar systems only [16, 17].

In other procedure, the cationic route for preparing mesoporous alumina has been described by Cabrera et al. [18]. Using hexadecyl trimethyl ammonium bromide in combination with triethanolamine in water , the authors claim that by varying the ratio of surfactant , water and triethanolamine is possible to adjust the pore size between 3.3 and 6.0 nm.

Mesoporous alumina has been prepared by several routes : cationic, anionic, nonionic and the nanocasting methods, often they have surfaces between 300 to 700 m^2/g and pore size of 2 to 10 nm or more. This type of alumina has application in several catalytic reactions , for example in the synthesis of hydrodesulfurization , epoxidation of styrene , partial oxidation of methane , oxidative dehydrogenation of ethane and metathesis catalysts [9]. The developing of mesoporous aluminas was preceded by the development of the mesoporous silicas .

p-aminobenzoic acid

The first successful synthesis of mesoporous a/umina was published several years later than the successful synthesis of silica and it is probably that this was due to the synthesis of mesoporous alumina is more complex compared to the synthesis of mesoporous silica [9].

This chapter aims to contribute to the update on the preparation of mesoporous aluminas using the cationic, anionic and nonionic methods. Some applications of this type of aluminas are mentioned.

2. General Aspects of Porous Solids

Porous materials have shown significant physicochemical properties for use in separation processes and heterogeneous catalysis [19, 20]. Ideally a porous material should have a narrow pore diameter distribution which is important for the design of a catalyst . They should also show chemical stability , thermal or hydrothermal stability and mechanical strength, high area, large pore volume and have particle sizes and morphology appropriate.

Porous materials are defined in terms of pore diameter and adsorption properties. According to the IUPAC definition [21], porous solids are divided into three classes : microporous (<2 nm), mesoporous (2-50 nm) and macroporous (> 50 nm) [22].

Well-known examples of microporous solids are zeolites . These solids are crystalline aluminosilicates (natural or synthetic) with periodic three-dimensional structure , are microporous materials formed by regular arrays of pores and channels with uniform pore size [23, 24].

The porous oxides are crystalline solids , however, offer distinct advantages in regard to their use for various materials such as membranes , monoliths, and fibers . They can have a suitable molecular size to the process needs.

Mesoporous solids can be obtained, such as amorphous gels, aerogels and glasses . Unlike zeolites , these materials have a disordered pore system and therefore present a wide distribution of pore diameter. These materials are typically used in separation processes and as catalyst support [25-28].

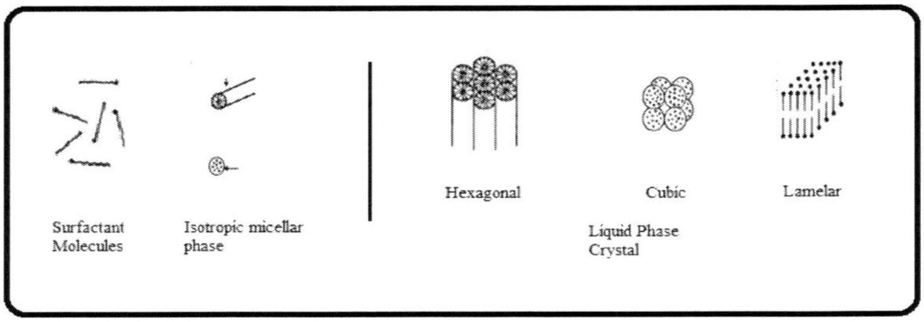

Figure 1. The role of the surfactant in aqueous solution to build the lamellar structure of mesoporous silica .

In the case of mesoporous solids based in silica (Figure 1), the role of surfactant in presence of water , the surfactant molecules are very active showing many structures in accordance with the increase of concentration. At low concentrations, the surfactant exists as mono-molecules and at high concentration is observed the formation of micelles . As the concentration increase, some arrangements named ensembles are organized in a close packing having a hexagonal shape (lamellar), but another geometric ensembles could be formed, for example cylinders which put in parallel arrangement produce a lamellar phase.

3. Alumina Using Anionic Surfactants

Vaudry et al. [29] conducted a synthesis of mesoporous alumina by anionic route, stearic acid and lauric acids used as surfactants , alcohol and diethyl ether as solvent. The aluminum source used was an aluminum alkoxide. The obtained mesoporous alumina showed specific surface s in the range of 500-700 m^2/g, with a pore size distribution around 2.0 nm (Table 2).

Sicard et al. [30] investigated the mechanism to remove the surfactant DSS from a mesoporous alumina using thermal analysis . The authors found that a strong interaction of the sulfate groups on the hydrophobic portion of the surfactant with the frame of the alumina during its formation was formed. This interaction probably caused sintering and collapse of alumina framework when the surfactant was removed. This explanation also has been reported by Valange [31].

In other studies Čejka et al. [32, 33] modified the procedure to synthesize the mesoporous alumina , using lauric acid and stearic acid as surfactants in a solution of 1-propanol. The samples prepared with lauric acid, calcined at 420°C showed a specific surface of 475 m^2/g. The samples prepared with stearic acid showed specific surfaces of 700 m^2/g. The pore diameter distribution showed a maximum of 3.5 nm. The same authors [32] showed that using stearic acid, the presence of micropores is not observed. These materials calcined at low temperatures do not exhibit crystalline structure by X-ray diffraction (XRD) and transmission electron microscopy (TEM) [33], however these aluminas were transformed into a crystalline phase when they were calcined at temperatures between 600°C and 1000°C.

Table 2. Textural Properties of mesoporous alumina prepared by some of the anionic surfactants (S-), Marquez et al. [8]

Al source	Solvent	Surfactant	H_2O/Al	S^-/Al	Temper. (°C)	pH	Ageing °C (H)	Decomp. (N_2) T(°C) Time(H)	Calcination(Air) T (°C) Time (H)	S_{BET} m^2/g	d_p nm	Ref.
Nitrate	water	Na-DDS[(a)]	640	0.2-1	20	7	---	450 3	450 2	443	3.4	31
Nitrate	water	Na-DDS[(a)]	850	2	25	7	25 24	---	500 4	27	--	41
Al_{13} Keggin	water	Na-laurate	640	0.2-1	20	5	-------	450 3	450 2	707	1	31
Al_{13} Keggin	water	Na-laurate	640	0.2-1	20	7	-------	450 3	450 2	336	1.5	31
Nitrate	water	Na-Palmi-tate [(b)]	640	0.2-1	20	6.5	-------	450 3	450 2	527	5.8	31
Al_{13} Keggin	water	Na-Palmi-tate [(b)]	640	0.2-1	20	6.5	-------	450 3	450 2	810	2.7	31
$Al(NO_3)_3$	water	SDS+urea	60	2:1	80	7	--------	--------	150 SCTA[(c)] 150 TCP[(d)]	580 20	150[(e)] 5[(e)]	30
$Al(NO_3)_3$	water	SDS+urea	60	2:1	80	7	--------	--------	150 SCTA 150 TCP	750 600	200[(e)] 170[(e)]	30
$Al(NO_3)_3$	water	SDS+urea	60	2:1	80	7	--------	--------	150 SCTA 150 TCP	600 200	150[(e)] 55[(e)]	30
$Al(NO_3)_3$	water	SDS+urea	60	2:1	80	7	--------	--------	150 SCTA 150 TCP	380 10	80[(e)] 5[(e)]	30
Sec-butoxide	1-Pro-panol	Lauric Acid	3.2	0.3	25	--	110 48	--------	430 1.5	710	1.9	15
Sec-butoxide	1-Pro-panol	Stearic Acid	3	0.32	25	--	100 50	410 2	420 9	758	3.7	32 33
Sec-butoxide	1-Pro-panol	Stearic Acid	3.1	0.32	25	--	100 50	410 2	450 19	397	4.7	32 33
Sec-butoxide	Pentane	Lauric Acid	3	0.32	25	--	25 15	430 ---	430 2	645	2.1 3.6	12
Nitrate	water	$(NH_4)_{12}W_{12}O_{41}$	35	0.01	25	8	25 24	110 12	500 6	240	4.5	96 97

[(a)]Na-DDS, [(b)] CTAB as contemplated, [(c)]SCTA=Sample controlled thermal analysis , [(d)] TCP = Traditional calcination protocol. [(e)]Pore Volume (cm^3/g).

Table 3. Textural Properties of some mesoporous alumina templated by cationic (S+) alkyltrimethylammonium surfactants, Marquez et al. [8]

Al source	Solvent	Surfactant	Solvent/ Al	S^-/Al	Temp. (°C)	pH	Ageing °C (H)	Decomp. (N_2) T(°C) Time(H)	Calcination(Air) T (°C) Time (H)	S_{BET} m^2/g	d_p [(b)] nm	Ref.
Chloride	water	C_{14}TMA[(a)]	50	1	80-90	7	--------	450 ---	450 2	486	5.5	37
Sec-butoxide	1-Buta-nol	C_{12}TMA	2	0.5	25	---	100 24	-----	500 4	429	4.5	40
Sec-butoxide	2-Buta-nol	C_{16}TMA	2	0.5	25	---	100 24	-----	500 4	480	5.0	41
Sec-butoxide	None	C_{16}TMA	2	0.5	25	---	150 72	-----	500 4	385	3.7	40
Nitrate	water	$C_{19}H_{42}BrN$ (CTAB)	213	-	60	8	60 24	100 24	550 4	331	5.0	42 43

(a) Trimethylammonium TMA.
(b) BJH Pore size distribution .

Table 4. Textural Properties of mesoporous alumina templated by non-ionic (N°) diblock copolymers , Marquez et al. [8]

Al source	Solvent	Copolymer	Solv/Al	N°/Al	Temp (°C)	Solvent Extract.	Ageing °C (H)	Decomp. (N_2) T(°C) Time(H)	Calcination(Air) T (°C) Time (H)	S_{BET} m^2/g	d_p nm	Ref.
Na-Alumi-nate	water	Tergitol 15-S-9	---	0.4	25	No	100 48	--------	500 20	360	--	87
Sec-buto-xide	water	Brij 56	----	0.5	70	Ethanol	80 0.5	--------	400 1	684	6.4	92
Sec-buto-xide	2-Buta-nol	Triton-X-114	2	0.1	25	Ethanol	25 24	550 1	550 4	555	4.0	52
Sec-buto-xide	2-Buta-nol	Triton-X-114	2	0.1	25	Ethanol	95 24	550 1	550 4	530	3	56
Sec-buto-xide	1,4 Dio xane	Tergitol 15-S-15	2	0.1	25	Ethanol	95 24	550 1	550 4	510	4.5	52

Table 5. Textural Properties of mesoporous alumina , Maekawa et al. [86]

Al source	Solvent	Surfactant	Cyclohexane/$AlO(Bt)_3$/ Surfactant	Temp. (°C)	Ageing °C (H)	Conductivity at 25 ^{0}C (S cm^{-1})	Calcination(Air) T (°C) Time (H)	S_{BET} m^2/g	$d_p^{(b)}$ nm
$AlO(Bt)_3$	water	SPAN 80[(a)]	2.70:0.30:0.08	25	--------	--------	500 6	360	3.0
$AlO(Bt)_3$	water	SPAN 85	1.35:0.15:0.10	25	45 15	$2.4x10^{-4}$	500 6	315	4.2
$AlO(Bt)_3$	water	TRITON X-114	2.70:0.30:0.15	25	45 15	$2.0x10^{-4}$	500 6	315	7.0
$AlO(Bt)_3$	water	TRITON X-114	0.68:0.15:0.08	25	45 15	$1.6x10^{-4}$	500 6	400	9.0
$AlO(Bt)_3$	water	TRITON X-114	1.35:0.15:0.08	25	45 15	$0.7x10^{-4}$	550 6	390	12.5

(a) SPAN.

(b) BJH Pore size distribution .

Mesoporous γ-alumina with relatively high surface area of 400 m^2/g and pore size of 3.8 nm was successfully prepared [34] employing the sol-gel route at room temperature and atmospheric pressure in the presence of conventional surfactant and fatty alcohol. The mesoporosity was retained when changing the surfactant to fatty alcohol.

In other study, mesoporous pseudo boehmite with novel pore properties was prepared [35] via the direct precipitation method using $Al(NO_3)_3$, as the inorganic alumina precursor and different surfactants containing bis (2-ethylhexyl) sulfosuccinate sodium salt, cetyltrimethylammonium bromide (CTAB), and tert-octylphenoxypolyethoxyethanol (Triton X-100) as the structure directing agents.

The presence of different surfactant micelles played an important role in the morphology and growth of pseudo boehmite. In addition, the pore properties could be enhanced significantly by the presence of transition metal ions .

Ordered mesoporous alumina with high surface area , large pore volume, and narrow pore-size distribution was synthesized [36] from organic aluminum precursors and anionic templates.

4. Alumina Using Cationic Surfactants

In the first case, the use alkyl trimethyl ammonium bromide (CTAB), Acosta et al. [37] published the first report on the synthesis of aluminum hydroxide mesophases with this cationic surfactant in aqueous solution (Table 3). $AlCl_3$ was used as a precursor, and the precipitation was controlled by the addition of urea . An alumina with poor crystallinity was produced when the sample was heated to remove the surfactant, with N_2 flow at 450°C followed by calcination in air.

The same authors explored the surfactant , tetradecyl trimethyl ammonium . The samples prepared showed an increase in the surface area (430-485 m^2/g) and also an increase in the pore volume (0.29 to 0.66 cm^3/g) [37].

Cabrera et al. [38, 39] prepared mesoporous aluminas using an aqueous solution of cetyl trimethyl ammonium bromide as the surfactant , and they calcined at 500°C. They have found a porous structure similar to a wormhole. Their results suggested that the porosity was caused by the agglomeration of small crystals of alumina .

In regard to solvents employed, the synthesis of mesoporous aluminas in aqueous solution is not well known compared with the synthesis using other solvents. The alumina prepared using 1-butanol in a molar ratio of aluminum sec butoxide/water equal to 2 and calcined at 500°C showed an ordered mesoporous structure and textural properties similar to those of samples synthesized in the aqueous medium made by Kim et al. [40].

In order to study the effect of the structure of the solvent, Ray et al. [41] modified the synthesis procedure used by Kim et al. [40] and showed that the change of solvent 1-butanol for 2-butanol leads to an increased area in the samples calcined at 500°C, from 401 to 475 m^2/g and a pore volume of 0.46 to 0.67 cm^3/g.

In another study, Gómez et al. [42, 43] studied five mesoporous aluminas by varying the concentration of cetyl-trimethyl ammonium bromide (CTAB) surfactant from 0 to 0.1Molar. The samples were named (C1-0, C1-0.01, C1-0.05, C1-0.075 and C1-0.1M) where the second number corresponds to the molar concentration of surfactant in aqueous solution of

$Al_2(SO_4)_3$. A solution with $Al_2(SO_4)_3$ was mixed with another solution containing the CTAB surfactant using an H_2O/Al (mole /atm) ratio of 213 (see Table 3).

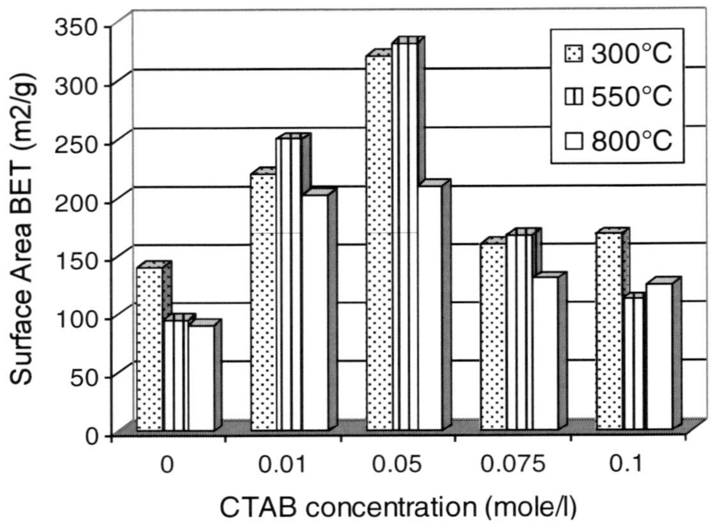

Figure 2. Effect of concentration of CTAB and calcination temperature over the surface area for the mesoporous alumina C1.

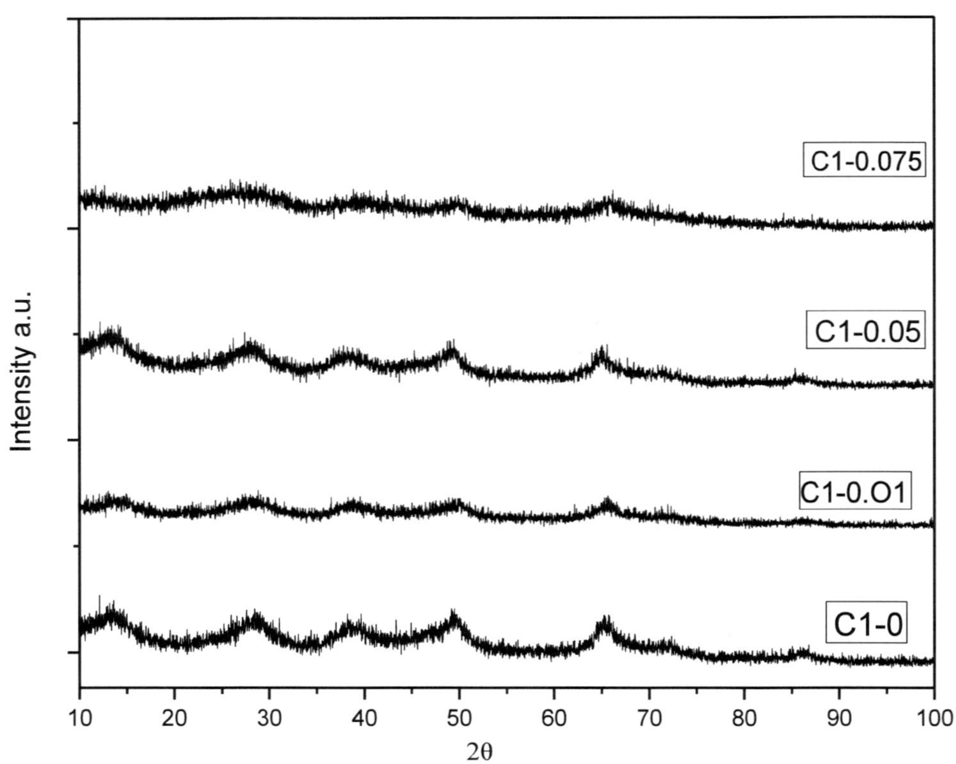

Figure 3. XRD of the samples calcined at 300°C (series C1).

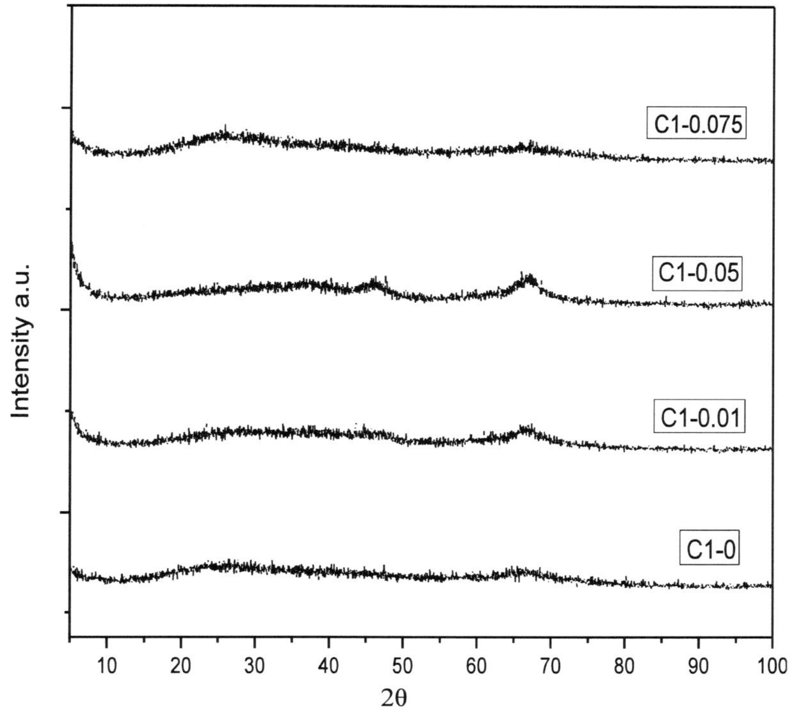

Figure 4. XRD of the samples calcined at 550°C (series C1).

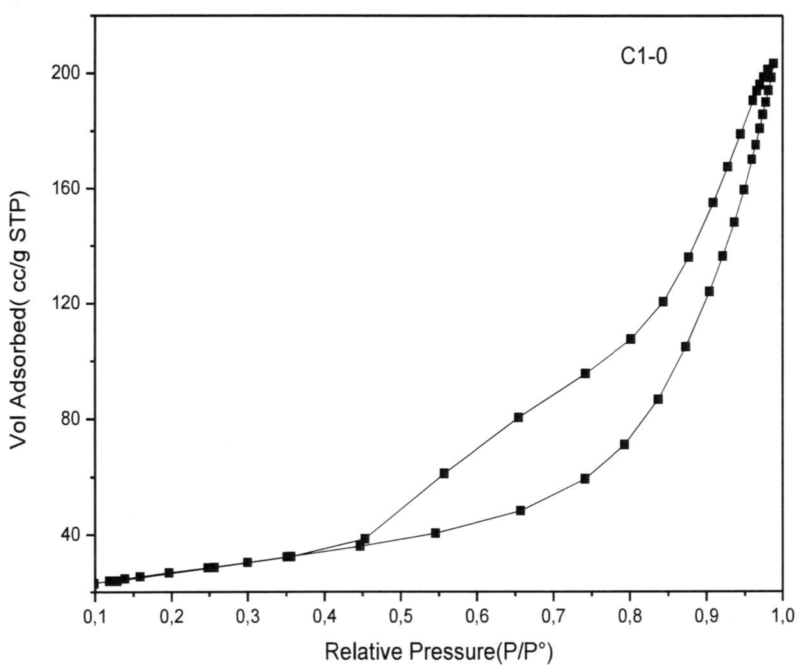

Figure 5(a) N2 isotherm for the sample C1-0.

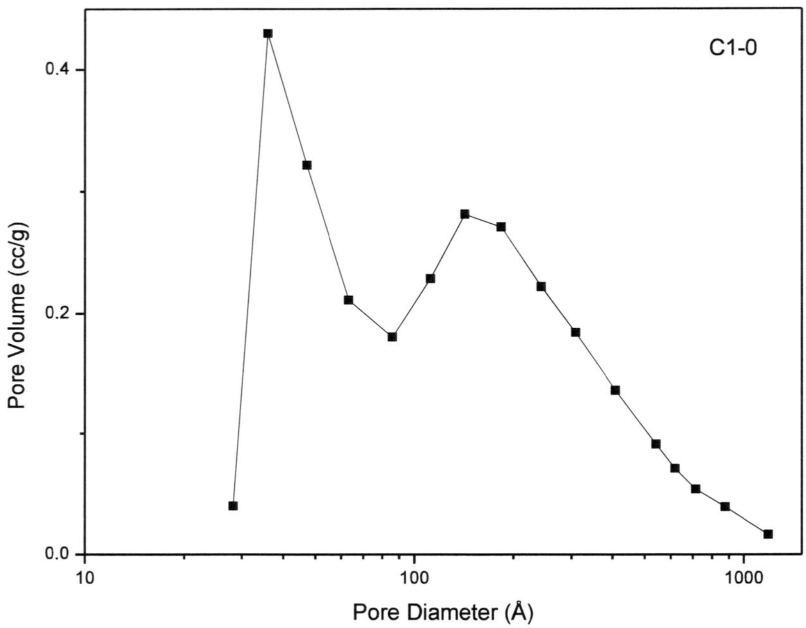

Figure 5(b). Pore distribution for the sample C1-0.

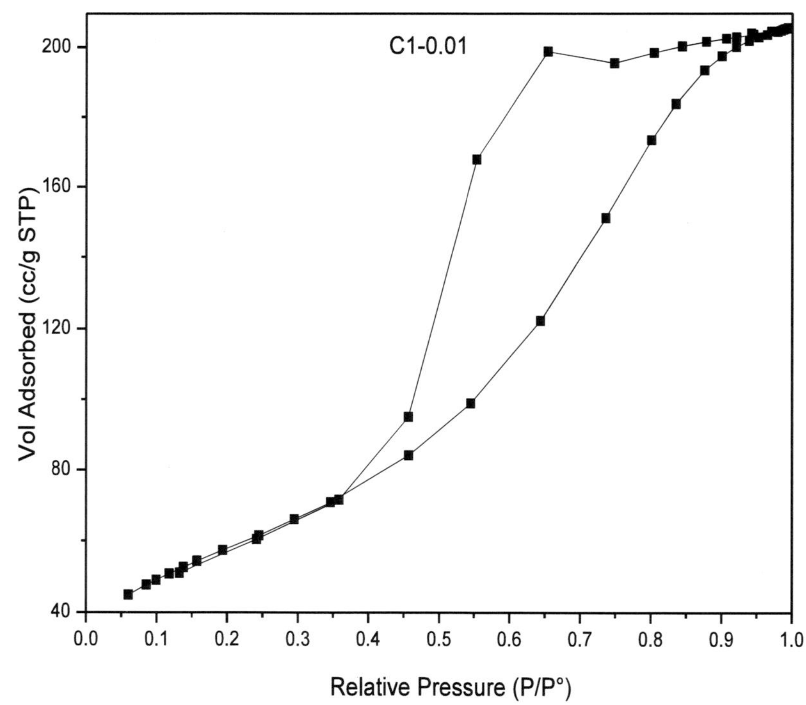

Figure 6(a). N_2 isotherm for the sample C1-0.01.

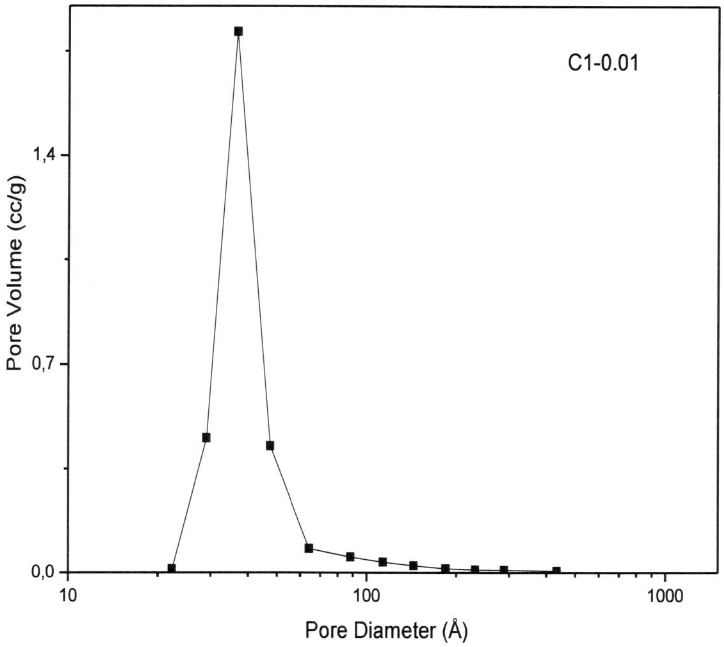

Figure 6(b). Pore distribution for the sample C1-0.01.

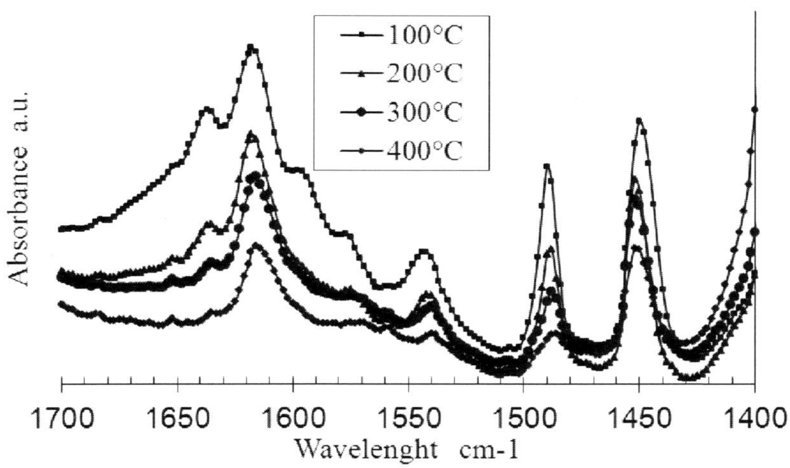

Figure 7. Infrared spectra of pyridine for the sample C1-0.075.

The surface area BET and pore volume increased as CTAB concentration increased passing through a maximum (Figure 2). After heating to 300°C boehmite was found by XRD (Figure 3) and after heating to 550°C, γ-Al_2O_3 was produced (Figure 4). The transition between γ to θ-Al_2O_3 was observed after calcination at 800°C, together with an increase of pore volume and pore diameter respect to the alumina without surfactant . A bimodal distribution of pore diameter of alumina (C1-0) without CTAB is shown in Figure 5 and a mono modal distribution for the sample C1-0.01, is observed in Figure 6. The combustion of CTAB was observed between 240 and 380°C. Brønsted and Lewis acidities were found on alumina samples having CTAB concentrations between 0.05 to 0.075M. The surface chemistry of mesoporous aluminas was affected by the presence of surfactant CTAB. At a concentration of CTAB of 0.075M, Lewis acid (1470cm^{-1}) and Brønsted sites (1485, 1500, 1540, 1610-1620 and 1640cm^{-1}) were observed by infrared spectroscopy of pyridine adsorbed (Figure 7). Conversely when the concentration of CTAB was 0.1 M (Figure 8) only Lewis acid sites were observed.

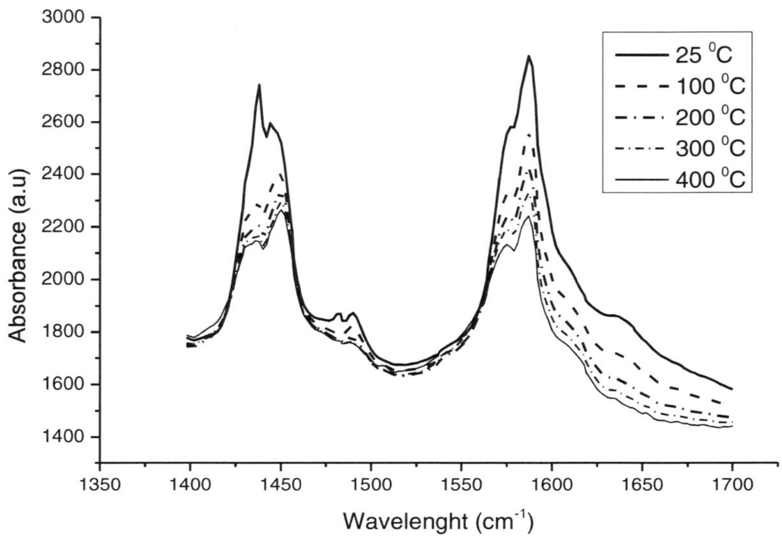

Figure 8. Infrared spectra of pyridine for the sample C1-0.1.

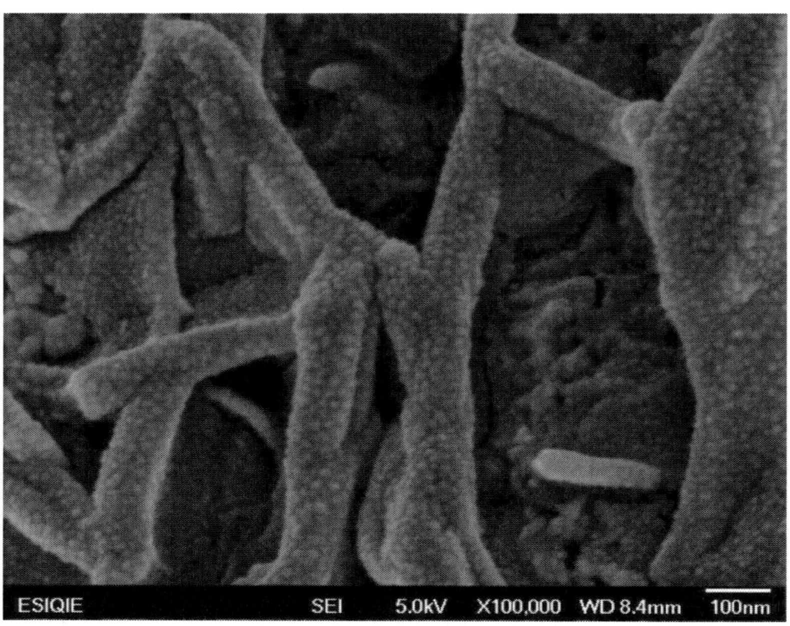

Figure 9. Mesoporous alumina prepared without CTAB (sample C1-0).

Using the scanning electron microscopy (SEM) the alumina prepared without CTAB (C1-0) showed a network of sheets and small circular fibers decorated with nanoparticles having less than 25 nm (Figure 9), the image suggests that each alumina sheet has small hemispherical protuberances on its surface.

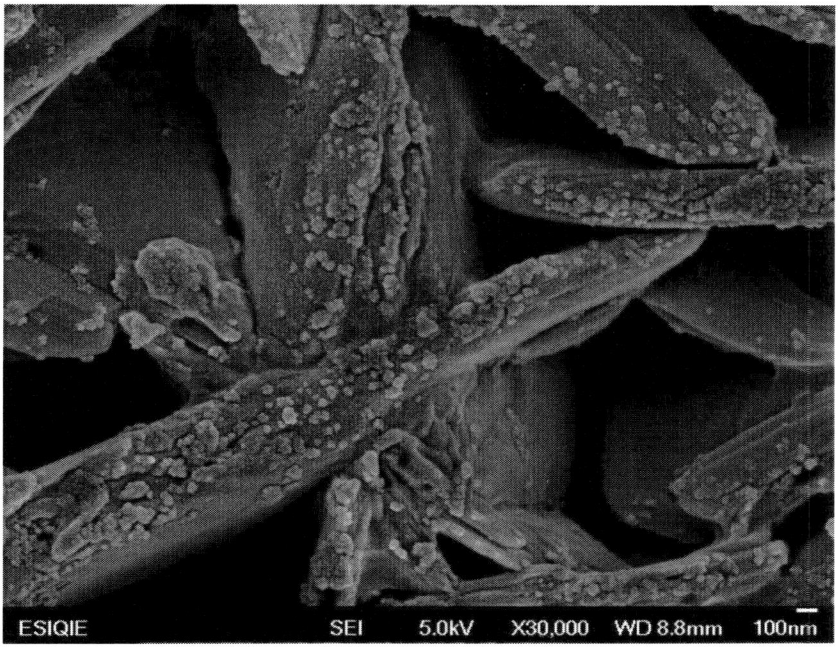

Figure 10. Image of mesoporous alumina prepared with 0.01M of CTAB (Sample C1-0.01).

By contrast, the sample prepared with 0.01 M CTAB showed a lamellar structure having only the small hemispherical nanoparticles on their edges (Figure 10). The image suggests that major changes occurred during the combustion of CTAB because there are not hemispherical nanoparticles on the sides of the alumina sheets.

In other study, the influence of the thermal treatment over the textural properties of a sol-gel mesoporous γ-Al_2O_3 synthesized in acid medium using cationic surfactants (e.g., hexadecyltrimethylammonium chloride/bromide) was investigated [44]. Among the different heating procedures tested, an isothermal step at 150°C under flowing N_2, led towards the most remarkable results (mesoporous γ-Al_2O_3 with BET area ~690 m^2/g). Different alkyltrimethylammonium bromide surfactants (alkyl ~ C_{10}-C_{18}) were used for the preparation of these aluminas, also leading towards meaning fully better materials under this thermal treatment (570-700 m^2/g).

In other paper, ordered mesoporous alumina was synthesized [45] (inserter en table 3) by the homogeneous precipitation method from aqueous $Al(NO_3)_3$ solution using CTAB as template. Surface area, and pore diameter of the alumina were 180 m^2/g, 5.4 nm respectively.

Other authors using CTAB prepared a mesoporous alumina with high specific surface area [46] by a sol-gel method using pelagic clay as the raw material. The alumina synthesized with a (1-hexadecyl) trimethylammonium bromide : urea mixed template showed a surface area of 385 m^2/g and a mean pore size of 3.6 nm.

5. Alumina Using Non-Ionic Surfactants

Non-ionic surfactant was introduced by Bagshaw et al. [47]. The surfactants used were copolymers (Tergitol, Igepal, and Triton), consisting of a chain of polyethylene oxide (PEO) hydrophilic with "n" units of ethylene oxide , and a hydrophobic hydrocarbon chain (R) with a structure R-(EO)n (Table 4).

Hydrolysis of aluminum sec-butoxide was carried out with 2-butanol in solution having a molar ratio of 2:1 water / alkoxide. The surfactant was removed by calcination in air flow at 500°C for 4h. The surface area was between 420-535 m^2/g pore size of 2.4 to 4.7 nm. Other researchers Luo et al. [48] also conducted the same type of research; they used these amphiphilic triblock copolymers to synthesize mesoporous aluminas. They found a surface area of 360 m^2/g and a pore diameter of 10 nm. Spherical particles with a size from 100 to 900 nm were obtained, and the shape depended on the synthesis conditions.

Meanwhile, González -Peña et al. [49, 50] prepared mesoporous alumina using aluminum sec-butoxide and the surfactants Tergitol and Triton in a 2-butanol solvent. TEM analysis showed a crystal structure having sheets or plate-shaped very similar to those observed in the alumina prepared by the sol-gel process without surfactant . The surfactants made of poly(ethylene oxide) chains seems to favor a partial breaking of the H_2 bonds that hold the boehmite-like layers together and contribute to limit the crystal growth as well as to have a control on the pore size . This method consisted of using amines along with the non-ionic surfactant [49, 50]. It has been reported that long-chain alkyl amines increase the size of the micelles , thereby increasing the pore size of siliceous MCM-41 [51].

Also, González-Peña [49] showed that the addition of amines to the synthesis gel improved thermal stability of mesoporous aluminas. Surface areas over 500 m^2/g were reported for materials calcined at 550°C. The surface area reported for samples calcined at 600°C was as high as 415 m^2/g, for a sample obtained using Triton X-114 and hexylamine. The authors also observed that pore size and volume increased with the number of ethylene oxide (EO) units of the surfactant when 1,4-dioxane was used as solvent [52]. The reported pore diameters for samples synthesized at room temperature with Triton X-114 (8 EO units), Tergitol 15-S-9 (9 EO units) and Tergitol 15-S-15 (15 EO units) were 3.5, 4.8 and 5.2 nm, respectively. In contrast, no correlation was observed when the solvent used was 2-butanol. This fact was attributed to the inability of the surfactants to form micelles in 2-butanol. It was reported that the mesoporous aluminas synthesized using both solvents exhibited an intricate stacking of amorphous aluminum oxyhydroxide platelets , from which the porous structure originates and that most of the aluminum was octahedrally coordinated [49, 52].

Indeed, it has been reported that poly(ethylene oxide) alkyl phenyl surfactants of Triton family do not aggregate into micelles in ethanol or methanol , whereas micelles are formed in solvents like, among others, ethylene glycol , 1,4-butanediol, ethylenediamine [53] or dioxane [54]. The authors concluded that in most of the syntheses reported in the literature (using non-ionic surfactants), the textural properties of the alumina were not determined by micellar chemistry. On the contrary, the surfactant would be located in the interlayer space, the oxygen atoms of the PEO hydrophilic chain interacting with the OH groups of the inorganic layers by H-bond. Several of these boehmite-surfactant *sandwiches* could be packed, which would partially collapse upon calcination.

To summarize, these results show that the boehmite-surfactant interaction is not strong enough to overcome the basic anisotropy of the Al-(O, OH)-Al network, and hence the basic layered morphology of boehmite is retained. Therefore, the control of hydrolysis and condensation reactions was also attempted by means of chemical modification of the alkoxide precursor with chelating ligands [52, 56].

It was found that the development of structural porosity was promoted, though at the expense of lower surface area s and smaller pore volumes. Simultaneously, a large increase in the concentration of tetra- and pentacoordinate aluminum was observed. N_2 sorption studies of samples calcined at 550°C showed that the ethylene glycol -modified precursor led to alumina samples containing slit mesopores while modification with ethyl acetoacetate and triethanolamine produced microporous aluminas with irregularly shaped pore channels. Calcination at 600°C led to the development of mesoporosity, but the pore structure was retained. TEM images combined with stochastic reconstruction methods [57- 59] showed that these microporous aluminas had a disordered but isotropic porous structure. The authors concluded that triethanolamine modification led to alumina samples with a network of cylindrical pores with very high connectivity , whereas modification with ethyl acetoacetate led to a material with lower accessibility of the pores due to their higher tortuosity.

Yang et al. [60] synthesized mesostructured aluminas and other metal oxides in ethanol solution with large mesopores using the triblock copolymer Pluronic P123, as structure directing agent. The alumina sample was prepared from $AlCl_3$ dissolved in a 10 wt% P123 ethanolic solution, and gelation was carried out at 40°C for several days. The recovered solids were calcined at 400°C to remove the surfactant . A notably large pore size (14 nm) and a relatively low BET surface area (300 m^2/g) were reported for this material (Table 4). The authors did not report any evidences of an ordered mesostructure.

In other work, monoolein-water bicontinuous cubic phase as a template for the synthesis of mesoporous alumina using $Al(NO_3)_3$ as precursor was used by Cruise et al. [61]. The cubic phase region was identified in a ternary phase diagram water-aluminum nitrate-monoolein by ^1H NMR . TEM and electron diffraction analysis of a sample calcined at 500°C showed the presence of γ-Al_2O_3 nanoparticles within an amorphous matrix .

In other work, Caragheorgheopol et al. [62] prepared mesoporous aluminas using the copolymer Pluronic (P123) and Tergitol (15-S-12) as templates in solution of acetonitrile and 2-butanol. The results showed that the samples washed with different solvents such as water (acetonitrile, ethanol and 2-butanol) at room temperature , changed the organization of surfactant aggregates, thereby changing the porous structure of alumina . The surface area s were in the range of 270-450 m^2/g, pore size 4.3 to 8.9 nm. The samples were dried at 120°C for 24h, before the washing treatment .

The first report of the synthesis of mesoporous alumina with ordered pore arrangement was developed by Somorjai and coworkers [63]. A sample having hexagonal mesoporous channels was prepared with the Pluronic P123 surfactant and a low proportion of surfactant /Al of 0.017. They showed by TEM images a mesoporous alumina with an organized order, very similar to the structures known as MCM-41.

In another paper, mesoporous alumina fibers were successfully synthesized [64] via an electrospinning technique combined with a sol-gel method in the presence of a triblock copolymer as a structure directing agent. The applications are in adsorption as well as related areas, such as catalysts and catalyst supports.

Ordered mesoporous alumina -doped titania thin films with anatase crystalline structure were prepared by using triblock copolymer Pluronic P123 as structure directing agent [65]. Uniform Al doping was realized by using aluminum isopropoxide as a dopant source which can be hydrolyzed together with titanium tetra isopropoxide.

Mesoporous molecular silicas and aluminas based materials were an excellent model supports for modification with catalytically active species for various organic reactions [66]. They offer well-ordered structures, although of amorphous type, high BET areas, large mesopores , and particularly the flexibility in the pore size adjustment . For an application in metathesis, two principal ways of incorporation of catalytically active site s were investigated. The first one represents the modification with inorganic oxides (W, Mo, Re) while the second one is based on immobilization of organ ometallic complexes (Mo, Ru, W).

In other case, employing P123 as a structure -directing agent and aluminum isopropoxide as aluminum source, Gd^{3+} or Eu^{3+} doped mesoporous alumina and co-doped mesoporous alumina were fabricated by a sol-gel method [67].

In other paper, alumina powder using P123/isopropoxide aluminum/HNO_3 reaction system, hierarchically macro/mesoporous alumina monoliths were fabricated [68] by introducing PEG8000 as the phase-segregation initiator to generate macropores . The mesopores size can be regulated by adjusting the PEG8000 content, while its shape is governed by P123 content.

Mesoporous aluminas with tailored adsorption , framework, and surface basic-acidic properties were prepared [69] using different structure -directing agents such as Brij78, Pluronic F108, poly(ethylene oxide)-poly(butylene oxide)-poly(ethylene oxide) B50-6600 and Vorasurf 504 block copolymers , which so far were not explored for the synthesis of alumina .

Employing phthalic acid (PA) as an interfacial protector and P123 as a structure - directing agent, ordered mesoporous γ-alumina with 2D hexagonal symmetry was success synthesized through the sol-gel method [70]. The best molar ratio of PA/Al^{3+} to synthesize the ordered alumina was 0.25. The mesoporous alumina with low apparent density was prepared using aluminum isopropoxide as raw material by surfactant template method [71]. Alumina was obtained under the conditions of mass fraction of polyethylene glycol 3%, a molar ratio of aluminum isopropoxide to H_2O 7:1. The product was γ-alumina with a specific surface of 498 m^2/g.

A general and easy method of synthesizing aluminas possessing hierarchical pore structure was presented [72]. This method is based on the integration of the sol-gel process with micelles of poly (ethylene oxide)-poly (propylene oxide)-poly(ethylene oxide) block copolymer and oil droplets of decahydronaftalene as pore templates.

A number of mesoporous nickel -alumina composite catalysts were prepared by a single-step surfactant -templating method using cationic, anionic, and non-ionic surfactant [73]. The effect of preparation method and surfactant identity on physicochemical properties , and catalytic activities of mesoporous nickel alumina catalysts in the steam reforming of liquid natural gas was investigated.

Mesoporous boehmite (γ-AlOOH) and gamma-alumina (γ-Al_2O_3) self supported films were prepared by a high-temperature transformation of γ-AlOOH precursor fibers [74]. A polymer , sodium polyacrylate, was used as a size/morphology controlling agent yielding to a stable colloidal suspension of boehmite ribbon-like fibers. The γ-Al_2O_3 film, obtained after calcination of the γ-AlOOH film, exhibited a high thermal stability up to 1000 °C and two distinct kinds of mesopores .

Starting from alumina with particle size of 0.3-0.4 μm, disk α-Al_2O_3 supports were prepared [75] via colloidal filtration with a mean pore size of 70 nm. The polished supports with mirror-like surface were used for mesoporous membrane preparation . A stable boehmite sol with a mean particle size of about 20 nm was successfully synthesized with aluminum sec-butoxide as a precursor through particle colloidal sol-gel route.

Using template polyethylene glycol (PEG) as structure directing agent and urea as the precipitant, alumina microfibers with mesoporous structures were synthesized, from a hydrothermal preparation and thermal-decomposition of precursor of ammonium aluminum carbonate hydroxide [76]. The results show that the microfibers with diameter of about 200-300 nm and length up to 8μm are prepared by using PEG as template. The mesoporous η-alumina microfibers with specific surface area 316 m^2/g and average pore diameter 2.5 nm are obtained after calcined at 900°C for 2 h, and the morphology is retained in the calcination process.

In other study, Cu catalysts supported on organized mesoporous alumina have been prepared [77]. Aluminum nitrate was used, PEG1540 as a template, mono ammonium phosphate as co-solvent and $Cu(NO_3)_2$. The results showed that Cu-Al_2O_3 with high surface area (>210m^2/g) and narrow pore size distribution (about 4nm). The pore structure was shown to be organized to some degree.

Using Al powder, mesoporous alumina samples with high surface area and comparatively narrow pore size distributions were synthesized for the first time using aluminum powder as an aluminum source and (PEG) as a structure directing agent via a sol-gel method in alkaline solution [78]. Alkali metal ion s used in the preparation play a significant role in determining the crystalline structure and morphology of the synthesized alumina.

The mesoporous γ-alumina samples were synthesized by the sol-gel method with P123 as the structure directing agent and different structural organic compounds as the interfacial protectors, such as phenylamine, o-phenylenediamine, p-aminobenzoic acid , benzoic acid and o-phthalic acid [79]. The functional group of interfacial protector had a great impact on the ordered structure and thermal stability of mesoporous γ-alumina.

Mesoporous γ-aluminas with large pore size (up to 19 nm) are prepared from dispersed pseudo-boehmite using pluronic P123 as a template [80]. It is found that these mesoporous alumina supported ReOx catalysts were more active and have far longer working life-span in gas-phase metathesis of 1-butene and 2-butene to propene than ReOx on conventional alumina with small pore size (5 nm).

In another investigation, organized mesoporous alumina has been synthesized by a precipitation method using $Al(NO_3)_3$ as source of aluminum, nonionic surfactant PEG1540 as template and $(NH_4)_2CO_3$ as precipitant [81]. The results showed that the organized mesoporous alumina had a high surface area (296 m^2/g) and a narrow pore size distribution (2.8-4.0 nm).

The synthesis of mesoporous polymer nanofibers by direct infiltration of microphase-separated block copolymers melts into self-ordered anodic aluminum oxide , and the subsequent removal of sacrificial domains from the nanofibers has been demonstrated [82]. This approach produces aluminum oxide hybrid membranes containing a mesoporous polymeric scaffold and thus, complements to the preparation of nanotubes by the self-assembly of amphiphilic block copolymers in solution .

In another work, ordered mesoporous alumina was synthesized by sol-gel method with aluminum isopropoxide as the source of alumina and PEO-PPO-PEO block copolymer F127 as the template [83]. The results showed a mesoporous alumina surface area (480 m^2/g) and worm-like pore was formed and ordered to some extent.

The one-pot synthesis of alumina -supported metal oxides (NiO) via self-assembly of a metal precursor and aluminum isopropoxide in the presence of triblock copolymer is described [84]. The resulting mesoporous mixed metal oxides possess hexagonal symmetry , well-developed mesoporosity, relatively high BET surface area , large pore widths, and crystalline pore walls. The generality of the recipe used for the synthesis of mesoporous nickel aluminum oxide was demonstrated by preparation of other alumina-supported metal oxides such as MgO, CaO, TiO_2, and Cr_2O_3.

Ordered mesoporous alumina has been synthesized [85] by a sol-gel method using aluminum isopropoxide as Al source, ethanol as solvent, and a mixed surfactant template (P123 and TritonX-100). The mesoporous alumina samples had surface area s of up to 500 m^2/g, pore volumes over 1.0 cm^3/g, and a narrow pore size distribution (2-12nm).

In accordance with Márquez et al. [8], a large number of synthesis of mesoporous aluminas assisted by nonionic surfactants have been reported using different aluminum sources, surfactants , solvents and heat treatments and due to the lack of detailed characterization makes it difficult to correlate the textural properties with the synthetic procedures used.

In other work, Maekawa et al.[86] prepared several mesoporous alumina synthesized by sol-gel method using nonionic copolymer as templates (Table 5). The pore size was controlled over the range of 2.8–12.5 nm by using different surfactant copolymers under different synthesis conditions. They found systematic dependence of conductivity upon pore size was observed, in which the conductivity increased with decreasing the pores size.

6. OTHER PREPARATIONS

Another synthesis of γ-Al_2O_3 nanofibers having partly crystalline walls were prepared by Zhu et al. [87]. They used non-ionic poly(ethylene oxide) surfactants to control the crystal growth of the boehmite. The colloids were obtained from sodium aluminate and surfactant mixture and then heated in autoclaves at 100°C. Fibrous γ-Al_2O_3 showed large pore volume (1.9 cm^3/g).

The same authors, [88] studied the thermal treatment and pH of aluminum hydrate of the surfactant cake, as well as the organic template. These parameters controlled the morphology and crystallinity of the boehmite nanoparticles .

Other authors have synthesized alumina with lathlike and scaffold-like framework structures [89, 90]. The synthesis method was based on the formation of a mesostructured surfactant -boehmite precursor that was transformed into mesostructured γ-Al_2O_3 by calcination at 550°C. These aluminas exhibited relatively high surface area s (over 300 m^2/g) and pore volumes up to 1.5 cm^3/g.

In another study, Hicks and Pinnavaia [91] also obtained mesoporous crystalline boehmite with a scaffold like structure by hydrolysis of aluminum sec-butoxide in ethanol - water solutions of amine surfactants .

In another study Ren et al. [92] reported the synthesis of hierarchical mesoporous-macroporous crystalline boehmite and γ-Al_2O_3, with a scaffold-like aggregation of nanofibers .

In another work, Lee et al.[93] obtained lathlike structures of γ-Al_2O_3 using cationic surfactants . Hydrolysis of aluminum sec-butoxide was carried out in the presence of alkyl trimethyl-ammonium surfactants in 1-butanol solution . The water /Al ratio had a major effect on the morphology . For water/Al ratios of 10 and 20, lathlike mesostructures were obtained while the ratio of 2 gave rise to the wormhole like type of pore structure .

Lee et al. [94] reported a strong effect on particle morphology when the hydrolysis of aluminum sec-butoxide in the presence of surfactants was carried out without addition of any organic solvent. Yao et al. [95] obtained mesoporous γ- Al_2O_3 with narrow pore size distribution without templates. The alumina sol was prepared at 25°C by ultrasonic treatment of an aqueous suspension of freshly precipitated $Al(OH)_3$, acidified with nitric acid . At lower calcination temperature s, part of the pseudo boehmite particles were retained, leading to the micro porosity and a low surface area .

Finally, the preparation of alumina using W ions (as ammonium tungstate) during the gel precipitation without surfactants has demonstrated high thermal stability . The Pt/alumina-WOx catalyst has demonstrated high resistance to sintering of Pt particles in the n-heptane hydroconversion [96, 97].

7. COMMERCIALIZATION

The alumina is used mainly as a raw material for the manufacture of metallic aluminum (90%) and is known as smelter grade alumina (SGA), the remaining 10% is devoted to non-metallurgical market, for example for industrial use. Some companies of market analysis [98] predict that the high purity alumina (99.99%) grow at a compound annual growth rate (CAGR) of 27.8% over the period 2013-2018.

Table 6. Applications of Mesoporous alumina reported in the patent literature

Application	Patent
Petroleum Hydrodesulfurization, Claus reaction, Dehydrogenation of butane to produce butenes and dehydration of alcohol to produce alkenes	[99]
Catalytic Cracking, hydrocracking and hydrodesulfurization of petroleum, steam reforming of natural gas and heavy naphthas to produce hydrogen, synthesis of ammonia and control automobile exhaust emissions.	[100]
Hydrogen storage, using an alumina nanotube. This storage method is more efficient, reliable, reproducible and economical than currently available H_2 storage methods.	[101]
Dehydration, Hydrotreating, Hydrogenation, Catalytic reforming, Steam reforming, Amination, Fischer Tropsch synthesis and Diels Alder synthesis.	[102]
Drying gases (Air, He, H_2, natural gas, etc.) drying liquids (hydrocarbons, petroleum l iquids, alcohols, etc.), Claus catalysts, Dehydration of alcohols, Isomerization, cracking and reforming of hydrocarbons, auto exhaust catalyst and chromatography.	[103]
Drug delivery vehicle. The alumina facilitates the in situ delivery of insoluble or poorly soluble drugs by using a vitamin E micelle as a host for the delivery of guest molecules.	[104]
Treatment of a fluid stream comprising a net hydrogen stream from a catalytic reforming process, where the hydrogen halide is HCl, or other halides (HBr, HF, and hydrogen iodide).	[105]
Fluid catalytic cracking, using rare earths for high thermal stability.	[106]

The global high purity alumina market can be categorized into four segments on the basis of application: LED applications, semiconductor applications, phosphor applications and other applications (in which the catalysts and adsorbents could be present). The key vendors are Alcoa Inc., Orbite Aluminae Inc., Sumitomo Chemicals Co. Ltd. and United Co. RUSAL plc.

There is not a market study of mesoporous alumina as such, but we can mention some of the applications reported in the patent literature [99-106] as catalysts or adsorbents (Table 6).

Conclusion

It was found that textural properties of the aluminas differ depending on the preparation method. It was identified mainly three preparation methods: anionic, cationic and non-ionic. The syntheses carried out with aqueous ionic surfactants produced aluminas with poorer textural properties than the aluminas prepared with nonionic methods in alcoholic solutions. It was found that the early stages of precipitation , aging, elimination of the surfactant and the thermal treatment s mainly calcination, more detailed experimental information is required to complete a thorough analysis in order to rationalize the scientific basis during the synthesis of mesoporous aluminas. In the open literature, many preparations of mesoporous alumina and physical characterization have been done while other aspects of the chemistry of surfaces, such as the determination of surface acidity have not been reported. The thermal stability of the aluminas in reaction conditions is still under development although it was clear that the increase in the calcination temperature produced lower areas and pore volumes and an increase in pore diameter.

References

[1] Ying, J.Y. Mehnert, C.P. Wong, M.S. *Angew. Chem. Int. Ed.*1999, 38, 56-77.

[2] Kresge, C.T. Leonowicz, M.E. W.J. Vartuli, J.C. Beck, J.S. *Nature* 1992,359, 710-712.

[3] Beck, J. S. Vartuli, J. C. Roth, W. J. Leonowicz, M. E. Kresge C. T. Schmitt K.D. Chu, C. T.W. Olson D. H. Sheppard, E. W. McCullen S. B. Higgins J. B Schlenker, J. L. *J. Am. Chem. Soc.* 1992, 114 10834-10843.

[4] Corma, A. *Chem. Rev.* 1997, 97, 2373-2419.

[5] Taguchi, A. Schüth, F. *Microporous Mesoporous Mater.* 2005, 77, 1-45.

[6] Misra, C., American Chemical Society Washington ACS Monograph 1986, 184.

[7] Schüth, F. Engineered Porous Catalytic Materials. *Ann Rev. Mater. Res.* 2005, 35, 209-238.

[8] Carlos Márquez-Alvarez, Naděžda Žilková, Joaquín Pérez-Pariente and Jiří Čejka, *Catal. Reviews Sci. and Engineering* 2008,50:2, 222-286.

[9] Čejka *J. Appl. Catal. A* 2003, 254,327-338.

[10] Pinnavaia, T.J. Zhang, Z.R. Hicks, R. W. In Nanoporous Materials IV, Stud. Surf. Sci. Catal. Sayari, A. Jaroniec, M., Eds. Elsevier Amsterdam, 2005, 156, 1-10.

[11] Li, C.R. Feng, Y.S. Yang, Q.H. The Research Progress of Mesoporous Alumina Progr. *Chem.* 2006, 18, 1482-1488 (in Chinese).

[12] F. Schüth, K. Unger, in: G. Ertl, H. Knözinger, J. Weitkamp (Eds.) *Preparation of Solid* Catalysts, Wiley-VCH, Weinheim 1999, 77-80.
[13] Agneta Caragheorgheopol, Adina Rogozea, Rodica Ganea *J. Phys. Chem. C.* 2010,114, 28-35.
[14] Yada, M. Machida, M, Kijima, T. *Chem. Commun.* 1996, 769-770.
[15] Vaudry, F. Khodabandeh, S. Davis, M.E. *Chem. Mater.* 1996, 8, 1451.
[16] Huo, Q.S. Margolese, D.I. Ciesla, U. Demuth, D.G. Feng, P.Y. Gier, T.E. Sieger, P. Firouzi, A, Chmelka, B.F.Schüth, F. Stucky, G.D. *Chem. Mater.*1994, 6, 1176-1191.
[17] Huo, Q.S., Margolese, D.I., Ciesla, U. Feng, P.Y. Gier, T.E. Sieger, P. Leon, R. Petroff,P.M, Schüth, F, Stucky, G.D. *Nature* 1994, 368, 317-321.
[18] Cabrera, S. El Haskouri, J. Alamo, J. Beltran, A. Beltran, D. Mendioroz, S. Marcos, M.D. Amorós, P. *Adv. Mater.* 1999,11, 379-381.
[19] Behrens P., *Adv. Mater.*1993, 5, 127.
[20] Barton T.J. Bull L.M., Klemperer W.G., Loy D.A., McEnaney B., Misono M., Monson P.A., Pez G., Scherer G.W., Vartuli J.C., Yaghi M. *Chem. Mater.* 1999, 11, 2633.
[21] Sing K.S.W., Everett D.H., Haul R.H.W., Moscou L., Pierotti R.A., Rouquerol J., Siemieniewska T., *Pure Appl. Chem.* 1985, 57, 603.
[22] Soler-Illia G., Sanchez C., Lebeau B., Patarin *J. Chem. Rev.*, 2002, 102, 4093.
[23] Meier W.M.& Olson D.H. Butterworths, London 1988
[24] Stucky G. D. MacDougall J. E. *Science* 1990, 247, 669
[25] Brinker C. J, Scherer G.W. Sol-Gel Science Academic Press New York, 1990
[26] Brinker C. J, Scherer G.W. *J. Non. Cryst. Solids* 1985, 70, 301.
[27] Hench L.L. and West J. K. *Chemical Reviews* 1990, 90, 33.
[28] Mackenzie J. D. J. *Non. Cryst. Solids* 1982, 1, 1.
[29] Vaudry, F.; Khodabandeh, S.; Davis, M.E. *Synthesis. Chem. Mater.* 1996, 8, 1451-1464.
[30] Sicard, L. Llewellyn, P.L. Patarin, J. Kolenda, F. *Microporous Mesoporous Mater.* 2001, 44, 195-201.
[31] Valange, S. Guth, J.L. Kolenda, F. Lacombe, S. Gabelica, Z. *Microporous Mesoporous Mater.* 2000, 35-36, 597-607.
[32] Čejka, J. Veselà, L. Rathouskỳ, J. Zukal, A. In Nanoporus Materials III Sayari, A., Jaroniec, M., Eds. Elsevier: Amsterdam Stud. *Surf. Sci. Catal.*, 2002, 141, 429-436.
[33] Čejka, J. Žilkovà, N. Rathouskỳ, J. Zukal, A. Jagiello, J. *Langmuir* 2004, 20, 7532-7539.
[34] Misran, H., Aminuddin,A.M., Zini, F.A.M., Ramesh, S. *ECS Transactions* 2010,28 (7) 57-65.
[35] Yang, F., Wang, Q., Yan, J., Fang, J., Zhao, J., Shen, W. *Ind. and Eng. Chem. Res.* 2012, 51 (47) 15386-15392.
[36] Li, R., Yuan, G.M., Chen, S.L., Hua, D.R., Zhou, Z., Sang, L., Chen, A.C. Ranliao Huaxue Xuebao/*Journal of Fuel Chemistry and Technology* 2012, 40 (7) 855-860.
[37] Acosta, S. Ayral, A. Guizard, C. Cot, L. *J. Sol-Gel. Sci. Technol.* 1996, 8, 195-199.
[38] Cabrera, S. El Haskouri, Alamo J., Beltran J., Beltran A., Medioroz D.S. Marcos, M.D. Amoròs, P. *Adv. Mater.* 1999, 11, 379-381.
[39] Cabrera, S. El Haskouri, Guillem J., Latorre C., Beltrán- Porter J., Beltrán A., Marcos D., Amoròs M.D.. *Solid State Sci.* 2000, 2, 405-420.

[40] Kim, H.J. Lee H.C. Choo, D.H. Lee, H.C. Chung, S.H. Lee, K.H. Lee, J.S. In Nanotechnology in Mesostructured Materials Park, S.E., Ryoo, R., Ahn, W.S., Lee C.W., Chang, J.S., Eds. Elsevier: Amsterdam Stud. *Surf. Sci. Catal.*, 2003; Vol. 146, 213-216.

[41] Ray, J.C. You, K.S. Ahn J.W. Ahn, *Microporous Mesoporous Mater.* 2007, 100, 183-190.

[42] Gómez G. P., J.L. Contreras, B. Zeifert, G.A. Fuentes, J. Navarrete, A. Vázquez, L. Nuño and J. Salmones, 18th Int. Symp. on Metast., *Amorph. and Nanostructurated Materials*, Gijón, Spain , Jun 26th -July 1st, 2011.

[43] Gómez G. P. J.L. Contreras, G.A. Fuentes, J. Navarrete, A. Vázquez,L. Nuño1, J. Salmones, B. Zeifert, Europacat, Glasgow, England August 29th to Sept 2nd (2011).

[44] Aguado, J., Escola,J.M., Castro, M.C. *Microporous and Mesoporous Materials* 128 (1-3) (2010) 48-55.

[45] Nie, R., Wang, J., Fei, J., Hou, Z., Zheng, X. CuihuaXuebao/*Chinese J. of Cataly.* (2011) 32 (2) 379-384.

[46] Zhang, P., Liu, J., Zhao, X., Wu, G. Russian J. of Physical Chemistry A 2010, 84 (13) 2351-2354.

[47] Bagshaw, S.A. Prouzet, E. Pinnavaia, T.J. *Science* 1995, 269, 1242-1244.

[48] Luo , Q. Li, L. Xue, Z, Zhao, D. In Nanoporous Materials II; Sayari, A., Jaroniec, M., Pinnavaia, T.J., Eds.; Elsevier: Amsterdam, *Stud. Surf. Sci. Catal.*, 2000; Vol. 129, 37-43.

[49] González-Peña, V. Dìaz, I. Màrquez-Alvarez, C. Sastre, E. Pérez-Pariente, *J. Therm. Microporous Mesoporous Mater.* 2001, 44, 203-210.

[50] González-Peña, V. Màrquez-Alvarez, C. Sastre, E. Pérez-Pariente J. Improved in Zeolites Mesoporous Materials at the Down of the 21st Century; Galameau, A., di Renso, F., Fajula, F., Vèdrine J., Eds.; Elsevier: Amsterdam Stud. *Surf. Sci. Catal.*, 2001, Vol. 135, 1072-1079.

[51] Sayari, A. Yang, Y. Kruk, M. Jaroniec, M. J. Phys. Chem. B. 1999, 103, 3651-3658.

[52] Gonzalez Peña,V., Márquez-Alvarez,C., Sastre E., Pérez-Pariente, *J. Stud. Surf. Sci. Catal.* 2002, V.142, 1283-1290

[53] Ray, A, Nature, 1971,231, 313-315; 49 and Ray, A, Nemethy, G., *J. Phys. Chem.*1971,75, 809-815.

[54] Becher, P. *J. Colloid Sci.* 1965,20,728-731.

[55] Becher, P., Trifiletti, S.E., *J. Colloid Interf. Sci.*1973,43,485-490.

[56] González Peña,V., Márquez-Alvarez,C., Diaz I., Grande M., Blasco, T. Pérez-Pariente *J. Microporous Mesoporous Mater.* 2005,80,173-182.

[57] Diaz I., González Peña,V., Márquez-Alvarez,C., Pérez-Pariente, *J. Collect. Czech. Chem. Commun.* 2003,68,1937-1948.

[58] Diaz I., González Peña,V., Márquez-Alvarez,C., Kikkinides,E,S. *Microporous Mesoporous Mater.* 2004,68,11-19.

[59] Diaz I., González Peña,V., Márquez-Alvarez,C., Pérez-Pariente,J., Kikkinides,E,S. *Stud. Surf. Sci. Catal.*, 2005,158,541-548.

[60] Yang, P.D., Zhao, D.Y., Margolese, D.I., Chmelka, B.F., Stucky, G.D. *Nature* 1998, 396, 152-155.

[61] Cruise, N. Jansson, K. Holmberg, K. *J. Colloid Interf. Sci.* 2001,241,527-529.

[62] Caragheorgheopol, A. Caldararu, H. Ionita, G. Savone F, žilkova, N. Zukal, A. Cejka, J. Solvent. *Langmuir* 2005, 21, 2591-2597.
[63] Niesz, K, Yang, P.D. Somorjai, G.A. *Chem. Commun.* 2005, 1986-1987.
[64] Wang, Y., Li, W., Jiao, X., Chen, D. *Journal of Materials Chemistry* A 2013,1 (36) 10720-10726.
[65] Jiang, X., Bastakoti, B.P., Weng, W., Higuchi, T., Oveisi, H. Suzuki, N., Chen, W.J., Yamauchi, Y. *Chemistry -A European Journal* 2013, 19 (33) 10958-10964.
[66] Balcar, H., Čejka, J. *Coordination Chemistry Reviews* 2013, 257 (21-22) 107-3124
[67] Yu, C.X., Liu, Y.L., Yang, Q., Sha, L., Wu, X. J. Chin. *J. of Inorg. Chemistry* 2012,28(2) 352-356.
[68] Zhang, K., Fu, Z., Nakayama, T., Suzuki,T., Suematsu, H., Niihara, K. Mater. *Research Bulletin* 2011, 46 (11) 2155-2162.
[69] Zhang Cai, W., Yu, J., Aroniec, M. *J. of Materials Chemistry* 2011, 21 (25) 9066-9072.
[70] Huang, F., Zheng, Y., Xiao, Y., Zheng, Y., Cai, G., Wei, K. *Materials Letters* 2011, 65 (2) 244-246
[71] Liu, D., Han, K., Ye, H., Tu, C. Huaxue FanyingGongcheng YuGongyi/ *Chemical Reaction Engineering and Technology* 2011, 27 (2) 160-165.
[72] Martins, L., Alves Rosa, M.A., Pulcinelli, S.H., Santilli, C.V. *Microporous and Mesoporous Materials* 2010,132 (1-2), pp. 268-275.
[73] Seo, J.G., Youn, M.H., Song, I.K. *Catalysis Surveys from Asia* 2010, 14 (1), 1-10.
[74] Mathieu, Y., Vidal, L., Valtchev, V., Lebeau, B. *New Journal of Chemistry* 2009, 33 (11) 2255-2260.
[75] Qi, H., Xing, W., Fan, Y. Huagong Xuebao/CIESC Journal 2009, 60 (10) 2628-2632.
[76] Zhu, Z.F., Sun , H.J., Liu, H., Yang, D., Zhang, J.Q., Guo, L.Y. Wuji Cailiao Xuebao/*Journal of Inorganic Materials* 2009, 24(5) 1003-1007.
[77] Wu, X., Zhao, R., Guo,F., Cui, W., Chen, J. Beijing Huagong Daxue Xuebao (Ziran Kexueban)/*J. Beijing Univer. of Chem. Technology* (Natural Science Edition) 2009,36 (1) 13- 17.
[78] Sun , Z.X., Zheng, T.T., Bo, Q.B., Vaughan, D., Warren, M. *J. of Materials Chemistry* 2008,18 (48) 5941-5947.
[79] Sun , Q., Zheng, Y., Li,Z., Zheng, Y., Xiao, Y., Cai, G. *Materials Letters* 2012, 84, 44-47
[80] Yuan, G., Zhou, Z., Li, R., Chen, A., Zheng, M., You, J. *J. Natural Gas Chemistry* 2012, 21 (3)352-359.
[81] Wen, J., Zhao, R., Guo,F., Chen, J. *J. Beijing Univer. of Chem. Tech.(Nat. Sci. Ed.* 2008,35 (6) 17-20.
[82] Wang, Y., Gösele, U., Steinhart, M. *Chemistry of Materials* 2008, 20 (2) 379-381.
[83] Li, Z.P., Zhao, R.H., Guo, F., Chen, J.F., Wang, G. Chem. *J. Chin. Universities* 2008, 29 (1) 13-17
[84] Morris, S.M., Fulvio,P.F., Jaroniec, M. *J. Amer. Chemical Society* 2008,130(45) 15210-15216.
[85] Li, Z., Zhao, R., Guo, F., Chen, J. J. Beijing Univer. *Chem. Techn.*(Nat. Sci. Ed.) 2008, 35 (3) 23- 26.
[86] Maekawa, H., Tanaka, R., Sato, T. Fujimaki, Y. Yamamura, T. *Solid State Ionics* 175 (2004) 281–285.
[87] Zhu, H.Y. Riches, J.D. Barry, J.C. *Chem. Mater.* 2002,14, 2086–2093.

[88] Zhu, H.Y. Gao, X.P. Song, D.Y. Ringer, S.P. Xi, Y.X. Frost, R.L. *Microporous Mesoporous Mater*. 2005, 85, 226–233.

[89] Zhang, Z. Hicks, R.W. Pauly, T.R. Pinnavaia, T.J. *J. Am. Chem. Soc*. 2002, 124, 1592–1593.

[90] Zhang, Z.; Pinnavaia, T.J. *J. Am. Chem. Soc*. 2002, 124, 12294–12301.

[91] Hicks, R.W. Pinnavaia, T.J. *Chem. Mater*. 2003, 15, 78–82.

[92] Ren,T.Z. Yuan, Z.Y. Su, B.L. *Langmuir* 2004, 20, 1531–1534.

[93] Lee, H.C. Kim, H.J. Rhee, C.H. Lee, K.H. Lee, J.S. Chung, S.H. *Microp. Mesop. Mater*. 2005, 79, 61–68.

[94] Lee, H.C. Kim, H.J. Chung, S.H. Lee, K.H. Lee, H.C. Lee, J.S. *J. Am. Chem. Soc*. 2003, 125, 2882–2883.

[95] Yao, N. Xiong, G.X. Zhang, Y.H. He, M.Y. Yang, W.S. *Catal. Today* 2001, 68, 97–109.

[96] J.L. Contreras, G.A. Fuentes, L.A. García, J. Salmones, B. Zeifert *J. Alloy Comp*. 483 (2009)450-452.

[97] J.L. Contreras, G.A. Fuentes, B. Zeifert, L.A. García and J. Salmones *J. Alloy Comp*.483(2009)371-373.

[98] http://www.marketwatch.com/story/global-haigh-purity -alumina -market-2014-2018.

[99] Davis, M., Vaudry, F.J.P., *US Patent* No. 5 863 515, Jen. (1999).

[100] Pinnavaia T., Zhang, Z., Hicks R. *US Patent* No.7,090,824 B2, Aug. (2006).

[101] Lee, J.S., Kyung, H.L., Lee H.C., Choo, D.H., *US P*atent No.7,399, 459 B2, Jul. (2008).

[102] Shan,Z., Jansen, J.C., Yeh, C.Y., Angevine, P.J., Maschmeyer, T. *US Patent* No.US 2007/0170096 A1, Jul. (2007).

[103] Lu G.Q., Zhu, H.Y., *Inter. Patent*: WO 01/74713 A1, Oct (2001).

[104] Ma, Y., Balkus, K.J., *US Patent*: 7,579,014 B2, Aug. (2009).

[105] Kanazirev, V.I., *US Patent*: 7,790,130 B2, Sept. (2010).

[106] Wang, K., Lemon, R.C., *US Patent*: 8,524,630 B2, Sept. (2013).

In: Comprehensive Guide for Mesoporous Materials. Volume 1 ISBN: 978-1-63463-958-3
Editor: Mahmood Aliofkhazraei

Chapter 2

Porous Clay Heterostructures (PCH) – Synthesis and Application in Catalysis and Adsorption

Lucjan Chmielarz and Andrzej Kowalczyk
Jagiellonian University, Faculty of Chemistry,
Kraków, Poland

Abstract

The process of cationic layered minerals intercalation with inorganic pillars leading to PILC and PCH materials was presented and discussed. PILCs are relatively well known group of clay-based materials, thus authors concentrated mainly on presentation of PCHs (Porous Clay Heterostructures), which are a relatively new group of porous materials that can be obtained from layered clay minerals by a surfactant directed method. PCHs are characterized by the combined micro-mesoporous structure , relatively high surface area , high thermal and hydrothermal stability , surface acidity and ion-exchange properties and therefore, these materials are very promising for application in various processes, including catalysis and adsorption . The synthesis of PCHs and their application in various catalytic and adsorption processes is presented and discussed.

1. Clay Minerals

Cationic layered clay minerals are still a very interesting group of materials for possible application in various areas, including catalysis and adsorption . It is related to their unique structures, which can be relatively easy modified in various ways to adjust properties of clay minerals to specific needs. There is a great variety of clay minerals with different layered structures and containing various chemical elements. An example could be montmorillonite , which is probably the most intensively studied clay mineral. The structure of montmorillonite is presented in figure 1. It can be seen that each monmorillonite layer consists of three sublayers. Central octahedral sublayer is sandwiched between two tetrahedral sublayers. Such structure is called 2:1 (two tetrahedral sheets and one octahedral sheet) or T-O-T (tetrahedral-octahedral- tetrahedral sheets). Si^{4+} are tetrahedral cations, while Al^{3+} are the main octahedral cations in montmorillonite sublayers. It must be mentioned that all tetrahedral positions are occupied by cations, while only two-thirds octahedral positions are occupied by Al^{3+} cations with vacancies in one-third octahedral positions. Clay minerals with such layered structure are classified as dioctahedral (two octahedral positions occupied and one vacancy). In montmorillonite part of trivalent Al^{3+} cations in octahedral sheets are replaced by divalent Mg^{2+} ions . Such substitution results in a negative charging of montmorillonite layers. The negative charge is compensated by common cations (e.g., Na^+ , K^+ , Mg^{2+}), which together with water molecules are located in the interlayer space of clay mineral. These interlayer cations can be relatively easy exchange for other cations, thus montmorillonite exhibits ion-exchange properties, which gives a possibility of its various modifications .

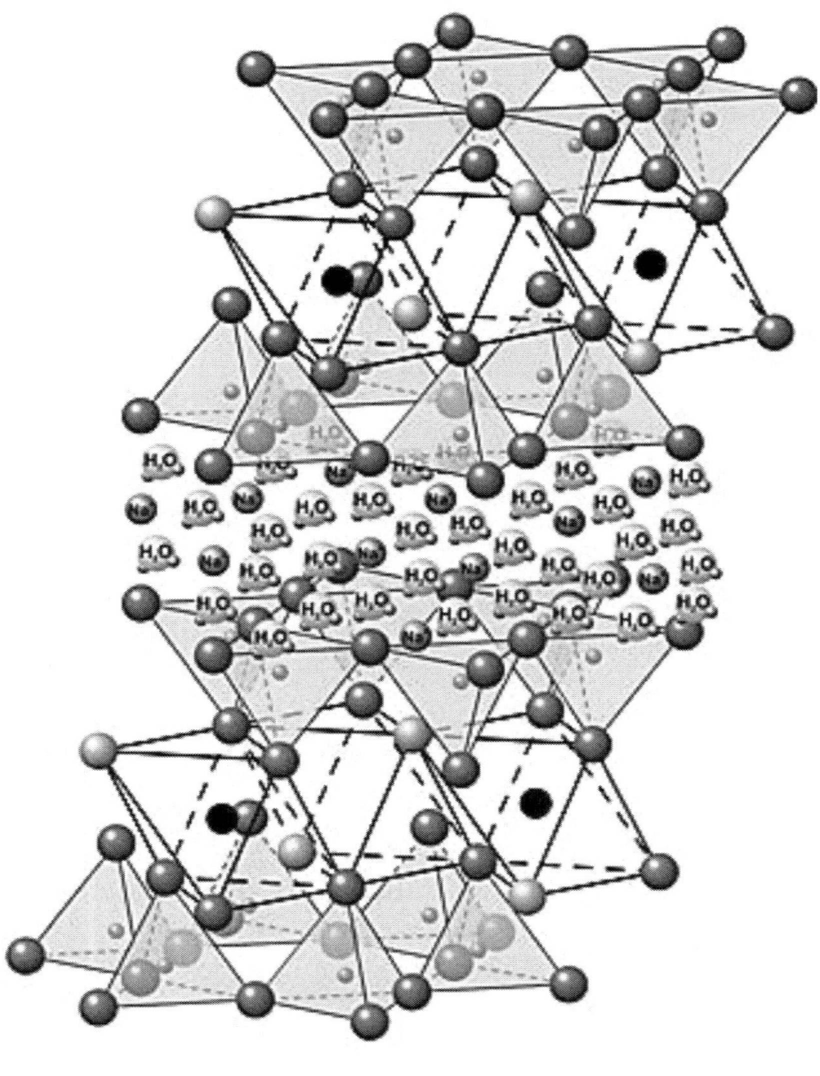

Figure 1. Model structure of montmorillonite .

Detailed description and discussion of the structure and properties of layered clay minerals can be found in [1].

2. Intercalation of Clay Minerals

Various methods were proposed for modification of cationic layered clay minerals to obtain materials for specific applications. Among them intercalation of cationic layered minerals with metal oxide pillars (e.g., Al_2O_3, TiO_2, ZrO_2) is considered as one of the most important methods. Clay minerals modified by this way, called pillared interlayered clays (PILCs), are characterized by a relatively high surface area (150-350 m^2 g^{-1}), developed microporous structure , ion-exchange properties and surface acidity . Porous clay heterostructures (PCHs) are another, very important group of materials, which can be produced from cationic layered clays [2]. The synthesis of PCHs is based on introduction of silica [3], silica-titania [4] or other silica-based multicomponent pillars into the interlayer space of clays by the surfactant directed method. The PCH materials are characterized by high surface area (up to 1000 m^2 g^{-1}), combined micro- and mesoporous structure, surface acidity and cation exchange properties.

Because PILCs are relatively well known group of clay-based materials , therefore in the next part of this chapter these materials will be only briefly presented, while more attention will be paid for various aspects related to the synthesis and application of the PCH materials.

2.1. Pillared Interlayered Clays (PILCs)

Increasing prices of oil in the mid-1970s resulted in intensification of studies focused on optimization of the catalytic cracking of crude oil . New catalysts with large pore size and good thermal and hydrothermal stability were required to increase the yield of lighter fractions from heavy crude oil. Cationic layered clay minerals intercalated with metal oxide pillars (PILCs) were one of the most important candidates to be such catalysts. The studies done in the mid-1970s were focused on the use of inorganic hydrated polyoxocations as pillaring agents [5]. Such pillaring agents, when calcined, dehydrate and dehydroxylate to form thermally stable metal oxide pillars, which permanently increase interlayer distance in clay mineral. Obtained PILC materials are characterized by microporous structure , high surface area , ion-exchange properties and surface acidity , however their hydrothermal stability was not good enough for catalytic cracking. Although, majority of the achievements related to the syntesis of PILCs was made in the 1970, these materials are still studied. The main directions of these studies are focused on syntesis of PILCs with specific properties (e.g., increased thermal stability [6, 7], modified surface acidity [8, 9]) and application of these materialas in various catalytic and adsorption processes [e.g., 10].

Procedure used for intercalation of cationic layered clay mineral with alumina pillars is schematicaly presented in figure 2. In the first step interlayer cations in clay mineral are exchanged for metal polyoxocations by ion exchange method. Introduction of large polyoxocations into the interlayer space of clay minereral results in an increase of dictance between clay layers.

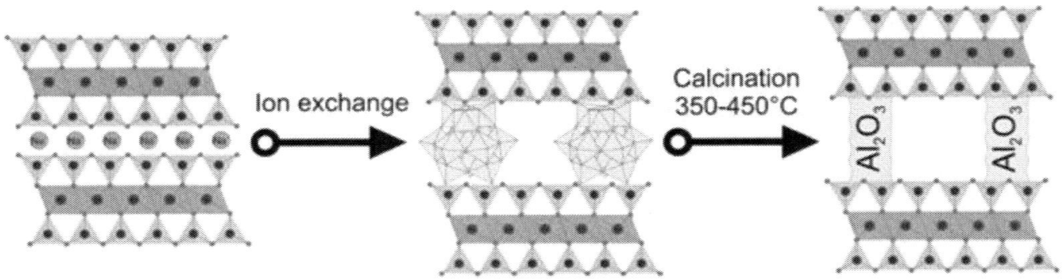

Figure 2. Intercalation of cationic layered clay mineral with alumina pillars.

In the next step the modified clay sample is calcined in order to decompose thermaly unstable polyoxocations into thermaly stable metal oxide pillars. Intercalation of cationic layered minerals with metal oxide pillars (e.g., Al_2O_3, TiO_2, ZrO_2) results in obtaining of materials characterized by a relatively high surface area (150-350 m^2 g^{-1}), developed microporous structure , ion-exchange properties and surface acidity .

There is a large number of scientific reports presenting various aspects of PILC materials , among them also some review papers [e.g., 11, 12], where reader can find more detailed information. Examples of papers presenting synthesis of PILCs and their applications in catalysis and adsorption are listed in tables 1 and 2, respectively. In the next part of this chapter new method of clay minerals intercalation based on a surfactant directed method will be presented and discussed.

2.2. Porous Clay Heterostructures (PCHs)

Since the development of the M41S family of ordered mesoporous silica materials in 1992 by researchers of Exon Mobile Company [33], an impressive progress in synthesis and modification of many new mesoporous solids based on the same templating mechanism has been reported [e.g., 34]. Based on these studies as well as experience related to the synthesis of PILCs, a new class of porous materials , based on layered clay minerals , was proposed by Galarneau et al. [2] in 1995. This new group of materials, called porous clay heterostructures or PCHs, is based on the formation of an MCM-like porous silica structure between clay layers. The synthesis of PCHs, presented schematically in figure 3, includes the following steps: (i) cationic alkylammonium templates and neutral alkylamine co-templates are intercalated in the interlayer space of host clay forming the micelle structures; (ii) the silica pillars are created by in-situ polymerization of silica source around the micelle structures in the interlayer space of clay; (iii) organic templates and co-templates are removed from the material by calcination or extraction with suitable solvents . The obtained material is characterized by high surface area and porosity , high thermal and hydrothermal stability and combined micro- and mesoporous structure. Additionally, PCH materials exhibit surface acidity and ion-exchange properties, which depend mainly on the type of parent clay used for their synthesis.

Up till now various types of cationic layered clays, including: fluorohectorite [2, 35], rectorite [2], montmorillonite [36], synthetic saponite [37], magadiite [38], vermiculite [39] and laponite [40, 41] were reported as host materials for the synthesis of PCHs.

Table 1. Examples of condition of PILC synthesis and their textural parameters

Host clay	CEC	Pillar composition	Pillar precursors	Calcination	*d* Spacings [nm]	BET surface area [m^2g^{-1}]	Ref.
Montmorillonite	76	Al_2O_3	$AlCl_3$	350	1.95	384	[17]
Montmorillonite	-	Al_2O_3	$AlCl_3$	200	1.88	126	[13]
Montmorillonite	116	Al_2O_3	$AlCl_3$	300	1.83	270	[23]
Montmorillonite	-	Al_2O_3	$AlCl_3$	400	-	254	[26]
Montmorillonite	84	Al_2O_3	Aluminium chlorohydrate	400	1.87	370	[21]
Montmorillonite	84	Al_2O_3	Aluminium chlorohydrate	400	1.84	250	[22]
Montmorillonite	90	Al_2O_3	Aluminium chlorohydrate	400	1.86	341	[15]
Montmorillonite	80	Al_2O_3	Aluminium chlorohydrate	500	1.93/1.54	183	[32]
Montmorillonite	81	Al_2O_3	Aluminium chlorohydrate	500	1.84	212	[19]
Montmorillonite	110	Al_2O_3	Aluminium chlorohydrate	500	1.77	304	[19]
Montmorillonite	-	Al_2O_3	$Al(NO_3)_3$	550	-	310	[24]
Bentonite F	97	Al_2O_3	$AlCl_3$	500	1.80	319	[20]
Bentonite F	97	Al_2O_3	$AlCl_3$	400	1.73	174	[27]
Bentonite P	92	Al_2O_3	$AlCl_3$	500	1.57	239	[20]
Bentonite	-	Al_2O_3	$AlCl_3$	400	1.70	142	[25]
Bentonite	-	Al_2O_3	$AlCl_3$	500	1.73	206	[31]
Bentonite	95	Al_2O_3	$Al(NO_3)_3$	300	1.80-1.82	215-258	[14]
Bentonite	-	Al_2O_3	$Al(NO_3)_3$	550	-	238	[24]
Saponite	80	Al_2O_3	Aluminium chlorohydrate	400	1.88	309	[21]
Saponite	56	Al_2O_3	Aluminium chlorohydrate	500	1.88	303	[19]
Saponite	61	Al_2O_3	Aluminium chlorohydrate	500	1.80	320	[19]
Synthetic beidellite	-	Al_2O_3	$AlCl_3$	200	1.88	251	[13]
Montmorillonite	76	TiO_2	$TiCl_4$	300	delaminated	383	[17]
Montmorillonite	84	TiO_2	$TiCl_4$	400	2.60	276	[22]
Montmorillonite	90	TiO_2	$TiCl_4$	450	2.30-2.70	282	[15]
Montmorillonite	-	TiO_2	$TiCl_4$	400	-	279	[26]
Montmorillonite	-	TiO_2	Ti-OEt	200	2.23	255	[13]
Synthetic beidellite	-	TiO_2	Ti-OEt	200	2.17	192	[13]
Montmorillonite	76	ZrO_2	$ZrOCl_2$	350	1.89	195	[17]
Montmorillonite	84	ZrO_2	$ZrOCl_2$	400	1.77	193	[22]
Montmorillonite	90	ZrO_2	$ZrOCl_2$	500	1.96	271	[15]
Montmorillonite Ch	81	ZrO_2	$ZrOCl_2$	500	1.65	171	[19]
Montmorillonite SB	110	ZrO_2	$ZrOCl_2$	500	1.64	216	[19]
Montmorillonite	-	ZrO_2	$ZrOCl_2$	400	-	242	[26]
Bentonite F	97	ZrO_2	$ZrOCl_2$	500	1.71	332	[20]

Table 1. (Continued)

Host clay	CEC	Pillar composition	Pillar precursors	Calcination	*d* Spacings [nm]	BET surface area [m^2g^{-1}]	Ref.
Bentonite F	97	ZrO_2	$ZrOCl_2$	400	1.75	192	[27]
Bentonite P	92	ZrO_2	$ZrOCl_2$	500	1.71	332	[20]
Smectite	103	ZrO_2	$ZrOCl_2$	500	1.65	165	[19]
Saponite	61	ZrO_2	$ZrOCl_2$	500	1.64	262	[19]
Montmorillonite	90	Fe_2O_3	$Fe(NO_3)_3$	400	1.21	134	[15]
Montmorillonite	86	Fe_2O_3	Fe(III) acetato complex	500	1.8	284	[18]
Bentonite F	97	Fe_2O_3	$FeCl_3$	500	2.30	245	[20]
Bentonite P	92	Fe_2O_3	$FeCl_3$	500	2.30	245	[20]
Montmorillonite	86	Cr_2O_3	Cr(III) acetato complex	500	1.78	286	[71]
Montmorillonite acid- activated	65	Cr_2O_3	Cr(III) acetato complex	500	1.75	270	[71]
Smectite clay	112	Al_2O_3/TiO_2	$AlCl_3$/Ti-OEt	450/600	1.96/1.88	178/178	[30]
Smectite clay	112	Al_2O_3/ZrO_2	$AlCl_3/ZrOCl_2$	450/600	2.06/1.96	224/198	[30]
Montmorillonite	-	Al_2O_3/Fe_2O_3	$Al(NO_3)_3/FeCl_3$	550	2.01	288	[24]
Montmorillonite	-	Al_2O_3/Fe_2O_3	$Al(NO_3)_3/Fe(NO_3)_3$	450	-	171	[28]
Bentonite	-	Al_2O_3/Fe_2O_3	$Al(NO_3)_3/FeCl_3$	550	1.94	247	[24]
Bentonite F	97	Al_2O_3/Fe_2O_3	$AlCl_3/FeCl_3$	400	1.80	206	[27]
Bentonite	-	Al_2O_3/Cr_2O_3	$AlCl_3/CrCl_3$	500	1.72	96	[31]
Bentonite	95	Al_2O_3/CuO	$Al(NO_3)_3/Cu(NO_3)_2$	300	1.85	208	[14]
Bentonite	-	Al_2O_3/CuO	$AlCl_3/CuCl_2$		1.81	164	[25]
Illite/smectite bentonite	75	Al_2O_3/CuO	$Al(NO_3)_3/Cu(NO_3)_2$	300	1.72	143	[16]
Montmorillonite	90	Fe_2O_3/ZrO_2	$Fe(NO_3)_3/ZrOCl_2$	400	2.01	295	[15]
Bentonite F	97	Fe_2O_3/ZrO_2	$ZrOCl_2/FeCl_3$	400	1.81	211	[27]

Ti-OEt - titanium tetraethoxide.

Table 2. Examples of PILC applications

Host clay	Pillar composition	Additional components	Applications in catalysis	Ref.
Montmorillonite	Al_2O_3	-	1-butene isomerisation, iso-propanol dehydratation	[21]
Montmorillonite	Al_2O_3	-	Reduction of NO by ethylene	[17]
Montmorillonite	Al_2O_3	CuO	Reduction of NO by ethylene	[17]
Montmorillonite	Al_2O_3	-	Cumene cracking	[13]
Montmorillonite	Al_2O_3	Pt	Hydroisomerisation–hydrocracking of heptane	[19]
Montmorillonite	Al_2O_3	-	Hydroxylation of phenol	[24]
Montmorillonite	Al_2O_3	-	NO reduction with ammonia	[26]
Montmorillonite	Al_2O_3	CuO	NO reduction with ammonia	[26]
Montmorillonite	Al_2O_3	-	Hydro isomerization of n-heptane	[32]
Bentonite	Al_2O_3	-	Hydroxylation of phenol	[24]

Host clay	Pillar composition	Additional components	Applications in catalysis	Ref.
Bentonite	Al_2O_3	-	Catalytic wet peroxide oxidation of reactive dyes	[25]
Bentonite F	Al_2O_3	-	Wet peroxide oxidation of phenol	[27]
Bentonite	Al_2O_3	-	CO oxidation	[31]
Saponite	Al_2O_3	-	1-butene isomerisation, iso-propanol dehydratation	[21]
Saponite	Al_2O_3	Pt	Hydroisomerisation–hydrocracking of heptane	[19]
Synthetic Beidellite	Al_2O_3	-	Cumene cracking	[13]
Smectite clay	Al_2O_3	Pt	Hydroisomerisation–hydrocracking of heptane	[19]
Montmorillonite	TiO_2	-	Reduction of NO by ethylene	[17]
Montmorillonite	TiO_2	CuO, Fe_2O_3	Reduction of NO by ethylene	[17]
Montmorillonite	TiO_2	-	Cumene cracking	[13]
Montmorillonite	TiO_2	-	NO reduction with ammonia	[26]
Montmorillonite	TiO_2	CuO, Fe_2O_3	NO reduction with ammonia	[26]
Synthetic Beidellite	TiO_2	-	Cumene cracking	[13]
Montmorillonite	ZrO_2	-	Reduction of NO by ethylene	[17]
Montmorillonite	ZrO_2	CuO	Reduction of NO by ethylene	[17]
Montmorillonite	ZrO_2	Pt	Hydroisomerisation–hydrocracking of heptane	[19]
Montmorillonite	ZrO_2	-	NO reduction with ammonia	[26]
Montmorillonite	ZrO_2	CuO	NO reduction with ammonia	[26]
Bentonite F	ZrO_2	-	Wet peroxide oxidation of phenol	[27]
Saponite	ZrO_2	Pt	Hydroisomerisation–hydrocracking of heptane	[19]
Smectite	ZrO_2	Pt	Hydroisomerisation–hydrocracking of heptane	[19]
Bentonite	Cr_2O_3	-	CO oxidation	[31]
Montmorillonite	Al_2O_3/ZrO_2	Pt	Hydroisomerisation–hydrocracking of heptane	[19]
Saponite	Al_2O_3/ZrO_2	Pt	Hydroisomerisation–hydrocracking of heptane	[19]
Smectite clay	Al_2O_3/ZrO_2	Pt	Hydroisomerisation–hydrocracking of heptane	[19]
Montmorillonite	Al_2O_3/Fe_2O_3	-	Hydroxylation of phenol	[24]
Montmorillonite	Al_2O_3/Fe_2O_3	V_2O_5, MnO, CoO, NiO, CuO, ZnO	Wet peroxide oxidation of phenol	[28]
Montmorillonite	Al_2O_3/Fe_2O_3	-	Wet peroxide oxidation of phenol	[28]
Bentonite	Al_2O_3/Fe_2O_3	-	Hydroxylation of phenol	[24]

Table 2. (Continued)

Host clay	Pillar composition	Additional components	Applications in catalysis	Ref.
Bentonite F	Al_2O_3/Fe_2O_3	-	Wet peroxide oxidation of phenol	[27]
Bentonite	Al_2O_3/Fe_2O_3	-	Wet hydrogen peroxide catalytic oxidation of model phenolic compounds (p-coumaric and p-hydroxybenzoic acids) and real olive oil milling wastewater	[29]
Bentonite	Al_2O_3/Cr_2O_3	-	CO oxidation	[31]
Bentonite	Al_2O_3/CuO	-	Catalytic wet peroxide oxidation of reactive dyes	[25]
Bentonite	Al_2O_3/CuO	-	Wet hydrogen peroxide catalytic oxidation of model phenolic compounds (p-coumaric and p-hydroxybenzoic acids) and real olive oil milling wastewater	[29]
Illite/Smectite bentonite	Al_2O_3/CuO	-	Catalytic wet peroxide oxidation of phenol	[16]
Bentonite F	ZrO_2/Fe_2O_3	-	Catalytic wet peroxide oxidation of phenol	[27]
			Applications in adsorptions	
Smectite clay	Al_2O_3/TiO_2	-	Zinc adsorption	[30]
Smectite clay	Al_2O_3/ZrO_2	-	Zinc adsorption	[30]

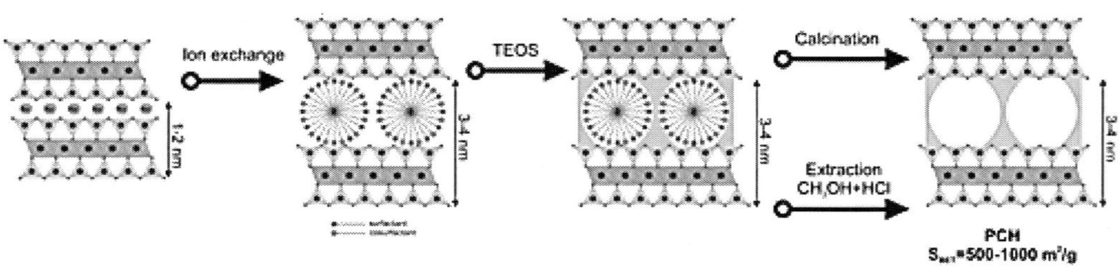

Figure 3. Synthesis of PCH materials .

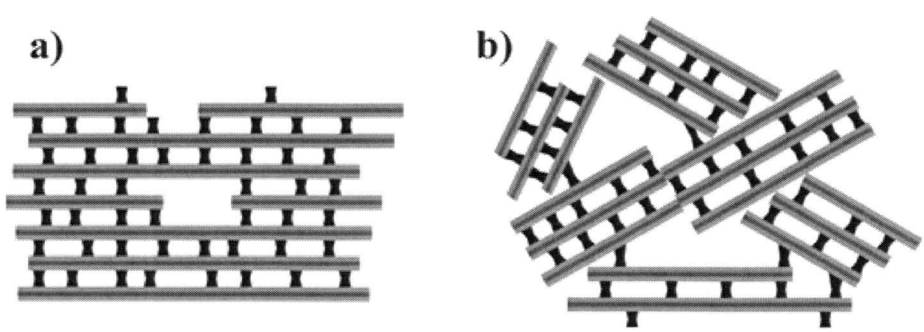

Figure 4. Orientation of clay layers in PCHs: face-to-face (a); face-to-face + edge-to-edge + edge-to-face (b) orientations.

Various features of layered clay minerals influence the nature of the obtained PCH materials, however the most important seems to be the ability for exchange of interlayer cations and layer charge density of host clays as well as their plate dimension. Exchange of interlayer inorganic cations, present in host clay, for alkyammonium cations is a necessary step for the formation of PCHs. Thus, only layered clays exhibiting ion-exchange properties can be used for the synthesis of PCH materials. Galarneau et al. [2] reported that for the clay samples with various layer charge density, including: rectorite, fluorohectorite, magadiite and vermiculite, the formation of the PCH structure was successful and the observed pore size s were invariant of these clays. Benjelloun et al. [42] reported that for large-sized host montmorillonite with a plate dimension of 1000 nm, the formation of micellar templates occurs mainly in the interlayer space and the orientation face-to-face is favored (Figure 4a). On the other hand for smaller-sized synthetic saponite with a plate dimension of 50 nm, the stacking of clay layers surrounding the micelles does not only occur parallel, due to the favored edge-to-edge and edge-to-face orientations (Figure 4b). Thus, the plate dimension in host clay influences the ordering of the clay layers in the PCH samples.

The next step in synthesis of PCHs is removal of organic matter from the pore system of as-synthetized materials. Calcination in air atmosphere and extraction with suitable solvents are the most popular methods used. Typically, calcination of as-synthetized PCHs is performed at temperature 550–650°C in air atmosphere. During this process the oxidative decomposition of organic surfactants occurs and gaseous products are evacuated from the pore systems of PCH. The appropriate choice of calcination parameters guarantees the complete removal of surfactants and products of their oxidative decomposition from pores of PCH materials. However, oxidation of organic surfactants results in very high local temperatures built up in the calcined samples, what may cause destabilization of the pillared structure. As it was reported by Benjelloun et al. [42] this effect is favored in case of small sized clay sheets. Burning of the organic surfactants in as-synthetized PCHs results in the formation of protons, which are necessary to compensate the clay layer charge. Thus, silica intercalated PCHs are acidic and still exhibit ion-exchange properties.

Extraction of the organic surfactants from as-synthetized PCHs is another important method used in the synthesis of this group of materials. There are two types of surfactants located in pores of PCHs – neutral and ionic. Strong electrostatic interaction between alkylammonium cations and negatively charged clay sheets is too strong for surfactant extraction with simple organic polar solvents [42]. Therefore, extraction solutions should contain small cations (e.g., H^+, Na^+, NH_4^+) that can exchange alkylammonium ions and compensate the negative charge of clay layers. Various extraction media, including: ethanol / HCl, ethanol/Na-acetate, ethanol/NH_4-acetate, ethanol/NH_4NO_3, acetone /HCl, methanol /HCl, were reported [3, 42, 43], however the best efficiency in the organic template extraction was found for the solution containing concentrated HCl diluted in methanol (volume ratio of 1 to 9), which extracted above 90% of organic surfactants from as-synthetized PCHs [42]. Thus, extraction, in contradiction to calcination, does not result in complete removal of the organic templates. On the other hand it was shown that extraction of the organic templates with methanol/HCl solution from as-synthetized PCH, obtained from synthetic saponite, resulted in a significant higher surface area and porosity compared to the calcined sample (1122 m^2 g^{-1} and 1.13 cm^3 g^{-1} for the extracted sample, 857 m^2 g^{-1} and 0.81 cm^3 g^{-1} for the calcined sample [42]). There was not such significant differences in textural parameters for calcined and extracted montmorillonite based PCHs [42]. It is probably related to the less thermally stable pillared structure obtained from the smaller-sized synthetic saponite, which can be destabilized during calcination, and the more thermally stable structure of pillared large-sized montmorillonite.

Guo et al. [44] reported that treatment of as-synthesized PCHs with ammonia solution (before calcination or extraction) results in a significant increase of their porosity (from 0.41 to 0.70 cm^3 g^{-1}) and pore diameter (from 1.7 to 3.1 nm). It was also found that temperature of ammonia solution used for the modification of PCHs significantly influenced their textural parameters.

Authors suggest that addition of ammonia to as-prepared PCHs introduces NH_4^+ and OH $^-$ into their interlayer space. Some of gallery surfactants is exchanged by NH_4^+, while OH^- redissolves SiO_2 . Similar process was also suggested by Kawi et al. [45]. The ammonia suspension of surfactant and polar head group NH_2 $^-$ of gallery co-surfactant should be concentrated and arranged in the gallery and hydrophobic alkyl chains focus on the core. Thus, micellar assemblies are formed, and more onium ions concentrated should result in the expansion of the interlayer distance in PCHs.

Hao et al. [46] reported the successful PCH synthesis in the medium of supercritical CO_2 . In the first step the sodium form of montmorillonite was modified with alkylammonium cations by ion exchange method. Then the modified clay sample was placed in a high-pressure autoclave, which was filled with CO_2 and finally TEOS was added. The reaction was performed at 80°C under a pressure of 12.0 MPa for 2 hours. The PCH samples obtained by this method were characterized by more uniform porous structure in comparison to PCHs synthetized by a classical method.

A multi-mesopores composite was successfully synthesized from natural montmorillonite milled with NaOH by Li et al. [47]. Two templating surfactants , cetyltrimethylammonium bromide (CTAB) and polyethylene glycol (PEG-200), were used in basic hydrothermal conditions in order to shape the pores in the clay samples. Templating surfactants were removed from the as-prepared samples by calcination. The general mechanism of the bi-modal mesopore structure formation is presented in figure 5. In the first step (a), clay layers of sodium form of montmorillonite were dispersed in a basic solution . In the second step (b), CTAB and PEG-200 were introduced into the suspension. Cetyltrimethylammonium (CTA) ions replace interlayer Na^+ cations and form micellar assemblies in the interlayer space of montmorillonite. Since the size of PEG-200 is too large to be accommodated into the interlayer space of montmorillonite, most of polyethylene glycol expands on the free surface of clay layer. In the following step (c), negatively charged amorphous SiO_2 (formed by the TEOS hydrolysis) is prone to be rearranged and condensates around the surface of micellar CTA assemblies, resulting in an ordered mesoporous structure in the interlayer spacing of montmorillonite.

At the same time irregular large pores are formed due to the presence of PEG-200. Finally, the organic templates are removed by calcination, which also completes the dehydroxylation and cross-linking of the assembled mesoporous silica structure (d). Authors reported that the obtained material is stable to up to 900°C, and its specific surface area was only slightly reduced by hydrothermal treatment (from 472 to 393 m^2 g^{-1} after boiling for 10 days).

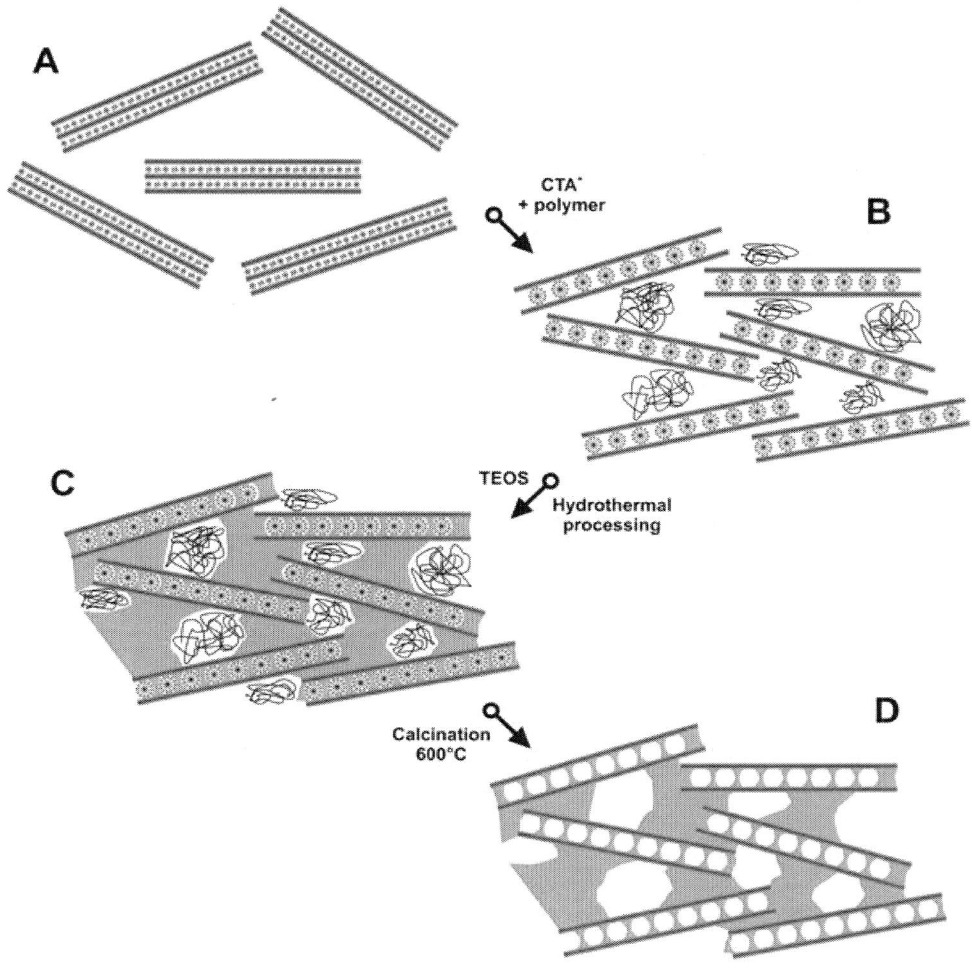

Figure 5. The proposed mechanism of formation of PCH with bimodal porosity .

The cation exchange capacity (CEC) of clay minerals is a very important parameter, which influences their unique properties such as the cation retention and diffusion process [3, 48]. In case of PCHs, obtained by calcination or extraction with acid -containing solvents , protons compensate the negative charge of clay layers. Such protons are strongly stabilized in the interlayer space of PCHs and their exchange for other cations is very difficult or even impossible. Moreover, in case of dioctahedral clays, protons can migrate to the vacant positions in dioctahedral sublayers, where they are inaccessible for the further cation exchange [3]. Thus, calcination or extraction with acid-containing solvents of as-prepared PCHs results in a CEC loss. Benjelloun et al. [3] suggested that exchange of protons for larger NH_4^+ cations, which are too large to migrate into vacant positions in octahedral sheets and are less stabilized in the interlayer space of clay, should protect CEC of PCHs. Three methods for the formation of NH_4^+-exchanged PCHs (obtained from montmorillonite and synthetic saponite) were proposed: (i) adsorption of ammonia under a gas flow on calcined and extracted PCHs; (ii) direct exchange in NH_4Cl solution ; (iii) solvent extraction with $NH_4Ac/EtOH/H_2O$ mixture. Authors reported that the most effective method was adsorption of ammonia, especially for the PCH samples obtained by extraction with a methanol /HCl solution. These results are supported by studies of Chmielarz et al. [49], who obtained about 2.5 times higher efficiency in introduction of Cu^{2+} cations by ion exchange method into NH_3-modified PCH in comparison to the unmodified PCH sample.

Thermal and hydrothermal stabilities of materials are crucial parameters necessary for designing catalysts or adsorbents for specific applications. Cassiers et al. [50] compared stability of various mesoporous silicas (MCM-41, MCM-48, KIT-1, FSM-16, HMS, SBA-15) and PCH obtained from synthetic saponite. Thermal treatment of these materials at various temperatures showed that the silica pillared structure of PCH is stable at temperature as high as 750°C. It must be mentioned that only MCM-41 and MCM-48 obtained from fumed silica presented better thermal stability , while other mesoporous silicas were less thermally stable. Studies of hydrothermal stability, conducted by exposing the samples to a N_2 stream containing 30 vol.% water vapor at 400°C for 48 and 120 hours, showed only small influence of the applied conditions on the textural parameters of the PCH sample.

Xia et al. [51] studied thermal and hydrothermal stability of silica intercalated montmorillonite . Prior to intercalation, the clay samples were mechanically treated in a laboratory high-performance planetary ball mill for 1, 2, 3 or 4 hours. Such treatment of the clay sample reduced its grain size from 34.5 μm (parent montmorillonite) to 0.38 μm (sample milled by 4 hours). It was reported that PCHs obtained from montmorillonite milled for 3 and 4 hours were characterized by lower surface area and pore volume than PCHs produced from clay that was not milled or milled only by 1 or 2 hours. The BET surface area of thermally treated PCHs (800°C, 2 h) decreased by 20-24% [51]. An increase in thermal treatment temperature to 850 and 900°C reduced surface area of PCH obtained from parent montorillonite from 619 to 368 and 70 m^2 g^{-1}, respectively [52]. Hydrothermal treatment of PCHs (samples dispersed in water , autoclaved at 100°C for 10 days) reduced the BET surface area by 15-23% with an exception of the PCH samples obtained from not milled montmorillonite, where BET surface area decreased by 62% [51].

Chmielarz et al. [53] studied titania -silica interlaced montmorillonites calcined at various temperatures. An increase in calcination temperature from 600 to 800°C significantly reduced their surface area . This effect depended on titanium content. For the sample intercalated with pure silica pillars an increase in calcination temperature reduced surface area by 21%, while for montmorillonite intercalated with pillars composed of silica-titania with molar ratio of 1.5, 7.5 and 15 the surface area decreased by 26, 26 and 42%, respectively. It was also reported that titanium incorporated into the pillars structure was present in the form of monomeric species and small polymeric chains . An increase in calcination temperature to 800°C only slightly modified titanium speciation in the sample with the highest titanium content.

Examples of condition used in the PCH synthesis and their textural parameters are shown in table 3.

Acidity of PCHs is a very important property and arises mainly from the host clay used for the synthesis of these materials . Various methods, including: (i) acid treatment of host clay [43, 54- 56]; (ii) incorporation of Al^{3+}, Ti^{4+} or Zr^{4+} cations into the structure of the silica pillars [e.g., 53, 57- 59]; (iii) post-synthesis impregnation or grafting of Al^{3+} cations; were reported in scientific literature [e.g., 37, 60].

Pichowicz and Mokaya [54] reported that PCHs with enhanced acidity can be prepared from acid -treated montmorillonite . It was suggested that acid-treatment of host clay generates mainly Brønsted acid sites. Moreover, PCHs obtained from acid-treated monmorillonite had significantly higher surface area than the samples prepared from raw clay. However, pore volumes were very similar for both series of the PCH materials . Kooli et al. [43] studied PCHs obtained from smectites treated with sulfuric acid solutions. It was shown that acid-treatment has improved the acidity and surface area of raw clays, however PCHs prepared from acid-treated clays were characterized by lower acidity and surface area comparing to the PCH samples obtained from raw clays. On the other hand Chmielarz et al. [55, 56] reported that treatment of vermiculite and phlogopite with nitric acid solution increased their surface area and acidity, however, an increase in duration of acid-treatment from 2 to 24 hours resulted in a decrease in surface acidity of vermiculite and phlogophite. Authors explained this effect by gradual leaching of Al^{3+} cations, responsible for the generation of acid sites, from clay layers. The successful synthesis of PCHs from acid-treated vermiculite was reported by Chmielarz et al. [59]. It was also shown that Lewis acid sites dominated over Brønsted acid centers and PCHs prepared from acid-treated vermiculite have lower surface area and acidity than the PCH samples obtained from montmorillonite and synthetic saponite (both without acid-treatment).

Another strategy for acid sites generation in PCHs is related to the modification of silica pillars composition . Chmielarz et al. reported successful incorporation of Ti^{4+} [53, 57] and Al^{3+}[58] cations into the structure of the silica pillars in PCHs produced from montmorillonite . Incorporation of titanium into the silica pillars resulted in a significant increase of surface acidity of the PCH samples, mainly by generation of weak Brønsted acid centers [53].

Titanium isopropoxide (TIP) and tetraethyl orthosilicate (TEOS) were used as precursors for the formation of the titania -silica pillars. The molar TIP/TEOS ratio used in the PCH synthesis was in the range of 0-0.11.

Titanium was present mainly in the form of isolated Ti^{4+} cations in the tetrahedral coordination and additionally, in the form of small polymeric chains . Authors showed the correlation between the content of titanium incorporated to the silica pillars and surface concentration of acid centers. Incorporation of aluminum into the silica pillars also significantly increased acidity of the PCH samples [58]. Aluminum isopropoxide (AIP) and tetraethyl orthosilicate (TEOS) were used as precursors for the formation of the alumina - silica pillars. The molar AIP/TEOS ratio used in the PCH synthesis was 0.05. Comparison of the montmorillonite based PCH samples intercalated with the titania -silica and alumina-silica pillars (in both cases the same mole Ti/Si and Al/Si ratios of 0.05) showed higher concentration of acid sites and larger contribution of stronger acid centers in case of the sample intercalated with alumina-titania pillars. The surface concentrations of acid centers for PCHs intercalated with silica, titania-silica and alumina-silica pillars were 99, 240 and 265 μmol g^{-1}, respectively [58].

Zhou et al. [61] applied a similar method for incorporation of alumna into the silica pillars and increasing their surface acidity . Similarly to described above studies, authors used aluminum isopropoxide (AIP) and tetraethyl orthosilicate (TEOS) as precursors for the formation of the alumina -silica pillars in the interlayer space of montorillonite. The surface concentrations of acid centers, mainly Lewis type, determined for PCH intercalated with the alumina-silica pillars (molar Al:Si ratio of 1:100) was 871 μmol g^{-1}.

Table 3. Examples of condition of PCH synthesis and their textural parameters

Host clay	Composition of pillars	Pillar precursors	Surfactant/ Co-surfactant	Surfactant removal	Interlayer distance [nm]	BET surface area [$m^2 g^{-1}$]	Pore volume [$cm^3 g^{-1}$]	Ref.
Montmorillonite	SiO_2	TEOS	HDTMACl/DDA	Cal: 550°C	n.d.	734	0.54	[3]
Montmorillonite	SiO_2	TEOS	HDTMABr/OA	Cal: 650°C	n.d.	795	0.75	[54]
Montmorillonite	SiO_2	TEOS	HDTMABr/DA	Cal: 650°C	n.d.	782	0.82	[54]
Montmorillonite	SiO_2	TEOS	HDTMABr/DDA	Cal: 550°C	n.d.	890	0.57	[43]
Montmorillonite	SiO_2	TEOS	HDTMABr/DDA	Cal: 550°C	n.d.	740	0.49	[43]
Montmorillonite	SiO_2	TEOS	HDTMABr/DDA	Cal: 550°C	n.d.	988	0.80	[43]
Montmorillonite	SiO_2	TEOS	HDTMABr/HDA	Cal: 550°C	1.75	1070	0.90	[37]
Montmorillonite	SiO_2	TEOS	HDTMABr/OA	Cal: 550°C	n.d.	589	0.41	[44]
Montmorillonite	SiO_2	TEOS	HDTMABr/HDA	Cal: 550°C	1.52	1022	0.72	[37]
Montmorillonite	SiO_2- Al_2O_3	TEOS- AIP	ODTMACl/DDA	Cal: 540°C	2.39	503	0.22	[61]
Montmorillonite	SiO_2	TEOS	HDTMACl/DDA	Cal: 600°C	2.75	816	0.54	[39]
Montmorillonite	SiO_2	TEOS	HDTMACl/HDA	Cal: 550°C	n.d.	579	0.47	[53]
Montmorillonite	SiO_2- TiO_2	TEOS- TIP	HDTMACl/HDA	Cal: 550°C	n.d.	501	0.37	[53]
Montmorillonite	SiO_2	TEOS	HDTMABr/DA	Cal: 650°C	n.d.	891	0.58	[59]
Montmorillonite	SiO_2- ZrO_2	TEOS- ZrEO	HDTMABr/DA	Cal: 650°C	n.d.	763	0.46	[59]
Montmorillonite	SiO_2- ZrO_2	TEOS- ZrEO	HDTMABr/DA	Cal: 650°C	n.d.	464	0.25	[59]
Montmorillonite	SiO_2- ZrO_2	TEOS- ZrOCl	HDTMABr/DA	Cal: 650°C	n.d.	847	0.45	[59]
Montmorillonite	SiO_2- ZrO_2	TEOS- ZrOCl	HDTMABr/DA	Cal: 650°C	n.d.	889	0.50	[59]
Montmorillonite	SiO_2- ZrO_2	TEOS- ZrOCl	HDTMABr/DA	Cal: 650°C	n.d.	525	0.64	[59]
Acid-treated	SiO_2	TEOS	HDTMABr/OA	Cal: 650°C	2.50	915	0.71	[54]
Acid-treated	SiO_2	TEOS	HDTMABr/DA	Cal: 650°C	n.d.	951	0.78	[54]
Bentonite	SiO_2	TEOS	HDTMABr/DDA	Cal: 550°C	n.d.	844	0.59	[43]
Bentonite (Na-form)	SiO_2	TEOS	HDTMABr/HDA	Cal: 600°C	delaminated	335	0.71	[76]
Bentonite (Na-form)	SiO_2	TEOS	HDTMABr/DDA	Cal: 550°C	delaminated	677	0.52	[60]
Bentonite	SiO_2- ZrO_2	TEOS- ZrPO	HDTMABr/HDA	Cal: 550°C	3.19	696	1.02	[40]

Host clay	Composition of pillars	Pillar precursors	Surfactant/ Co-surfactant	Surfactant removal	Interlayer distance [nm]	BET surface area [m^2 g^{-1}]	Pore volume [cm^3 g^{-1}]	Ref.
Synthetic lithium	SiO_2	TEOS	HDTMABr/DA	Cal: 550°C	2.35	750	n.d.	[35]
Synthetic lithium	SiO_2	TEOS	HDTMA+/DA	Cal: 550°C	2.35	750	n.d.	[35]
Fluorohectorite	SiO_2	TEOS	HDTMABr/DA	Ext: HCl/ethanol	2.14	550	n.d.	[68]
Synthetic saponite	SiO_2	TEOS	HDTMACl/DDA	Ext: HCl/ethanol	n.d.	997	0.76	[3]
Synthetic saponite	SiO_2	TEOS	HDTMACl/DDA	Ext: HCl/ethanol	2.60	1118	0.97	[3]
Synthetic saponite	SiO_2	TEOS	CTMABr/DDA	Cal: 550°C	n.d.	899	0.50	[50]
Synthetic saponite	SiO_2	TEOS	HDTMABr/HDA	Cal: 550°C	1.68	999	0.94	[37]
Synthetic saponite	SiO_2	TEOS	HDTMABr/HDA	Cal: 550°C	2.23	910	0.62	[37]
Synthetic saponite	SiO_2	TEOS	HDTMACl/DDA	Cal: 550°C	2.63	906	0.74	[39]
Acid-treated	SiO_2	TEOS	HDTMACl/DDA	Cal: 600°C	delaminated	612	0.40	[39]
Acid-treated	SiO_2	TEOS	HDTMABr/DDA	Cal: 550°C	delaminated	463	0.57	[41]
Smectite (Volclay	SiO_2	TEOS	HDTMABr/OA	Ext: HCl/ethanol	n.d.	620	0.46	[84]

ZrOCl - zirconium (IV) oxychloride.
TEOS – tetraethyl orthosilicate.
TIP - titanium isopropoxide .
AIP - aluminium isopropoxide.
ZrEO - zirconium (IV) ethoxide.
n.d. – not determined.
$HDTMA^+$ - hexadecyltrimethylammonium ion.
HDTMACl - hexadecyltrimethylammonium chloride.
HDTMABr - hexadecyltrimethylammonium bromide.
DDTMABr - dodecyltrimethylammonium bromide.
ODTMACl - octadecyltrimethylammonium chlorid.

Similar strategy was applied by Cecilia et al. [62] to prepare silica -zirconia intercalated PCH from natural bentonite. TEOS and ZIP (zirconium isopropoxide) were cocondensated to form interlayer pillars. Centyltrimethylammonium bromide and hexadecylamine were used as surfactants and co-surfactants, respectively. Organic templates were removed from the as-prepared PCH samples by calcination at 550°C. The samples containing pillars with the Si/Zr molar ratios of 1, 2, 5, 10 and 15 were prepared. It was found that surface area and pore volume decreased with an increase in Zr-content (696 m^2 g^{-1} and 1.02 cm^3 g^{-1} for Si/Zr=15 and 210 m^2 g^{-1} and 0.37 cm^3 g^{-1} for Si/Zr=1), while the surface concentration of acid sites increased with an increase in zirconium loading (150 μmol g^{-1} for Si/Zr=15 and 533 μmol g^{-1} for Si/Zr=1). It was also reported that incorporation of zirconium into the silica pillars structure generated mainly Brønsted-type acidity .

Ahenach et. al. [37] reported that Brønsted acidity of PCHs can be enhanced upon Al-grafting. Aluminum acetylacetonate complex, which interacts with the surface silanol groups of the PCH support, was used as aluminum precursor. Al-grafting was followed be a calcination step. The grafting of AlO_x-species onto the PCH surface resulted in the formation of Si-(OH)-Al bonds , which gave rise to a strong Brønsted acidity. Such grafting method was successfully implemented for the generation of Brønsted acid sites in montmorillonite and saponite based PCHs.

Polverejan et al. [63] increased surface acidity of PCHs by deposition of aluminum from $AlCl_3$ and $NaAlO_2$ solutions. In the first step silica intercalated PCH was prepared from synthetic saponite, which was used as host material for aluminum deposition. PCH was dispersed in a solution of $AlCl_3$ or $NaAlO_2$ to give the gallery Si/Al molar ratios of 5, 10 and 20. The ^{27}Al MAS NMR analysis showed that aluminum was deposited in the form of tetrahedraly coordinated Al^{3+} cations grafted into gallery silica. However, it was also reported that for the Si/Al ratio below 10 the formation of undesired octahedral aluminum was detected for PCH modified with a $AlCl_3$ solution. Such effect was not observed for the samples modified with a $NaAlO_2$ solution and the formation of octahedral aluminum was not detected for the Si/Al ratio as low as 5. Deposition of aluminum increased the surface concentration of acid centers from 630 to 1820 μmol g^{-1} (value obtained for the sample with the Si/Al ratio of 5 modified with $NaAlO_2$).

Mao et al. [64] reported a similar procedure of aluminum deposition . Silica intercalated clay was dispersed in a solution of $NaAlO_2$ to give the gallery Si/Al ratios of 7.2 and 14.5. To form the protonated forms of alumina modified PCHs, sodium ions in the as-made aluminated products were exchanged for ammonium ions by equilibrating in a solution of NH_4NO_3 and then calcined at 600°C. Alumina modified PCHs exhibited interlayer distance in the range of 2.2-2.4 nm (determined by XRD), while pore size s in the large micropore to small mesopore range (1.5–1.6 nm) was determined by low-temperature nitrogen adsorption . An increase in alumina loading decreased surface area (807 and 445 m^2 g^{-1} for Si/Al ratios of 14.5 and 7.2, respectively) and increased surface acidity 1.46 and 1.78 mmol g^{-1} for Si/Al ratios of 14.5 and 7.2, respectively) of the samples. It was suggested that the basic $NaAlO_2$ solution may promote the cleavage of some Si-O bonds resulting in a partial rearrangement of the interlayer mesostructure. Authors related the observed decrease in interlayer pore size, surface area and pore volume to an increase in density of the interlayer framework upon aluminum incorporation.

Zimowska et al. [60] effectively increased surface acidity of silica intercalated laponite by post synthesis deposition of aluminum. First series of the PCH samples was modified by impregnation method using a solution of $Al(NO_3)_3$. Second series was obtained by grafting of Al-species using aluminum isoporopoxide. It was reported that deposition of aluminum generated mainly Lewis-type acidity. Brønsted acid centers were less abundant and rather weak. An increase in aluminum loading resulted in a higher concentration of Brønsted and Lewis acid sites. For a similar level of aluminum doping , more acid sites of both types were formed when $Al(NO_3)_3$ was used for its deposition.

Another way for functionalization of PCHs for catalysis and adsorption is generation of organic functional groups attached into the silica interlayer pillars. The most important method, called "one-pot" synthesis , is based on cocondensation of a tetraalkoxysilane and an organoalkoxysilane to form interlayer pillars [e.g., 65, 66].

Ishii [65] reported successful synthesis of PCHs containing methyl group s attached into the silica pillars. Authors used tetraethoxysilane (TEOS) and methyltriethoxysilane (MTS) as silica source (separately and mixtures). Then, the organic surfactants were extracted with a HCl-ethanol solution . Calcination of the as-prepared PCH samples is not useful in this case, because apart from burning of the organic surfactants it could result in the complete oxidation of methyl groups attached into the silica pillars. It was reported that surface area of the samples decreased with an increase of the MTS/TEOS ratio used for their synthesis. A water adsorption study showed that the samples obtained using MTS/TEOS mixture had a hydrophobic surface as a result of the successful incorporation of methyl groups to the surface of the silica pillars [66].

Wei et al. [36] successfully synthesized a series of PCHs with phenyl groups attached to the silica pillars. TEOS and phenyltriethoxysilane (PhOS) were used as silica precursors for montmorillonite intercalation, with the PhOS/TEOS molar ratios of 0:1, 0.25:0.75, 0.35:0.65 and 0.5:0.5.

The organic surfactants were removed from as-synthesized PCHs by extraction with a HCl-ethanol solution or by calcination at 600°C (only in case of the sample prepared without using PhOS). It was found that surface area , pore volume and pore size significantly decreased with an increase of the PhOS/TEOS molar ratio. PCHs containing phenyl groups attached into the silica pillars exhibited hydrophobic properties , while PCH without such functional groups was hydrophilic. Additionally, it was reported that the samples containing phenyl groups were more acidic in comparison to PCH without such functional groups.

Pires at al. [67] studied a synthesis of clay intercalated with silica pillars containing anchored amine group s. The PCH samples were obtained by cocondensation methods using TEOS and 3-aminopropyltriethoxysilane (APTES) as precursors of the interlayer silica pillars.

Two molar APTES/TEOS ratios of 2.5 and 25.0 were used for the clay intercalation. The air-dried solids were solvent extracted with an ethanol solution of HCl to remove organic surfactants from their pore system. The PCH samples functionalized with amine group s exhibited increased hydrophilic properties in comparison to PCHs without such surface functional groups, what was shown by water adsorption measurements .

The successful grafting of mercaptopropyl functional groups on PCH was reported by Mercier and Pinnavaia [68] and Diaz et al. [69]. In the first step fluorohectorite was intercalated with the silica pillars. TEOS was used as silica source, while hexadecyltrimethylammonium chloride and decylamine were used as surfactants and co-surfactants, respectively. The organic templates were removed from the porous structure of PCH in two stages. Firstly, by treatment of the as-prepared sample with a solution of HCl to protonate decylamine and exchange hexadecyltrimethylammonium cations. The exchange reaction was followed by refluxing of the PCH sample in ethanol . Grafting of mercaptopropyl groups was done by refluxing of PCH in dry toluene containing mercaptopropyltrietoxysilane (MPTMS).

The obtained product was filtered, washed with toluene and finally washed with ethanol to remove ungrafted MPTMS. The method applied for grafting of mercaptopropyl groups significantly reduced surface area of PCH (from 550 to 175 $m^2 g^{-1}$) and pore size (2.1 nm - before grafting; no mesoporosity was observed for the MPTMS modified sample). This effect was explained by deposition of significant amounts of MPTMS in the pores of the PCH sample. It was estimated that the mercaptopropyl group density on the pore surface is 1.5 groups per nm^2 (1.1 mol g^{-1}).

An alternative way for the PCH functionalization with mercaptopropyl functional groups was applied by Tchinda at al. [70, 71]. The synthesis was based on cocondensation of TEOS and mercaptopropyltrietoxysilane (MPTMS) to form the interlayer pillars with attached mercaptopropyl groups.

A series of the PCH samples were prepared using mixtures of TEOS and MPTMS with the molar ratio of MPTMS/(TEOS + MPTMS) in the range from 0 to 35%. Surfactants were extracted using refluxing with a HCl/methanol mixture. The surface area , pore volume and mean pore diameter gradually decreased with an increase in amount of MPTMS used for the PCH synthesis . However, it should be mentioned that for all PCHs synthesized by cocondensation method these parameters were significantly higher comparing to the sample prepared by the mercaptopropyl groups grafting. The density of mercaptopropyl groups increased with an increase in MPTMS used for the PCH synthesis and was in the range from 0 to 3.0 mmol g^{-1}.

Thiol (-SH) functional groups can be relatively easy oxidized into sulfonic groups (-SO_3H) using a solution of H_2O_2. This method was used for oxidation of –SH to -SO_3H grafted on various mesoporous silica materials [e.g., 72, 73].

3. Modified Clays as Catalysts and Adsorbents

Porous clay heterostructures (PCHs), due to their unique properties, have been studied as materials for various applications, including also catalysis and adsorption .

3.1. Catalysis

Selective catalytic reduction of nitrogen oxides (NO and NO_2) with ammonia (NO-SCR or DeNOx) is one of the most intensively studied processes, in which PCHs were recognized as effective catalysts.

$$4\,NO + 4\,NH_3 + O_2 = 4\,N_2 + 6\,H_2O \quad (1)$$

$$2\,NO_2 + 4\,NH_3 + O_2 = 3\,N_2 + 6\,H_2O \quad (2)$$

Chmielarz et al. [39] studied silica intercalated PCHs obtained from natural montmorillonite , acid -treated vermiculite and synthetic saponite as catalysts of the DeNOx process. It was reported that PCH obtained from acid-treated vermiculite was significantly more active than silica intercalated montmorillonite and saponite. Authors related high catalytic activity of vermiculite based PCH with a relatively high content of iron , which is a natural component of this clay and is well known as element active in the DeNOx process. Significant catalytic activation of PCHs was observed after their modification with copper and iron by ion exchange method. Deposition of copper activated PCHs for the low-temperature DeNOx process, while modified clays doped with iron were active at higher temperatures [39, 49]. The PCH based catalysts were found to be stable in the presence of water vapor and SO_2, which are typical components of flue gas ses emitted by electric power stations. Chmielarz et al. [4] reported that incorporation of some titania into the silica pillars of PCHs results in a significant increase of their catalytic activity in the DeNOx process (Figure 6). Montmorillonite intercalated with the silica-titania pillars and doped with iron or copper (by ion exchange method) effectively operated in the DeNOx process in a broad temperature range from 300 to 450-500°C, so at temperature much higher than the commercial catalysts based on V_2O_5-TiO_2 metal oxide system. Taking into account their high thermal and hydrothermal stability and resistance for SO_2 poisoning, montmorillonite intercalated with the silica-titania pillars could be an interesting candidate for the commercial high temperature DeNOx catalyst . However, procedures used for intercalation of clay minerals needs to be optimized in order to decrease the costs of the PCH synthesis as well as to improve its catalytic properties .

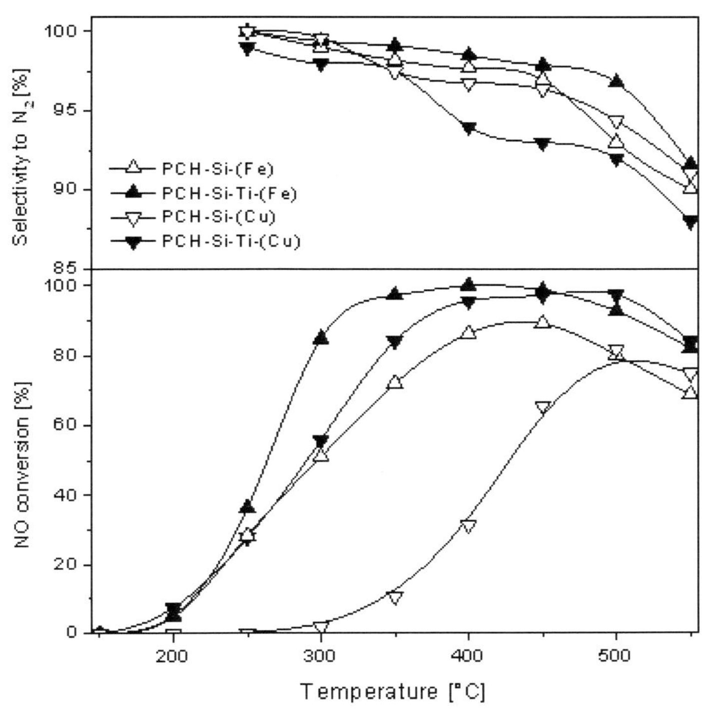

Figure 6. Catalytic performance of PCH intercalated with silica and silica-titania pillars doped with copper or iron .

Selective catalytic oxidation of ammonia to N_2 and H_2O (SCO -NH_3) is another environmental process, in which the PCH based catalysts were tested.

$$4\ NH_3 + 3\ O_2 = 2\ N_2 + 6\ H_2O \quad (3)$$

Chmielarz et al. [74] studied silica intercalated saponite modified with copper and iron by ion exchange method in the role of catalysts for the SCO -NH_3 process.

It was reported that the Cu-containing catalysts effectively operated at temperatures above 400°C, while PCHs doped with iron were active at temperatures higher by about 100°C. For both series of the catalysts the selectivity to N_2, which is a desired product of ammonia oxidation , was above 90% in the studied temperature range (T<550°C).

Park et al. [75] studied silica intercalated PCHs obtained from kenyaite and magadiite as catalytic supports for partial oxidation of methane (POM).

$$2\ CH_4 + O_2 = 2\ CO + 4\ H_2 \quad (4)$$

The pillared structure of PCHs was stable up to temperature as high as 800°C, thus application of such materials for the high temperature POM was possible. Transition metals , Ni and Pd, which played a role of active components of the POM catalysts, were deposited by impregnation method. The studied catalysts showed a high methane conversion and remained stable for a reaction time of almost 100 hours. PCH modified with palladium selectively converted methane into hydrogen . Authors related the excellent selectivity of the reaction to hydrogen with a high and uniform dispersion of palladium in the interlayer space of PCHs.

Hao et al. [46] reported high catalytic activity of PCHs modified with cobalt (20 wt.%) by impregnation method in the Fischer-Tropsch process . Authors compared two types of the PCH supports – obtained by classical method (ncPCH) and in medium of supercritical CO_2 (scPCH). It was shown that the PCH based catalysts presented higher selectivity to C_5-C_{12} hydrocarbons and considerably lower selectivity to heavy hydrocarbons (C_{21+}) than silica deposited cobalt (Co/SiO_2). This effect was explained by a possible cracking of long-chain hydrocarbons over acidic sites present in PCHs. The catalyst based on scPCH presented higher selectivity to higher hydrocarbons than the catalyst based on ncPCH. Authors related these differences in selectivity to the more ordered pore structure of scPCH than ncPCH.

Mao et al. [64] studied PCHs modified with alumina by post-synthesis method (Al-PCH) in the role of the catalysts for cumene cracking. It was shown that deposition of alumina increased surface acidity of the PCH samples and activated them in the cracking process. Al-PCHs were found to be more active in the cumene cracking than Al-MCM-41and Al-ZSM-5 (figure 7), and what is very important, were less affected by the coke formation and therefore significantly more stable than the other studied catalytic system s. Al-PCHs presented remarkable hydrothermal stability , which could be attributed to a unique combination of two factors: (i) the presence of tetrahedral aluminium grafted onto the interlayer silica framework and (ii) the stable pillared structure . Moreover, limited deactivation of the Al-PCH catalysts is possibly related to the open interlayer porous structure, that greatly improves access to the acid sites located on the pore walls and minimizes the coke formation in pores.

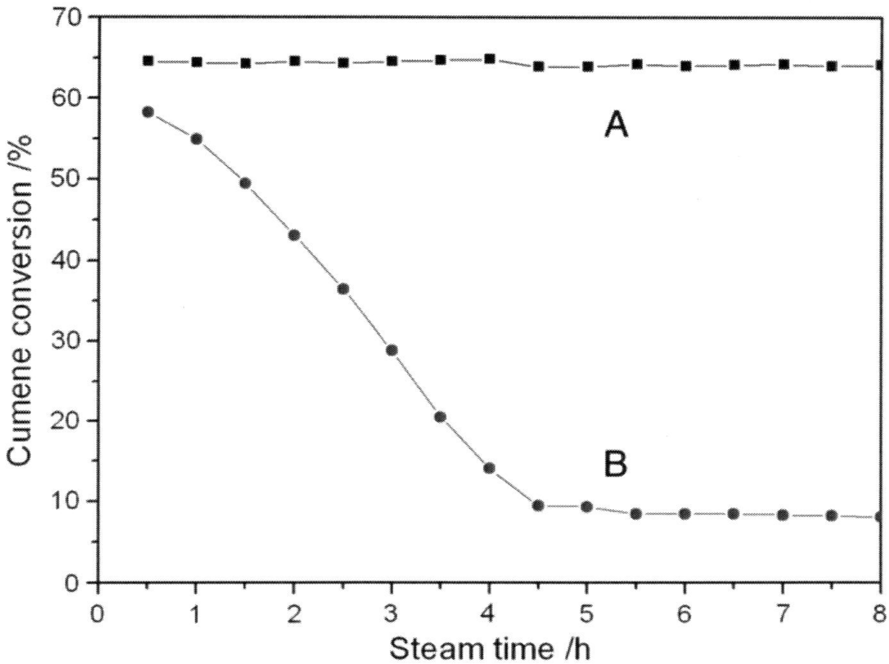

Figure 7. Stability tests of Al-PCH (A) and Al-MCM-41(B) in the process of cumene cracking [64].

Similar results were obtained by Polverejan et al. [63], who showed that modification of PCHs with alumina increased their surface acidity , and therefore activated them in the process of cumene cracking. Authors reported that alumina modified PCHs were significantly more active in the studied process than ordered mesoporous silicas MCM-41 and HMS doped with alumina.

Jeenpadiphat et al. [76] studied silica intercalated PCHs in the role of the catalysts for the esterification reaction of methanol with oleic acid or high acid content refined palm oil . Surfactants were removed from as-prepared PCHs by calcination or extraction with a HCl/ethanol solution . The obtained material was treated with H_2SO_4 solution. Such modification of PCH significantly reduced its surface area (from 412 to 212 m^2 g^{-1}) and generated surface acid sites (from 0.10 to 1.96 mmol g^{-1}). Activation of PCHs with a H_2SO_4 solution significantly increased their catalytically activity for oleic acid or refined palm oil esterification with methanol, due to an increase in the number of catalytic acid sites. Moreover, these modified clay catalysts were found to have a higher deactivation tolerance, in terms of stability and reusability , than the commercial Amberlyst-15 based catalyst . Thus, H_2SO_4-activated PCHs seem to be very interesting candidates for the catalysts of high acid content oils esterification for sustainable biodiesel production.

Li et al. [77] studied the mesoporous montmorillonite -based materials modified with $H_3PW_{12}O_{40}$ (HPW) heteropolyacid as catalysts for oxidative desulfurization of dibenzothiophene-containing model oil . Sol-gel and impregnation methods were used for HPW deposition . The catalysts obtained by sol-gel method showed a homogeneous dispersion of HPW even at 25 wt.% loadings. In this case, HPW was encapsulated into the gallery silica framework structure of pillared montmorillonite. The catalysts obtained by sol-gel method presented better catalytic performance than the catalysts containing HPW deposited by impregnation method. Under the experimental condition s (25 mL of model oil, 0.12 mL of 30 vol.% of H_2O_2 aqueous solution , 0.15 g of catalyst , 60°C), the sulfur content in the model oil was reduced from 500 to 7.1 ppm (in the presence of the catalyst containing 15wt.% HPW).

Kuźniarska-Biernacka et al. [78] reported the studies of PCHs functionalized with two chiral manganese (III) salen complexes in the role of catalysts for the asymmetric epoxidation of olefins . The first type of the salen complex was anchored onto PCH derivatized with phenoxyl groups by axial coordination between the metal center and support, while the second type of the manganese(III) salen complex, containing vinyl functionalities, was grafted onto mercapto-derivatized PCH through the ligand . The obtained materials were studied as the catalysts of asymmetric epoxidation of olefins using m-chloroperoxybenzoic acid /N-methylmorpholine-N-oxide as oxygen source. The studied materials presented a very good catalytic performance with respect to epoxide selectivity and enantioselectivity as well as conversion and stability .

3.2. Adsorption

Mercier and Pinnavaia [68] reported a succesfull appliction of fluorohectorite based PCH functionalized with 3-mercaptopropyltrimethoxysilane (post synthesis grafting) as effective and selective adsorbent for Hg^{2+} trapping. The adsorption capacity was 74 mmol g^{-1} and up to 67% of the immobilized thiol groups were accessible for Hg^{2+} trapping. Authors related such high efficiency of PCH modified with thiol groups to their uniform open-framework mesoporosity, which facilitates access to the reactive sites attached to the pore walls.

On the other hand Diaz et al. [69] studied PCHs obtained from synthetic beidelite as well as natural montmorillonite , hectorite and vermiculite functionalized with thiol groups (post synthesis 3-mercaptopropyltrimethoxysilane grafting) as adsorbents for removal of cadmium and lead from solutions. It was reported that grafting of –SH groups activated PCHs for adsorption of studied heavy metals . Additionally, tests of Cd^{2+} and Pb^{2+} plant uptake were pursued by ray-grass cultivation on polluted soil amended or not with functionalized synthetic beidellite during 3 months were done. It was reported that the cadmium uptake by ray-grass grown on polluted soil has been quite depressed by amendments with 0.17 or 0.5%, however these amendments were inefficient for lead.

Tchinda et al. [70] studied thiol functionalized PCHs obtained from smectite clay. The mesoporous framework was formed by surfactant directed co-condensation of tetraethoxysilane (TEOS) and 3-mercaptopropyltrimethoxysilane (MPTMS), at various MPTMS/TEOS ratios, in the interlayer region of clay. In the next step thiol functionalized PCHs were deposited as thin films onto the surface of glassy carbon electrodes and applied to the voltammetric detection of Hg^{2+} subsequent The PCH mesostructure offered a favorable environment for fast diffusion rates , resulting in great sensitivity , while the layered morphology ensured good mechanical stability .

Arellano-Cardenas et al. [79] presented studies of PCH (silica intercalated bentonite) as adsorbent of phenol and dichlorophenols (DCPs) from water solutions. Authors reported the following adsorption capacities of PCH: 14.5 mg g^{-1} for phenol; 48.7 mg g^{-1} for 3,4 DCP and 44.5 mg g^{-1} for 2,5 DCP. The values of maximal adsorption capacity for DCPs were higher than those reported for alumina pillared clays and comparable with some ionic exchange resins .

Authors proposed adsorption mechanism of phenol molecules on the PCH surface, through the formation of hydrogen bonds between the hydroxyl group of phenol and surface silanol groups of PCH.

PCHs functionalized with phenol and propylamine groups were studied as selective adsorbents for the separation of the selected biogas generated in landfills (CO_2 , CH_4, C_2H_6) [80]. It was shown that modification of the PCH surface with organic groups changed the affinity of the surface for different gases. Introduction of propylamine groups on the PCH surface significantly increased the separation of CO_2 from CH_4 and C_2H_6. Moreover, the functionalization of PCH with phenyl groups, improved the selectivity of the separation of C_2H_6 from CH_4.

Nunes et al. [81] studied PCHs functionalized with benzene moieties as adsorbents of the selected volatile organic compounds (VOCs): methanol , methyl ethyl ketone, toluene and trichloroethylene. The PCH samples were prepared by cocondensation method using TEOS and bis(triethoxysilyl)benzene (BTEB) as precursors of the organo-silica pillars. The adsorption of VOCs was favored over water vapour. The best results were obtained for adsorption of toluene and methyl ethyl ketone. Also Qu et al. [82] presented the adsorption studies of the selected VOCs (acetone , toluene, ethylbenzene, *o*-xylene, *m*-xylene and *p*-xylene) by PCH (silica intercalated bentonite, non-functionalized with organic species). It was reported that PCH had a higher adsorption affinity for the aliphatic hydrocarbon compound (acetone) than for aromatic compounds , what was attributed to the HOMO energy effects of VOCs.

Interesting studies of the selected organic compounds separation in the presence of PCHs functionalized with zirconium were reported by Pinto et al. [59]. The PCH based adsorbents presented high selectivity in the separation of binary systems of aromatic molecules such as toluene /benzene or p-xylene/benzene, as well as for benzene/n-hexane . The separation p-xylene/toluene was effective in the presence of the studied clay adsorbents. Authors related high separation efficiency of the aromatic molecules with the presence of zirconium, which increased acidity of PCHs. The PCH based adsorbents exhibited more favorable selectivity values for the studied binary mixture separations than the majority of other adsorbent materials presented in the literature. In a group of inorganic materials, only some zeolites presented higher selectivity values but at higher temperatures, so in less favorable working conditions .

Pinto and Pires [83] compared a series of PCHs with respect to their ability to the adsorption and separation of the selected organic compounds (n-hexane , benzene , cyclohexane, 3-methylpentane, toluene and ethylbenzene). Silica intercalated PCH as well as its modifications functionalized with phenyl groups (cocondensation method, TEOS and phenyltriethoxysilane were used as pillars precursors) and amine group s (grafting of 3-aminopropyltriethoxysilane on silica intercalated PCH) were used as adsorbents . Separation of the binary benzene/n-hexane system was most effective in the presence of silica intercalated PCH, while anchoring of amine groups significantly increased separation efficiency of the ethylbenzene/benzene system.

Examples of PCH applications in catalysis and adsorption are listed in table 4.

Table 4. Examples of PCH applications

Host clay	Pillar composition/ Pillar precursors	Additional components	Applications in catalysis	Ref.
Montmorillonite	SiO_2 /TEOS Al_2O_3/AIP		The alkylation of catechol with tert-butyl alcohol to synthesize 4-tertbutylcatechol	[61]
Montmorillonite	SiO_2/ TEOS		The alkylation of catechol with tert-butyl alcohol to synthesize 4-tertbutylcatechol	[61]
Montmorillonite	SiO_2/ TEOS		Reduction of NO with ammonia	[39]
Montmorillonite	SiO_2/ TEOS	CuO, Fe_2O_3	Reduction of NO with ammonia	[39]
Bentonite (montmorillonite >90%)	SiO_2/ TEOS		1-Butene isomerization	[52]
Bentonite (montmorillonite >90%)	SiO_2 / TEOS ZrO_2/ ZrPO		1-Butene isomerization	[52]
Bentonite	SiO_2/ TEOS		Esterification of oleic acid w ith methanol Esterification refined palm oil	[76]
Vermiculite	SiO_2/ TEOS		Reduction of NO with ammonia	[39]
Vermiculite	SiO_2/ TEOS	CuO, Fe_2O_3	Reduction of NO with ammonia	[39]
Synthetic Fluorohectorite	SiO_2/ TEOS		Dehydration of 2-methylbut-3-yn-2-ol (MBOH)	[35]
Synthetic Saponite	SiO_2/ TEOS	Al_2O_3	Cumene Cracking	[63]
Synthetic Saponite	SiO_2/ TEOS		Oxidation of ammonia, reduction of NO with ammonia	[39], [49], [74]
Synthetic Saponite	SiO_2/ TEOS	CuO, Fe_2O_3	Oxidation of ammonia, reduction of NO with ammonia	[39], [49], [74]
Synthetic Laponite	SiO_2/ TEOS		Ethanol decomposition	[60]
Synthetic Laponite	SiO_2/ TEOS	Al_2O_3	Ethanol decomposition	[60]
			Applications in adsorptions	
Fluorohectorite	SiO_2/ TEOS	MPTMS	Adsorption of Hg^{2+}	[68]
Bentonite	SiO_2/ TEOS		Adsorption of phenol and dichlorophenol	[79]
Bentonite	SiO_2/ TEOS		Adsorption of methane, ethane and carbon dioxide	[84]
Bentonite	SiO_2 / TEOS Ph/ PhOS		Adsorption of methane, ethane and carbon dioxide	[84]
Natural Portuguese clay	SiO_2/ TEOS		Adsorption: methanol , methylethylketone, trichloroethylene, methanol, toluene	[81]
Natural Portuguese clay	SiO_2 / TEOS -C_6H_4-/ BTEB		Adsorption: methanol , methylethylketone, trichloroethylene, methanol, toluene	[81]

ZrPO - zirconium (IV) propoxide.
BTEB -bis(triethoxysilyl)benzene .
PhOS - Ph - Phenyl group Phenyltriethoxysilane.
MPTMS – mercaptopropyltrimethoxysilane.

Conclusion

Layered clay minerals are still very promising raw materials for the preparation of advanced catalysts and adsorbents . Intercalation of clays with interlayer pillars by surfactant directed method results in the combined micro-mesoporous mesoporous structure , relatively high surface area , high thermal and hydrothermal stability , surface acidity and ion-exchange properties. Such materials have been tested in various catalytic and adsorption processes. Some of the results reported in literature are very promising but its industrialization is limited due to relatively high cost of the PCH synthesis . Thus, preparation procedure still needs optimization to decreases the synthesis cost as well as to adjust PCH properties to the specific applications.

References

[1] M. F. Brigatti, E. Galan, B. K. G. Theng in Handbook of Clay Science, (F. Bergaya, B. K. G. Theng, G. Lagely, Eds.) Elsevier, Amsterdam, 2006 pp. 19-86.

[2] A. Galarneau, A. Barodawalla, T. J. Pinnavaia, *Nature,* 374 (1995) 529-531.

[3] M. Benjelloun, P. Cool, T. Linssen, E. F. Vansant, *Micropor. Mesopor. Mater.,* 49 (2001) 83-94.

[4] L. Chmielarz, Z. Piwowarska, P. Kustrowski, B. Gil, A. Adamski, B. Dudek, M. Michalik, *Appl. Catal. B,* 91 (2009) 449–459.

[5] R. Burch, *Catal. Today*, 2 (1988) 185-186.

[6] F. J. del Rey-Perez-Caballero, G. Poncelet, *Micropor. Mesopor. Mater.*, 41 (2000) 169-181.

[7] F. J. del Rey-Perez-Caballero, G. Poncelet, *Micropor. Mesopor. Mater.,* 37 (2000) 313-327.

[8] L. Chmielarz, R. Dziembaj, T. Grzybek, J. Klinik, T. Łojewski, D. Olszewska, H. Papp, *Catal. Lett.,* 68 (2000) 95-100.

[9] L. Chmielarz, R. Dziembaj, T. Grzybek, J. Klinik, T. Łojewski, D. Olszewska, A. Węgrzyn, *Catal. Lett.*, 70 (2000) 51-56.

[10] J. M. Adams, R. W. McCabe in Handbook of Clay Science, (F. Bergaya, B. K. G. Theng, G. Lagely, Eds.) Elsevier, Amsterdam, 2006 pp. 541-582.

[11] A. Vaccari, *Appl. Clay Sci.,* 14 (1999) 161-198.

[12] A. Gil, S. A. Korili, R. Trujillano, M. A. Vicente, *Appl. Clay Sci.,* 53 (2011) 97-105.

[13] R. Swarnakar, Kerstin B. Brandt, Ronald A. Kydd, *Appl. Catal. A*, 142 (1996) 61-71.

[14] N. Frini, M. Crespin, M. Trabelsi, D. Messad, H. Van Damme, F. Bergaya, *Appl. Clay Sci.,* 12 (1997) 281-292.

[15] N. Maes, I. Heylen, P. Cool and E. F. Vansant, *Appl. Clay Sci.,* 12 (1997) 43-60.

[16] J. Barrault, C. Bouchoule, K. Echachoui, N. Frini-Srasra, M. Trabelsi, F. Bergaya, *Appl. Catal. B,* 15 (1998) 269-274.

[17] R. T. Yang, N. Tharappiwattananon, R. Q. Long, *Appl. Catal. B,* 19 (1998) 289-302.

[18] K. M Parida, T Mishra, D. Das, S. N Chintalpudi, *Appl. Clay Sci.,* 15 (1999) 463-475.

[19] S. Moreno, R. Sun Kou, R. Molina, G. Poncelet, *J. Catal.*, 182, (1999) 174–185.

[20] P. Cañizares, J. L. Valverde, M. R. Sun Kou, C. B. Molina, *Micropor. Mesopor. Mater.,* 29 (1999) 267-281.
[21] M. Trombetta, G. Busca, M. Lenarda, L. Storaro, R. Ganzerla, L. Piovesan, A. J. Lopez, M. Alcantara-Rodrìguez, E. Rodriguez-Castellón, *Appl. Catal. A,* 193 (2000) 55-69.
[22] K. Bahranowski, A. Kielski, E. M. Serwicka, E. Wisła-Walsh, K. Wodnicka, *Micropor. Mesopor. Mater.,* 41 (2000) 201-215.
[23] J.M Guil, J.A Perdigón-Melón, M Brotas de Carvalho, A.P Carvalho, J Pires, *Micropor. Mesopor. Mater.,* 51 (2002) 145-154.
[24] S. Letaïef, B. Casal, P. Aranda, M. A. Martin-Luengo, E. Ruiz-Hitzky, *Appl. Clay Sci.,* 22 (2003) 263–277.
[25] S. C. Kim, D. K. Lee, *Catal. Today*, 97 (2004) 153–158.
[26] L. Chmielarz, P. Kuśtrowski, M. Zbroja, W. Łasocha, R. Dziembaj, *Catal. Today*, 90 (2004) 43–49.
[27] C. B. Molina, J. A. Casas, J. A. Zazo, J. J. Rodríguez, *Chem. Eng. J.,* 118 (2006) 29–35.
[28] M. Kurian, S. Sugunan, *Chem. Eng. J.,* 115 (2006) 139–146.
[29] S. Caudo, G. Centi, Ch. Genovese, S. Perathoner, *Appl. Catal. B,* 70 (2007) 437–446.
[30] D. L. Guerra, C. Airoldi, V. P. Lemos, R. S. Angélica, *J. Hazard. Mater.,* 155 (2008) 230–242.
[31] F. Tomul, S. Balci, *Appl. Clay Sci.*, 43 (2009) 13–20.
[32] F. Kooli, *Micropor. Mesopor. Mater.,* 167 (2013) 228–236.
[33] C. T. Kresge, M. E. Leonowicz, W. J. Roth, J. C. Vartuli, J. S. Beck, *Nature,* 359 (1992) 710-712.
[34] J. Y. Ying, Ch. P. Mehnert, M. S. Wong, *Angew. Chem. Int. Ed.,* 38 (1999) 56-77.
[35] A. Galarneau, A. Barodawalla, T. J. Pinnavaia, *Chem. Comm.,* (1997) 1661-1662.
[36] L. Wei, T. Tang, B. Huang, *Micropor. Mesopor. Mater.*, 67 (2004) 175-179.
[37] J. Ahenach, P. Cool, E. F. Vansant, *Phys. Chem. Chem. Phys.,* 2 (2000) 5750-5755.
[38] O. Y. Kwon, H. S. Shin, *Chem. Mater.*, 12 (2000) 1273 -1278.
[39] L. Chmielarz, Z. Piwowarska, P. Kuśtrowski, B. Dudek, B. Gil, M. Michalik, *Appl. Catal. B,* 88 (2009) 331-340.
[40] H. Palkova, J. Madejova, M. Zimowska, E. Bielanska, Z. Olejniczak, L. Litynska-Dobrzynska, E. M. Serwicka, *Micropor. Mesopor. Mater.,* 127 (2009) 228–236.
[41] H. Palkova, J. Madejova, M. Zimowska, E. M. Serwicka, *Micropor. Mesopor. Mater.,* 127 (2010) 237–244.
[42] M. Benjelloun, P. Cool, P. Van Der Voort, E. F. Vansant, *Phys. Chem. Chem. Phys.*, 4 (2002) 2818–2823.
[43] F. Kooli, P. C. Hian, Q. Weirong, S. F. Alshahateet, F. Chen, *J. Porous Mater.*, 13 (2006) 319-324.
[44] Q. Guo, Z. Zhang, X. Zhang, F. Ling. W. Sun , R. Li, K. Xie, *J. Porous Mater.,* 16 (2009) 209-213.
[45] S. Kawi, Y. Z. Yao, *Micropor. Mesopor. Mater.*, 28 (1999) 25-34.
[46] Q. Q. Hao, Z. W. Liu, B. Zhang, G. W. Wang, C. Ma, W. Frandsen, J. Li, Z. T. Liu, Z. Hao, D. S. Su, *Chem. Mater.*, 24 (2012) 972−974.
[47] F. Li, Y. Jiang, M. Xia, M. Sun , B. Xue, X. Ren, *J. Porous Mater.,* 17 (2010) 217–223.
[48] P. L. Meier, G. Kahl, *Clays Clay Miner.,*47 (1999) 386-388.
[49] L. Chmielarz, P. Kuśtrowski, R. Dziembaj, P. Cool, E. F. Vansant, *Catal. Today*, 119 (2007) 181-186.

[50] K. Cassiers, T. Lssen, M. Mathieu, M. Benjelloun, K. Schrijnemakers, P. Van Der Voort, P. Cool, E. F. Vansant, *Chem. Mater.*, 14 (2002) 2317-2324.

[51] M. Xia, Y. Jiang, L. Zhao, F. Li, B. Xue, M. Sun , D. Liu, X. Zhang, *Colloids and Surfaces A: Physicochem. Eng. Aspects,* 356 (2010) 1–9.

[52] M. Xia, Y. Jiang, F. Li, M. Sun , B. Xue, X. Chen, *Colloids and Surfaces A: Physicochem. Eng. Aspects,* 338 (2009) 1–6.

[53] L. Chmielarz, P. Kuśtrowski, Z. Piwowarska, B. Dudek, M. Michalik, *J. Solid State Chem.,* 182 (2009) 1094-1104.

[54] M. Pichowicz, R. Mokaya, *Chem. Comm.,* (2001) 2100-2101.

[55] L. Chmielarz, A. Kowalczyk, M. Michalik, B. Dudek, Z. Piwowarska, A. Matusiewicz, *Appl. Clay Sci.,* 48 (2010) 156-162.

[56] L. Chmielarz, M. Wojciechowska, M. Rutkowska, A. Adamski, A. Węgrzyn, A. Kowalczyk, B. Dudek, P. Boroń, M. Michalik, A. Matusiewicz, *Catal. Today,* 191 (2012) 25-31.

[57] L. Chmielarz, Z. Piwowarska, P. Kuśtrowski, A. Węgrzyn, B. Gil, A. Kowalczyk, B. Dudek, R. Dziembaj, M. Michalik, *Appl. Clay Sci.,* 53 (2011) 164-173.

[58] L. Chmielarz, A. Kowalczyk, M. Wojciechowska, P. Boroń, B. Dudek, M. Michalik, *Chem. Papers,* 68 (2014) 1219–1227.

[59] M. L. Pinto, J. Marques, J. Pires, *Sep. Purif. Technol.*, 98 (2012) 337-343.

[60] M. Zimowska, H. Palkova, J. Madejova, R. Dula, K. Pamin, Z. Olejniczak, B. Gil, E. M. Serwicka, *Micropor. Mesopor. Mater.,* 175 (2013) 67–75.

[61] C. Zhou, X. Li, Z. Ge, Q. Li, D. Tong, *Catalysis Today*, 93–95 (2004) 607–613.

[62] J. A. Cecilia, C. Gracia-Sancho, F. Franco, *Mircopor. Mesopor. Mater.*, 176 (2013) 95-102.

[63] M. Polverejan, Y. Liu, T. S. Pinnavaia, *Chem. Mater.,* 14 (2002) 2283-2288.

[64] H. Mao, B. Li, L. Yue, L. Wang, J. Yang, X. Gao, *Appl. Clay Sci.,* 53 (2011) 676-683.

[65] R. Ishii, M. Nakatsuji, K. Ooi, *Micropor. Mesopor. Mater.*, 79 (2005) 111-119.

[66] M. Nakatsuji, R. Ishii, Z.-M. Wang, K. Ooi, *J. Coll. Inter. Surf. Sci.,* 272 (2004) 158-166.

[67] J. Pires, M. L. Pinto, J. Estella, J. C. Echeverría, *J. Coll. Inter. Sci.,* 317 (2008) 206-213.

[68] L. Mercier, T. J. Pinnavaia, *Micropor. Mesopor. Mater.*, 20 (1998) 101-106.

[69] M. Diaz, P. Cambier, J. Brendlé, R. Prost, *Appl. Clay Sci.,* 37 (2007) 12-22.

[70] A. J. Tchinda, E. Ngameni, A. Walcarius, *Sens. Actuators B.,* 121 (2007) 113-123.

[71] A. J. Tchinda, E. Ngameni, I. T. Kenfack, A. Walcarius, *Chem. Mater.,* 21 (2009) 4111-4121.

[72] P. F. Siril, N. R. Shiju, D. R. Brown, K. Wilson, *Appl. Catal. A*, 364 (2009) 95–100.

[73] M. Sasidharan, A. Bhaumik, *J. Mol. Catal.,* 367 (2013) 1– 6.

[74] L. Chmielarz, P. Kuśtrowski, M. Drozdek, R. Dziembaj, P. Cool, E.F. Vansant, *Catal. Today,* 114 (2006) 319–325.

[75] K. W. Park, J. H. Jung, H. J. Seo, O. Y. Kwon, *Micropor. Mesopor. Mater.*, 121 (2009) 219-225.

[76] S. Jeenpadiphat, D. N. Tungasmita, *Powder Technol.*, 237 (2013) 634–640.

[77] B. Li, Z. Liu, J. Liu, Z. Zhou, X. Gao, X. Pang, H. Sheng, *J. Coll. Inter. Surf. Sci.,* 362 (2011) 450–456.

[78] I. Kuźniarska-Biernacka, C. Pereira, A. P. Carvalho, J. Pires, C. Freire, *Appl. Clay Sci.*, 53 (2011) 195–203.

[79] S. Arellano-Cardenas, T. Galardo-Valazquez, G. Osorio-Revilla, M. S. Lopez-Cortez, *Water Environ. Res.*, 80 (2008) 60-67.

[80] M. L. Pinto, J. Pires, J. Rocha, *J. Phys. Chem. C.*, 112 (2008) 14394–14402.

[81] C. D. Nunes, J. Pires, A. P. Carvalho, M. J. Calhorda, P. Ferreira, *Micropor. Mesopor. Mater.*, 111 (2008) 612–619.

[82] F. Qu, L. Zhu, K. Yang, *J. Hazard. Mater.*, 170 (2009) 7–12.

[83] M. L. Pinto, J. Pires, *Micropor. Mesopor. Mater.*, 151 (2012) 403–410.

[84] J. Pires, M. Bestilleiro, M. Pinto, A. Gil, *Sep. Purif. Technol.*, 61 (2008) 161-167.

In: Comprehensive Guide for Mesoporous Materials. Volume 1 ISBN: 978-1-63463-958-3
Editor: Mahmood Aliofkhazraei

Chapter 3

MESOPOROUS FILMS AND MEMBRANES FROM ASSEMBLY OF SILICA COLLOIDAL SPHERES

Amir Khabibullin and Ilya Zharov*
Department of Chemistry, University of Utah,
Salt Lake City , UT, US

ABSTRACT

This chapter describes the preparation and surface modification of mesoporous films and membranes based on silica colloidal spheres. Silica spheres form close-packed fcc lattice with ordered arrays of interconnected mesopores upon vertical deposition from colloidal solutions. Silica colloidal films can be prepared on various solid supports and suspended in glass or silicon support. Free-standing silica colloidal membranes can be obtained by sintering above 1000 °C. Silica colloidal membranes have a highly ordered geometry ; they are capable of size-selective transport and can be surface-modified in a well-defined and controlled manner with small molecules and polymers . The surface modification can be achieved using silanol chemistry, and polymer brushes with narrow mass distribution and controlled length can be formed on the mesopore surface using atom transfer radical polymerization and ring-opening polymerization . The transport in the resulting mesoporous membranes can be controlled by pH and ionic strength, temperature , light , and small molecule binding. Proton conducting silica colloidal membranes can be prepared by filling the mesopores with sulfonated polymer brushes.

Keywords: Silica, colloidal crystal, membrane, surface, responsive, separations, proton conductivity

1. INTRODUCTION

Porous films and membranes are important in fundamental studies of molecular transport and molecule-surface interactions in confined spaces [1], and are used in separations [2, 3] and sensing [4, 5]. They have been prepared using a variety of methods including lithography [6], anodic oxidation of aluminum membranes [7], track etching of polymers [8], surfactant -directed self-assembly [9], and by templating colloidal crystals [10, 11]. Responsive mesoporous membranes are produced by modifying the mesopore surfaces with a molecular layer containing chargeable moieties that are capable of electrostatic interactions [12] with the diffusing species . Alternatively, controlled molecular transport has been achieved by modifying the mesopore surfaces with polymer molecule s that respond to environmental stimuli [13]. A number of responsive mesoporous membranes have been prepared using self-assembled polymeric membranes [9], zeolites [14], silicon nitride membranes [6], and nanotubes [15, 16]. Such membranes can be used in separations of biomacromolecules [17] and drug molecules [18], in controlled release and drug delivery systems [19, 20], and in chemical sensors [4]. These applications result from the ability to control the molecular transport through the mesopores . This work is summarized in Table 1.1

Table 1.1. Summary of materials , preparation methods, and applications of responsive nanoporous membranes

Materials	Methods of the preparation	Applications
Copolymers [9]	Lithography [6]	Separations [17, 18]
Zeolites [14]	Anodic oxidation of aluminum [7]	Controlled release and drug delivery [19, 20]
Silicon nitride [6]	Track-etching of polymers [8]	Chemical sensors [4]
Carbon nanotubes [15, 16]	Surfactant-directed self-assembly [9]	
	Templating colloidal crystals [10, 11]	

Silica colloidal crystals are promising materials for mesoporous films and membranes because they form through self-assembly , allow straightforward control of mesopore size in a broad range, possess exceptionally high molecular throughput, and offer a large variety of surface chemistries.

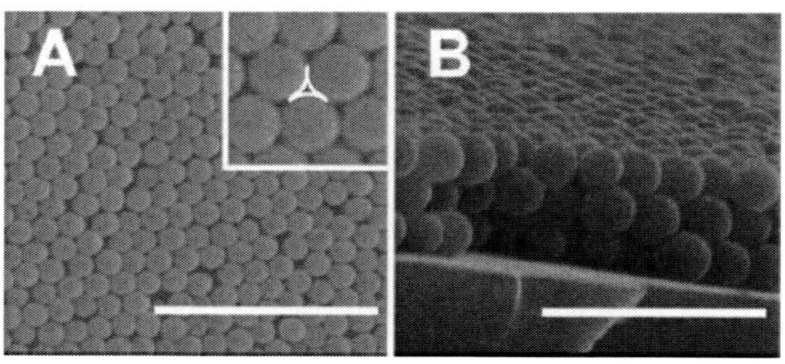

Figure 1.1. SEM images of silica colloidal membranes prepared on glass from 440 nm diameter silica spheres deposited (A) top view (size bar 4 µm); and (B) side view (size bar 2 µm). The geometric projection of a pore observed from the (111) plane is outlined in the inset in (A). Reproduced with permission from ref. [36]. Copyright : 2005 The American Chemical Society.

Silica colloidal crystals comprise a close-packed face-centered cubic (fcc) lattice of amorphous non-porous silica spheres of a sub-micrometer diameter (Figure 1.1) with ordered arrays of interconnected three-dimensional nanoscale voidse [21]. The preparation of silica spheres (Scheme 1.1) is straightforward [22], self-assembly of the spheres is well developed [23], and pore size in the crystals can be readily controlled by selecting the sphere size (the distance from the center of the mesopore projection to the nearest silica sphere surface is ca. 15% of the sphere radius). Surface silanol groups can be directly modified by nucleophilic silylation to introduce a variety of functional groups [24]. Alternatively, silica surface can be first modified with 3-aminopropyltriethoxysilane (Scheme 8.1.1), followed by treatment with organic molecules carrying electrophilic moieties such as acyl chloride, isocyanate, isothiocyanate, carboxylic acid , sulfonyl chloride [25], or succinimidyl ester [26]. 3-Aminopropyl-triethoxysilane-treated silica can also be modified with 2-bromoisobutyrylbromide [27], which can serve as atom transfer radical polymerization initiator [28]. This provides the possibility of growing various polymer brushes on the silica surface [29, 30, 31, 32].

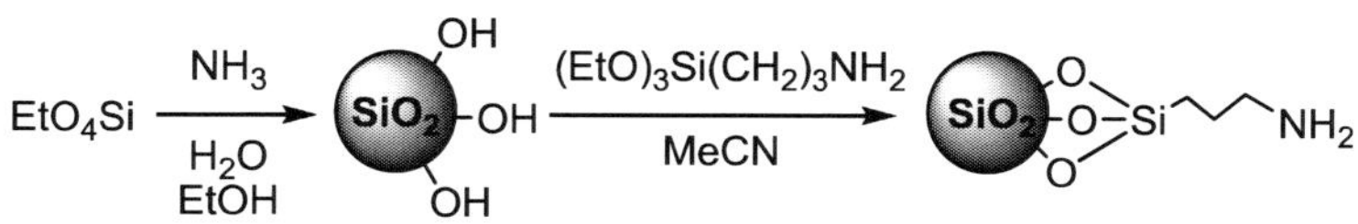

Scheme 1.1. Preparation and surface amination of silica spheres.

An important advantage of colloidal crystals as mesoporous membranes is their highly ordered structure , which allows using accurate mathematical descriptions of the molecular transport [33, 34, 35]. The effective diffusivity of molecules in the fcc lattice, $D_{colloid}$, can be expressed as $(\varepsilon/\tau)D_{sol}$, where D_{sol} is the diffusivity of molecules in free solution , and the void fraction ε (0.26) and the tortuosity τ (~3.0) are intrinsic geometrical parameters independent of the size of the silica spheres used to prepare the colloidal crystal. An estimate of the molecular flux J_{fcc} can be obtained using equation (1) [33], where ΔC is the concentration gradient and L is the thickness of the membrane [33]:

$$J_{colloid} = (\Delta C/L) \times (\varepsilon/\tau)D_{sol} \quad (1)$$

For a typical D_{sol} of 10^{-5} $cm^2 \cdot s^{-1}$, L of 10 μm, and ΔC of 10 mM, $J_{colloid}$ is ~10^{-8} $mol/cm^2 \cdot s$. Importantly, the diffusive flux of small molecules normal to the (111) plane of a semi-infinite colloidal crystal is only ca. 10 times smaller relative to the free solution value, independent of the size of the spheres used to assemble the crystal [35]. Thus, the rate of molecular transport remains significant, even when the pore size is reduced to the nanoscale to impart molecular transport selectivity .

In 2005 we introduced, for the first time, the concept of permselective silica colloidal mesoporous films by describing pH -responsive amine -modified colloidal membranes with controlled transport of positively charged species [36]. Later, we reported a detailed study of transport through amine-modified colloidal membranes [37], as well as membranes modified with sulfonic acids [38, 39] and spiropyran moieties [40]. We developed several methods for the preparation of silica colloidal membranes, including suspended colloidal membranes [41, 42] and mechanically robust free-standing silica membranes [43]. We showed that unmodified free-standing silica colloidal membranes are capable of size-selective transport [44]. We also developed methods to modify the colloidal mesopores with acrylamide [45], and introduced temperature responsive poly(*N*-isopropylacryl amide), PNIPAAM [46], pH- and ion-responsive poly(2-(dimethylamino)ethyl methacrylate), PDMAEMA [47, 48], temperature-responsive poly(L-alanine) [49], small molecule-responsive aptamer [50], and studied the molecular transport in these polymer -modified membranes as a function of the environmental conditions . Finally, we prepared proton conducting silica colloidal membranes by filling the mesopores with sulfonated polymers , and studied the properties of these membranes [51]. This work is summarized in Table 1.2.

Table 1.2. Summary of responsive silica colloidal membranes

Stimulus	Silica surface modification
pH and ionic strength	Amines [36,37] Sulfonic acid groups [38,39] PDMAEMA [47,48]
Temperature	PNIPAAM [46] Poly(L-alanine) [49]
Small molecule binding	Aptamer [50]
Light	Spiropyran [40]

2. Preparation of Silica Colloidal Films and Membranes

2.1. Supported Silica Colloidal Films

Silica colloidal films can be assembled on a hydrophilic solid support (glass, oxidized silicon , metal, etc.) using the vertical deposition technique [23]. In this technique silica colloidal films are deposited onto the support, which is placed vertically into colloidal solution of silica spheres in ethanol . The solutions are allowed to evaporate for 2-3 days in a vibration -free environment . The thickness of the colloidal film is controlled by varying the weight percent of silica spheres in the colloidal solution.

2.2. Suspended Silica Colloidal Membranes

Suspended silica colloidal membranes can be prepared in silicon [41] and glass [42] supports. For example, silica colloidal membranes suspended in silicon have been prepared using a 0.3 mm thick Si(100) wafer, with an array of openings shaped as a truncated square pyramid with 50 ×50 μm smaller dimension (2a, Figure 2.1), or a single 40 × 40, 100 × 100, 250 × 250 or 500 × 500 μm opening. A wafer was suspended vertically in a colloidal solution of silica spheres in absolute ethanol . The solvent evaporated at room temperature , resulting in the formation of suspended silica colloidal membranes (Figure 2.2). Silica spheres are packed in fcc orientation as seen in Figure 2.3B).

To measure the thickness of the suspended membranes and to test them for the absence of cracks and pinholes, diffusion of Rhodamine B through these suspended silica membranes was studied. Rhodamine B diffusion rate increased linearly with increasing the membranes area. The experimental diffusion rates were in good agreement with the calculated values. Silica colloidal membranes suspended in glass can be prepared in a similar way [42] using a 150-μm-thick glass slide. Figure 2.4 shows an SEM image of an unfilled opening in a glass microscope slide anda light microscope image of a silica colloidal membrane suspended in openings in glass microscope slide covers.

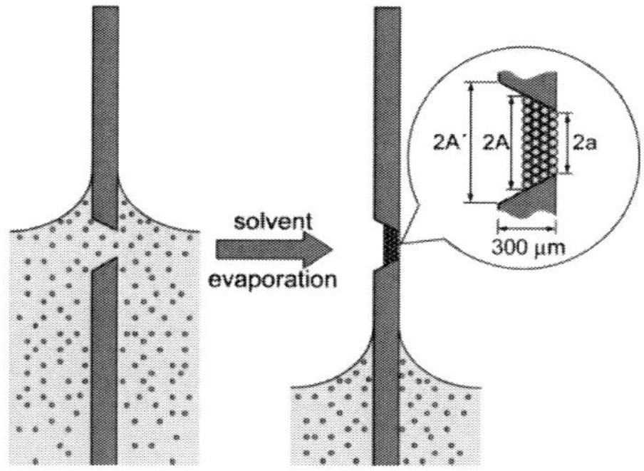

Figure 2.1. Scheme of formation of a suspended opal membrane in a frustum-shaped opening in a 300 μm-thick silicon wafer. Reproduced with permission from ref. [41]. Copyright : 2006 The American Chemical Society.

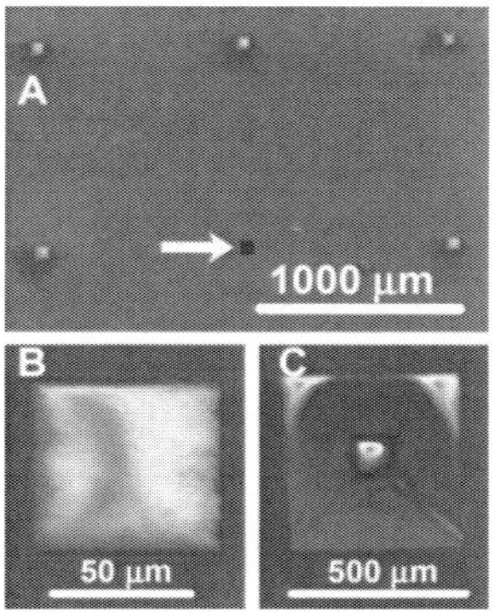

Figure 2.2. SEM images of a 0.3 mm thick silicon wafer containing frustum-shaped openings with 50 × 50 μm smaller dimension after the opal membrane formation using 440 nm silica spheres. (A) Front side of the wafer with five filled and one unfilled (marked with white arrow) opening. (B) Front and (C) back side of a filled opening. Reproduced with permission from ref. [41]. Copyright : 2006 The American Chemical Society.

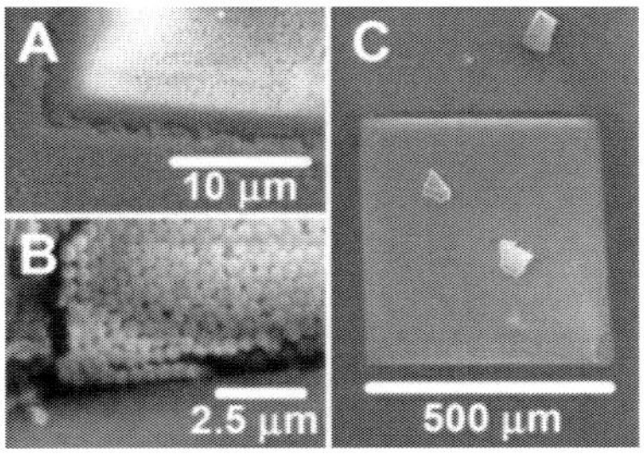

Figure 2.3. SEM images of an opal membrane peripherally supported (A) in 50 × 50 μm opening in a 0.3 mm thick silicon wafer, showing (B) close-packed fcc lattice of the 440 nm silica spheres in the supported opal membrane; and (C) an opal membrane in a 500 × 500 μm opening in a 0.3 mm thick silicon wafer. Reproduced with permission from ref. [41]. Copyright : 2006 The American Chemical Society.

The images show that the colloidal membrane contained no major cracks through the middle of the membrane. The silica spheres formed fcc lattice (Figure 2.5) with point defects but no major cracks or other mechanical defects that would extend through the membrane. The SEM images of both sides of the slides showed the membranes to be flush with the glass or slightly elevated above the glass substrate surface [42]. Diffusion of $Fe(bpy)_3^{2+}$ through silica membranes suspended in glass was in good agreement with predicted values suggesting that the membranes contain no major mechanical defects, although small defects have been observed at the periphery of the membranes.

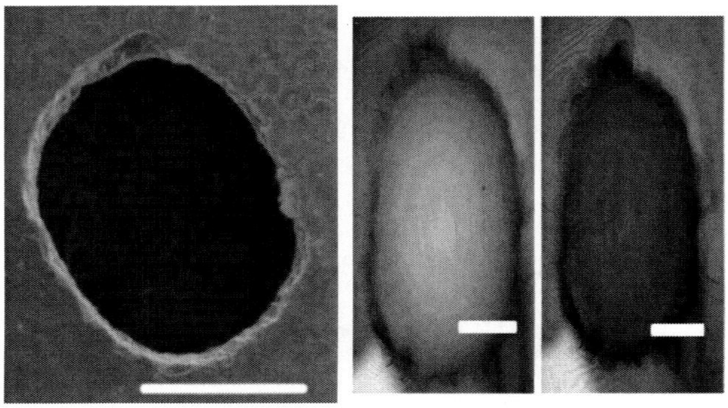

Figure 2.4. Left: SEM image of a representative glass microscope slide with a 455 × 364 μm opening (size bar is 250 μm). Right: Light microscope image of a representative colloidal membrane suspended in glass with a 455 × 364 μm opening (size bar is 100 μm). Reproduced with permission from ref. [42]. Copyright : 2011 Springer.

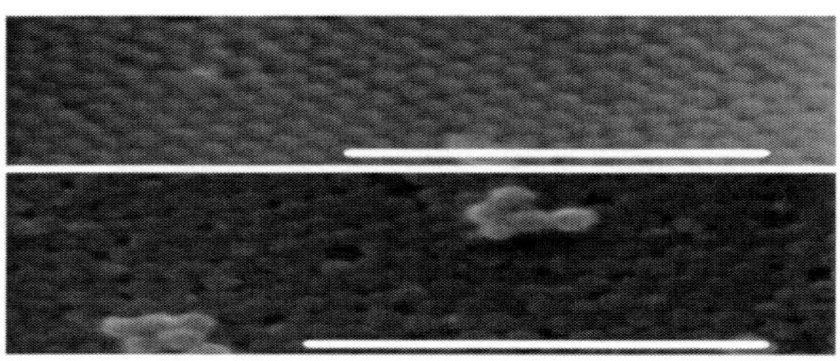

Figure 2.5. SEM image of suspended colloidal membranes in a glass microscope slide cover comprised of 239 nm silica spheres (size bar is 2 μm). Reproduced with permission from ref. [42]. Copyright : 2011 Springer.

2.3. Free-Standing Silica Colloidal Membranes

The free-standing silica colloidal membranes can be prepared as a material that is mechanically durable and possesses large area (Figures 2.6 and 2.7). To prepare such membranes, the silica spheres assembled into a colloidal crystal are physically bonded together by sintering at 1050 °C, at which temperature silica flows at the surface. The preparation process is time and cost -effective. The sintering process can be followed by rehydroxylation with tetrabutylammonium hydroxide in water to restore the surface hydroxyl groups [52].

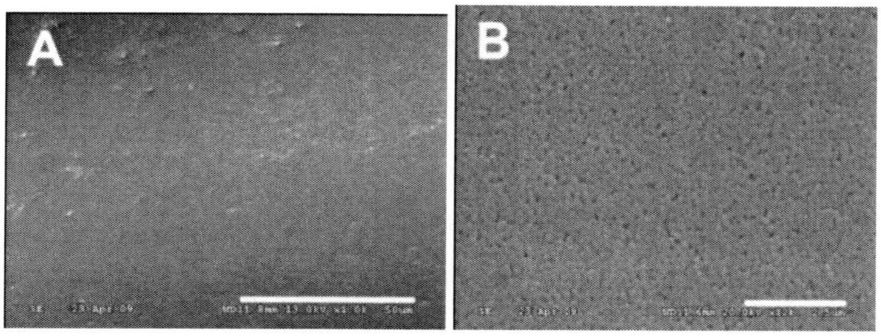

Figure 2.6. SEM images of sintered colloidal crystals comprised of 180 nm silica spheres: A. SEM image showing no major cracks over a large area (size bar = 50 μm); B. Magnified image displaying the close-packed fcc lattice (size bar = 2.5 μm). Reproduced with permission from ref. [43]. Copyright : 2009 The American Chemical Society.

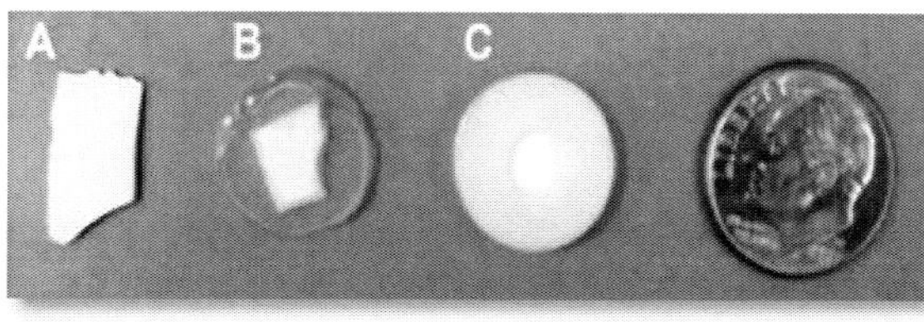

Figure 2.7. Photographs of sintered silica colloidal membranes . (A) as-sintered; (B) without PTFE washers showing the sintered colloidal membrane in the epoxy; (C) with PTFE washers. Reproduced with permission from ref. [43]. Copyright : 2009 The American Chemical Society.

The diffusional flux of the $Fe(bpy)_3^{2+}$ ion through free-standing silica colloidal membranes in acetonitrile is on the order of 3.6×10^{-10} $mol \cdot s^{-1} \cdot cm^{-2}$ and is in good agreement with the calculated flux, confirming that the membranes are crack-free and have no major defects [43].

Table 2.1. Types of silica colloidal membranes

Silica colloidal membrane	Design
Supported	Supported on glass-shrouded Pt electrode [47,48]
Suspended	Suspended in silicon [41] or glass [42] openings
Free-standing	No support required [43,44], can be sealed in gasket

3. Size-Selectivity in Silica Colloidal Membranes

Free-standing sintered silica colloidal membrane show size selectivity for the transport of macromolecules [44], in a way similar to the size selectivity for other mesoporous membranes [53, 54]. Polyamidoamine (PAMAM) dendrimers were used as diffusion probes due to their high monodispersity, well-defined molecular architecture [55] and size ranging from 2-20 nm. Rhodamine B isothiocyanate-labeled PAMAM dendrimers were prepared via the amide bond formation . The extent of dye-labeling evaluated using UV absorbance (Table 3.1) was purposefully kept low to avoid altering the size of the diffusion probes.

Table 3.1. Characteristics of diffusing species [44]

PAMAM	Diameter, nm[a]	No. of -NH_2 end groups per dendrimer	No. of dyes per dendrimer
G1	1.9	8	2.87
G2	2.6	16	2.79
G3	3.6	32	2.72
G4	4.4	64	3.00

[a] Rhodamine B diameter was estimated as 1.6 nm, the diameters of the dendrimers are given as provided by the manufacturer (Dendritic Nanotechnologies, Inc.).

Table 3.2 summarizes the calculated diffusion coefficient D_{eff} values for the dye-labeled PAMAM dendrimers used. Comparison of the calculated D_{eff} values for the same membrane showed that the diffusion rate decreased with increasing size of the diffusing species . The results for the membranes sintered from 100 nm spheres (7.5 nm 'radius ' mesopores) showed an abrupt change in diffusion rates from G2 to G3. While G2 dendrimer diffused about three-fourths as fast as G1, G3 has a diffusion rate *ca.* three times lower, and G4 and G5 have roughly the same diffusion rates, about four times slower than G1. This is noteworthy considering the fact that there is only about 3 - 4 nm difference between the diameters of G4 and G5 and of G1 dendrimers (Table 3.1). Moreover, the size of the colloidal mesopores is significantly larger than the dendrimer diameter, suggesting an additional effect of tortuosity on the size selectivity . The D_{eff} is dependent on the mesopore size, with the smallest D_{eff} values corresponding to diffusion through the 7.5 nm mesopores (Table 3.2). The selectivity for G4 dendrimer is lower by a factor of ca. 2 for the membranes with the larger mesopores.

Table 3.2. Calculated average effective diffusion coefficients through as-sintered and rehydroxylated membranes made using 100 and 362 nm silica spheres [44]

Pore "diameter", nm	Diffusing PAMAM species	D_{eff} (cm^2/s) × 10^{-6}	% D_{eff} ratio relative to G1
15	G1	2.35 ± 0.42	100
	G2	1.79 ± 0.31	76 ± 19
	G3	0.82 ± 0.15	35 ± 9
	G4	0.57 ± 0.10	24 ± 6
54	G1	1.68 ± 0.27	100
	G4	0.70 ± 0.13	42 ± 10

In order to confirm that the observed transport selectivities are due to the size discrimination and not electrostatic effects, the diffusion experiments were also performed using the rehydroxylated membranes . The difference between as-sintered and rehydroxylated membranes is that the former have a greater number of terminal siloxanes (Si-O-Si) on the surface [56]. If transport selectivity of the dendrimers through the sintered membranes is affected by electrostatic effects, different diffusion rates through the rehydroxylated membranes would be expected due to the interactions between the negatively charged mesopore surface and the cationic dye-labeled dendrimers. However, a comparison of both calculated D_{eff} and selectivity values showed little difference in diffusion rates between the as-sintered and the rehydroxylated membranes. This confirmed that the molecular transport selectivity exhibited by the sintered membranes results exclusively from the size discrimination.

4. Responsive Pore-Filled Silica Colloidal Films and Membranes

4.1. pH - and Ionic Strength-Responsive Films and Membranes

To demonstrate that silica colloidal membranes can be functionalized to acquire pH - responsive permselectivity we performed surface-initiated atom transfer radical polymerization (ATRP) of 2-(dimethylamino)ethyl methacrylate (DMAEMA) inside the mesopores of the silica colloidal films comprised of 255 nm silica spheres (Scheme 4.1) and studied the molecular transport across the resulting PDMAEMA-filled mesoporous films as a function of pH and salt concentration for positively charged ($Ru(NH_3)_6^{3+}$) and neutral ($Fc(CH_2OH)_2$) redox-active species [47]. The polymer brush thickness inside the mesopores after 20 hours of polymerization was estimated as 10-15 nm, thus filling most of the void inside the colloidal crystal [47].

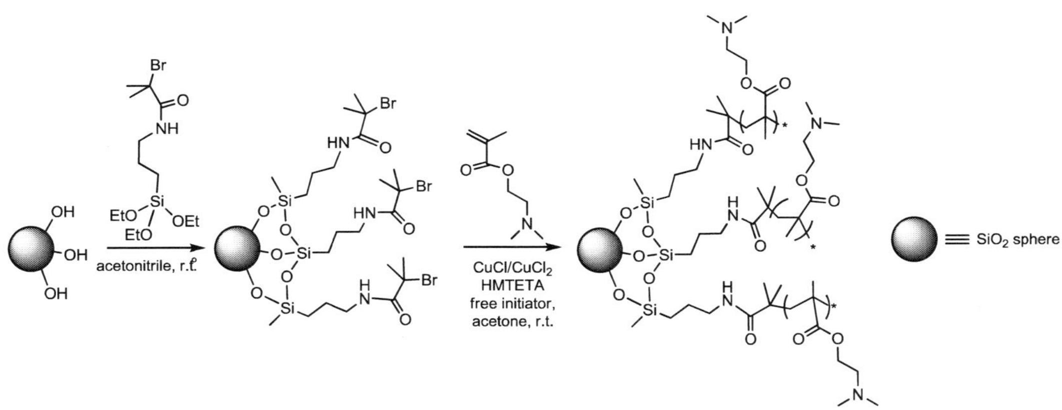

Scheme 4.1. Modification of silica spheres with PDMAEMA brushes. Reproduced with permission from ref. [47]. Copyright : 2008 The American Chemical Society.

PDMAEMA is a well-studied environmentally responsive polymer [57] whose behavior is governed by both electrostatic and hydrophobic interactions. It has been found that at low pH electrostatic repulsions between the protonated tertiary amine group s lead to PDMAEMA swelling [58]. In contrast, at high pH most of the amine groups are deprotonated and neutral . As a result, interactions between nonpolar groups in a polar solvent (hydrophobic interactions) are increased, which leads to a more compact PDMAEMA conformation. In addition, PDMAEMA exhibits ion-responsive behavior. Increasing ionic strength of the solution leads to a collapsed conformation of PDMAEMA at acidic pH as a result of charge screening in the protonated polymer [59].

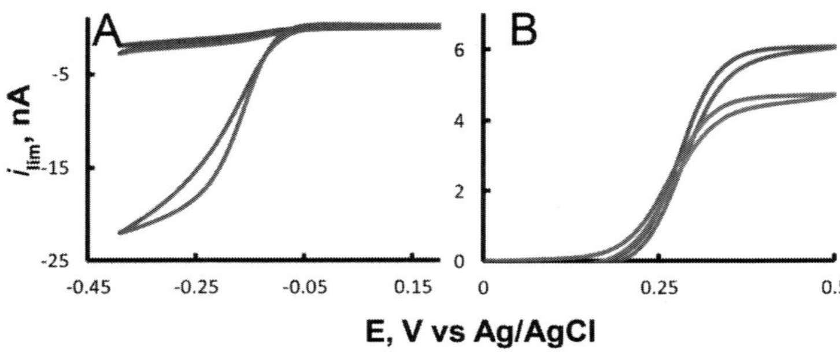

Figure 4.1. Representative voltammetric responses for PDMAEMA-colloidal film Pt electrodes (20 h polymerization) at different pH for $Ru(NH_3)_6^{3+}$ (A) and for $Fc(CH_2OH)_2$ (B). Voltammograms recorded above pH 5 are at the bottom in (A) and at the top in (B). Voltammograms recorded below pH 4 are at the top in (A) and at the bottom in (B). Reproduced with permission from ref. [47]. Copyright : 2008 The American Chemical Society.

First, we studied the pH response for PDMAEMA-filled colloidal membranes using positively charged $Ru(NH_3)_6^{3+}$. As can be seen in Figure 4.1A, the limiting current for PDMAEMA-modified colloidal membrane electrodes is highly dependent on pH. This effect is summarized in Figure 4.2. The limiting current increases with increasing pH by ~80% with an abrupt change at pH~4-5.

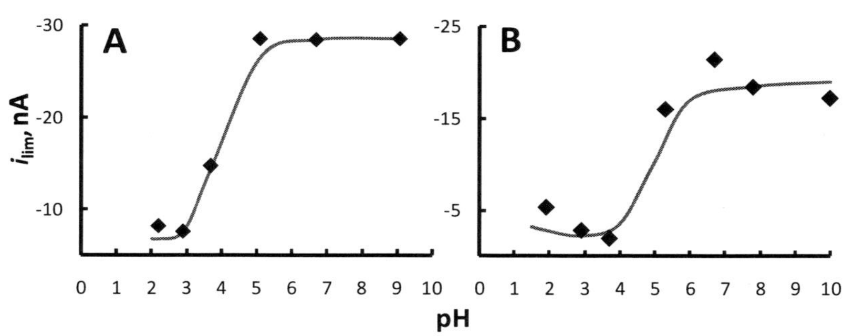

Figure 4.2. Limiting current ($Ru(NH_3)_6^{3+}$) as a function of increasing pH for PDMAEMA-colloidal membrane Pt electrodes for 5 h (A) and for 20 h (B) polymerizations. Reproduced with permission from ref. [47]. Copyright : 2008 The American Chemical Society.

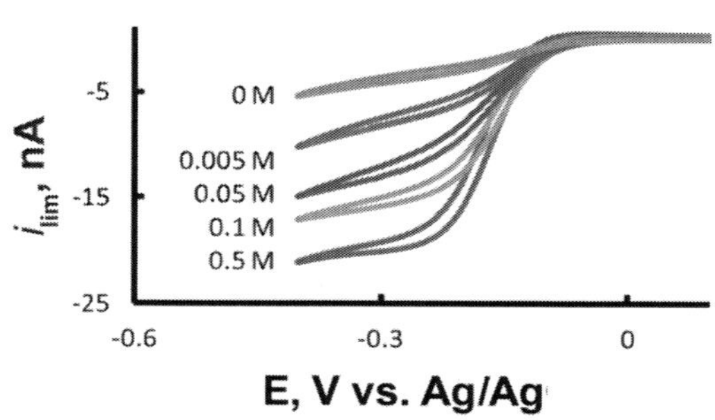

Figure 4.3. Voltammetric responses of the protonated PDMAEMA-modified colloidal film Pt electrodes for $Ru(NH_3)_6^{3+}$ reduction in water as a function of KCl concentration. KCl concentrations

are shown adjacent to each voltammogram. Reproduced with permission from ref. [47]. Copyright : 2008 The American Chemical Society.

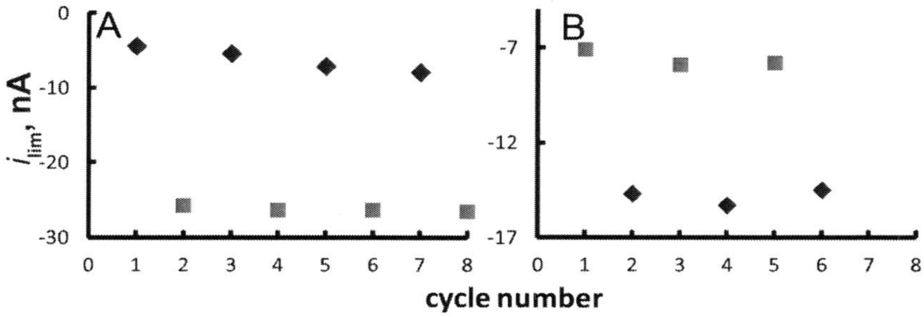

Figure 4.4. (A) Limiting current ($Ru(NH_3)_6^{3+}$) at pH 2 (squares) and at pH 7 (diamonds). (B) Limiting current ($Ru(NH_3)_6^{3+}$) at 0.05 M KCl (squares) and at 0 M KCl (diamonds) at pH=2 for PDMAEMA-colloidal film Pt electrodes after polymerization for 5 hours. Reproduced with permission from ref. [47]. Copyright : 2008 The American Chemical Society.

For polyelectrolyte brushes, the chain conformation is governed by the electrostatic interactions between the charged monomer units [60]. At high pH , the amine group s of the polymer chain s are deprotonated, and the polymer is considered to be neutral . As a result, the polymer chains tend to attain a collapsed conformation due to the hydrophobic interactions, and the diffusion of the positively charged $Ru(NH_3)_6^{3+}$ is not sterically hindered by the polymer brush, nor it is repelled electrostatically. In contrast, at low pH, the polymer brush becomes protonated and stretches away from the surface as a result of electrostatic repulsions between the charged monomer units and between the polymer chains. The diffusion of $Ru(NH_3)_6^{3+}$ in this case is blocked as a result of both electrostatic repulsion from the positively charged polymer chains and as a result of the steric hindrance.

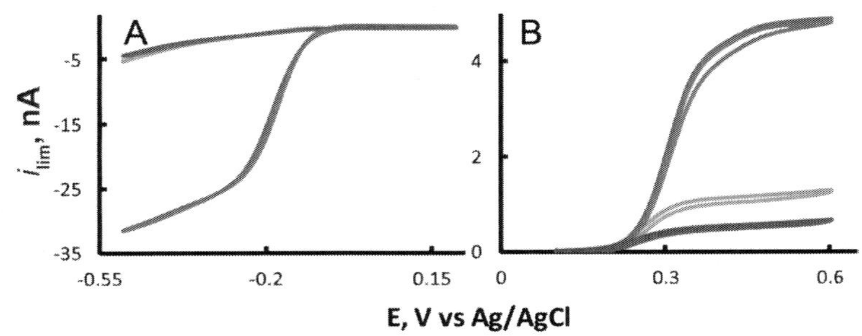

Figure 4.5. Representative voltammetric responses of $Ru(NH_3)_6^{3+}$ (A) and $Fc(CH_2OH)_2$ (B) of PDMAEMA-colloidal film Pt electrodes (5 h polymerization). Before quaternization (bottom), after quaternization (top), and after quaternization in the presence of 0.5 M KCl (top) in (A). Before quaternization (top), after quaternization (bottom), and after quaternization in the presence of 0.5 M KCl (middle) in (B). Reproduced with permission from ref. [47]. Copyright : 2008 The American Chemical Society.

To isolate the pH effect on the polymer conformation, we examined the diffusion of a neutral redox-active molecule, $Fc(CH_2OH)_2$, assuming that it does not electrostatically interact with the polymer chain s . The limiting current of $Fc(CH_2OH)_2$ decreased only by ~30% (Figure 4.1B) with no abrupt change when pH was lowed from neutral to acidic , which should result exclusively from the conformational changes in the PDMAEMA chains.

The influence of the solution ionic strength on the diffusion across the protonated PDMAEMA-filled colloidal films was studied by measuring the limiting current of $Ru(NH_3)_6^{3+}$ and $Fc(CH_2OH)_2$ as a function of KCl concentration. Figure 4.3 shows the dependence of the limiting current of $Ru(NH_3)_6^{3+}$ on the salt concentration at pH 2.4 where the PDMAEMA chains are protonated. The limiting current increases with increasing KCl concentration. The addition of KCl progressively screens the charge within the polymer brush. As a result, the diffusion of $Ru(NH_3)_6^{3+}$ becomes easier. Similar results were observed for the colloidal membranes obtained after both 5 and 20 h polymerization . These results are in good agreement with previously reported studies for PDMAEMA-modified surfaces [60, 61, 62, 63]. No significant effect of the salt concentration on the diffusion of $Fc(CH_2OH)_2$ at low pH was observed, which suggests that, under these conditions, the conformation of the PDMAEMA chains is not significantly affected by the salt concentration [62, 63]. Both pH- and ion-dependent changes in limiting current were reversible (Figure 4.4). The mesoporous PDMAEMA-filled colloidal films can be cycled between the low and high pH and between the low and high ionic strength regimes without apparent loss of responsiveness .

To demonstrate that the "weak" polyelectrolyte PDMAEMA can be converted into a "strong" polyelectrolyte brush with a fixed, pH -independent number of charges, the quaternization of the polymer by treating the PDMAEMA-colloidal film electrodes with ethyl bromide was performed [64]. Figure 4.5 shows the limiting current of $Ru(NH_3)_6^{3+}$ and $Fc(CH_2OH)_2$ of these quaternized electrodes at neutral pH.

The limiting current of both neutral and positively charged species decreased significantly after the quaternization. This suggests that the polymer brush carrying a large number of quaternary ammonium ions blocks the diffusion of ions and molecules almost completely regardless of pH as a result of both strong electrostatic repulsion and steric hindrance. Next, the limiting current for $Ru(NH_3)_6^{3+}$ and $Fc(CH_2OH)_2$ at 0.5 M concentration of KCl was measured. No significant change in the limiting current for $Ru(NH_3)_6^{3+}$ with increased ionic strength (Figure 4.5A) was observed, indicating that the positive charge of the quaternized polymer brush was not significantly screened under these conditions. At the same time, for $Fc(CH_2OH)_2$ a ~10% increase in the limiting current was observed (Figure 4.5B), suggesting small conformational changes in the polymer chain s . This result is in agreement with the previously reported data [60], which demonstrated that at moderate salt concentration the thickness of strong polyelectrolyte brushes does not strongly depend on the salt concentration due to the osmotic pressure of the counterions trapped inside the brush (osmotic brush regime).

PDMAEMA brushes were also prepared on the surface of mesopores in free-standing silica colloidal membranes [43], which showed similar pH -responsive transport behavior [48]. The silica surface was functionalized using the reaction of the surface amines with 2-bromoisobutyryl bromide, in order to attach the ATRP initiator moieties to the mesopore surface, followed by the ATRP of DMAEMA inside the mesopores of the free-standing membranes, as described above. In order to estimate the molecular weight of the polymer chain s formed on the mesopore surface, we carried out the polymerization of DMAEMA in the presence of a small amount of sacrificial initiator in solution . This resulted in the formation of a free polymer, which we isolated and analyzed by gel permeation chromatography (GPC). We assumed that the chains formed at the surface and those formed in solution have similar molecular weights. The M_w PDMAEMA after 16 hours of polymerization was 19,900 Da, after 22 hours was 23,300 Da, and after 44 hours was 30,500 Da.

We used $Fe(bpy)_3^{2+}$ diffusion measurements to determine the responsive behavior of the PDMAEMA-filled membranes with and without trifluoroacetic acid present in the diffusing solution . Figure 4.6 shows the diffusion rate of $Fe(bpy)_3^{2+}$ with and without trifluoroacetic acid present in solution through the membranes polymer -modified at different polymerization time s. For the membrane modified with the polymer for 16 hours, the diffusion rate of $Fe(bpy)_3^{2+}$ was by 42% (2.4 times) lower in the presence of 50 mM trifluoroacetic acid than in the absence of the acid. After a longer polymerization time (22 hours) the acid-controlled response of the membrane increased greatly (Figure 4.6 B) and the diffusion blockage reached 95%. The switching behavior of the polymer-modified colloidal membranes was reversible as was shown by cycling between "on" and "off" states . In addition to the polymer length, increasing the membrane thickness as well as decreasing the mesopore size enhanced the gating behavior and allowed to achieve the complete acid-controlled gating.

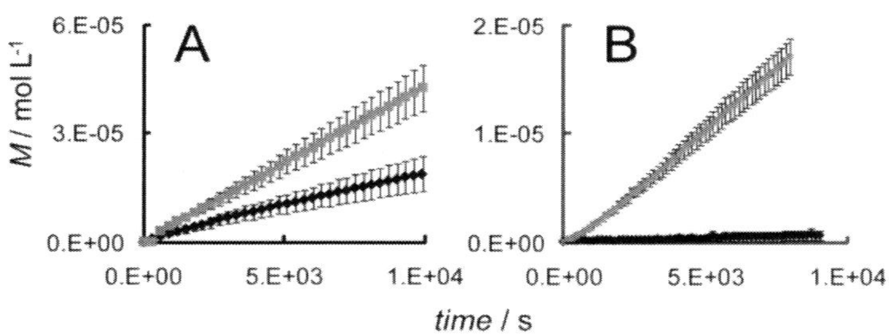

Figure 4.6. Diffusion rates of $Fe(bpy)_3^{2+}$ through PDMAEMA-modified colloidal membranes (180 nm silica spheres) with (black) and without (grey) 50 mM trifluoroacetic acid after (A) 16 hours and (B) 22 hours of polymerization . Reproduced with permission from ref. [48]. Copyright : 2010 Wiley, Inc.

4.2. Temperature-Responsive Silica Colloidal Films

PNIPAM-filled films . Poly(*N*-isopropylacrylamide), PNIPAAM, is a well-known temperature responsive polymer [65] that has been used in the preparation of thermoresponsive membranes [66, 67]. We modified colloidal film mesopores of 15 nm "radius " with PNIPAAM brushes (Scheme 4.2) [68] of 9.3 nm estimated thickness, and measured the temperature response for the membranes modified at different polymerization time s. As can be seen in Figure 4.7, two types of response are observed. For mesoporous films modified with a thin polymer brush, the limiting current increases with temperature, with a gradual change at ca. 29 °C (Figure 4.7A). For colloidal films filled with thick polymer brush a reverse change is observed, where the limiting current decreases with increasing temperature, with an abrupt change at ca. 29 °C (Figure 4.7B).

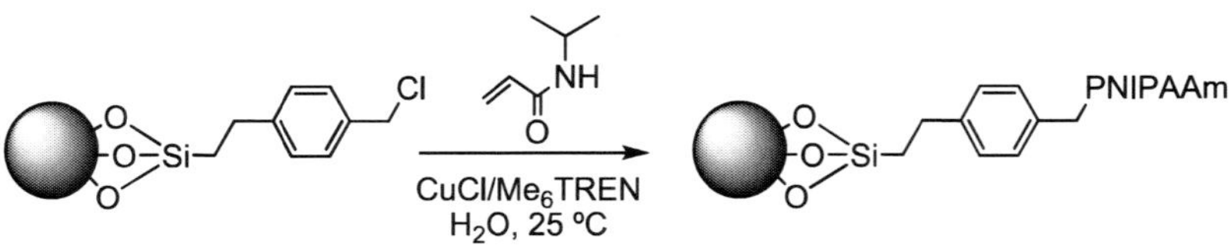

Scheme 4.2. Preparation of PAAM and PNIPAAM brushes on silica sphere surface.

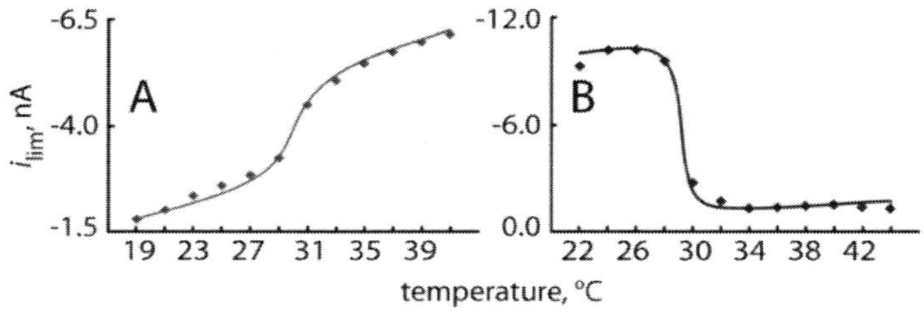

Figure 4.7. Limiting current (Ru(NH3)63+) as a function of increasing temperature for PNIPAAM-opal membrane Pt electrodes after polymerization for (A) 15 min, and (B) 90 min. Reproduced with permission from ref. [46]. Copyright : 2007 The American Chemical Society.

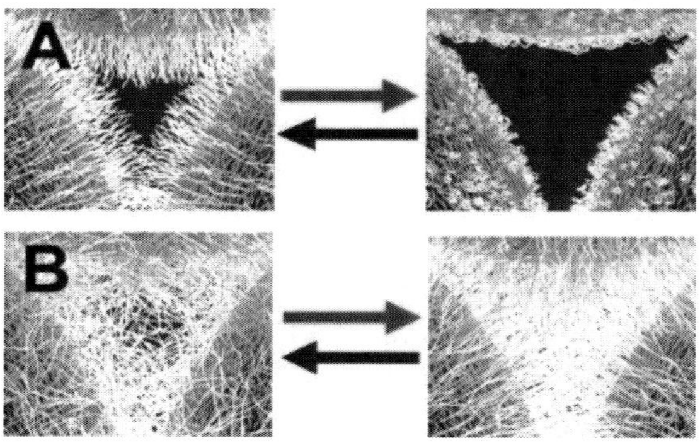

Figure 4.8. Schematic representation of the processes that occur upon heating and cooling of a (A) PNIPAAM brush (15 min polymerization) and (B) PNIPAAM gel (90 min polymerization) inside a colloidal mesopore. Reproduced with permission from ref. [46]. Copyright : 2007 The American Chemical Society.

These results are consistent with two types of PNIPAAM morphologies [66, 67] inside the mesopores , which lead to two types of molecular transport mechanisms through these mesopores. We believe that when *N*-isopropylacrylamide is polymerized for a short period of time, it forms a dense brush (Figure 4.8A). Transport through such mesopores mainly happens in the polymer -free volume of the mesopores. With rising temperature the conformation of the polymer chain s inside the brush changes in such a way that the brush shrinks, providing a larger volume for diffusion (Figure 4.8A), which results in the observed increase in molecular transport. When polymerization is conducted for a sufficiently long time, polymer chains in the brushes growing from the opposite mesopore wall become long enough to meet and interpenetrate. This leads to a highly porous and permeable hydrogel structure (Figure 4.8B). When the temperature is increased, the hydrogel becomes dehydrated and impermeable to aqueous permeants (Figure 4.8B), but it does not shrink to open the mesopores [66, 67].

Polyalanine-filled films . It has been shown that the poly(L-alanine) conformation responds to solvent polarity , ionic strength, and temperature [69]. Thus, poly(L-alanine) brushes with 12 nm thickness were grown on the surface of the 19 nm "radius " mesopores in silica colloidal films assembled on Pt electrodes (opal electrodes) according to Scheme 4.3 [49]. Polymerization time was varied from 7 minutes to 6 hours to create polymer brushes of various lengths.

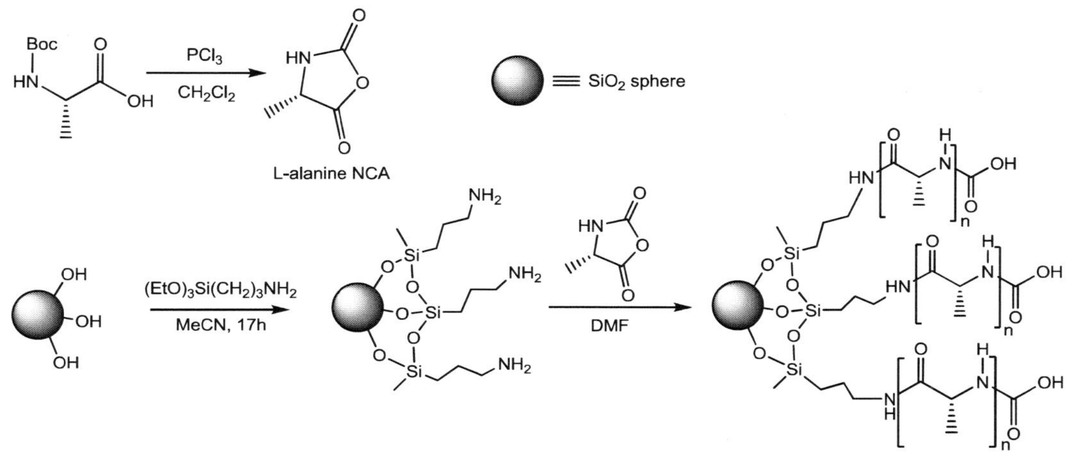

Scheme 4.3. Modification of silica spheres by poly(L-alanine) brushes. Reproduced with permission from ref. [49]. Copyright : 2008 The Royal Society of Chemistry.

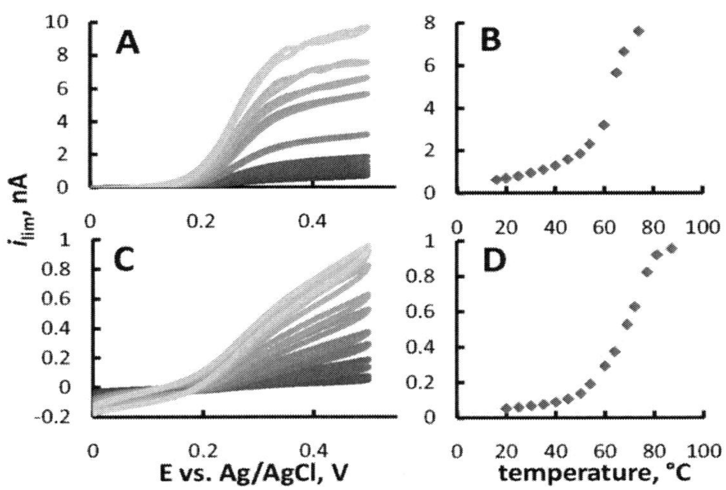

Figure 4.9. (A-C) Representative voltammetric responses, and (B-D) plots of $Fc(CH_2OH)_2$ limiting current as a function of temperature for poly(L-alanine)-colloidal membrane Pt electrodes after polymerization for 1 hour and 3 hours, respectively. Reproduced with permission from ref. [49]. Copyright : 2008 The Royal Society of Chemistry.

We investigated the temperature -response for poly(L-alanine)-filled colloidal films using cyclic voltammetry. To exclude the possibility that the observed changes in the molecular transport would result from electrostatic effects [36, 70], the temperature-response of colloidal film electrodes for a neutral molecule, $Fc(CH_2OH)_2$ was examined. As can be seen in Figure 4.9, the limiting current for the modified colloidal film electrodes is indeed affected by temperature. For all colloidal film electrodes the limiting current slightly increased with increasing temperature as a result of increasing diffusion coefficient [37]. We observed the effect of temperature for both thin and thick polymer brushes. For mesoporous films modified with a thinner polymer brush (polymerization time of 1 hour), the limiting current increased with increasing temperature, with a transition temperature of ca. 65 °C (Figure 4.9A and 4.9B). For colloidal films filled with thick polymer brushes (polymerization time of 3 hours) a temperature response was observed at a higher temperature of ca. 75 °C (Figure 4.9C and 4.9D). The temperature-dependent change in the limiting current was reversible for both types of polymer brush. The poly(L-alanine)-modified films could be cycled between low and high temperatures without apparent loss of responsiveness (Figure 4.10).

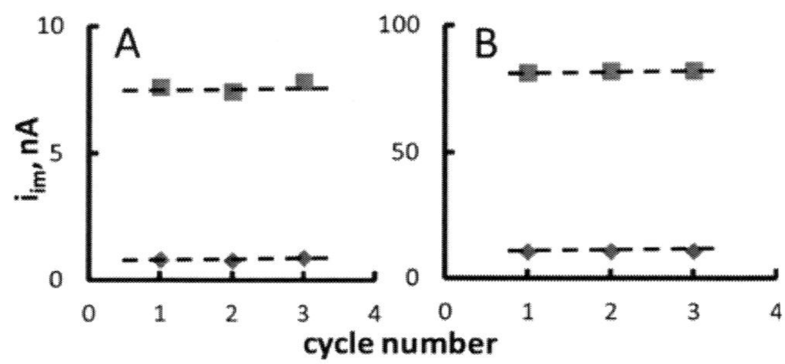

Figure 4.10. Limiting current at low (diamonds) and high (squares) temperature for poly(L-alaniny)-colloidal membrane Pt electrodes ($Fc(CH_2OH)_2$) after polymerization for (A) 1 hour, and (B) 3 hours, as a function of temperature cycling . Reproduced with permission from ref. [49]. Copyright : 2008 The Royal Society of Chemistry.

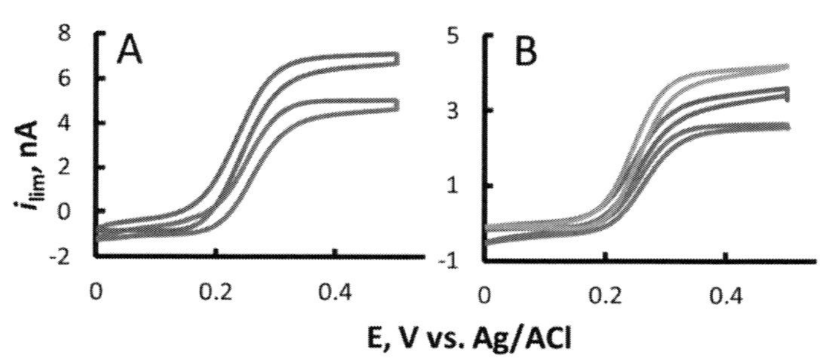

Figure 4.11. ($Fc(CH_2OH)_2$) voltammetric responses for colloidal film Pt electrodes surface-modified with poly-L-alanine for 3 hours (A) and 6 hours (B) at pH 3 (bottom line), pH 6 (middle line), and pH 8 (top line). Reproduced with permission from ref. [49]. Copyright : 2008 The Royal Society of Chemistry.

pH responsive behavior of poly(L-alanine)-modified silica colloidal films was also investigated. As can be seen in Figure 4.11, the limiting current for the poly(L-alanine)-modified colloidal film electrodes is affected by pH, where the limiting current increases with increasing pH. This suggests that as pH is increased, the polymer brush shrinks, increasing the mesopore size. This reverse is also true, as the limiting current decreases for decreasing pH indicating an elongation of the polymer brush. This trend, while smaller for colloidal films carrying thinner polymer brushes, is consistent for the entire range of polymer brush sizes studied. The pH-dependent change in the limiting current is reversible for polymer brushes of various lengths. The mesoporous polymer-modified colloidal films could be cycled between low and high pH. The observed pH responsive behavior may originate from the electrostatic interaction between the amine end-groups of the poly(L-alanine) brush and the residual amines on the silica surface in a manner similar to that described earlier for other types of surface-immobilized polymer brushes [33, 71]. To confirm this hypothesis we estimated the grafting density of poly(L-alanine) chains on the silica spheres using thergogravimetric analysis (TGA). Based on its results we estimated the grafting density as 0.61 chains per nm^2, well below the maximum grafting density that can be achieved for the silica surface [72] thus supporting our explanation of the pH-responsive behavior observed for the poly(L-alanine)-modified colloidal mesopores .

4.3. Light-Responsive Silica Colloidal Films

A spiropyran moiety that undergoes a reversible transformation from the neutral form to the charged merocyanine forms and back upon irradiation with UV and visible light , respectively, (Scheme 4.4) [73] can be used to control the ionic permselectivity of a mesoporous colloidal film by light. The succinimidyl ester derivative [74] of the spiropyran was prepared and used to treat the amine -modified colloidal films assembled on a Pt electrode (Scheme 4.5) [40]. The successful silica surface modification with spiropyran moieties was confirmed using UV spectroscopy .

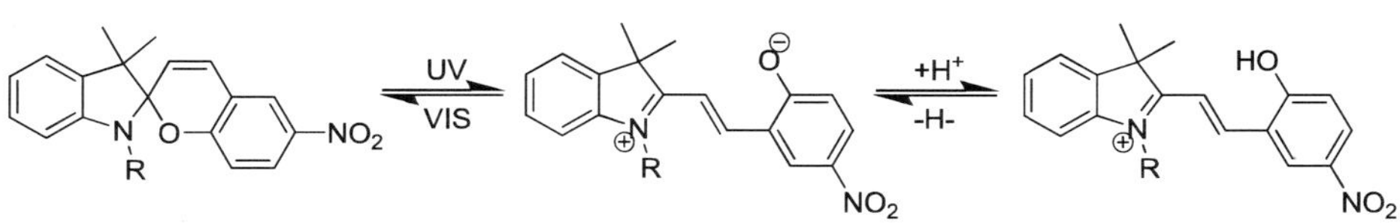

Scheme 4.4. Interconversions of neutral , zwitterionic and charged spiropyran forms.

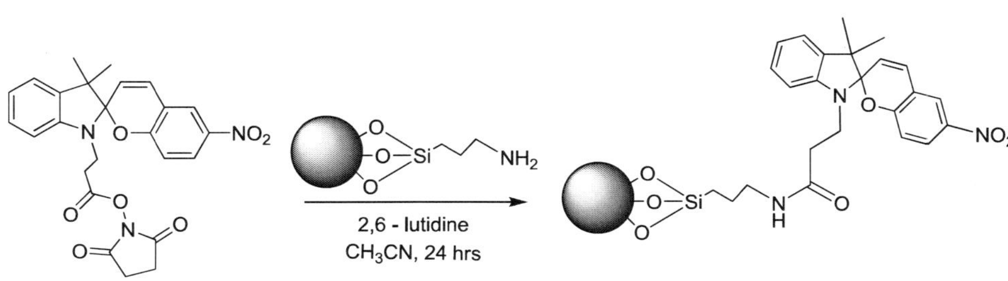

Scheme 4.5. Attachment of spiropyran moiety to the silica sphere surface. Reproduced with permission from ref. [40]. Copyright : 2011 Springer.

Next, the flux of charged molecules through the spiropyran-modified film was measured using cyclic voltammetry, in the same way as for the amine -modified membranes . It was found that upon irradiation with UV light the transport of positively charged species decreased, and that it can be restored to its original value upon irradiating the membrane with visible light (Figure 4.12).

The permselectivity for the molecular transport of cations was only about 15% for the spiropyran-modified films , compared to nearly 100% in the case of amine -modified films. This may result from having fewer spiropyran molecules on the surface, or from incomplete conversion of the spiropyran molecules to the merocyanine form.

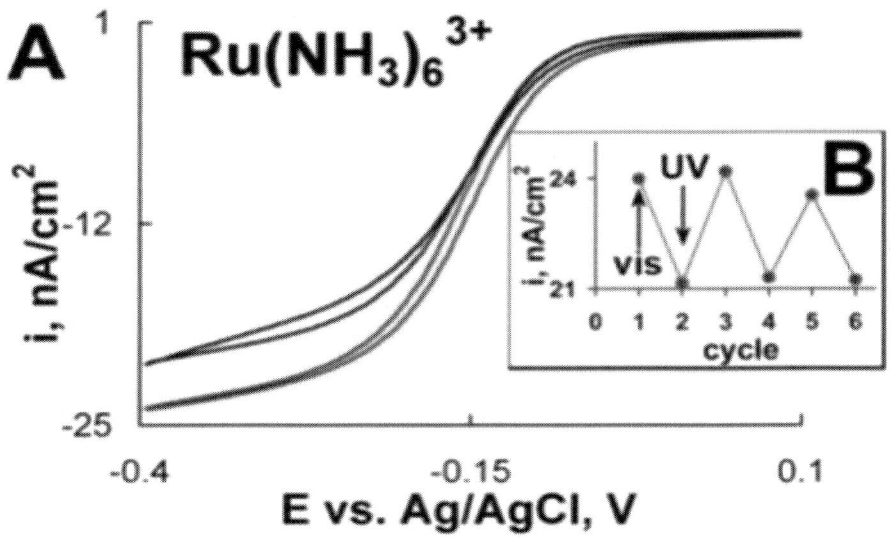

Figure 4.12. (A)Voltammetric response of a spiropyran-modified opal Pt electrode after irradiation with UV (top curve) and visible light (bottom curve). (B) Time profile of the response. Reproduced with permission from ref. [40]. Copyright : 2011 Springer.

4.4. Small Molecule-Responsive Silica Colloidal Films

We demonstrated that silica colloidal crystals with reversible permselectivity controlled by the presence of a small molecule can be prepared by surface modification of the mesopores with a responsive DNA aptamer [50]. Aptamer oligonucleotides exhibit selective and specific binding properties towards molecules [75] and undergo conformational changes in response to small molecule binding. The aptamer used in our work was based on an oligonucleotide binder selective for the detection of cocaine which was previously engineered by Stajonovic et al., [75]. The secondary structure for the 32-base cocaine aptamer possesses a three-way junction, in the middle of which there is a cavity which binds the target molecule (Figure 4.13). In the absence of a target, the aptamer is thought to remain partially unfolded, with only one of the three junctions folded. We expected that the conformational change of the oligonucleotide inside the mesopores in the presence of cocaine would cause a reversible increase in the rate of diffusion through the mesopore.

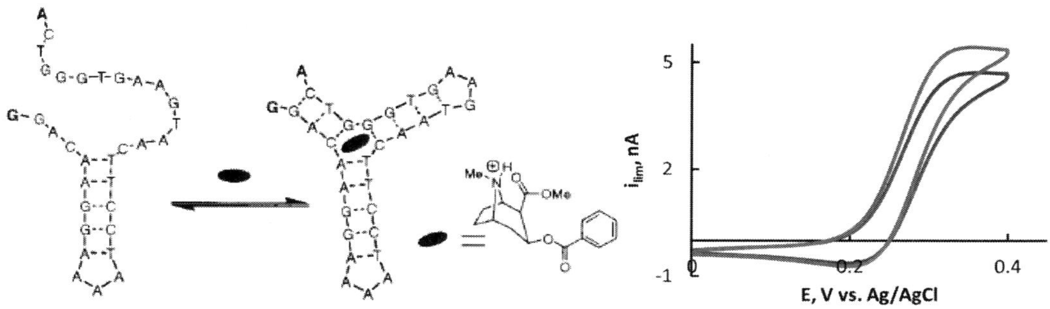

Figure 4.13. Left: Cocaine-sensing aptamer binding to cocaine . Right: Representative $Fc(CH_2OH)_2$ voltammetric response for an aptamer-modified opal electrode made of 290 nm silica spheres in the absence (bottom) and in the presence of cocaine (top). Reproduced with permission from ref. [50]. Copyright : 2011 Institute of Organic Chemistry and Biochemistry AS CR.

The aptamer described above was attached to the surface of silica colloidal film comprising 290 nm silica spheres (resulting in 22.5 nm radius mesopores) via maleimide-activated support. The transport rates of a redox-active probe molecule (ferrocene dimethanol) through the resulting mesoporous films was measured as a function of cocaine concentration using cyclic voltammetry. A neutral redox probe, $Fc(CH_2OH)_2$, was used to exclude the possibility that the observed changes in the molecular transport would result from electrostatic effects [36, 37].

As can be seen in Figure 4.13, the limiting current for the aptamer-modified colloidal film electrodes is affected by the addition of cocaine to the solution . A $9.0 \pm 3.5\%$ change in the limiting current was observed as a result of the cocaine addition (measurements were performed in triplicate to ensure reproducibility), corresponding to 0.6 nm increase in effective mesopore radius . The cocaine-dependent change in the limiting current is reversible and the mesoporous aptamer-modified colloidal films could be regenerated by immersing the electrode into a potassium phosphate buffer. This behavior was attributed to the conformational change in the secondary structure where the aptamer changes from partially unfolded conformation with only one of the three junctions folded to a conformation with a three-way junction containing the cocaine molecule in the internal cavity (Figure 4.13). With the formation of the three-way junction, the space that the aptamer occupies inside the mesopores is reduced, allowing for increased transport through the mesopores.

We also investigated the response of the aptamer-modified colloidal membranes with smaller mesopores (silica spheres of 100 nm in diameter resulted formation of 7.8 nm mesopores) to the presence of cocaine using cyclic voltammetry, as described above. We found that the change in limiting current resulting from cocaine binding is indeed affected by the size of the mesopore. For the mesopores that are ca. 2.9 times smaller the limiting current increase is ca. 2.6 times higher compared to the larger mesopores. This observation was rationalized by assuming that the aptamer change in size as the result of cocaine binding remains constant regardless of the mesopore size, while having a greater effect for the smaller mesopores. Indeed, the calculated change in the organic layer thickness inside the mesopores resulting from cocaine binding is essentially identical for both mesopore sizes studied (0.4 nm for the smaller and 0.6 nm for the larger pores). This thickness change is presumably proportional to both the conformational change the aptamer undergoes in the process of the small molecule binding and its packing density on the mesopore surface.

5. Proton Conducting Silica Colloidal Membranes

We studied the proton transport in free-standing membranes prepared from surface-sulfonated silica nanospheres. The silica sphere surface was sulfonated using 1,3-propanesultone in refluxing toluene followed by washing [38]. To evaluate the proton transport in assemblies of sulfonated silica colloidal spheres we prepared two types of samples [39]. First, both unmodified and sulfonated 200 nm silica spheres were pressed into disk-shaped pellets using a hydraulic press. Second, we prepared ordered colloidal membranes by vertical deposition of both unmodified and sulfonated silica spheres from 4 wt% ethanol colloidal solutions onto the glass substrate . This process led to mechanically robust ~1-mm-thin free-standing colloidal membranes. Both materials were free of mechanical defects such as cracks , and both possess similar porosity . However, their structure at the nanoscale was quite different. Based on SEM images (Figure 5.1) it appears that the pellet has multiple disordered regions compared to the self-assembled material, which seems to possess more structural order.

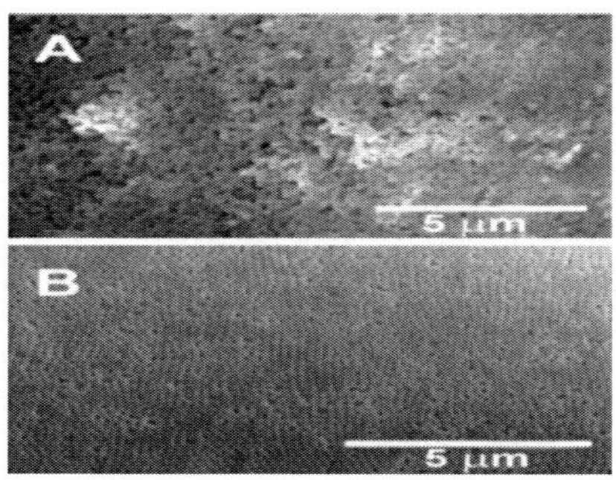

Figure 5.1. SEM images of the disordered (A) and self-assembled close-packed (B) materials prepared from ca. 200-nm-diameter sulfonated silica spheres as described in the text. Reproduced with permission from ref. [39]. Copyright : 2008 The American Chemical Society.

Table 5.1. Proton conductivity of disordered and close-packed colloidal assemblies made of sulfonated silica spheres [39]

Material	Proton conductivity (S/cm) at 100 °C			
	dry air		98% R.H	
	unmodified	sulfonated	unmodified	sulfonated
disordered	$<10^{-7}$	0.5×10^{-6}	2.2×10^{-6}	7.0×10^{-5}
close-packed	$<10^{-7}$	0.8×10^{-6}	5.0×10^{-6}	1.4×10^{-2}

We used electrochemical impedance spectroscopy to measure the proton conductivity of the above materials (Table 5.1). The proton conductivity of unmodified pellets at 100 °C in dry air is below 10^{-7} S/cm (the lower limit of the instrument sensitivity). Next, we measured the proton conductivity of the self-assembled colloidal membranes . We found that at 100 °C in dry air the proton conductivity of unmodified self-assembled colloidal materials was also very low. In contrast, under these conditions the proton conductivity of the sulfonated self-assembled colloidal materials was 0.82×10^{-5} S/cm, nearly two times higher compared to the disordered sulfonated membranes. We believe that this may result from a higher number of interconnected mesopores and/or a shorter distance the protons have to travel, leading to more efficient proton transport in the self-assembled colloidal material. The lower proton conductivity of the disordered membranes comprising the sulfonated silica spheres might result from the disordered structure of the membrane with a small number of continuous mesoporous channels and a poor contact between the silica spheres. Indeed, it is well known that increasing the network organization of a proton conducting material enhances the proton transport through that material [76, 77, 78].

The proton conductivity of the unmodified and sulfonated dirordered membranes at 100°C and 98% R.H. is given in Table 1. The proton conductivity of the sulfonated membranes increases ca. 14-fold (to 7.0×10^{-5} S/cm) with increasing relative humidity from zero to 98%. Increasing the relative humidity has a much more pronounced effect on the proton conductivity of self-assembled sulfonated colloidal membranes . At 98% R.H. and 100 °C its proton conductivity was ca. 10^{-2} S/cm, more than two orders of magnitude higher than that of the sulfonated disordered pellets. Since in dry air the difference between the ordered and disordered colloidal membranes was much smaller, we speculate that the high proton conductivity of the ordered membrane at high humidity results from its better ability to retain water .

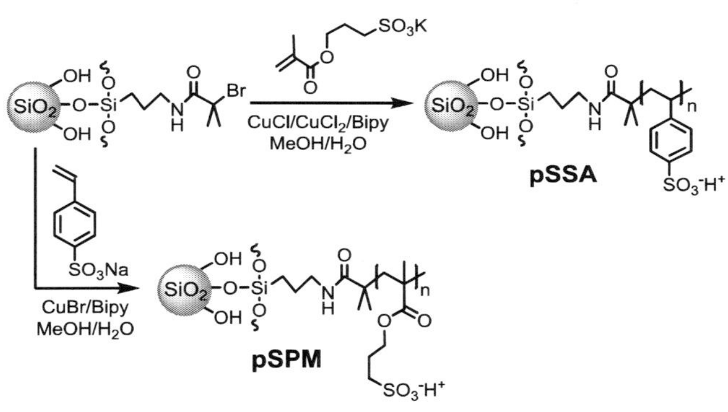

Scheme 5.1. Preparation of poly(3-sulfopropyl-methacrylate) (pSPM) and poly(stryrenesulfonic acid) (pSSA) brushes on silica surface. Reproduced with permission from ref. [51]. Copyright : 2008 The American Chemical Society.

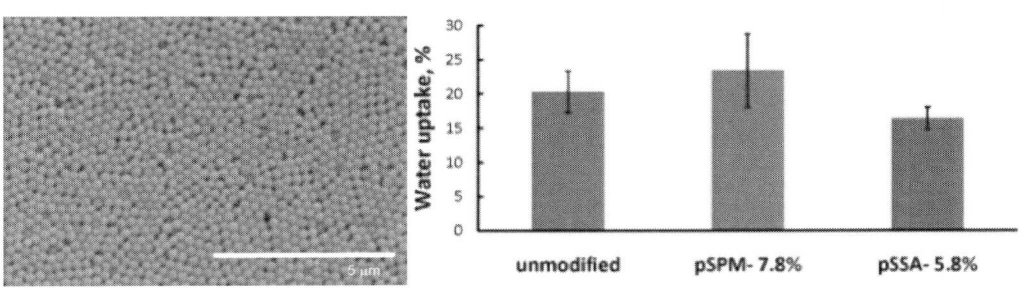

Figure 5.2. Left: SEM image of sintered silica colloidal crystal comprised of ~300 nm diameter silica nanoparticles . Size bar = 5 μm. Right: water uptake for sintered silica membranes after soaking in water at room temperature for 24 hours. Reproduced with permission from ref. [51]. Copyright : 2008 The American Chemical Society.

We also prepared silica colloidal membranes filled with poly(3-sulfopropyl-methacrylate) (pSPM) and poly(stryrenesulfonic acid) (pSSA) brushes covalently attached to the mesopore surface [51]. Figure 5.2 shows an SEM image of a sintered silica colloidal membrane, comprised of ~300 nm diameter silica spheres. The membranes ~300 μm thick and ~1 cm across were filled with pSPM and pSSA sulfonated polymer brushes by surface-initiated ATRP (Scheme 5.1). TGA was used to characterize the polymer-filled membranes. The polymer weight % for pSPM membranes is ~8 wt% and the for pSSA membranes is ~6 wt% after 10 hours of polymerization . Water uptake of the membranes (Figure 5.2) was measured by soaking the membranes in water at room temperature for 24 h. Given that the void fraction of the colloidal crystal is 26%, it appears that both polymers are significantly hydrated, with pSPM being hydrated almost completely, and pSSA falling slightly below. The water uptake of pSPM- and pSSA-filled silica membranes is lower than that of Nafion® (~38% water uptake measured under the same conditions [79]). The polymer-filled membranes did not swell in the course of these experiments.

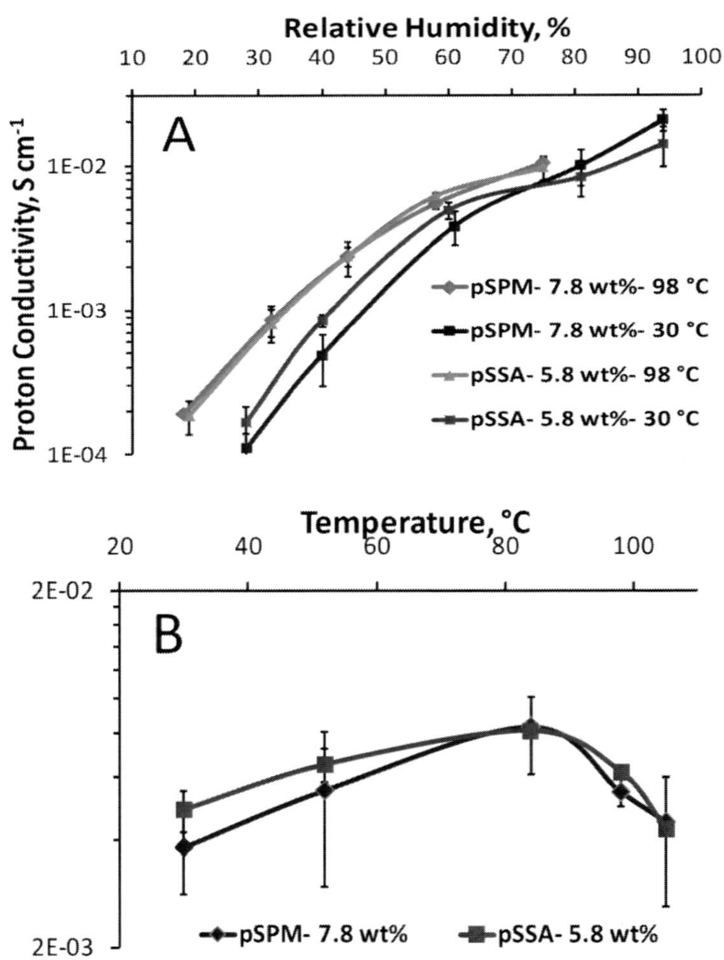

Figure 5.3. Proton conductivity of pSPM and pSSA sintered colloidal membranes as a function of (A) relative humidity at 30 °C and 98 °C and (B) temperature at 60% relative humidity. Reproduced with permission from ref. [51]. Copyright : 2008 The American Chemical Society.

We measured the proton conductivity of the polymer -filled colloidal membranes using electrochemical impedance spectroscopy . Figure 5.3A shows the proton conductivity of pSPM- and pSSA-filled membranes at 30 °C and 98 °C as a function of relative humidity . It is clear that both membranes have similar proton conductivities which increase with relative humidity, with a maximum value of ~2 × 10^{-2} S/cm achieved at 30 °C and 94% R.H. The proton conductivity of the polymer-filled colloidal membranes is affected by temperature , as shown in Figure 5.3B. The proton conductivity for both membranes at 60% R.H. increases gradually with increasing temperature up to 85 °C, and then decreases.

Overall, the proton conductivity values obtained for the polymer -filled mesoporous silica colloidal membranes are comparable to those of Nafion (0.002–0.07 S/cm). The most extensively sulfonated polystyrene membranes reported (20 mol% sulfonation) had the proton conductivity of 5 × 10^{-2} S/cm at room temperature in their fully hydrated state [80].

CONCLUSION

In this chapter, we described the preparation , surface modification and transport properties of mesoporous silica colloidal films and membranes prepared by assembly of silica spheres into close-packed fcc lattice with arrays of mesopores interconnected. It is important to note that the mathematical description of the flux through these membranes based on fcc packing is an approximation, albeit a good one, but their overall characteristics such as pore size and porosity are nevertheless accurately described assuming the perfect fcc packing. The films are formed upon vertical deposition from colloidal solution onto various solid supports. The memabranes can be suspended in a glass or silicon support. Free-standing silica colloidal membranes can be prepared by sintering at temperature above 1000 °C. The quality of the fcc packing depends on the thickness of the films and membranes, with thinner materials showing much better ordering. Most importantly, in all cases the films and membranes can be prepared without mechanical defects .

Silica colloidal membranes are capable of size-selective transport and can be surface-modified in a well-defined and controlled manner with small molecules and polymers . We achieved surface modification with small molecules using silanol chemistry, and formed polymer brushes with narrow mass distribution and controlled length on the mesopore surface using atom transfer radical polymerization and ring-opening polymerization . The transport in the resulting mesoporous films and membranes can be controlled by pH and ionic strength, temperature , light , and small molecule binding. We also prepared proton conducting silica colloidal membranes by filling the mesopores with sulfonated polymer brushes, and studied the properties of these membranes.

There are various potential applications for mesoporous silica colloidal films and membranes . Due to their ordered mesoporous structure and mechanical durability , they can be applied in nanofluidics, ultrafiltration, and as catalyst support. Reversible control of transport via external stimuli may be useful in drug-release devices, in size-, charge- and structure-selective separations, and in microfluidic and sensing devices. Presently, these application are under investigation in research laboratory, and no commercial applications are available.

Mesoporous silica colloidal films membranes described in this chapter broaden the growing field of inorganic mesoporous materials containing ordered arrays of mesopores and prepared via self-assembly , such as anodized alumina [81], containing highly ordered pores with columnar structure , and mesoporous silica, containing multiple cylindrical channels arranged in a 2D network of honeycomb-like structure [82, 83, 84]. While the pore size s and geometries are different for these materials , their surface modification and the transport phenomena are similar, in all cases allowing to produce responsive materials [85].

REFERENCES

[1] P.T. Tanev, J.-R. Butruille and T.J. Pinnavaia in *Chem. Adv. Mater.*, L.V. Interrante, M.J. Hampden-Smith, eds., Wiley-VCH: New York, N. Y, 1998, p. 329.

[2] M.E. Davis, *Nature* 417, 813–821 (2002).

[3] G. Artu, I. Roosmasari, K. Richau and J. Hapke, *Separation Sci. Tech. 4*2, 2947–2986 (2007).

[4] H. Bayley and C.R. Martin, *Chem. Rev.* 100, 2575–2594 (2000).

[5] A. Piruska, M. Gong, J.V. Sweedler and P.W. Bohn, *Chem. Soc. Rev.* 39, 1060–1072 (2010).

[6] H.D. Tong, H.V. Jansen, V.J. Gadgil, C.G. Bostan, C.G.E. Berenschot, C.J.M. van Rijn and M. Elwenspoek, *Nano Lett.* 4, 283–287 (2004).

[7] C.-S. Toh, B.M. Kayes, E.J. Nemanick and N.S. Lewis, *Nano Lett.* 4, 767–770 (2004).

[8] M. Yoshida, M. Asano, T. Suwa, N. Reber, R. Spohr and R. Katakai, *Adv. Mater. 9,* 757–758 (1997).

[9] N.G. Liu, D.R. Dunphy, P. Atanassov, S.D. Bunge, Z. Chen, G.P. Lopez, T.J. Boyle and C.J. Brinker, *Nano Lett.* 4, 551–554 (2004).

[10] H. Xu and W.A. Goedel, *Angew. Chemie Int. Ed.* 42, 4694–4696 (2003).

[11] H. Xu and W.A. Goedel, *Langmuir,* 18, 2363–2367 (2002).

[12] M. Nishizawa, V.P. Menon and C. R. Martin, *Science,* 268, 700–702 (1995).

[13] M.W. Urban, Ed. *Stimuli-responsive polymeric membranes and coatings , ACS Symposium Series, Vol. 912*, ACS, Washington , 2005.

[14] S. Kallus, J.-M. Condre, A. Hahn, G. Golemme, C. Algieri, Ph. Dieudonne, P. Timmins and J.D.F. Ramsay, *J. Mat. Chem.* 12, 3343–3350 (2002).

[15] L. Sun and R.M. Crooks, *J. Am. Chem. Soc.* 122, 12340–12345 (2000).

[16] Z. Siwy, E. Heins, C.C. Harrell, P. Kohli and C. R. Martin, *J. Am. Chem. Soc.* 126, 10850–10851 (2004).

[17] S. Yu, S.B. Lee, M. Kang and C.R. Martin, *Nano Lett.* 1, 495–498 (2001).

[18] H.H. Ramaile and J.B. Schlenoff, *J. Am. Chem. Soc.* 125, 6602–6603 (2003).

[19] T.A. Desai, D.J. Hansford, L. Kulinsky, A.H. Nashat, G. Rasi, J. Tu, Y. Wang, M. Zhang and M. Ferrari, *Biomed. Microdev.* 2, 11–40 (1999).

[20] S. Sershen and J. West, *Adv. Drug Delivery Rev.* 54, 1225–1235 (2002).

[21] S. Wong, V. Kitaev and G.A. Ozin, *J. Am. Chem. Soc.* 125, 15589–15598 (2003).

[22] W. Stöber, A. Fink and E. Bohn, *J. Colloid. Interface Sci.* 26, 62–69 (1968).

[23] P. Jiang, J.F. Bertone, K.S. Hwang and V.L. Colvin, *Chem. Mater.* 111, 2132–2140 (1999).

[24] T. Hanni in P. Brown and E. Grushka, eds., *Advances in Chromatography*, Vol. 40, Marcel Dekker, New York, 2000, pp. 315–357.
[25] S. Flink, F.C.J.M. van Veggel and D.N. Reinhoudt, *J. Phys. Org. Chem.* 14, 407–415 (2001).
[26] N. Kanoh, S. Kumashiro, S. Simizu, Y. Kandoh, S. Hatakeyama, H. Tashiro and H. Osada, *Angew. Chem. Int. Ed.* 42, 5584–5587 (2003).
[27] K. Matyjaszewski and J. Xia, *Chem. Rev.* 101, 2921–2990 (2001).
[28] J.-S. Wang and K. Matyjaszewski, *Macromolecules* 28, 7901–7910 (1995).
[29] X. Kong, T. Kawai, J. Abe and T. Iyoda, *Macromolecules* 34, 1837–1844 (2001).
[30] S. Minko, D. Usov, E. Goreshnik and M. Stamm, *Macromol. Rapid. Commun.* 22, 206–211 (2001).
[31] Y. Tran, P. Auroy and L.-T. Lee, *Macromolecules* 32, 8952–8964 (1999).
[32] M. Biesalski, D. Johannsmann and J. Ruhe, *J. Chem. Phys.* 117, 4988–4994 (2002).
[33] E.L. Cussler, *Diffusion. Mass Transfer in Fluid Systems.* 2nd Ed., University Press, Cambridge, 1997.
[34] J. Crank, *The Mathematics of Diffusion.* 2nd Ed., Oxford University Press, New York, 1975.
[35] H. Brenner and D. A. Edwards, *Macrotransport Processes*; Butterworth-Heinemann: Boston, 1993.
[36] M.R. Newton, A.K. Bohaty, H.S. White and I. Zharov, *J. Am. Chem. Soc.* 127, 7268–7269 (2005).
[37] M.R. Newton, A.K. Bohaty, H.S. White and I. Zharov, *Langmuir* 22, 4429–4432 (2006).
[38] J.J. Smith and I. Zharov, *Langmuir* 24, 2650–2654 (2008).
[39] J.J. Smith, R. Abbaraju and I. Zharov, *J. Mater. Chem.* 18, 5335–5338 (2008).
[40] A.K. Bohaty M.R. Newton and I. Zharov, *J. Porous Mater.* 17, 465–473 (2010).
[41] A.K. Bohaty and I. Zharov, *Langmuir* 22, 5533–5536 (2006).
[42] Bohaty, A. Abelow and I. Zharov, *J. Porous Mater.* 18, 297-304 (2011)
[43] K. Bohaty, J.J. Smith, and I. Zharov, *Langmuir,* 25, 3096-3101 (2009).
[44] P. Ignacio-de Leon and I. Zharov, *Chem. Commun.* 47, 553-555 (2011).
[45] O. Schepelina and I. Zharov, *Langmuir*, 22, 10523–10527 (2006).
[46] O. Schepelina and I. Zharov, *Langmuir*, 23, 12704–12709 (2007).
[47] O. Schepelina and I. Zharov, *Langmuir* 24, 14188–14194 (2008).
[48] O. Schepelina, N. Poth and I. Zharov, *Adv. Funct. Mater.* 20, 1962–1969 (2010).
[49] A.E. Abelow and I. Zharov, *Soft Matter*, 5, 457–462 (2009).
[50] A.E. Abelow, R.J. White, K.W. Plaxco and I. Zharov, *Coll. Czech Chem. Commun.* 76, 683–694 (2011).
[51] J.J. Smith, I. Zharov, *Chem. Mater.* 21, 2013-2019 (2009).
[52] P. K. Jal, S. Patel and B. K. Mishra, *Talanta,* 62, 1005-1028 (2004)
[53] C.-H. Hou, X. Wang, C. Liang, S. Yiacoumi, C. Tsouris and S. Dai, *J. Phys. Chem. B* 112, 8563–8570 (2008).
[54] X. S. Peng, J. Jin and I. Ichinose, *Adv. Funct. Mater.* 17, 1849–1855 (2007).
[55] D.A. Tomalia, H. Baker, J. Dewald, M. Hall, G. Kallos, S. Martin, J. Roeck and P. Smith, *Macromolecules* 19, 2466–2468 (1986).
[56] P.K. Jal, S. Patel and B.K. Mishra, *Talanta* 62, 1005–1028 (2004).

[57] G.B. Webber, E.J. Wanless, V. Butun, S.P. Armes and S. Biggs, *Nano Lett.* 2, 1307–1313 (2002)..
[58] J.I. Amalvy, E.J. Wanless, Y. Li, V. Michailidou, S.P. Armes and Y. Duccini, *Langmuir* 20, 8992–8999 (2004).
[59] J. Gao, G. Zhai, Y. Song and B. Jiang, *J. Appl. Polymer Sci.* 107, 3548–3556 (2008).
[60] S. Sanjuan, P. Perrin, N. Pantoustier and Y. Tran, *Langmuir* 23, 5769–5778 (2007).
[61] L. Zhou, W. Yuan, J. Yuan and X. Hong, *Mater. Lett.* 62, 1372–1375 (2008).
[62] M. Zhang, L. Liu, H. Zhao, Y. Yang, G. Fu and B. He, *J. Colloid. Inter. Chem.* 301, 85–91 (2006).
[63] M. Zhang, L. Liu, H. Zhao, Y. Yang, G. Fu and B. He, *Polymer* 48, 1989–1997 (2007).
[64] J. Huang, H. Murata, R.R. Koepsel, A.J. Russel and K. Matyjaszewski, *Biomacromolecules* 8, 1396–1399 (2007).
[65] M. Heskins and J.E. Guillet, *J. Macromol. Sci., Chem. A*, 2, 1441–1455 (1968).
[66] T. Peng and Y.-L. Cheng, *J. Appl. Polym. Sci.* 70, 2133–2142 (1998).
[67] Y. Li, L.-Y. Chu, J.-H. Zhu, H.-D. Wang, S.-L. Xia and W.-M. Chen, *Ind. Eng. Chem. Res.* 43, 2643–2649 (2004).
[68] N. Idota, A. Kikuchi, J. Kobayashi, Y. Akiyama, K. Sakai and T. Okano, *Langmuir* 22, 425–430 (2006).
[69] T. Sun , G. Wang, L. Feng, B. Liu, Y. Ma, L. Jiang and D. Zhu, Angew. *Chem. Int. Ed.* 43, 357–360 (2004).
[70] W.N.E. van Dijk-Wolthuis, L. van de Water, P. van de Wetering, M.J. van Steenbergen, J.J. Kettenes-van den Bosch, W.J.W. Schuyl and W.E. Hennink, *Macromol. Chem. Phys.* 198, 3893–3906 (1997).
[71] M. Ballauff and O. Borisov, *Curr. Op. Colloid & Interface Sci.* 11, 316–323 (2006).
[72] P.K. Jal, S. Patel and B.K. Mishra, Talanta 62, 1005–1028 (2004), and references therein.
[73] R.C. Bertelson, Spiropyrans. In *Organic Photochromic and Thermochromic Compounds.* J. C. Crano and R. J. Guglielmetti, Eds., Plenum Publishing Corp.: New York, 1999.
[74] Fissi, O. Pieroni, G. Ruggeri and F. Ciardelli, *Macromolecules* 28, 302–309 (1995).
[75] M.N. Stojanovic, P. Prada and D.W. Landry, *J. Am. Chem. Soc.* 123, 4928–4931 (2001).
[76] T. Teuka, K. Tadanaga, A. Hayashi and M. Tatsumisago, *J. Am. Chem. Soc.* 128, 16470–16471 (2006).
[77] G. Gao, Y. Yang, D. Lee, S. Holdcroft and B. Frisken, *J. Macromolecules* 39, 8060–8066 (2006).
[78] R. Subbaraman, H. Ghassemi and T.A. Zawodzinski Jr., *J. Am. Chem. Soc.* 129, 2238–2239 (2007).
[79] F. Pereira, K. Vallé, P. Belleville, A. Morin, S. Lambert and C. Sanchez, *Chem. Mater.* 20, 1710–1718 (2008).
[80] N. Carreta, V. Tricoli and P. Picchioni, *J. Membr. Sci.* 166, 189–197 (2000).
[81] S. Kipke and G. Schmid, *Adv. Funct. Mater.* 14, 1184–1188 (2004).
[82] J. Vivero-Escoto, I. Slowing, B.G. Trewyn and V.S.-Y. Lin, *Small* 6, 1952–1967 (2010).
[83] M. Vallet-Regi, A. Ramila, R.P. del Real and J. Perez-Pariente, *Chem. Mater.* 13, 308–311 (2001).

[84] G.J.A.A. Soler-Illia and O. Azzaroni, *Chem. Soc. Rev.* 40, 1107-1150 (2011).
[85] A.E. Abelow and I. Zharov, *J. Mater. Chem.* 22, 21810-21818 (2012).

In: Comprehensive Guide for Mesoporous Materials. Volume 1 ISBN: 978-1-63463-958-3
Editor: Mahmood Aliofkhazraei

Chapter 4

MESOPOROUS CATALYTIC MATERIALS AND FINE CHEMISTRY

***Elena Pérez-Mayoral,*[*,1] *Elena Soriano,*[2] *Rosa M. Martín-Aranda*[1] *and Francisco J. Maldonado-Hódar*[3]**

[1]Departamento de Química Inorgánica y Química Técnica,
Facultad de Ciencias, Universidad Nacional de Educación a Distancia, Madrid, Spain
[2]Instituto de Química Orgánica General (CSIC), Madrid, Spain
[3]Departamento de Química Inorgánica, Facultad de Ciencias,
Universidad de Granada, Granada, Spain

ABSTRACT

The development of new materials , such as those based on mesoporous structures involved in green processes, is a current and hot issue with special relevance on the design of novel, more efficient and environmental-friendly catalysts. In fact, in the last decade, the number of reported novel mesoporous materials with application in catalysis has been notably increased.

In this sense, this chapter is intended as an update of catalytic processes of interest involving mesoporous materials such as functionalized mesoporous silicates and metallosilicates , metal-organic-frameworks and mesoporous carbons. This overview covers the most recent achievements in the application of mesoporous structures as catalysts and even as catalyst supports in acid -base reactions related to Fine Chemistry.

1. INTRODUCTION

Zeolites are the most traditional microporous materials with applications in the petrochemical and fine chemical industries but also including the gas separation and purification and ion exchange [1, 2]. Among the new application fields of zeolites are the

* Corresponding author: Elena Pérez Mayoral, mail: eperez@ccia.uned.es, Phone: (34) 91 3989047, Fax: (34) 91 3986697

intensification process [3], green chemistry [4], hybrid materials [5], medicine [6], sensing microsystems [7], and nanotechnology [8].

Noteworthy, the zeolite term is often used in a broad sense. In fact, it can be extended to the porous-tailored materials including the oxide molecular sieves, metal organic frameworks (MOFs), porous carbons, sol–gel derived oxides or even porous heteropolyanion salts [9].

The main limitation of using microporous zeolites in catalysis is the small pore size which blocks the catalytic transformation of large molecules . Then, the mesoporous zeolitic materials emerge as alternative materials; the development of those have experimented an impressive progress since the discovery of MCM-41 by Scientists of Mobil Corporation in 1992 [10]. The mesoporous silicas constitute a fascinating class of materials useful for many technological applications because of their porous and morphological characteristics. In fact, the development of new modified mesoporous silicas is an area of extensive research being a hot issue of current interest in nanochemistry . Interestingly, mesoporous silicates find their application in adsorption , separation, as biomolecules immobilizing supports and even as templates for the preparation of mesoporous carbons. It is particularly relevant the application of mesoporous silicas in catalysis mainly due to their well-defined surface chemical properties , high surface area s, and narrow pore size distributions which allow the transformation of bulky substrates [11, 15].

Likewise, zeolites show also structural similarities with metal–organic frameworks (MOFs) such as the pore size and topology . MOFs are new type of hybrid nanoporous materials which have recently attracted much attention as interesting porous materials exhibiting a wide variety of potential applications in chemical engineering , chemistry, and materials science , including gas storage , gas/vapor separation, adsorbents of VOCs, luminescence and drug delivery [16-19].

More recently, MOFs have emerged as the golden opportunities in the field of catalysis [20]. The huge structural variety of MOFs with pore size ranging from ultramicropores to mesopores ones allows the rational design of new and appropriated catalysts with size and shape selectivity . It makes them the union bridge between zeolites and mesoporous silicas .

In other sense, the production of porous carbon materials is of great interest from an economic and technological point of view. Today, there is a wide collection of carbon materials with very different origin, chemical and textural properties , and consequently, covering a large variety of applications. In particular, mesoporous carbon materials present a high interest in catalysis , medical applications and drug delivery , immobilization of biomolecules to develop biosensors or decontamination of big molecular size pollutants . It is noteworthy that while microporosity is an intrinsic characteristic of carbon materials, because more or less micropores are generated during carbonization by the release of gases formed from the heteroatoms present in any carbon precursor (mainly H,O, N, S), the experimental condition s and precursors to obtain mesopores should be carefully fitted.

As can be deduced, the development and characterization of novel and well-defined mesoporous catalytic materials that are active and selective for new environmentally friendly chemical processes is an urgent need and a challenging scientific target . To this end, in the last years the computational methodology has emerged as a useful tool for understanding the properties, structures of active site s and catalytic behavior of these mesoporous materials [21]. The following three contemporary computational methods are suitable for studying such aspects and chemical reactions as they describe reactants , products and active sites with their immediate environments , while employing differing levels of approximation to describe the effect of the extended environment : i) molecular cluster approaches, ii) periodic methods, and iii) embedding techniques .

This chapter is, hence, intended as an update of the most relevant and interesting catalytic processes involving mesoporous materials , mainly functionalized mesoporous silicates and metallosilicates , MOFs and mesoporous carbons. The subject is presented from a multidisciplinary point of view, covering the research fields of material science and catalysis , and emphasizing on the optimization of chemical processes useful in Fine Chemistry. Mesoporous carbons and their combinations (composites) are promising materials in different catalyzed areas. In the appropriated section different approaches used for the preparation of mesoporous carbon materials are presented; among them, activation of carbon precursors, nanocasting and synthesis of designed polymeric precursors are described. Finally, the performance of these materials as catalysts or catalysts support is shortly presented for different applications.

2. Ordered Mesoporous Silicates and Metallosilicates

Several reviews concerning to the structural characteristics and properties of mesoporous zeolitic materials have been reported [2, 22].

In general, mesoporous silicas exhibit a weak acidity ; therefore, the introduction of active functions is a fundamental topic for their application in catalysis [23]. Activation of mesoporous silicates is performed by using two different synthetic approaches: i) direct synthesis in one-pot, where the active phase is added to the initial reaction mixture, and (ii) post-synthesis methods, which firstly require the preparation of the support and subsequent modification with the active species . However, both of them show some disadvantages; while the incorporation of active catalytic sites is considered easier in the one-pot approach, the mesoporous system often results on a less ordered material. In this sense, the post-synthesis modification allows the preparation of the well-ordered structures. In other way, the second approximation allows the incorporation of higher amount of the active phases but also the functionalities often can block the channels.

Among the catalytic active mesoporous silicates, it should be cited i) the catalytic system s containing heteroatoms, and ii) metal oxides and iii) the inorganic-organic hybrid materials . Mesoporous silicates containing single heteroatoms are frequently used in acid -base promoted reactions (materials incorporating aluminum) [24] or even as redox catalysts (including tin or niobium) [25, 26]. Special mention merits the simple mesoporous structures modified with molybdenum oxide and others catalyzing the olefin metathesis [27, 28]. Finally, the modification of the mesoporous surfaces by immobilizing organic active species provides inorganic-organic hybrid structures which are especially relevant, as can be deduced from the huge number of papers daily reported [29-31].

Silica is considered as one of the most useful matrix to make heterogeneous organocatalysts. The most important requirements are a highly accessible surface area , a huge number of silanols for anchoring organic linkers and a hydrothermal stability . In this sense, this section is mainly addressed to review firstly some of the most studied silica -bound organocatalysts involved in green processes avoiding the use of metals and, hence, considered as environmentally benign catalysts (Scheme 1). Secondly, the modification of the surface of mesoporous silicas with heteroatoms, as an alternative to the covalent anchoring of active catalytic sites, is revised. Finally, the development of multifunctional mesoporous silicas modified with at least two different active species which can operate in cooperation is a topic of current interest. It is known as catalyst design by bio-inspiration, imitating the action mode of enzymes , the most efficient biocatalysts [32].

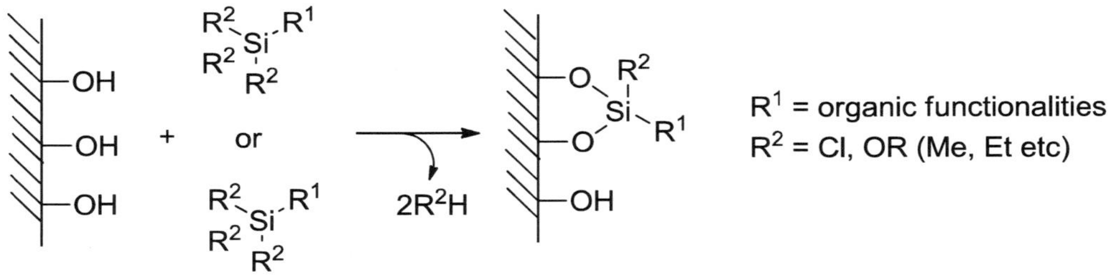

Scheme 1. Surface reactions leading to covalent anchoring of organic R groups.

As mentioned above, the attention of this section is mainly paid to the revision of mesoporous silicates of special relevance in the green catalytic synthesis of fine chemicals . Considering the large number of the reactions which use this type of catalysts, we have focused on the revision of those catalytic system s useful in the preparation of biologically active compounds , emphasizing on the mesoporous silicas which promote the condensation reactions leading to bioactive heterocycles.

Some of the most relevant results are summarized in Tables 1-4 including the used substrates and catalysts, reaction products and the reaction conditions.

2.1. Acid/Base Mesoporous Silicates

Among the most widely used reactions in the synthesis of fine chemical products and catalyzed by acid and/or basic mesoporous silicas are acetalizations [33, 34], acylation reactions [35], Michael additions and condensation reactions such as Henry or Claisen–Schmidt reactions and Knoevenagel condensations [36].

2.1.1. Surface Modification with Sulfonic Acids

The creation of acidic catalytic sites on mesoporus silicas is frequently performed by binding of sulfonic groups using one or two steps for their functionalization [37]. In this regard, Das et al. [38, 39] modified the MCM-41 and MCM-48 materials with propylthiol groups, transforming them into propylsulfonic acid groups under mild oxidative conditions using H_2O_2 (Scheme 2). This synthetic strategy was also reported for the functionalization of FSM-16 and SBA-15 materials.

Interestingly, Dufaud et al. [40] described the synthesis of acidic SBA-15 in which two sulfonic acids are spatially separated from the other ones; this approach consists in the functionalization of SBA-15 matrix with a bissilylated disulfide reagent, subsequent cleavage of the disulfide bridges, and finally the oxidation of the resulting thiols moieties into sulfonic acid groups (Scheme 3). A second method used by the authors for the preparation of SBA-15 mesoporous silica functionalized with aryl sulfonic groups involved the hydrolysis of the corresponding sulfonate ester in an aqueous solution of sulfuric acid (Scheme 4).

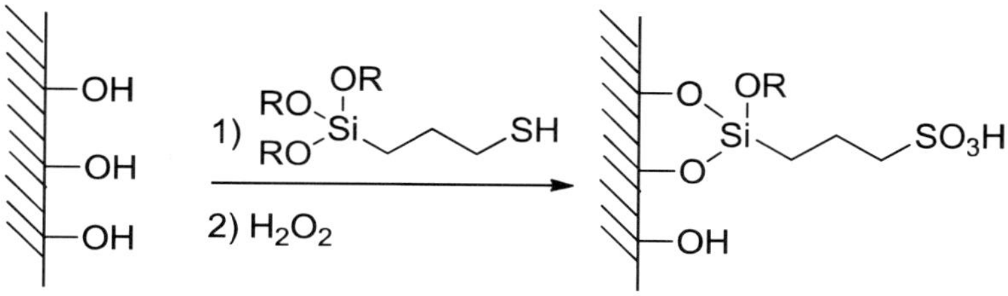

Scheme 2. Synthetic approach for the preparation of mesoporous silicas functionalised with propylsulfonic acid groups.

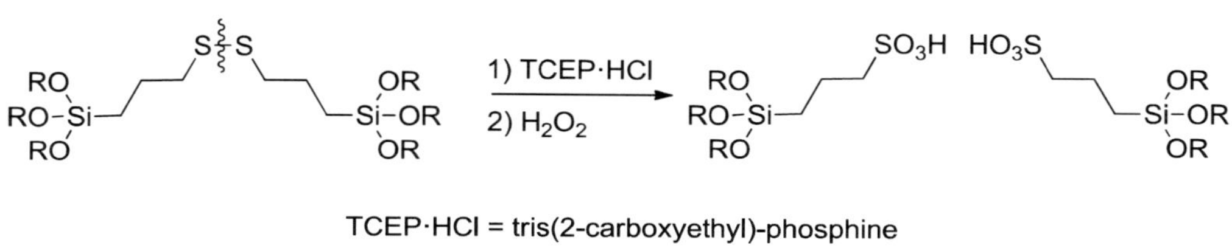

Scheme 3. Modification of SBA-15 matrix with a bissilylated disulfide reagent, followed by cleavage of the disulfide bridges and oxidation of the thiols moieties.

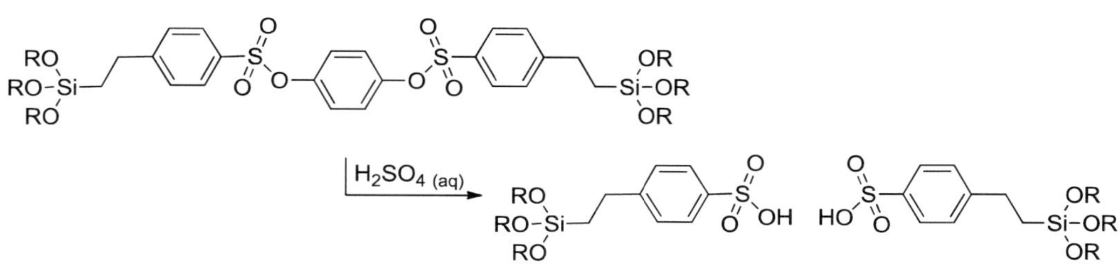

Scheme 4. Modification of SBA-15 matrix with aryl sulfonic acids.

These porous acidic materials were tested in the synthesis of *bisphenol A*, an interesting compound useful in the preparation of polycarbonate plastics and epoxy resins ; thus, the condensation of phenol and acetone preferentially led to the *p,p'*-isomer of *bisphenol A* with higher selectivity and good yields under mild reaction conditions [39] (Scheme 5).

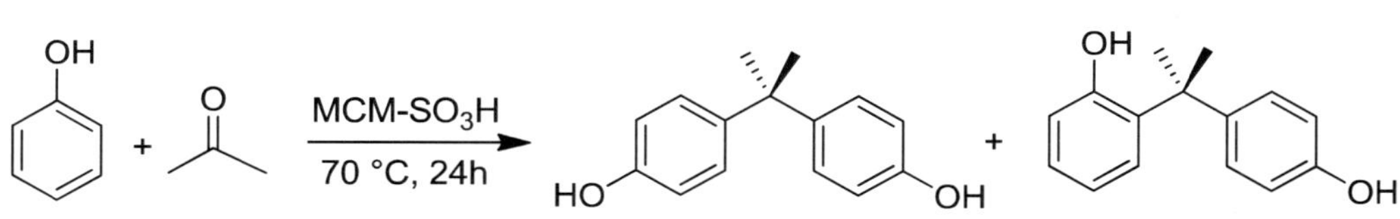

Scheme 5. Synthesis of the *p,p'*-isomer of bisphenol A by condensation of phenol and acetone .

Shimizu et al. [41] reported the acetalization of carbonyl compounds with ethylene glycol catalyzed by propyl-sulfonic acid -functionalized FSM-16 mesoporous silica prepared by a conventional post-synthetic method. The authors reported the formation of the corresponding 1,3-dioxolane with higher rate and yield than other traditional porous material such as zeolites or clays (Scheme 6).

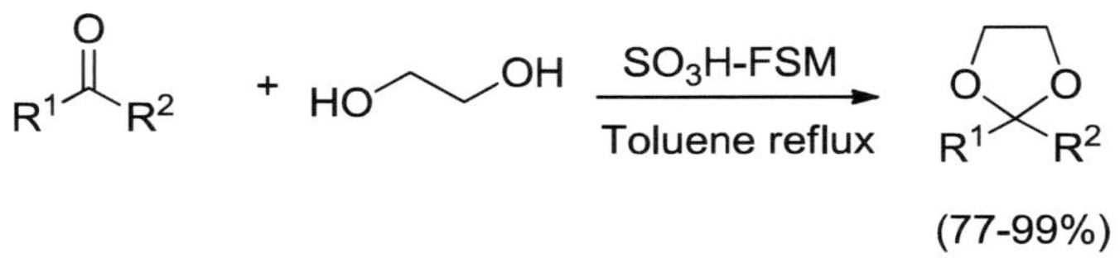

R^1	R^2
Ph	Me
Ph	H
p-Cl-C_6H_4	H
p-MeO-C_6H_4	H
o-OH-C_6H_4	H
$CH_3(CH_2)_6$-	Me
$CH_3(CH_2)_6$-	H

Scheme 6. Acetalization of carbonyl compounds from different carbonyl compounds and ethylene glycol .

The synthesis of heterocyclic compounds is a subject of great and current interest derived from their pharmacological applications. In this sense, the Friedländer condensation is one of the most used and cited reactions in Organic Synthesis and the method of choice for the synthesis of a large variety of nitrogen heterocyclic ring systems such as quinolines, naphthyridines, phenanthrolines, quindolines, or acridones [42].

Under this topic, the synthesis of quinolines, *via* Friedländer condensation , catalyzed by SBA-15 modified with propyl sulfonic acids was reported by Lopez-Sanz et al. [43]. They carried out the condensation between 2-aminoaryl ketones and ethyl acetoacetate which afforded mixtures of quinolines /quinolones (Scheme 7). The selectivity of the reaction was dependent on the starting substrate and differences in the 2-aminoaryl ketone behavior might be related to the structure and stability of the presumed key intermediate, as theoretical calculations suggested, that would control the evolution of the reaction to the corresponding reaction products.

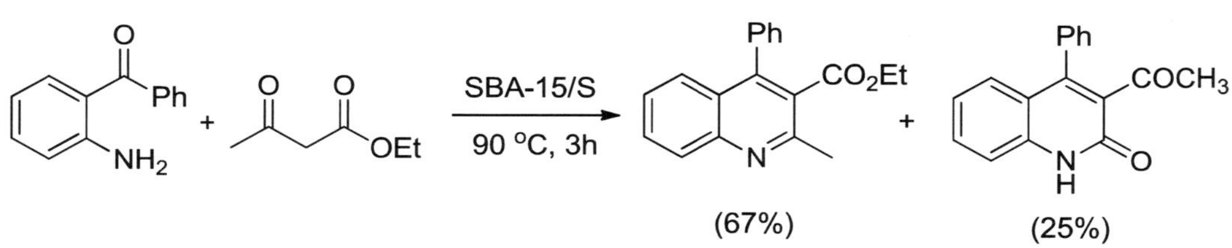

Scheme 7. Friedländer condensation of 2-aminoaryl ketones and ethyl acetoacetate.

More recently, Rostamnia et al. [44] described a new technology which combines the use of ultrasound activation in a nanoreactor system, constituted by SBA-15 bearing covalently bonded propyl-sulfonic groups, for the ultrafast synthesis and waste -free of some biologically interesting organic molecules , such as 2*H*-indazolo[2,1-*b*]phthalazine-triones (Scheme 8A), polyhydroquinolines (Scheme 8B), and α-aminophosphonates (Scheme 8C) through multi-component reactions . This green methodology provides a synergistic means of an efficient approach of the reactants to acidic sites, and suitable mesochannels to drive out the products and water for next recycles.

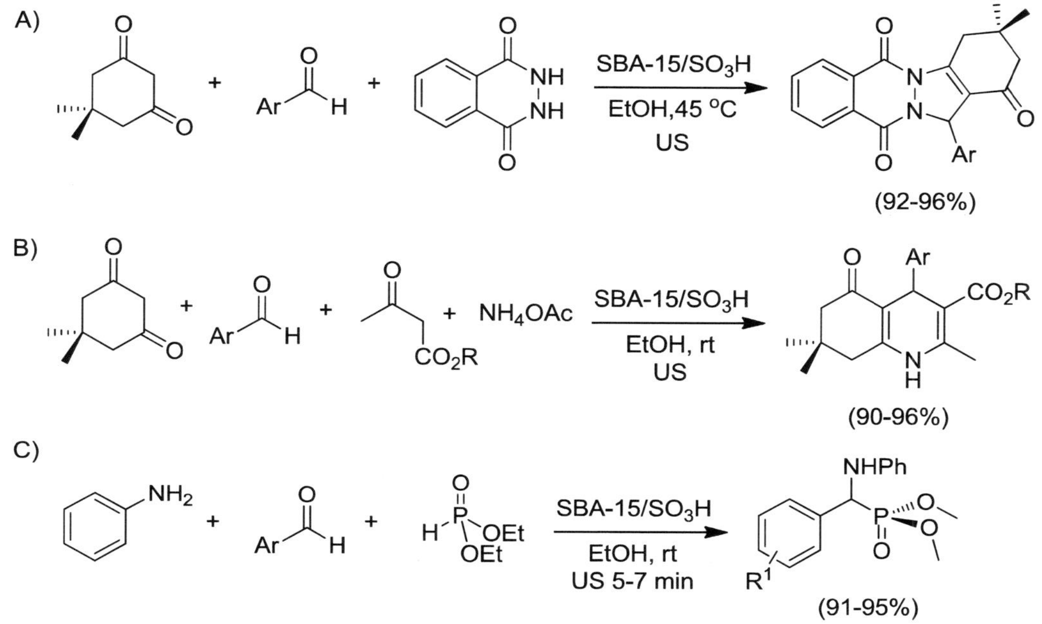

Scheme 8. Three-component reactions catalyzed by SBA-15 functionalized with propyl sulfonic acids.

In other way, sulfuric acid loaded on MCM-41 by using the impregnation method has been reported. This acidic mesoporous solid was tested in the coupling reaction between totally *O*-unprotected *D*-glucuronic acid with alcohols , under monomode microwave irradiation or conventional heating and solvent-free conditions, predominantly affording the corresponding monosubstituted β-D-glucofuranosidurono-6,3-lactones in good yields (Scheme 9) [45].

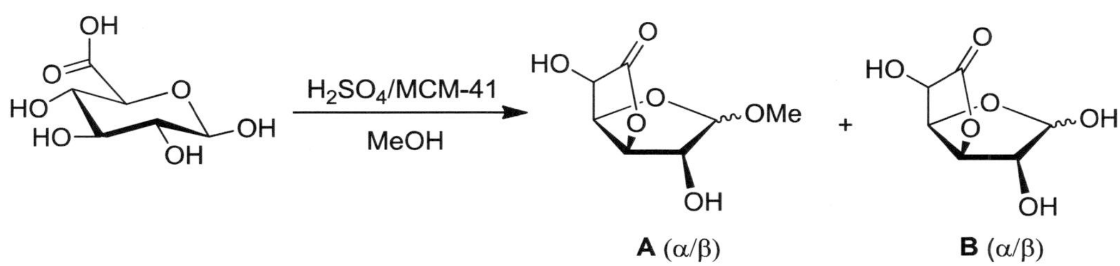

Catalyst	MW	Thermal	
	A (%)	A (%)	B(%)
H_2SO_4/MCM-41	9 (α/β 20:80)	30 (α/β 10:90)	45 (α/β 20:80)
HPWA/MCM-41	68 (α/β 50:50)		

Scheme 9. Synthesis of monosubstituted β-D-glucofuranosidurono-6,3-lactones from D-glucuronic acid with alcohols under thermal and microwave irradiation .

Table 1. Reactions catalyzed by ordered mesoporous silicates modificated with sulfonic acids

Catalyst	Substrate	Product	Reaction conditions	Reference
SO_3H-FSM	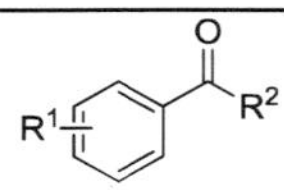	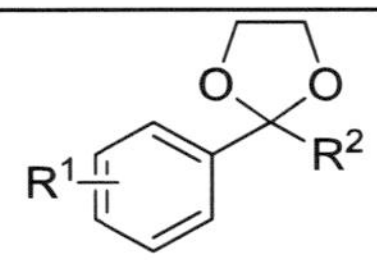 (77-99%)	Carbonyl compound (5 mmol), ethylene glycol (6 mmol), catalyst (0.1 mol%) in toluene at reflux in air	41
SO_3H-FSM	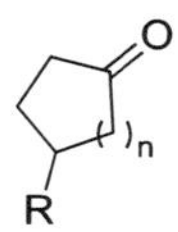	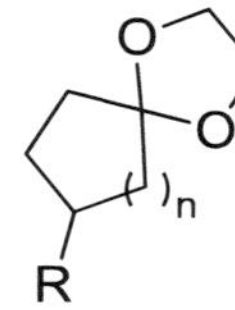 (98-99%)	Carbonyl compound (5 mmol), ethylene glycol (6 mmol), catalyst (0.1 mol%) in toluene at reflux in air	41
MCM-SO_3H	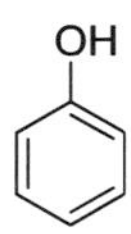	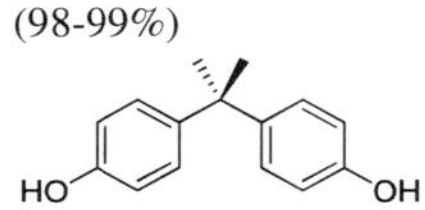 (30%, Selectivity = 92%)	Phenol/acetone molar ratio 5:1, catalyst (50 mg) at 70 °C, 24h	38
SBA [Si]-CH_2CH_2aryl-SO_3H SBA [Si]-propyl-SO_3H SBA [Si]-propyl-SO_3H HO_3S-propyl-[Si]	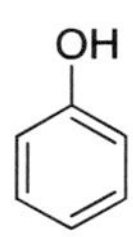	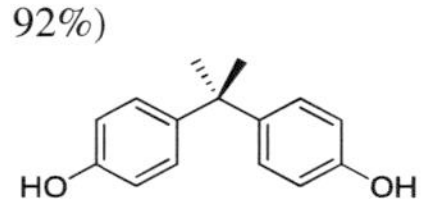	Phenol/acetone molar ratio 7/1 at 85 °C, 24 h	40
SBA-15/S	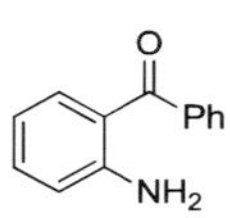	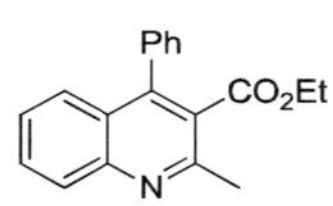 (67%)	2-Aminoaryl ketone (1 mmol), ethyl acetoacetate (5 mmol), catalyst (25 mg) at 90 °C, 3h	43
SBA-15/SO_3H	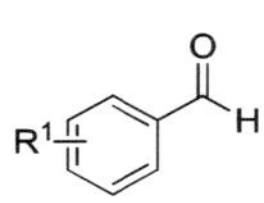	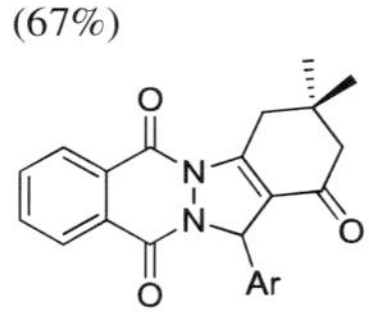 (92-96%)	Aldehyde (3.3 mmol), dimedone (3 mmol), phthalhydrazide (3 mmol), catalyst (0.07 g, ~4 mol%) in EtOH at 45 °C,15–20 min under US irradiation	44
SBA-15/SO_3H	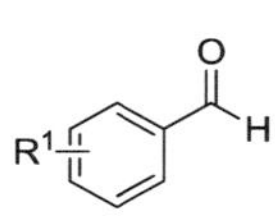	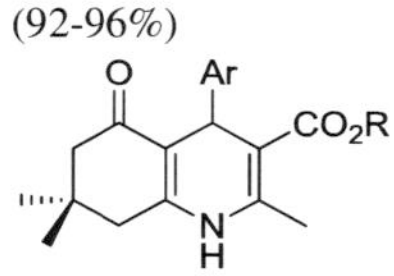 (90-96%)	β-Diketon (3 mmol), aldehyde (3.3 mmol), dimedone (3 mmol), ammonium acetate(4 mmol), catalyst (0.07 g, 4 mol%) in EtOH at rt under US irradiation	44
SBA-15/SO_3H	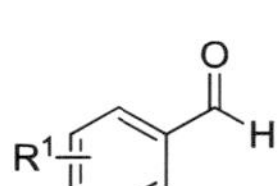	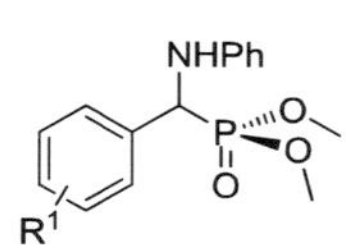 (91-95%)	Aldehyde (5 mmol), amine (5 mmol), dimethyl phosphite (6 mmol), catalyst (0.06 g,~2 mol%) in EtOH at rt under US irradiation	44
H_2SO_4/MCM-41 HPWA/MCM-41	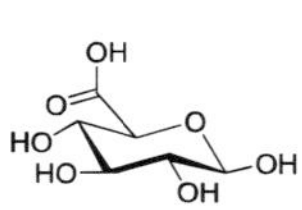	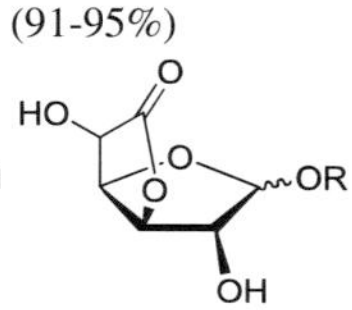 (32-76%)	D-Glucuronic acid (1.03 mmol), MeOH (6 equiv), catalyst (50 mg) at 85 °C, 10 min, under thermal and MW activation (80 W)	45

2.1.2. Mesoporous Silicas Incorporating Heteroatoms

Mesoporous silicas incorporating other single atoms and exhibiting acidic properties have been also widely studied. (Al)SBA-15 with hexagonal channel ordering was found to exhibit higher catalytic activity as compared with its analogue (Al)MCM-41 [46, 47]. Different strategies have been developed for the synthesis of these types of materials . Hu et al. [48] reported a study concerning the nature of acid sites in different series of (Al)SBA-15 showing various Si/Al ratios by using NMR techniques ; the alumination process of the samples was carried out using sodium aluminate at room temperature . The results reported by the authors indicated that the Brønsted acid sites in (Al)SBA-15 samples are due to the generation of terminal silanol groups in the vicinity of aluminum atoms, acid strength of (Al)SBA-15 materials being comparable to that of microporous zeolites . Remarkably, no Lewis acid sites can be found in the (Al)SBA-15 materials.

Recently, Zukal et al. [49] described the alumina -grafted SBA-15 materials prepared by post-synthesis alumination of SBA-15 matrix by using aluminum chlorhydrol. The most important conclusion of this study is that the successive alumination process occurs by gradual filling of the corona surrounding the pores, the mesoporous structure of the resulting materials being preserved without observing blockage of the mesopores . More recently some of the authors completed the study concerning to the effect of the gradual alumination of SBA-15 on the structure and adsorption properties by using ^{1}H and variable-temperature ^{15}N NMR techniques [50]. In this sense, the aluminum film is formed in the confinement of the cavities of rough mesopore walls resulting in aluminum islands on the inner surfaces of mesopores. The alumination process produces Brønsted and Lewis acidity ; the first one is attributed to silanol groups interacting with aluminum, whereas the Lewis acid sites are depending of the Si/Al ratios. Samples showing high Si/Al ratios exhibited Lewis acidity attributed to tetracoordinated aluminum, while at lower Si/Al ratios, the acidity is likely due to the presence of pentacoordinated aluminum.

Regarding to the catalytic applications, Srivastava and Srivastava [51] described a novel method for the protection /deprotection of amino alcohols (Scheme 10) and carbonyl compounds by employing (Al)SBA-15, as environmental-friendly and reusable catalyst .

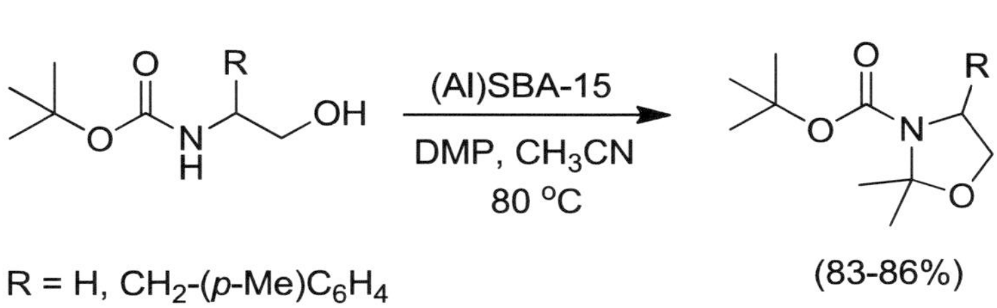

Scheme 10. Protection of N(Boc) amino alcohols using Al-SBA-15.

Almost at the same time Pérez-Mayoral et al. [52] reported the acetalation/ketalation of carbonyl compounds catalyzed by acid or basic aluminated mesoporous SBA-15, under free-solvent conditions and thermal activation , in the presence of ethanol and/or triethyl orthoformate (Scheme 11). The procedure therein reported was applied to the preparation of key intermediate compounds in the synthesis of a pH indicator for Proton Magnetic Resonance Spectroscopic Imaging, ^{1}H MRSI (Scheme 12).

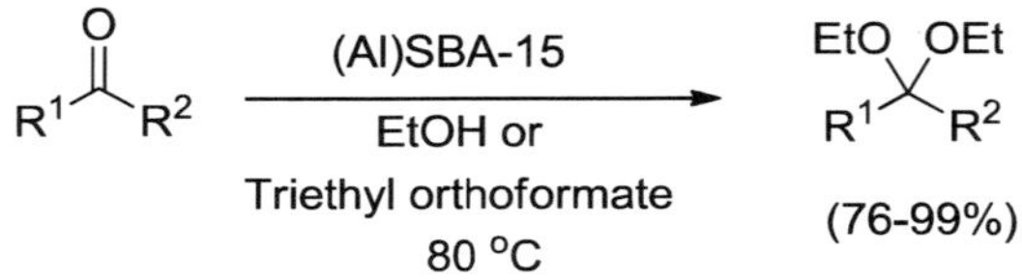

R^1	R^2
$CH_3(CH_2)_5$	H
Ph	H
$-(CH_2)_5-$	

Scheme 11. Acetalation/ketalation reaction from carbonyl compounds catalyzed by (Al)SBA-15.

(Al)SBA-15 was also reported as efficient catalyst in the synthesis of quinolines /quinolones, *via* Friedländer reaction, preferentially leading to the corresponding quinoline but with a lesser selectivity than H-BEA zeolite (Scheme 7) [53, 54].

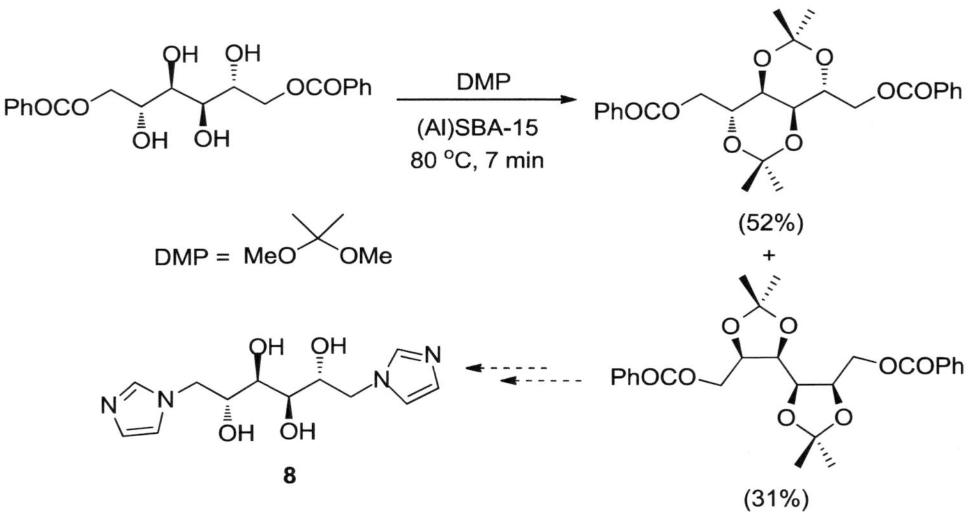

Scheme 12. Synthesis of intermediate for the preparation of the pH indicator for ^{1}H MRSI.

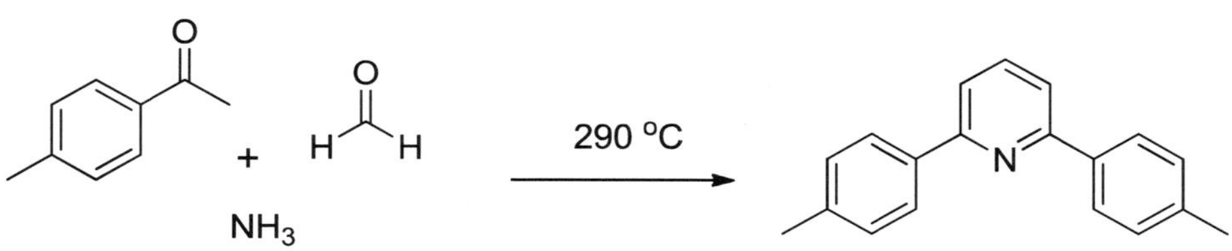

Catalyst (Si/Al ratio)	Conversion (%)	Selectivity (%)
(Al)MCM-41(21)	78	82
Zn(Al)MCM-41(75)	86	99
Zn(Al)MCM-41(151)	82	97
Zn(Al)MCM-41(228)	76	91
Zn(Al)MCM-41(380)	68	81

Scheme 13. Synthesis of 2,5-aryl pyridines.

The presence of transition metal s on molecular sieves also enhances their acid properties. In this sense, novel mesoporous materials showing different Si/(Zn + Al) ratios were found to be efficient and selective acid catalysts for the aminocyclization reaction of 4-methyl acetophenone with formaldehyde and ammonia , under vapor phase reaction conditions, (Scheme 13) [55]; higher selectivity than that for (Al)MCM-41 or even different zeolites was observed. The selective synthesis of 1,4-diazabicyclo (2.2.2)octane , an amine useful in the synthesis polyurethane foam s, from ethanolamine and catalyzed by Zn(Al)MCM-41 was also reported by some of the authors [56]. The Lewis acid properties of microporous and mesoporous materials with Sn sites in different frameworks and the catalytic performace and diastereoselectivity of these materials in the cyclization of *citronellal* to *isopulegol*, have been analyzed by Corma et al. [57] in a combined experimental and computational study (Scheme 14). They concluded that tin species with one or two hydrolyzed Sn–O–Si bridges act as strong Lewis acid sites, and are highly active in several reactions , whereas tin species in framework positions coordinated to four framework siloxy groups are not active.

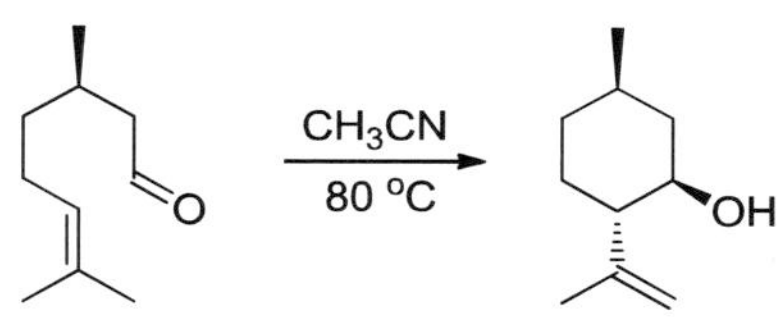

Scheme 14. Cyclization of citronellal into isopulegol.

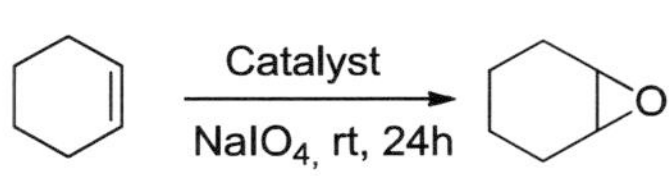

Catalyst	Conversion (%)	Selectivity (%)
Mn-NH-HMS	92	100
Mn-NH-Hβ	76	100
Mn-NH-HY	54	100

Scheme 15. Epoxidation of cyclohexene by $NaIO_4$.

Mesoporous silica containing heteroatoms have been reported not only as interesting catalysts with acid -base properties. In this sense, silica-supported titanium -based heterogeneous catalysts are employed industrially in the production of epoxides using alkyl hydroperoxides (typically, *tert*-butylhydroperoxide, TBHP) as the oxidant [58]. Variants of such catalysts prepared from titanocene dichloride and the mesoporous silica MCM-41 possess active centers that are "single-site" in the sense that they are spatially well separated at the inner surfaces of the mesopores and consist of Ti(IV)-OH groups tripodally anchored to the silica via covalent bond s to oxygen . A combined experimental (XAFS measurements) and computations study revealed a quantitative three-dimensional picture of the six-coordinated Ti(IV)-centered active site during production of epoxides from alkenes in the presence of hydroperoxide as oxidants [59].

More recently, Jiang et al. have reported the Mn(III)-porphyrin grafted mesoporous silica through amide bond. These materials have shown structural durable properties and they behave as steadily reusable catalysts for epoxidation , showing promising properties over other zeolites (Scheme 15). In addition, the computed energy profiles for epoxidation of alkenes by $NaIO_4$ and H_2O_2 were also determined to explain the different catalytic efficiency [60].

Table 2. Reactions catalyzed by ordered mesoporous silicates incorporating heteroatoms

Catalyst	Substrate	Product	Reaction conditions	Reference
(Al)SBA-15	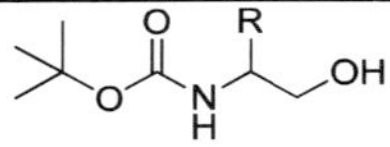	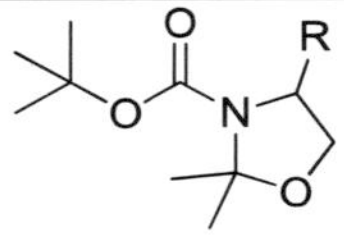 (83-86%)	*Protection*: Amino alcohols (5 mmol), DMP (15 mmol), acetonitrile , catalysts (10% w/w), at 80 °C, 8 h *Deprotection*: acetonide (5 mmol), methanol (5 ml), catalyst (10% w/w) at 40 °C, 3h	51
(Al)SBA-15	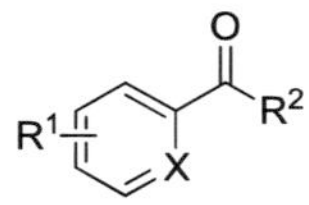	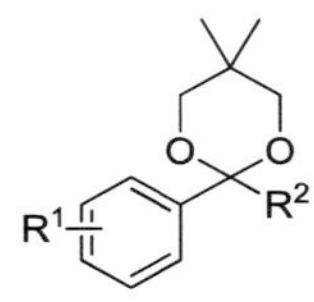 (53-96%)	Carbonyl compounds (5 mmol), alcohol (7.5 mmol), catalyst (10 wt%) in 1,2 dicholoro ethane at 85 °C, 5h	51
(Al)SBA-15	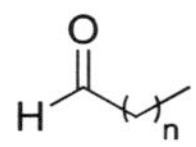	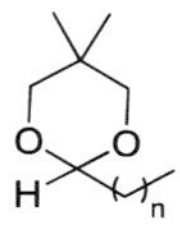 (85-87%)	Carbonyl compounds (5 mmol), alcohol (7.5 mmol), catalyst (10 wt%) in 1,2 dicholoro ethane at 85 °C, 5h	51
(Al)SBA-15	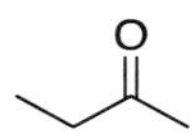	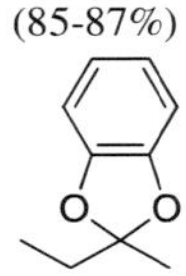 (96%)	Carbonyl compounds (2 mmol), catechol (2 mmol), toluene, catalyst (10 wt%) at 120 °C, 2h	51
(Al)SBA-15	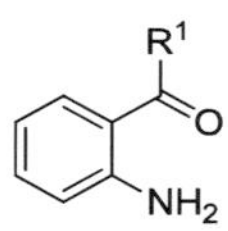	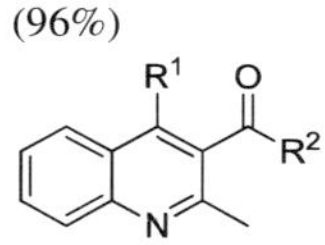 (31-77%)	*Procedure A:* 2-aminoaryl ketone (1 mmol), 1,3-dicarbonyl compound (1.5 mmol), catalyst (50 mg), in toluene , at 363 K. *Procedure B:* 2-aminoaryl ketone (1 mmol), 1,3-dicarbonyl compound (1.5 mmol), catalyst (50 mg) at 363 K.	53, 54
(Al)MCM-41 Zn(Al)MCM-41	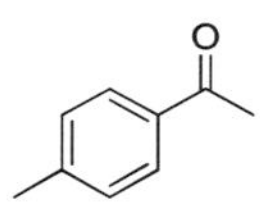	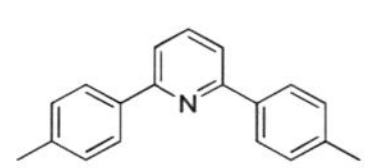 (81-99%)	4-MAP/NH_3 mol ratio = 2:5 2 mol of formaldehyde (2 mol), catalyst (2g) at 290 °C	55
MCM-41-Sn(OH) Sn-MCM-41 MCM-41-$SnMe_2$	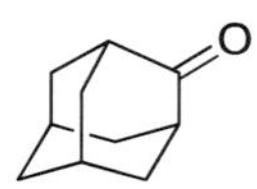	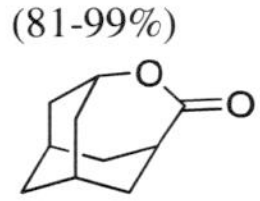	Adamantanone (150 mg), aqueous hydrogen peroxide 35% (150 mg), catalyst (50 mg) in dioxane at 90 °C	57
Ti-MMM-2 Ti-MMM	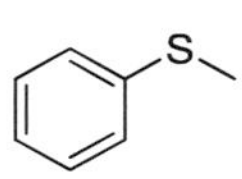	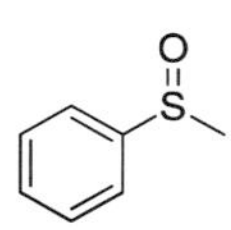 (74-75%)	0.1M MPS, MPS/H_2O_2 = 1/1.2, 20 °C, MeCN, 35 min	58b
Ti-MMM-2 Ti-MMM	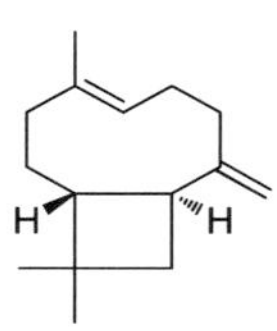	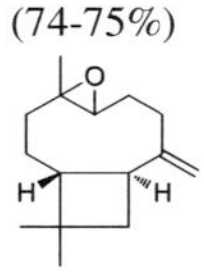 (52-56%)	0.1M CP, CP/H_2O_2 = 1/1, MeCN, 50 °C, 4 h	58b
Ti-MMM-2 Ti-MMM	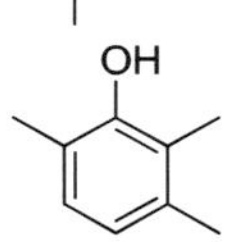	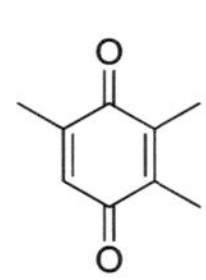	0.1M TMP, TMP/H_2O_2 = 1/3.5, 80 °C, MeCN, 25 min	58b

Catalyst	Substrate	Product	Reaction conditions	Reference
Mn-NH-HMS	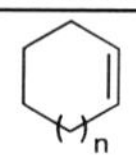	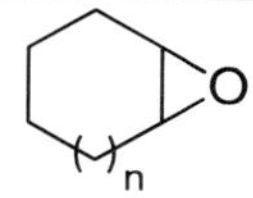 (92-100%)	Alkene (1 mmol), catalysts (the amount of Mn(TCPP)Cl was adjusted to 0.002 g), imidazole (2.8 mmol), $NaIO_4$ (2 mmol; H_2O, 10 mL) in CH_3CN at rt	60
Mn-NH-HMS	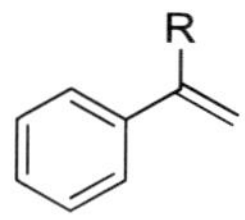	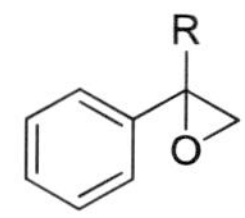 (76, 89%, Selectivity 76, 80%)	Alkene (1 mmol), catalysts (the amount of Mn(TCPP)Cl was adjusted to 0.002 g), imidazole (2.8 mmol), $NaIO_4$ (2 mmol; H_2O, 10 mL) in CH_3CN at rt	60
Mn-NH-HMS	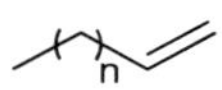	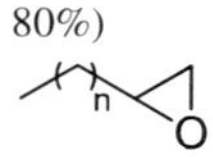 (12-86%, Selectivity 100%)	Alkene (1 mmol), catalysts (the amount of Mn(TCPP)Cl was adjusted to 0.002 g), imidazole (2.8 mmol), $NaIO_4$ (2 mmol; H_2O, 10 mL) in CH_3CN at rt	60

2.1.3. Surface Modification with Basic Centers

The development of environmentally friendly basic catalysts which can replace alkaline metals or hydroxides is an interesting issue in the production of fine chemicals . In general, solid acid catalysts have been extensively studied in chemical technology ; however much less attention has been paid to the basic materials . Basicity in mesoporous silicates can be introduced by impregnation of basic metal oxides or covalent binding with amines , the main disadvantage of both being the leaching of active species and low hydrothermal stability of amine group s, respectively.

Regarding this subject, Michalska et al. [61] reported the synthesis of basic MCM-41, by impregnation of (Nb)MCM-41 and (Al)MCM-41 with alkali metal acetates. Basicity of the MCM-41 type mesoporous silica can be controlled by selecting the appropriate alkali metal and the chemical composition of the matrix . These types of mesoporous silica, MCM-41 and (Nb)MCM-41 modified with alkaline metals (from Li to Cs), were lately reported as useful catalysts for the synthesis of the key intermediate compounds for the synthesis of 1,4-dihydropyridines *via* Knoevenagel condensation , the most useful C=C bond forming strategy in Organic Synthesis (Scheme 16) [62]. It is interesting to note that niobium atoms located in the mesoporous material plays a role of structural promoter preventing the structure disordering caused by the treatment with alkali metals.

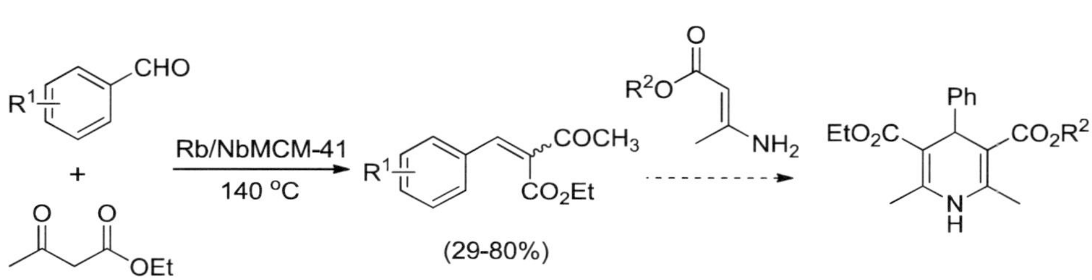

Scheme 16. Knoevenagel condensation in the synthesis of the key intermediate for the preparation of 1,4-dihydropyridines.

Other basic mesoporous silicas commonly used are those functionalized with amine group s. Basicity of amines is lower than that for the tetraalkylammonium hydroxides. It was possible to prepare organic ammonium quaternary salt by grafting to the pure silica MCM-41 as useful catalysts for aldol and Knoevenagel condensations and Michael additions, under mild conditions (Scheme 17) [63, 64]. The reported methodology was used in the selective synthesis of chromenes, from salicylaldehyde and diethyl 2-pentenedicarboxylate (Scheme 18); chromenes are oxygen -containing heterocycles widely distributed in nature, and also synthetic compounds with commercial interest in the preparation of pharmaceuticals .

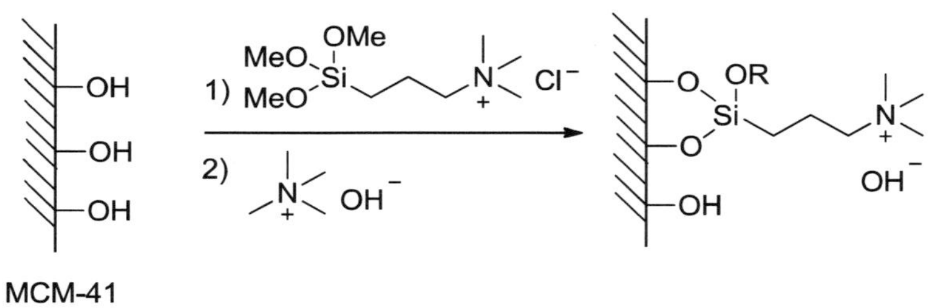

Scheme 17. Synthesis of organic ammonium quaternary salt by grafting to MCM-41.

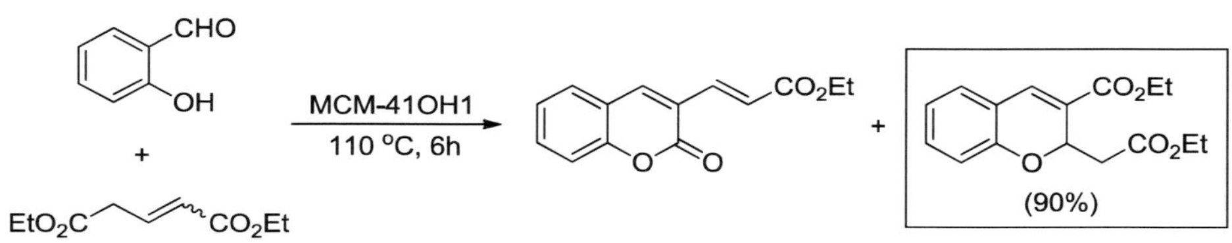

Scheme 18. Selective synthesis of chromenes from salycilaldehyde and diethyl 2-pentenedicarboxylate.

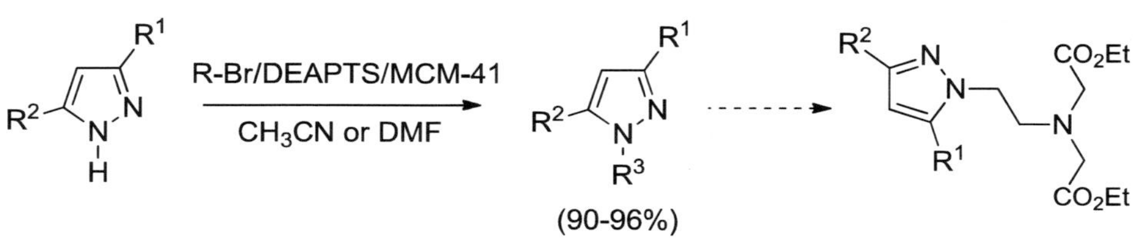

Scheme 19. *N*-Alquilation of pyrazole catalyzed by amino -grafted mesoporous silica .

A large variety of amino -functionalized mesoporous materials have been reported because of its application for capture and separation of gases [65-67]. These basic materials are often useful in green processes although these solids show lower thermal stability , as mentioned above. Among the most recent results are the selective *N*-alkylation of pyrazole catalyzed by different mesoporous basic molecular sieves (Cs)Al-SBA-15 or APTS/ MCM-41 and DEAPTS/MCM-41 (where APTS and DEAPTS are aminopropyl and diethylaminopropyl groups, respectively) [68]; DEAPTS/MCM-41 with tertiary amine group s proved the best catalyst for this transformation . The pyrazole alkylation reaction has been clarified by a computational study of the reaction mechanism , with two different cluster model s to simulate the solid. The authors demonstrated that the alkylation of pyrazole often using stoichiometric amounts of bases is, in this case, a catalytic process. The investigated methodology was used in the preparation of *N*-alkyl pyrazoles as key intermediates for the preparation of ligands which exhibit interesting chelating properties (Scheme 19).

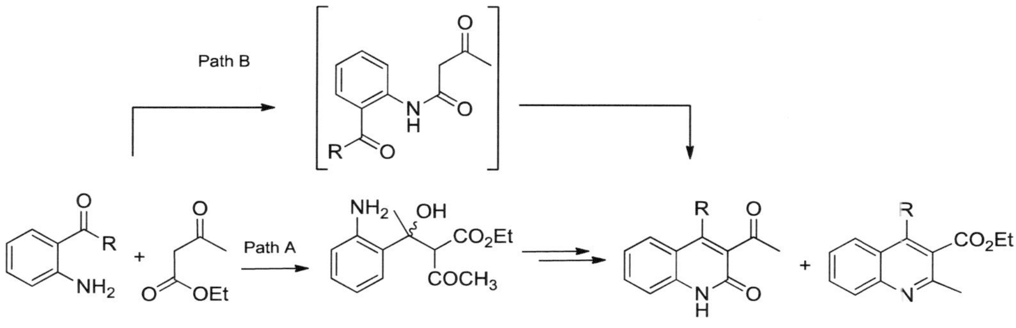

Scheme 20. Friedländer condensation of 2-aminoaryl ketones and ethyl acetoacetate. Path A catalyzed by (Al)SBA-15 and Path B when using amino -grafted mesoporous molecular sieve s.

Concerning the preparation of heterocyclic compounds with synthetic interest, Dominguez-Fernandez et al. [53] investigated the Fiedländer condensation between 2-aminoaryl ketones and ethyl acetoacetate catalyzed by amino -grafted MCM-41 and other acid mesoporous molecular sieve such as (Al)SBA-15. While the condensation catalyzed by acid catalyst yielded mixtures of corresponding quinolin-2(1*H*)-ones and quinolines (Path A), dramatic selectivity changes were observed when using amino-grafted MCM-41 materials , which exclusively led to quinolones, isolated with quantitative yield and total selectivity (Path B) (Scheme 20).

Aider et al. [69] have recently published the synthesis of coumarins catalyzed by amino -grafted SBA-15 (Scheme 21); the reaction takes places by a Knoevenagel condensation followed by a non-catalytic lactonization step. Additionally, the authors reported therein a theoretical DFT study which demonstrates that the amino moiety of MAP/SBA-15 (where MAP is methyl aminopropyl) shows a dual acid -base catalytic behavior (as H-donor/H-acceptor) when used in the synthesis of coumarins *via* Knoevenagel condensation.

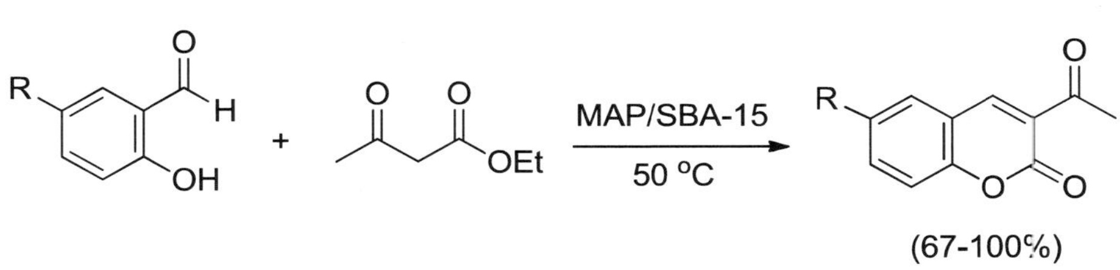

Scheme 21. Synthesis of coumarins from salycilaldehydes and ethyl acetoacetate.

Claisen-Schmidt condensations followed by Michael addition reaction over aminopropyl-functionalized SBA-15 materials to synthesize flavanones with good yields and high selectivities were published by Wang et al. [70]. It was observed that the catalytic activity decreased with increasing loading of amine group s on SBA-15 probably because of the decreasing in the surface area and pore volume (Scheme 22).

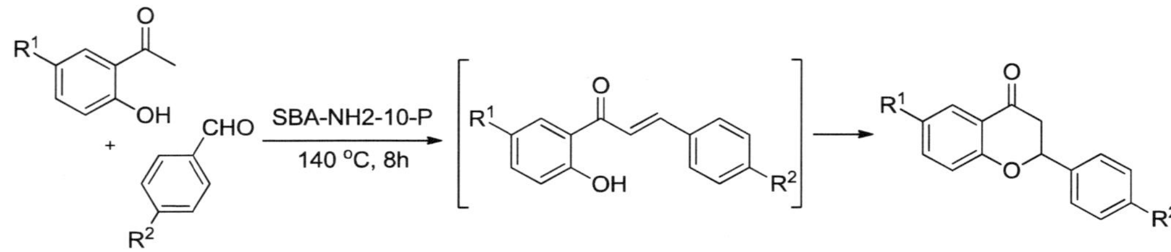

Scheme 22. Synthesis of flavanones catalyzed by aminopropyl-functionalized SBA-15 molecular sieves.

Table 3. Reactions catalyzed by basic ordered mesoporous silicates

Catalyst	Substrate	Product	Reaction conditions	Reference
MCM-41OH1	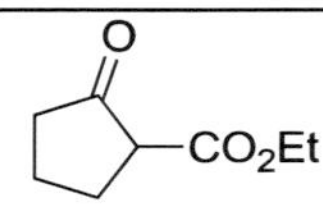	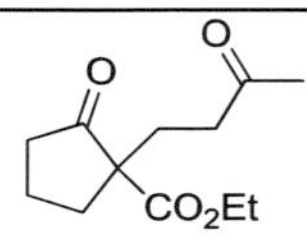 (85%)	2-Oxocyclopentane 2-carboxylate (8 mmol), methyl vinyl ketone (8 mmol), catalyst (60mg) in CH_3CN at rt under N_2, 1h	64
MCM-41OH1	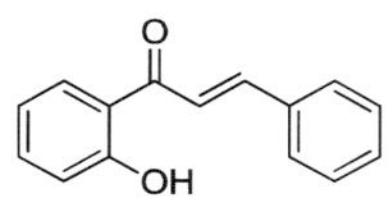	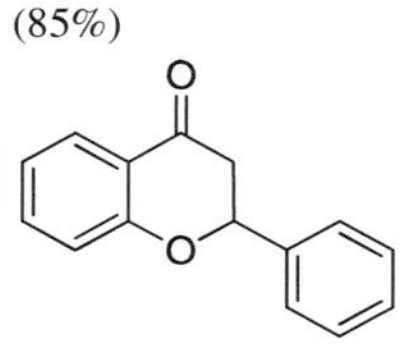 (65%, Selectivity 78%)	Benzaldehyde (10 mmol), 2-hydroxyacetophenone (8 mmol), catalyst (15% m/m) at 130 °C, 8 h	63, 64
SBA-NH2- 10-P	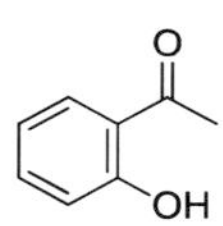	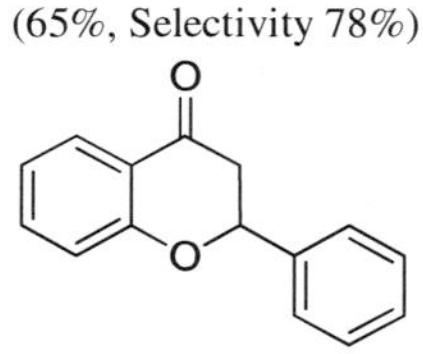 (84-92%, Selectivity 69-71%)	Benzaldehyde (10 mmol), 2-hydroxyacetophenone (10-15 mmol), catalyst (0.15 g) at 140 °C, 8 h	70
DEAPTS/MCM-41	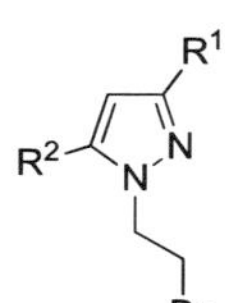	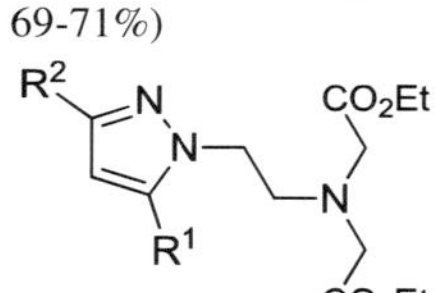 (79-81%)	Alkyl bromide (0.5mmol), diethyl iminodiacetate (0.5mmol), catalyst (20 wt% with regard to the amine) in DMF at 100 °C, 4-4.5h	68
DEAP/MCM-41 MAP/MCM-41 AP/MCM-41	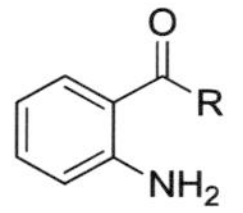	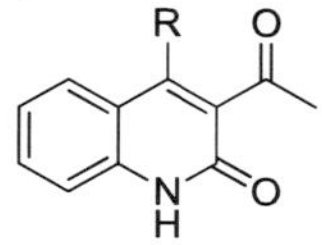 (42-100%)	*o*-Aminoaryl ketone (1 mmol), ethyl acetoacetate (1.5 mmol), catalyst (20 wt.% with respect to the *o*-aminoaryl ketone), in toluene at 100 °C, 1-7 h	53
DEAP/SBA-15 MAP/ SBA-15	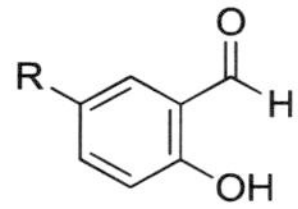	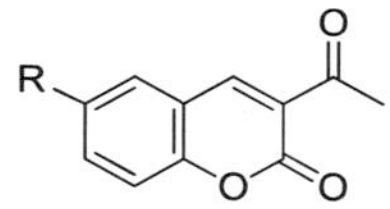 (67-100%)	2-Hydroxibenzaldehyde (2 mmol), ethyl acetoacetate (2 mmol), catalyst (50 mg), at 50 °C	69

2.1.4. Surface Modification with Ionic Liquids

On the other hand, the immobilization of ionic liquids (ILs) over mesoporous silica combines the advantages of homogeneous and heterogeneous catalysis , such as high catalytic activity and selectivity , easy product separation and catalyst recyclability [71]. A recent example of ILs immobilized over SBA-15 was reported by Nargis Parvin et at. [72]. The authors prepared samples of SBA-15 functionalized with 1-methyl-3-[(3-triethoxysilyl) propyl] imidazolium chloride by using the co-condensation method under microwave irradiation . The samples under study was found to be efficient to catalyze the Knoevenagel condensation of different aromatic and heteroaromatic aldehydes with ethyl cyanoacetate, in solvent free conditions, and also under microwave irradiation at short reaction time s.

2.2. Bifunctional Materials

The specific and selective chemical reactions in living organisms are often processes taking place in cascade and promoted by multifunctional catalysts, *the enzymes* . During the last two decades, part of the efforts in the heterogeneous catalysis field has been addressed to the development of inorganic–organic hybrid materials which can mimic the enzymatic behavior [73]. The preparation of silica materials containing different functional groups which can act in cooperation frequently provides both catalytic activity and selectivity superior to the monofunctional analogues materials or even more than the homogeneous catalyst s [74, 75].

Numerous examples of bifunctional mesoporous silicas useful as catalysts for the synthesis of fine chemicals have been recently reported. Blasco-Jiménez et al. [76] described different amino -grafted mesoporous silicas based on MCM-41 and (T)MCM-41 matrixes (where T is Nb or Al) useful for the production of *N*-alkyl imidazoles, with antiviral activities, by Michael addition between imidazole and ethyl acrylate , under ultrasound activation, and also in the Knoevenagel condensation reactions (Scheme 23). The most important finding from these works was that the chemical composition of the mesoporous materials under study influences the amine basicity of the materials giving rise to the higher basicity for the samples containing Nb or Al [77a].

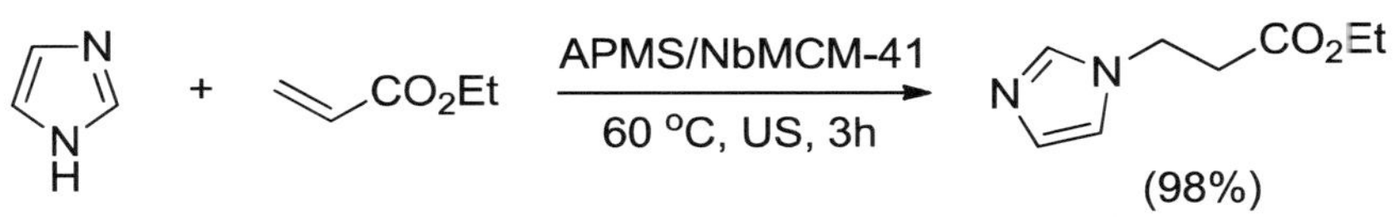

Scheme 23. Condensation of imidazole and ethyl acrylate .

This type of metallosilicates finds also applications for the isomerization of *Safrole*, a naturally occurring heterocyclic compound used as flavoring agent in drugs and in the manufacture perfumes or soaps (Scheme 24) [77b].

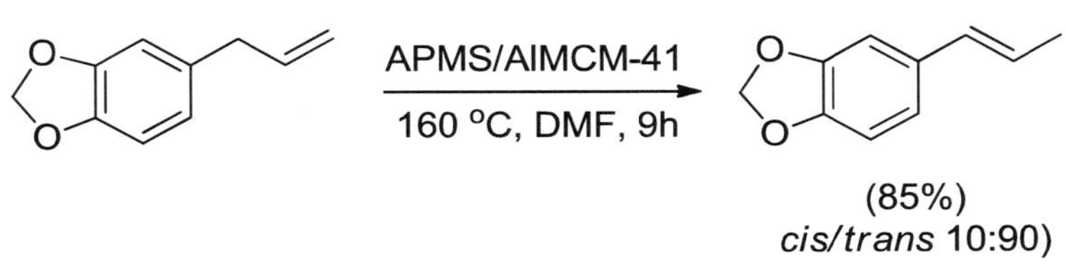

Scheme 24. Isomerization of safrole over amino -grafted MCM-41 materials .

More recently Smuszkiewicz et al. [78, 79] have reported a new series of ordered amino - grafted mesoporous metallosilicates based on MCF structures which have been found to successfully catalyze the selective synthesis of quinolines, *via* Friedländer reaction, from 2-amino-5-chlorobenzaldehyde and ethyl acetoacetate (Schemes 25 and 26). As an example of the interest of these structures, the quinoline (where Cl is H) is known as histone acetyltransferase (HAT) inhibitor and also a building block for the synthesis of lavendamycin (Schemes 25). Experimental and theoretical studies demonstrated that the presence of water in non-activated samples plays an important role in the formation of the corresponding quinoline by stabilizing key transition structures and intermediate species and enhancing the regioselectivity of the reaction [78].

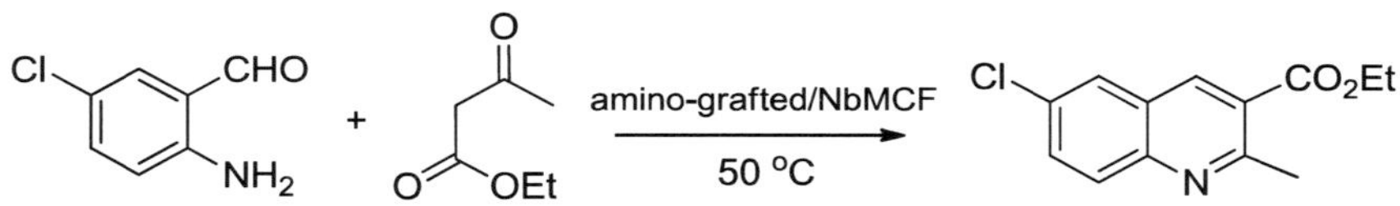

Catalyst	Time (min)	Yield (%)
APMS/MCF	300	100
2APMS/MCF	15	99
APMS/NbMCF	30	98
2APMS/NbMCF	120	87

Scheme 25. Synthesis of quinolines through Friedländer reaction.

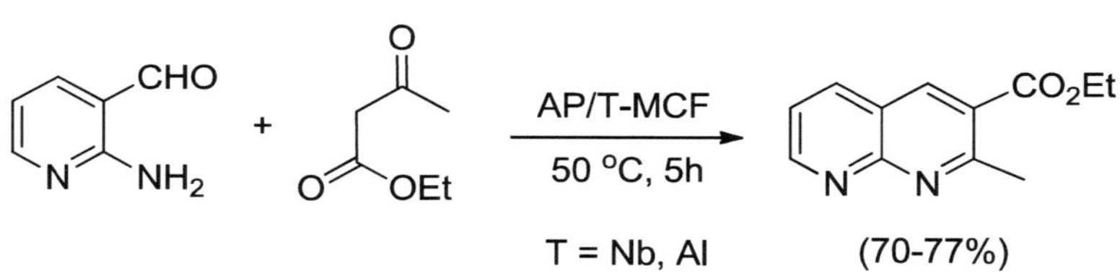

Scheme 26. Synthesis of naphthyridines through Friedländer reaction.

Additionally, a computational study to analyze in detail the first step of the condensation was reported. In this sense, the calculations confirmed cooperative effects of the basic amino group and the metal center of these bifunctional catalysts that give rise to moderate energy barriers for the C-C bond formation .

Corma et al. [80] reported series of bifunctional organic catalysts, simultaneously containing acid and basic active site s with ionic liquid structure , useful for Claisen-Schmidt and Knoevenagel condensations. The authors studied also some mechanistic insights by means of DFT calculations. The distance between acid and base sites was the unique factor which controls the catalytic activity observed for the Knoevenagel condensation . In contrast, the organocatalyst spatial orientation is the key parameter to induce high activity and selectivity in the Claisen-Schmidt condensation where the transition state for the rate-determining step involves two simultaneous or concerted proton transfers. The authors reported detailed mechanistic insights into the cooperative role played by the basic and acid sites concluding that the synthesis of *trans*-chalcones follows a mechanism inspired in natural aldolases.

Table 4. Reactions catalyzed by bifunctional mesoporous silicates

Catalyst	Substrate	Product	Reaction conditions	Reference
NH_4-NbMCM-41 APMS/NbMCM-41 APMS/MCM-41	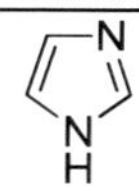	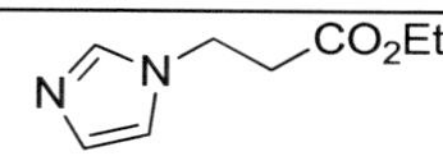 Thermal (81-95%) US (93-98%)	Imidazole (5 mmol), ethyl acrylate (5 mmol) at 60 °C, 3h US:40 kHz of frequency, 550W of power	77a
APMS/MCM-41 APMS/NbMCM-41 APMS/AlMCM-41	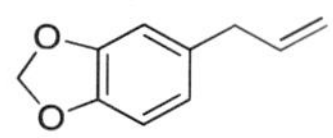	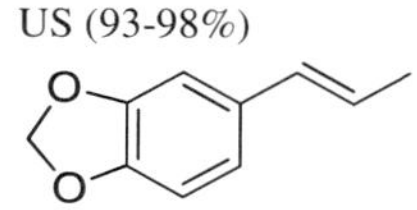 (35-85%)	Safrole (0.4 g), catalyst (0.25 g) in DMF at 160 °C	77b
APMS/MCF 2APMS/MCF APMS/NbMCF 2APMS/NbMCF	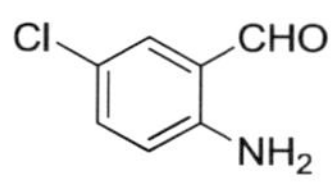	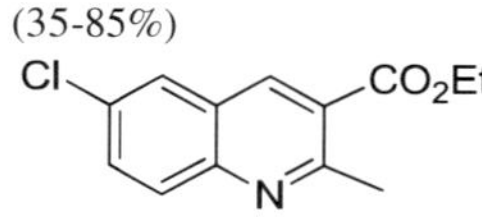 (87-100%)	2-Amino-5-chlorobenzaldehyde 7 (0.5 mmol) and ethylacetoacetate 2 (5 mmol), catalyst (25 mg) at 50°C	78
AP/MCF AP/Nb-MCM AP/Al-MCM AP/Nb-MCM AP/Al-MCM	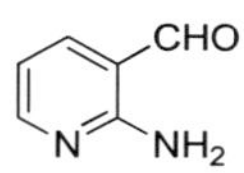	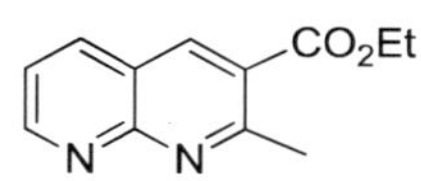 (70-77%)	2-Amino-5-chlorobenzaldehyde 7 (0.5 mmol) and ethylacetoacetate 2 (5 mmol), catalyst (25 mg) at 50°C	79

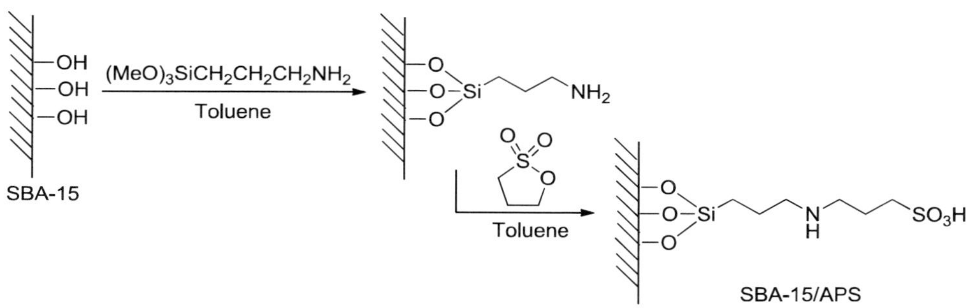

Scheme 27. Synthetic approch for the preparation of SBA-15/APS.

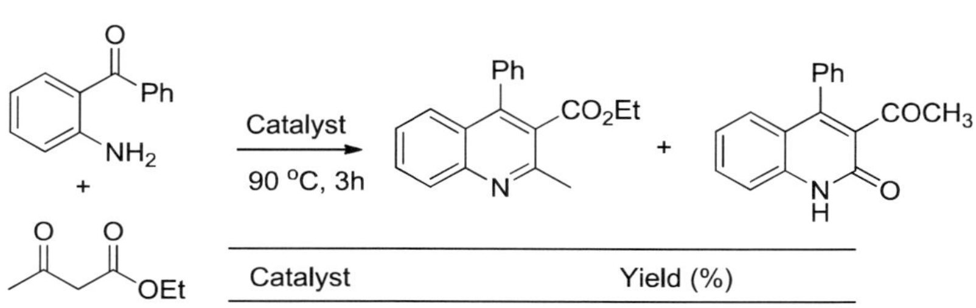

Catalyst	Yield (%)	
SBA-15/APS	86	14
SBA-15/AP	3	63
SBA-15/S	67	25

Scheme 28. Friedländer condensation catalyzed by SBA-15/APS.

Concerning mesoporous bifunctional organocatalysts, Lopez-Sanz et al. [43] also reported the first mesoporous catalyst based on SBA-15 incorporating simultaneously basic and acid functions able to promote the Friedländer reaction leading to quinolines with high yields and selectivities (Schemes 27 and 28). Although SBA-15/S (Scheme 2) was more active than SBA-15/APS, this last afforded the corresponding quinoline with the highest selectivity .

3. METAL-ORGANIC-FRAMEWORKS

MOFs are porous coordination polymers constructed from the assembly of organic ligands with metal ion s or metal clusters particularly interesting because of their high surface area , permanent porosity and stability , as well as the ease synthesis based on predictable designs. Analogously to zeolites , MOFs are considered a type of porous materials with tunable pore size s ranging from non-pores to mesopores . A review particularly interesting regarding the strategies in the design of mesoporous MOFs and some of their applications in gas storage [81], catalysis and sensors among others was recently reported by Song et al. [82, 83].

MOFs have been recently revised as catalysts for the synthesis of fine chemicals [84]. The most important advantage of MOFs in catalysis is the large density of active site s as metal coordination vacancies as well as their high porosity . From an environmental point of view, both characteristics make them improved catalytic system s in comparison with the homogeneous analogues, the transition metal catalysts. In this context, Dhakshinamoorthy et al. [85] reviewed the catalytic activity of commercially available MOFs for a variety of acid - promoted reactions or even for transformations requiring redox centers.

Table 5. Reactions catalyzed by MOFs

Catalyst	Substrate	Product	Reaction conditions	Reference
Fe(BTC) $Al_2(BDC)_3$ $Cu_3(BTC)_2$ Zn-MOF	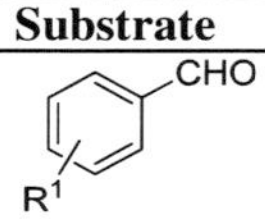	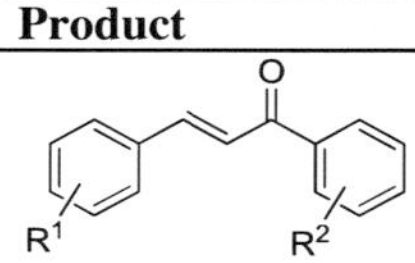 (6-90%)	Benzaldehyde (0.47 mmol), acetophenone (0.466 mmol), catalyst (150 mg) at 110 °C, 24 h	86
Fe(BTC) $Cu_3(BTC)_2$ $Al_2OH(BDC)_3$	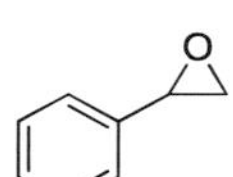	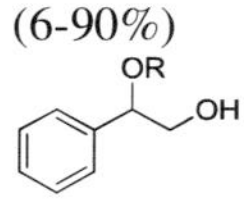 (14-99%)	Epoxide (0.250 g, 2 mmol), alcohols (5 mL), catalyst (50 mg)	87
Fe(BTC) $Cu_3(BTC)_2$ $Al_2(BDC)_3$	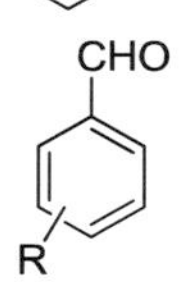	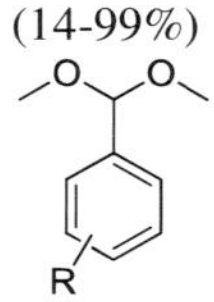 (30-94%)	Aldehyde (1 mmol), methanol, $Cu_3(BTC)_2$ (100 mg) at rt, 24 h	88
$Cu_3(BTC)_2$	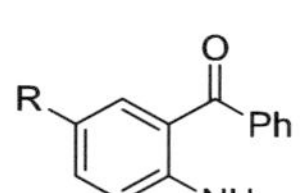	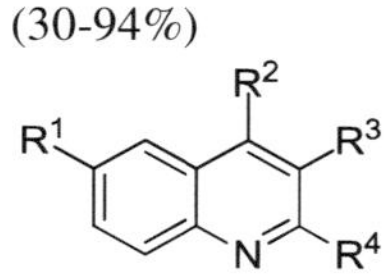 (41-99%)	Acetylacetone (5 mmol), 5-chloro- 2-amino- benzophenone (1 mmol), $Cu_3(BTC)_2$ (0.083 mmol, 50 mg) at 90 °C, 2 h	91, 92a
Fe(BTC) $Cu_3(BTC)_2$	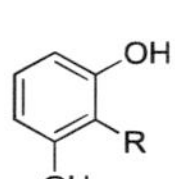	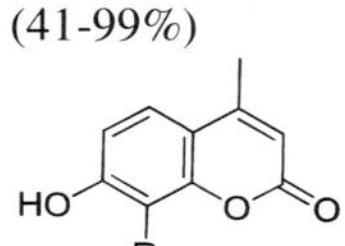	Phenol (8.5 mmol), ethyl acetoacetate (10 mmol), catalyst (0.2 g) in $PhNO_2$, at 130 °C, 23h	93

Although there are reported a lot of MOFs and they have been widely used in redox reactions , this section is only focused on reviewing the most studied mesoporous catalytic system s promoting some of the most relevant acid -base catalyzed transformations useful in Fine Chemistry; some of them are summarized in Table 5.

3.1. Acid-Base Promoted Reactions

One of the most known MOFs is the termed *Basolite materials* commercialized by BASF (Table 6). In general, these MOFs exhibit Lewis-acid properties. Particularly, $Cu_3(BTC)_2$ and Fe(BTC) show one coordination vacancy often occupied by a solvent molecule which can be removed by activation at the optimal temperature or exchanged by other ligands acting as a coordinative unsaturated site. These sites are typical Lewis centers which accept one electron lone pair from the reagents , thus acting as solid acids.

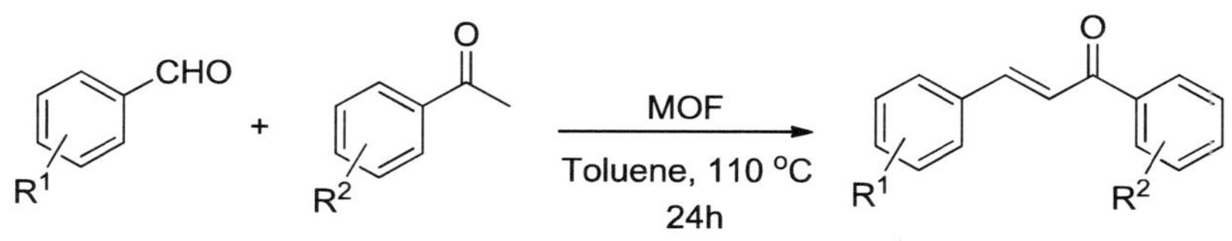

Catalyst	Yield (%)
Fe(BTC)	90
$Al_2OH(BDC)_3$	71
$Cu_3(BTC)_2$	6
Zn-MOF	<2

Scheme 29. Claisen–Schmidt condensation reaction between aromatic aldehydes and ketones .

Table 6. Textural parameters and physical properties of MOFs commercially available from BASF (*from* Ref. 85)

MOF	Ligand	Formula	S_{BET} ($m^2 g^{-1}$)	Pore dimensions (Å)
Basolite F 300	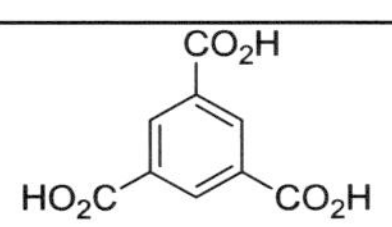	Fe(BTC)	840	21
Basolite C 300 (HKUST-1)	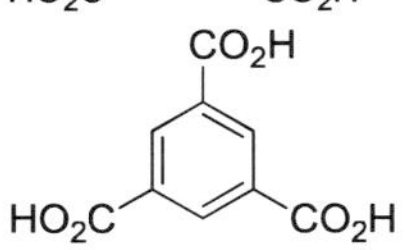	$Cu_3(BTC)_2$	1019	11, 16, and 6
Basolite A 100 (MIL-53(Al))	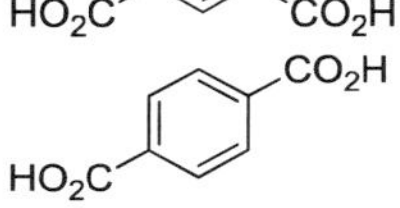	Al(OH) (BDC)	950	8.5
Basolite Z 1200 (ZIF-8)	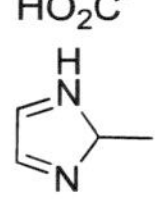	$Zn(MeIm)_2$	1947	11.6

Fe(BTC), $Cu_3(BTC)_2$ and Al(OH)(BDC) have been reported as acid heterogeneous catalysts in aldolic condensations affording chalcones, intermediate compounds of pharmacological interest (Scheme 29) [86]. The highest yield to the corresponding chalcone is indicative of the advantage of using Fe(BTC) instead of others iron salts such as iron(III) chloride hexahydrate as homogeneous catalyst s. Remarkably, Al(OH)(BDC) apparently lacking of exchangeable coordination position, proved to be an even more active catalyst than $Cu_3(BTC)_2$.

Basolite materials have been also reported in epoxide ring opening reactions with alcohols or amines . In this context, Fe(BTC) and $Al_3OH(BDC)_3$ were found to be efficient heterogeneous catalysts leading to 2-methoxy-2-phenylethanol [87] with almost quantitative yield and excellent selectivity (Scheme 30 and 31). Fe(BTC) also promotes the ring opening of styrene oxide with aniline following a $S_N{}^1$ mechanism.

Catalyst	Conversion(%)	Selectivity (%)
Fe(BTC)	99	94
$Cu_3(BTC)_2$	93	95
$Al_2OH(BDC)_3$	67	96

Scheme 30. Ring opening of styrene oxide with methanol catalyzed by MOFs.

Fe(BTC), $Cu_3(BTC)_2$ and $Al_2(BDC)_3$ are also active and selective catalysts for the poly-*N*-methylation of aromatic primary amines with dimethyl carbonate (Scheme 32) [88].

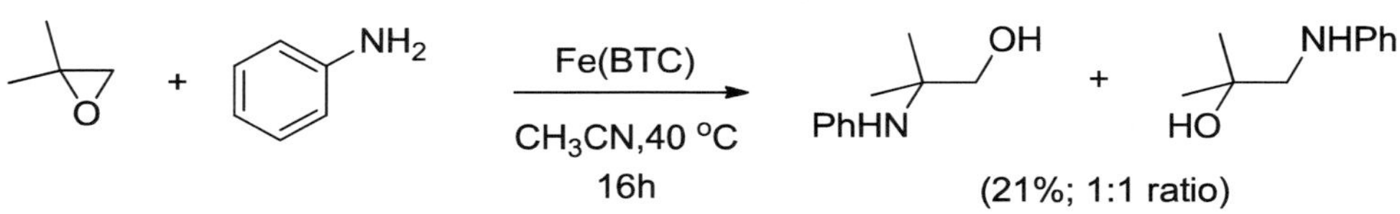

Scheme 31. Ring opening of 1,2-epoxy-2-methylpropane with aniline .

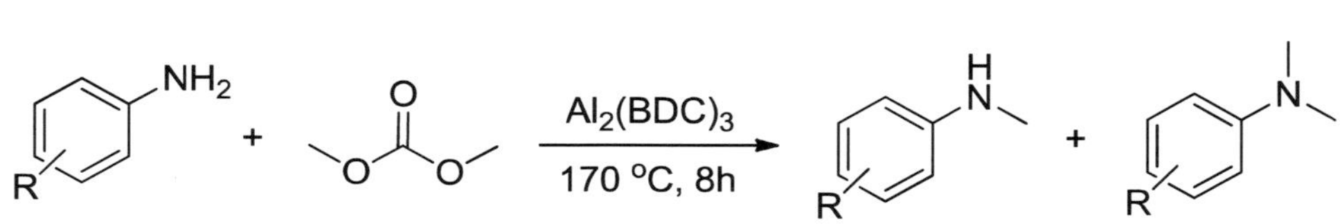

Scheme 32. Methylation of primary aromatic and aliphatic amines with DMC promoted by $Al_2(BDC)_3$.

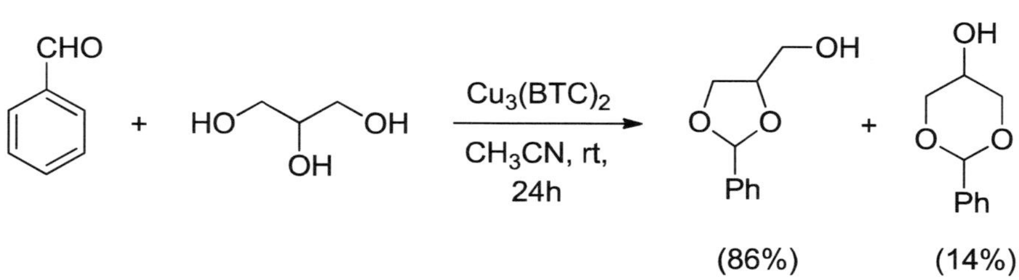

Scheme 33. Acetalization reaction of benzaldehyde with glycerol catalyzed by $Cu_3(BTC\ )_2$.

The $Cu_3(BTC)_2$ catalyzed synthesis of dimethyl acetals using methanol as reagent, an incompatible alcohol with Lewis-acid catalyst, has been interestingly reported [89]. $Cu_3(BTC)_2$ efficiently catalyzed the acetalization of benzaldehyde, at room-temperature yielding the corresponding dimethyl acetal in 88% and with almost total selectivity. Fe(BTC) and Al(OH)(BDC) selectively afforded the dimethyl acetal of benzaldehyde with lower conversions, 71 and 66%, respectively. The catalytic activity and selectivity in the acetalization of benzaldehyde with glycerol, at room temperature and using acetronitrile as solvent, catalyzed by $Cu_3(BTC)_2$ was also reported. In this case, a mixture of (2-phenyl-1,3-dioxolan-4-yl)methanol and the corresponding six-membered ring isomer, 2-phenyl-1,3-dioxan-5-ol, was yielded (Scheme 33).

MOFs have also been reported in condensation reactions for the preparation of heterocyclic compounds, widely used in Organic Synthesis [90]. In this sense, $Cu_3(BTC)_2$ also constitutes an efficient catalyst for the synthesis of quinolines *via* Friedländer reaction. Pérez-Mayoral and Cejka [91] reported for the first time the synthesis of 3-acetyl-5-chloro-2-methyl-4-phenylquinoline, under solvent-free conditions, at 80 °C in the presence of 8.3 mol% $Cu_3(BTC)_2$ resulting in 80% of chemical yield after only 1h of reaction time (Scheme 34). Catalyst under study exhibiting hard Lewis acid character led to highly improved catalytic activity when compared with other molecular sieves showing high concentration of Lewis acid sites, e.g., in BEA and (Al)SBA-15. High concentration of active sites in $Cu_3(BTC)_2$ together with the concerted effect of a pair of adjacent Cu^{2+} coordinatively unsaturated active sites are behind the very high quinoline yield reached at short reaction time [92]. The experimental investigation was accompanied by a computational study of the reaction mechanisms catalyzed by Brønsted and Lewis acid centers. The periodic DFT calculations identified two effects of the catalyst on the course of the Friedländer reaction: (i) lowering of energy barriers, especially for the annulation reaction step and (ii) favorable geometry of the reaction precursor formed by adsorption of reactants on two adjacent Cu^{2+} sites in $Cu_3(BTC)_2$.

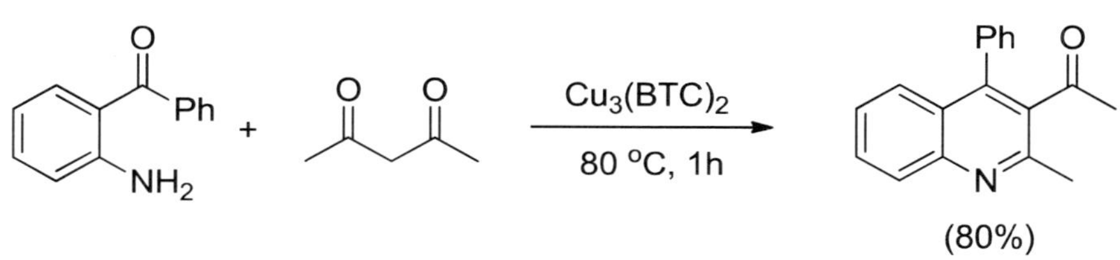

Scheme 34. Friedländer reaction between 2-aminobenzophenone and acetylacetone.

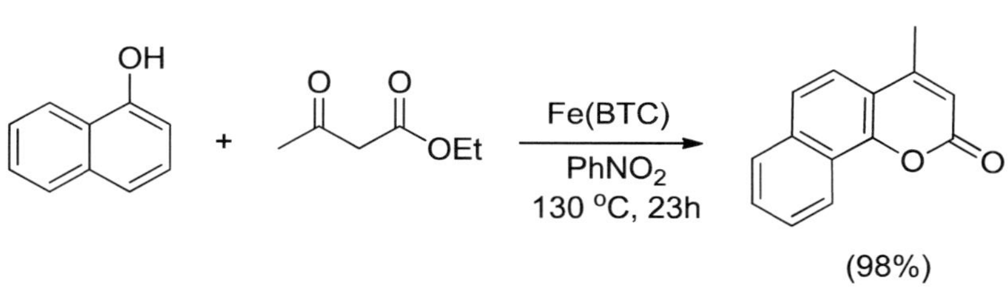

Scheme 35. Pechmann condensation between α-naphthol and ethyl acetoacetate.

Pechmann reaction promoted by $Cu_3(BTC)_2$ and Fe(BTC) to yield coumarins has been reported by Opanasenko et al. [93]. Both of them were found to be efficient catalysts for the synthesis of coumarin derivatives from α-naphthol and ethyl acetoacetate, at 130 °C, as compared with the traditional zeolites, 4-methyl-2H-benzo[*h*]chromen-2-one being selectively isolated after 23h of reaction time (Scheme 35).

However, when using resorcinol and pyrogallol as starting materials the destruction of the MOF framework was observed because of the strong adsorption of the substrates on the active site s. Then, the Pechmann reaction catalyzed by MOFs is strongly dependent on the type of substrate used. The better activity of MOFs under study in comparison with zeolites in the transformation of the most bulky substrate was rationalized in terms of the regularity in the arrangement of active sites within the framework of MOFs, resulting in the existence of two active centers in close proximity to each other, as well as in the optimum pore size.

The catalysis in post-synthesized MOFs also represents an attractive issue. The presence of coordinatively unsaturated metal sites in MOFs can provide a powerful way for the immobilization of electron-rich functional groups such as amines grafted to the surface. In this context, amine-grafted MIL-101 functionalized with ethylenediamine and diethylenetriamine have shown high activities in the Knoevenagel condensation [94].

Hartmann and Fischer [95] recently reported a novel MOFs, Fe-MIL-101-NH_2 and Al-MIL-101-NH_2, providing non-coordinated primary amines as active solid Bronsted-basic catalyst highly active for the Knoevenagel condensation of benzaldehyde with ethyl cyanoacetate. The novel amino containing MOFs proved substantially more active than the traditional solid bases such as MgO and hydrotalcite.

4. Mesoporous Carbon Materials

The major production of carbon materials correspond to activated carbon s (ACs), used in many processes including catalysis, adsorption, purification, separation, medical applications and decontamination, among others. However, together ACs there is a large family of carbon materials prepared for specific applications by controlled synthesis routes (fullerenes, nanofibers, nanotubes, aerogels, foams, etc). Today the unique properties of graphene and their derivative materials focus a large part of the researcher's attention.

4.1. Activated Carbons (ACs)

Activated carbons are prepared from a large variety of cheap precursors because many residues from agricultural (almond shell, coconuts, olive stones, wood, pruning rest, etc) industrial (coal and coal residues, petroleum refineries, plastics, etc) or even quotidian live activities (sewage sludge, packages, spend tires, etc) are used. In general, these precursors are first pyrolyzed in inert atmosphere to form a char that is thereafter selectively burned (activated) to develop the adequate porosity range. The activation processes are divided into physical and chemical activation. The physical activation, also called thermal activation, implies that the carbonized previously obtained is gasified in oxidant atmosphere, typically CO_2 or stream flows, at high temperature (between 700-900 °C). In the chemical activation, the carbon precursor or their derivative chars are intimately impregnated with an activating agent (typically H_3PO_4, $ZnCl_2$, or Na or K hydroxides) and then thermally treated in inert atmosphere. The material obtained is then washed with acid solutions and finally with distillated water to remove the chemical rests. Both procedures are applicable to any carbon precursor, and thus, a large variety of ACs can be prepared depending on the raw materials and the experimental condition s used in the preparation .

Many studies in the literature compare the characteristics of materials prepared by physical and chemical activation [96]. Physical activation permits to obtain ACs with a narrower and homogeneous pore size distribution (PSD). However, ACs obtained by this method are mainly microporous materials with very high surface area values, and consequently, mainly useful as adsorbents . Obviously, the microporosity can be enlarged by increasing the activation degree (decreasing the ACs yield), looking for the mesoporosity development. Alternatively, chemical activation produced a more opened porosity (meso and macropores) but also more heterogeneous PSD.

Due to the interest of mesoporous materials in many fields regarding the microporous ones, some studies are focused on the experimental condition s needed to obtain cheap mesoporous ACs . In this sense, it was reported that mesoporous ACs (mesoporosity between 70-80% of the total pore volume) with a mesopore volume up to 1 cm^3 g^{-1} are obtained from olive stones by combining physical activation in CO_2 and chemical activation with $CaCl_2$ [97].

Mesoporous ACs were prepared also from other biomass residues like sugar cane bagasse or sunflower seed hull [98]. In this case, mesoporous ACs are directly obtained by chemical activation using H_3PO_4 or $ZnCl_2$. Nevertheless, a base -leaching pretreatment of the raw lignocellulosic precursors (boiling solution of NaOH 0.5 M) is necessary to favor the development of mesopores . After activation, samples were first acid -washed (HCl 3M) to remove the phosphorous and zinc compounds and finally with distilled water until neutralization. A greater porosity and surface area are obtained by using $ZnCl_2$. Similar results than those previously exposed are obtained: i) the mesoporosity ranges between 74-81% of the total porosity and ii) the values of mesopore volumes are around 1.1 cm^3 g^{-1}.

Chemical activation seems to be therefore preferable to the physical activation processes in order to obtain highly mesoporous materials from different types of biomass residues . Nevertheless, chemical activation procedure is longer and consumes great amounts of chemicals and water for activation and cleaning , being therefore susceptible to generate pollution . Moreover, the corrosion in the equipment has also an important economic impact in the production of ACs .

Some alternatives are consequently needed in order to maintain the economics and technological advantages associated to the ACs productions. It is well known the capacity of some types of plants to capture metal compound from the soils, in such a way that they can be used for phyto-remediation of heavy metal polluted soils [99]. Recently it was proposed the self-activation of pokeweed (Phytolacca americana L.) because the high content of K_2O of this plant (approximately 6.5 % in a dry weight basis) [100]. In this case, a mesopore volume of around 0.4 cm^3 g^{-1} with a mean diameter of 3.1 nm was obtained. In spite of the mesopore volume is smaller than in the previous cases, this procedure can be economical and environmentally favorable; anyway, these ACs presents a high capacity of phenol removal. The mesopore volume become negligible if the pokeweed raw material is de-mineralized by treatment with 0.5 M HCl for 24 h and washed with distilled water until neutral pH before activation, showing the importance of the nature and content of the inorganic components (ash) of the different ACs precursors.

Although biomass is an attractive carbon resource, mesoporous ACs are also prepared from other raw precursors. Zhank et al. [101] prepared activated carbon by activation with KOH of residues from coal liquefaction reaching samples with a S_{BET} larger than 2000 m^2g^{-1} and mesopore volumes at around 1.0 cm^3g^{-1}. Lin and Teng [102] obtain mesoporous activated carbons from waste tires. In this case, the mean pore size obtained is 50 nm (i.e., very large mesopores , in the frontier of macropores) and a total pore volume ranged between 0.4-0.7 cm^3 g^{-1} depending on the activation degree, where between 66-81% correspond to mesopores.

The main advantage of the ACs is the low production cost and the available technology of production at large scale. This allows industrial applications of ACs. However, and despite the high mesopore volume that can be obtained, the main drawback of these materials is the heterogeneity of the samples, either from a PSD or chemical composition point of view.

4.2. Ordered Mesoporous Carbons: Nanocasting Techniques

The preparation of carbon materials by this procedure includes four steps: i) the preparation of templates with well-defined nanostructure, ii) preparation of a carbon precursor / template mixture, iii) carbonization , and iv) the template removal. Depending on the nature of the template used, the procedure is called hard or soft templating (endo or exotemplating). The hard templating procedure use solid porous templates, which are infiltrated with a carbon precursor before carbonization. After removing the template, a negative replica of the template porosity is obtained. The soft templating method use surfactant type organic molecules that form micelles in a solution of the carbon precursor, which normally polymerizes around these micelles before carbonization. Others classifications refers to hard templates as thermally stables inorganic compounds (silica , Na_2CO_3, $Ni(OH)_2$, etc) while thermally decomposable polymers as polyurethane , polystyrene , polymethil metacrilate, and amphiphilic block polymers poly(ethylene oxide) (PEO) and poly(propylene oxide) (PPO) (PEO-PPO-PEO) are considered as soft templates.

Nanocasting technology was already patented by Knox and Gilbert in 1982 [103]. They obtain porous glassy carbons (PGC) using silica gel as template and a phenol -hexamine mixture as carbon precursor. When carbonized below 1000 °C the PGS obtained contain micro and mesopores [104]. However, the synthesis of ordered mesoporous carbons (OMC) was strongly developed after 1999 [105, 106] when Ryoo used aluminosilicate MCM-48 as template and sucrose as carbon source. The subsequent removal of the inorganic framework leads to a hexagonal arrangement of cylindrical carbon nanorods (CMK-1) leaving a homogenous mesoporosity. Lately, different mesoporous silica materials (MCM-41, SBA-15, etc) were also used as template [107]. The PSD of this OMC can be fitted by using the same silica precursor with different wall thicknesses, by using different types of mesoporous templates or by using templates with different nanoparticle size.

Even nowadays, nanocasting procedures present important economic and technological limitations for the production of carbon materials at large scale, and should be therefore strongly improved. The synthesis of silica SBA-15 structures inside a carbon precursor solution (glucose , soluble starch , starch from wheat, arabic gum or furfuryl alcohol) was recently proposed as "one- pot" procedure to avoid the initial synthesis of the template and the complicated infiltration process [108]. In this case, infiltration occurs simultaneously to SBA-15 formation and the micro and mesopore volumes obtained are 0.35 and 1.13 cm^3 g^{-1}, respectively. Also, semi-graphitized OMC were obtained using SBA-15 as hard template and boric acid as pore expanding agent [109]. The samples possess extensive graphitic domains as evidenced by TEM , Raman and XRD techniques , which can improve their applications in electrochemical devices. In the same line, the combination of templating / activation procedures can also favor the preparation of highly porous materials with hierarchical porosity , because the macro-mesoporosity can be induced by the template and microporosity by the activation procedures. Using this combination [110], materials up to S_{BET} of 2700 m^2g^{-1} and mesopore volume of 5.4 cm^3 g^{-1} were obtained.

Mesoporous carbon materials were prepared also using graphite as precursor and an intermediate templating / intercalation procedure [111, 112]. Graphite is first oxidized (GO) and expanded with the appropriate surfactant molecules before react with silica precursor (TEOS). TEOS is hydrolyzed between the GO layers and the GO intercalated surfactant-silica materials carbonized at different temperatures, treated with HF and finally washed carefully. Ultrafine carbon microtextures with three to four carbon layers stacking are assemble to form large wrinkled layers which overlap with each other leaving between them homogeneous mesopores of 1.8 – 2.0 nm in diameter and mesopore volume of 0.4 - 0.7 cm^3 g^{-1} were obtained, being the micropore size negligible in any case [111]. Pt, Pd or Ru catalysts were prepared by a similar procedure by doping the GO with the corresponding complexes precursors [112].

4.3. Carbon gels: Nanomaterials with Hierarchical Porosity

Alternatively, polymeric materials can be designed as precursors of mesoporous carbons. Carbon aerogels were first synthesized by Pekala [113] using aqueous solutions of resorcinol - formaldehyde and Na_2CO_3 as polycondensation catalyst . This preparation route has several advantages regarding those previously exposed, thus, pure materials (in contrast with ACs) are obtained with highly developed, well-defined and hierarchical porosity [114] and avoid the long and difficult procedure of preparation, infiltration and removal of templates. Nevertheless, because materials are prepared from pure reactants , they are not so cheap as ACs, and the preparation of aerogels present the drawback of using non-conventional techniques as the supercritical drying , difficult to be applied at industrial scale.

This procedure is carried out also in different steps: preparation of the initial solutions, gelation , drying , carbonization and alternatively, activation. The sol – gel technique is very flexible, but the experimental condition s should be carefully controlled in order to obtain the desired porosity and composition [115]. Hydrogels structure is constructed by interconnected primary particles formed during gelation; the shape , dimension and packing of these primary particles determine therefore the morphology , porosity and mechanical properties of materials , being these parameter defined by the nature and concentration of monomers , solvent and catalysts used. Solvent remains between the three-dimensional arrangement of primary particles, and the drying method selected is a key point; in fact, materials are named according to the pertinent drying method: aerogels (supercritical drying), xerogels (thermal drying) or cryogels (cryogenic drying). The direct drying of hydrogels in the oven leads to cheap xerogels avoiding supercritical technologies, but provokes the stronger collapse of the mesoporosity. After drying, the PSD depend on the carbonization temperature : carbonization also produces certain shrinkage but the weight loss increase up to 800-900 °C favoring a pore opening. Finally, this carbon gels can be physically or chemically activated [116] according the procedures previously exposed.

Mesoporous materials are obtained by applying the original Pekala method, where the role of Na_2CO_3 catalysts is based on the ability of carbonate anions to deprotonate the resorcinol molecule [115]. The role of the counter-cation was initially neglected, but it was pointed out [117] that the PSD shifted to smaller mesopores with increasing the polarizing power of the alkali metal used as counter-ion (M_2CO_3). Similarly, it was revealed that transition metal s are active in all the synthesis process, including polymerization , carbonization and activation, modifying textural properties and favoring the graphitization of the carbon aerogels [118]. Highly mesoporous Co and Ni-doped carbon aerogels were obtained [116, 118] directly by carbonization. Doping carbon gel with transition metals is an interesting alternative to prepare very stable heterogeneous catalysts [119, 120] with tunable porosity and surface chemistry .

Therefore, the preparation method requires controlling several parameters, but this fact simultaneously favors the possibility of use of cheap monomers like phenol , catechol, etc. [121] or alternative drying method [122] like microwaves , that significantly shorten the preparation time. Materials are obtained in powder, grains, pellets, monoliths, films , coatings of ceramic honeycomb or foams structures [123], their surface composition can be modified by using monomers containing heteroatoms [124] like melamine or urea (for N-doping) of after synthesis treatments [125]. The resorcinol -formaldehyde polymerization was also combined with the soft templating method, using Pluronic F127 solution followed by the addition of acid catalysts to favor the condensation reaction [126]. After activation with CO_2 , samples showed a S_{BET} of 2700 m^2 g^{-1} and a pore volume of 2.01 g cm^{-3} (81 % corresponding to mesopores). Thus, very different experimental condition s can be selected to obtain carbon materials with fitted shape , dimensions, surface chemistry and porosity by applying the sol-gel polymerization of organic monomers. The flexibility of the process, the available technology and the high quality of the samples prepared can favor the preparation at large scale of mesoporous carbon materials for different applications by applying this method or their combinations.

4.4. Catalytic Performance

In spite that carbon materials are used mainly as catalysts support of metallic phases they are also actives in many catalyzed reactions . Their applications are based on economic aspects, as previously described, but are also favored by their improved properties regarding inorganic catalysts like greater stability on acid and basic media , higher thermal stability (no phase transitions or dimensional changes are produced) in inert atmosphere , large and adjustable porous texture , surface chemistry and hydrophilic - hydrophobic character, etc. These advantages were summarized by Rodríguez-Reinoso in a classical manuscript [127]. The relationship between carbon surface properties and catalytic activity and the perspectives of nanocarbons in catalysis are continuously analyzed and discussed [128, 129].

Thus, nanocarbons were used as catalysts in reaction as: oxidative dehydrogenation , dehydration of alcohols , SO_x and SH_2 oxidation , NO_x reduction, catalytic ozonation or catalytic wet air oxidation [130]. Acid mesoporous carbons found also its application in biodiesel production [131, 132] although other functionalities (sulfonic groups, imidazole, etc) can be used to favor various transesterification , Knoevenagel or Aldol reactions [133].

Matos et al. [134] recently reported the ring opening reaction of epoxides by alcohols and amines using mesoporous activated carbon as efficient and environmentally friendly heterogeneous catalysts (Scheme 36). Mesoporous carbons under study were synthesized by polymerization of resorcinol and formaldehyde , followed by the oxidation of the surface with HNO_3 and then functionalized with H_2SO_4. Mesoporous activated carbon proved to be effective catalysts for the ring opening reaction of epoxides affording the corresponding substituted alcohols with excellent conversions and selectivity , both over 95%. Remarkably, the carbon material can be re-used by regeneration and thus restoring the original activity.

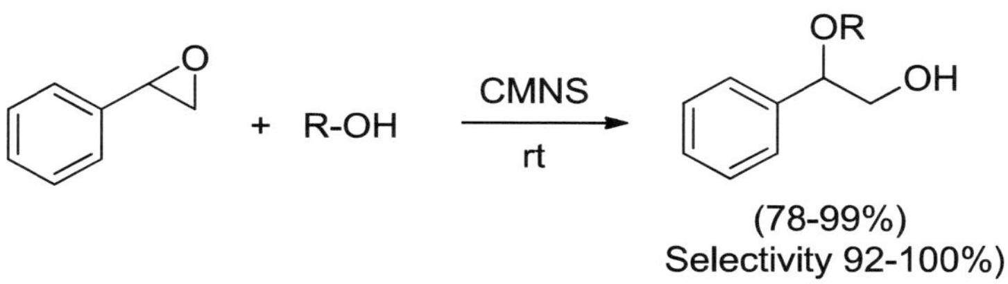

Scheme 36. Ring opening reaction of epoxides with alcohols .

More recently, López-Sanz et al. [135] have described an interesting study regarding to the catalytic behavior of acidic carbon materials in the Friedländer condensation for the synthesis of quinolinas. The authors reported different series of acidic activated carbon materials , which can be considered as cheaper alternatives to the traditional acidic mesoporous silicates or even zeolites for the synthesis of quinolines/quinolones. Textural parameters of the acidic activated carbon materials together with their acidic properties are important factors that affect the reaction selectivity . In addition, computational calculations have provided mechanistic insights into the catalytic behaviour of different types of acidic active site s.

ACs , ordered carbons, carbon aerogels are used in the development of electro- or photo-catalysts. Their combination with inorganic phases producing composites (C/Al_2O_3, C/SiO_2 or C/TiO_2), develops the porosity , increases the hydrophobic character and moderate the strong acidity of the inorganic oxides [136, 137] largely improving the catalytic performance of each phase. The synergism between the phases is the basis of the preparation of specific photocatalysts [138]. When used as support of metallic phases, they were applied as selective hydrogenation of α,β-unsaturated aldehydes , nitrobenzene or fatty acids [139, 140], as an example. They can be also be functionalized by anchoring organ ometallic compounds improving the product yields and reuse of these compounds used as homogenous catalysts [141, 142]. Mesoporous carbon material can be used as supports of metal oxides . In this sense, highly basic CaO nanoparticles immobilized on mesoporous carbon materials , prepared by impregnation , have been investigated by Raja et al. [143] in the base -catalyzed transesterification of ethylacetoacetate by aryl, aliphatic and cyclic primary alcohols . They proved highly active with higher conversion within a very short reaction time .

Mesoporous carbon xerogels were also used as enzyme supports [144] and it was suggested that the mesopore volume prevent the enzyme denaturalization being their specific activity between 2 – 13 times higher than for soluble enzyme.

CONCLUSION

This chapter provides an up-to date regarding the catalytic applications of porous materials with different composition and structural nature such as mesoporous silicas , MOFs and carbons.

It is well known that homogeneous catalyst s, often used for the production of fine chemicals , frequently decompose or undergo deactivation under the reaction conditions. Therefore, the immobilization of that in a mesoporous framework often improve the stability of the active catalytic sites, although some diffusion restrictions could apply to large substrates depending on the pore size . In other sense, when comparing the catalytic performance of mesoporous silicas and MOFs or with homogeneous analogues as soluble salts , it is often observed that MOFs exhibit higher catalytic activity mainly due to the higher concentration of active species even acting in cooperation . Remarkably, the lower thermal and chemical stability of MOFs does not imply that they cannot be useful catalysts, but also that the reaction conditions must be appropriately selected; low reaction temperature s in the presence of non-coordinating solvents , if any, represent some of the especially relevant parameters for MOF catalytic applications.

As summarized in this chapter, many studies concerning the synthesis and activation of mesoporous carbons have been reported. However, only a few examples of this type of materials are involved in the synthesis of fine chemicals . It is possible to prepare mesoporous carbons exhibiting different acid -base properties and porosity or even as supports for active species as anticipated. In this sense, mesoporous carbons could emerge as environmental and cheaper alternative catalysts for the production of valuable chemicals being that an area in expansion. Metals-loading mesoporous carbons could even compete, why no, with MOFs related to the higher stability and concentration of active catalytic sites.

REFERENCES

[1] (a) Corma, A; Martinez, A. *Stud. Surf. Sci. Catal.* 2005, *157*, 337-366; b) Clerici M. G. *Top. Catal.* 2000, *13*, 373–386.
[2] Martínez, C.; Corma A. *Coord. Chem. Rev.* 2011, *255*, 1558–1580.
[3] Stankiewicz, A. *Chem. Eng. Proc.* 2003, *42*, 137–144.
[4] Anastas, P. T.; Kirchhoff, M. M.; Williamson, T. C. *Appl. Catal. A: Gen.* 2001, *221*, 3–13.
[5] Choi, S.; Coronas, J.; Jordan , E.; Oh, W.; Nair, S.; Onorato, F.; Shantz, D. F.; Tsapatsis, M. *Angew. Chem. Int. Ed.* 2008, *47*, 552–555.
[6] Galownia, J.; Martin, J.; Davis, M. E. *Micropor. Mesopor. Mater.* 2006, *92*, 61–63.
[7] Urbiztondo, M. A.; Pellejero, I.; Villarroya, M.; Sese, J.; Pina, M. P.; Dufour, I.; Santamaria, J.; *Sens. Actuators B: Chem.* 2009, *137*, 608–616.

[8] Tsapatsis, M. *AIChE J.* 2002, *48*, 654–660.
[9] Coronas, J. *Chem. Eng. J.* 2010, *156*, 236–242.
[10] Beck, J. S.; Vartuli, J. C.; Roth, W. J.; Leonowicz, M. E.; Kresge, C. T.; Schmitt, K. D.; Chu, C. T. W.; Olson, D. H.; Sheppard, E. W.; McCullen, S. B.; Higgins, J. B.; Schlenker, J. L. *J. Am. Chem. Soc.* 1992, *114*, 10834-10843.
[11] Sayari, A. *Chem. Mater.* 1996, *8*, 1840-1852.
[12] Taguchi, A.; Schüth, F. *Microp. Mesop. Mater.* 2005, *77*, 1-45.
[13] Ying, J. Y.; Mehnert, C. P.; Wong, M. S. *Angew. Chem. Int. Ed.* 1999, *38*, 56-77.
[14] Corma, A; Garcia, H. *Adv. Synth. Catal.* 2006, *348*, 1391-1412.
[15] Martín-Aranda, R. M.; Cejka J. *Top. Catal.* 2010, *53*, 141–153.
[16] Kuppler, R. J.; Timmons, D. J.; Fang, Q.-R.; Li, J.-R.; Makal, T. A.; Young, M. D.; Yuan, D.; Zhao, D.; Zhuang, W.; Zhou, H.-C. *Chem. Rev.* 2009, *253*, 3042–3066.
[17] Mueller, U.; Schubert, M.; Teich, F.; Puetter, H.; Schierle-Arndta, K.; Pastré J. *J. Mater. Chem.* 2006, *16*, 626-636.
[18] Keskin, S.; Kızılel, S. *Ind. Eng. Chem. Res.* 2011, *50*, 1799–1812.
[19] Valvekens, P.; Vermoortele, F.; De VosCatal, D. *Sci. Technol.* 2013, *3*, 1435-1445.
[20] Farrusseng, D.; Aguado, S.; Pinel, C. *Angew. Chem. Int. Ed.* 2009, *48*, 7502-7513.
[21] Catlow, C. R. A.; French, S. A.; Sokol, A. A.; Thomas, J. M. *Phil. Trans. R. Soc. A* 2005, *363*, 913-936; b) Rimola, R.; Costa, D.; Sodupe, M.; Lambert, J.-F.; Ugliengo, P. *Chem. Rev.* 2013, *113*, 4216−4313.
[22] Taguchi, A.; Schüth F. *Microp. Mesop. Mater.* 2005, *77*, 1–45.
[23] Hoffmann, F.; Cornelius, M.; Morell, J.; Froba, M. *Angew. Chem. Int. Ed.* 2006, *45*, 3216-3251.
[24] Karthikeyana, G.; Pandurangan, A. *J. Mol. Catal. A: Chem.* 2012, *361– 362*, 58-67.
[25] Nowak, I.; Feliczak, A.; Nekoksova, I.; Cejka J. *Appl. Catal. A* 2007, *321*, 40-48.
[26] Feliczak-Guzik, A.; Nowak I. *Catal. Today* 2009, *142*, 288–292.
[27] Oikawa, T.; Ookoshi, T.; Tanaka, T.; Yamamoto, T.; Onaka, M. *Micropor. Mesopor. Mater.* 2004, *74*, 93-103.
[28] Aguado, J.; Escola J. M.; Castro, M. C.; Paredes, B. *Appl. Catal. A* 2005, *284*, 47-57.
[29] Pal, N.; Bhaumik, A. *Adv. Colloid Interface Sci.* 2013, *189*, 21-41.
[30] Melero, J. A.; van Grieken, R,; Morales, G. *Chem. Rev.* 2006, *106*, 3790-3812.
[31] Campelo, J. M.; Luna D.; Luque, R.; Marinas, J. M.; Romero, A. A. *ChemSusChem* 2009, *2*, 18-45.
[32] Díaz, U.; Brunel, D.; Corma, A. *Chem. Soc. Rev.* 2013, *42*, 4083-4097.
[33] Vinu, A.; Justus, J.; Balasubramanian, V. V.; Halligudi, S. B.; Ariga, K.; Mori, T. *Collect. Czech. Chem. Commun.* 2008, *73*, 1112-1131.
[34] Rat, M.; Zahedi-Niaki, M.H.; Kaliaguine, S.; Do, T.O. Micropor. Mesopor. Mater. 2008, *112*, 26–31.
[35] Bejblová, M.; Prochazková, D.; Cejka, J. *ChemSusChem* 2009, *2*, 486-499.
[36] Park, S. E.; Prasetyanto, E. A. *Top. Catal.* 2009, *52*, 91-100.
[37] El Kadib, A.; Finiels, A.; Brunel, D. *Chem. Commun.* 2013, *49*, 9073-9076.
[38] Das, D.; Lee, J.-F.; Cheng, S. *Chem. Commun.* 2001, 2178 – 2179;
[39] Das, D.; Lee, J.-F.; Cheng, S. *J. Catal.* 2004, 223, 152 – 160.
[40] Dufaud, V.; Davis, M. E. *J. Am. Chem. Soc.* 2003, *125*, 9403 –9413.
[41] Shimizu, K.; Hayashi, E.; Hatamachi, T.; Kodama, T.; Higuchi, T.; Satsuma, A.; Kitayama, Y. *J. Catal.* 2005, *231*, 131 – 138.

[42] Shiri, M.; Zolfigol, M. A.; Kruger, H. G.; Tanbakouchian, Z. *Adv. Heterocyclic Chem.* 2011, *102*, 139-227; b) Marco-Contelles, J.; Pérez-Mayoral, E.; Samadi, A., do Carmo Carreiras, M.; Soriano, E. *Chem. Rev.* 2009,*109*, 2652–2671.

[43] López-Sanz, J., Pérez-Mayoral, E.; Soriano, E.; Sturm, M.; Martín-Aranda, R. M.; López-Peinado, A. J.; Cejka, J. *Catal. Today* 2012, *187*, 97– 103.

[44] Rostamnia, S.; Xin, H.; Liu, X.; Lamei, K. *J. Mol. Catal. A: Chem.* 2013, *374– 375*, 85– 93.

[45] Richel, A.; Laurent, P.; Wathelet, B.; Wathelet, J.-P.; Paquot, M. *Tetrahedron Lett.* 2010, *51*, 1356–1360.

[46] Vinu, A.; Devassy, B. M.; Halligudi, S. B.; Böhlmann, W.; Hartmann, M. *Appl Catal A: Gen.* 2005, *281*, 207-213.

[47] Vinu, A.; Sawant, D. P.; Ariga, K.; Hartmann, M.; Halligudi, S. B. *Micropor. Mesopor. Mater.* 2005, *80*, 195-203.

[48] Hu, W.; Luo , Q.; Su, Y.; Chen, L.; Yue, Y.; Ye, C.; Deng, F. *Micropor. Mesopor. Mater.* 2006, *92*, 22-30.

[49] Zukal, A.; Siklová, H.; Cejka J. *Langmuir* 2008, *24*, 9837-9842.

[50] Gurinov, A. A.; Rozhkova, Y. A.; Zukal, A.; Cejka, J.; Shenderovich, I. G. *Langmuir* 2011, *27*, 12115–12123.

[51] Srivastava, P.; Srivastava, R. *Catal. Commun.* 2008, *9*, 645-649.

[52] Pérez-Mayoral E.; Martín-Aranda, R. M.; López-Peinado, A. J.; Ballesteros, P.; Zukal, A.; Cejka, J. *Top. Catal.* 2009, *52*, 148–152.

[53] Domínguez-Fernández, F.; López-Sanz, J.; Pérez-Mayoral, E.; Bek, D.; Martín-Aranda, R. M.; López-Peinado, A. J.; Cejka, J. *ChemCatChem* 2009, *1*, 241–243.

[54] López-Sanz, J.; Pérez-Mayoral, E.; Procházková, D.; Martín-Aranda, R. M.; López-Peinado, A. J. *Top. Catal.* 2010, *53*, 1430–1437.

[55] Selvaraj, M.; Lee, T. G. *Micropor. Mesopor. Mater.* 2005, *85*, 52–58.

[56] Selvaraj, M.; Min, B. R.; Shul, Y. G.; Lee, T. G. *Micropor. Mesopor. Mater.* 2004, *74,* 157–162.

[57] Boronat, M.; Concepción, P.; Corma, A.; Navarro, M. T.; Renz, M.; Valencia , S. *Phys. Chem. Chem. Phys.*, 2009, *11*, 2876–2884.

[58] Murugavel, R.; Roesky, H. W. *Angew. Chem., Int. Ed. Engl.* 1997, *36*, 477-479; b) Hutchings, G. J. *Chem. Commun.* 1999, 301-306; c) Kholdeeva, O. A.; Melgunov, M. S.; Shmakov, A. N.; Trukhan, N. N.; Kriventsov, V. V.; Zaikovskii, V. I.; Malyshev, M. E.; Romannikov, V. N. *Catal. Today* 2004, *91-92*, 205-209.

[59] Sankar, G.; Thomas, J. M.; Catlow, C. R. A.; Barker, C. M.; Gleeson, D.; Kaltsoyannis, N. *J. Phys. Chem. B* 2001, *105*, 9028-9030; b) Thomas, J. M.; Catlow, C. R. A.; Sankar, G. *Chem. Commun.* 2002, 2921–2925. c) For embedded cluster calculations, see: French, S. A.; Sokol, A. A.; To, J.; Catlow, C. R. A.; Phala, N. S.; Klatt, G.; van Steen, E. *Catal. Today* 2004, *93–95*, 535–540.

[60] Zhang, W.-J.; Jiang, P.-P.; Zhang, P.-B.; Zheng, J.-W.; Li, H. *Bull. Korean Chem. Soc.* 2012, *33*, 4015-4022.

[61] Michalska, A.; Daturi, M.; Saussey, J.; Nowak, I.; Ziolek, M. *Micropor. Mesopor. Mater.* 2006, *90*, 362-369.

[62] Zienkiewicz, Z.; Calvino-Casilda, V.; Sobczak, I.; Ziolek, M.; Martín-Aranda, R. M.; López-Peinado, A. J. *Catal. Today* 2009, *142*, 303–307.

[63] Rodriguez, I.; Iborra, S.; Corma, A.; Rey, F.; Jordá, J. L. *Chem. Commun.* 1999, 593–594.

[64] Rodriguez, I.; Iborra, S.; Rey, F.; Corma, A. *Appl. Catal. A: Gen.* 2000, *194-195* 241-252.

[65] Loganathan, S.; Tikmani, M.; Ghoshal, A. K. *Langmuir* 2013, *29*, 3491-3499.

[66] Olea, A.; Sanz-Pérez, E. S.; Arencibia, A.; Sanz, R.; Calleja, G. *Adsorption* 2013, *19*, 589-600;

[67] Feng, X.; Hu, G.; Hu, X.; Xie, G.; Xie, Y.; Lu, J.; Luo , M. *Ind. Eng. Chem. Res.* 2013, *52*, 4221-4228.

[68] Matos, I.; Pérez-Mayoral, E.; Soriano, E.; Zukal, A.; Martín-Aranda, R. M.; López-Peinado, A. J.; Fonseca, I.; Cejka, J. *Chem. Eng. J.* 2010, *161*, 377-383.

[69] Aider, N.; Smuszkiewicz, A.; Pérez-Mayoral, E.; Soriano, E.; Martín-Aranda, R. M.; Halliche, D. Menad, S. *Catal. Today* 2014, *227*, 215–222.

[70] Wang, X.; Cheng, S. *Catal. Commun.* 2006, *7*, 689-695.

[71] Li, H.; Bhadury, P. S.; Song, B.; Yang, S. *RSC Adv.* 2012, *2*, 12525–12551.

[72] Parvin, M. N.; Jin, H.; Ansari, M. B.; Oh, S.-M.; Park, S.-E. *Appl. Catal. A: Gen.* 2012, *413– 414*, 205– 212.

[73] Shylesh, S.; Thiel W. R. *ChemCatChem* 2011, *3*, 278 – 287.

[74] Díaz, U.; Brunel, D.; Corma, A. *Chem. Soc. Rev.* 2013, *42*, 4083-4097.

[75] Margelefsky, E. L.; Zeidan, R. K.; Davis, M. E. *Chem. Soc. Rev.* 2008, *37*, 1118–1126.

[76] Blasco-Jiménez, D.; López-Peinado, A. J.; Martín-Aranda, R. M.; Ziolek, M.; Sobczak, I. *Catal. Today* 2009, *142*, 283–287.

[77] a) Blasco-Jiménez, D.; Sobczak, I.; Ziolek, M.; López-Peinado, A. J.; Martín-Aranda, R. M. *Catal. Today* 2010, *152*, 119–125; b) Sobczak, I.; Ziolek, M.; Pérez-Mayoral, E.; Blasco-Jiménez, D.; López-Peinado, A. J.; Martín-Aranda, R. M. Catal. Today 2012, *179*, 159– 163

[78] Smuszkiewicz, A.; López-Sanz, J.; Pérez-Mayoral, E.; Soriano, E.; Sobczak, I.; Ziolek, M.; Martín-Aranda, R. M.; López-Peinado, A. J. *J. Mol. Catal. A: Chem.* 2013, *378*, 38–46.

[79] Smuszkiewicz, A.; Pérez-Mayoral, E.; Soriano, E.; Sobczak, I.; Ziolek, M.; Martín-Aranda, R. M.; López-Peinado, A. J. *Catal. Today* 2013, *218–219*, 70–75.

[80] Corma, A.; Boronat, M.; Climent, M. J.; Iborra, S.; Monton, R.; Sabater, M. J. *Phys. Chem., Chem. Phys.* 2011, *13*, 17255-17261.

[81] For a recent review on computational studies, *see* Yang, Q.; Liu, D.; Zhong, C.; Li, J.-R. *Chem. Rev.* 2013, *113*, 8261−8323

[82] Song, L.; Zhang, J.; Sun , L.; Xu, F.; Li, F.; Zhang, H.; Si, X.; Jiao, C.; Li, Z.; Liu, S.; Liu, Y.; Zhou, H.; Sun, D.; Du, Y.; Caoh, Z.; Gabelica, Z. *Energy Environ. Sci.* 2012, *5*, 7508-7520.

[83] Kuppler, R. J.; Timmons, D. J.; Fang, Q.-R.; Li, J.-R.; Makal, T. A.; Young, M. D.; Yuan, D.; Zhao, D.; Zhuang, W.; Zhou H.-C. *Coord. Chem. Rev.* 2009, *253,* 3042–3066.

[84] Dhakshinamoorthy, A.; Opanasenko, M.; Cejka, J.; Garcia, H. *Catal. Sci. Technol.* 2013, *3*, 2509-2540.

[85] Dhakshinamoorthy, A.; Alvaro, M.; García, H. *Chem. Commun.* 2012, 11275–11288.

[86] Dhakshinamoorthy, A.; Alvaro, M.; Garcia, H. *Adv. Synth. Catal.* 2010, *352*, 711-717.

[87] Dhakshinamoorthy, A.; Alvaro, M.; Garcia, H. *Chem. Eur. J.* 2010, *16*, 8530-8536.

[88] Dhakshinamoorthy, A.; Alvaro, M.; Garcia, H. *Appl. Catal. A* 2010, *378*, 19-25.
[89] Dhakshinamoorthy, A.; Alvaro, M.; Garcia, H. *Adv. Synth. Catal.* 2010, *352*, 3022-3030.
[90] Dhakshinamoorthy, A.; Opanasenko, M.; Cejka, J.; Garcia, H. *Adv. Synth. Catal.* 2013, *355*, 247 – 268.
[91] Pérez-Mayoral, E.; Cejka J. *ChemCatChem* 2011, *3*, 157–159.
[92] a) Pérez-Mayoral, E.; Musilová, Z.; Gil, B.; Marszalek, B.; Položij, M.; Nachtigall, P.; Čejka, J. *Dalton Trans.* 2012, *41*, 4036–4044; b) Polozij, M.; Pérez-Mayoral, E.; Cejka J.; Hermann, J.; Nachtigall, P. *Catal. Today* 2013, *204*, 101– 107.
[93] Opanasenko, M.; Shamzhy, M.; Cejka, J. *ChemCatChem* 2013, *5*, 1024-1031.
[94] Hwang, Y. K.; Hong, D. Y.; Chang, J. S.; Jhung, S. H.; Seo, Y. K.; Kim, J.; Vimont, A.; Daturi, M.; Serre, C.; Férey, G. *Angew. Chem. Int. Ed.* 2008, 47, 4144-4148.
[95] Hartmann, M.; Fischer, M. *Micropor. Mesopor. Mater.* 2012, *164*, 38–43.
[96] Prauchner, M. J.; Rodríguez-Reinoso, F. *Micropor. Mesopor. Mater.* 2012, *152* 163–171.
[97] Juárez-Galán, J. M.; Silvestre-Albero, A.; Silvestre-Albero, J.; Rodríguez-Reinoso, F. *Micropor. Mesopor. Mater.* 2009, *117*, 519–521.
[98] Liou, T.-H. *Chem. Eng. J.* 2010, *158*, 129–142.
[99] Ji, P.; Sun , T.; Song, Y.; Ackland, M. L.; Liu, Y. *Environm. Pollution* 2011, *159*, 762-768.
[100] Chen, Y.-D.; Huang, M.-J.; Huang, B.; Chen, X.-R. *J. Anal. Appl. Pyrol.* 2012, *98*, 159–165.
[101] Zhang, J.; Jin, L.; Zhu, S.; Hu, H. *J. Natur. Gas Chem.* 2012, *21*, 759–766.
[102] Lin, Y.-R.; Teng, H. *Micropor. Mesopor. Mater.* 2002, *54*, 167–174.
[103] Knox, J. H.; Gilbert, M. T. U.K. Put., 7939449 (1982); U.S. Put., 4,263,268 (1982); W. German Pat., P 2946688-4 (1982).
[104] Knox, J.H.; Kaur, B.; Millward, G. R. *J. Chromatography* 1986, *352*, 3-25.
[105] Ryoo, R.; Joo, S. H.; Jun, S. *J. Phys. Chem. B* 1999, *103*, 7743–6.
[106] Lee, J.; Yoon, S.; Hyeon, T.; Oh, S. M.; Kim, K. B. *Chem. Commun.* 1999, 2177-2178.
[107] Enterría, M.; Suárez-García, F.; Martínez-Alonso, A.; Tascón, J. M D. *Micropor. Mesopor. Mater.* 2012, *151*, 390–396.
[108] Enterría, M.; Suárez-García, F.; Martínez-Alonso, A.; Tascón, J. M. D. *Carbon* 2013, *54*, 365–377.
[109] Li, Y.; Zhong, J.; Yang, X. Z.; Lan, G. J.; Tang, H. D.; Liu, H. Z. *New Carbon Mater.* 2011, *26*, 123-129.
[110] Górka, J.; Jaroniec, M. *Carbon* 2011, *49*, 154-160.
[111] Wang, Z.-M.; Hoshinoo, K.; Yamagishi, M.; Yoshizawa, N.; Kanoh, H.; Hirotsu, T. *Micropor. Mesopor. Mater.* 2006, *93*, 254–262.
[112] Gotoha, K.; Kawabata, K.; Fujii, E.; Morishige, K.; Kinumoto, T.; Miyazaki, Y.; Ishida, H. *Carbon* 2009, *47*, 2112 –2142.
[113] Pekala, R.W. US Patent, 4997804, (1991).
[114] Baumann, T. F.; Worsley, M.A.; Han, T. Y.-J.; Satcher, J. H. Jr. *J. Non-Cryst. Solids* 2008, *354*, 3513-3515.
[115] Al-Muhtaseb S.A.; Ritter, J. A. *Adv. Matter.* 2003, *15*, 101-114.
[116] Maldonado-Hódar, F. J.; Moreno-Castilla, C.; Pérez-Cadenas, A. F. *Micropor. Mesopor. Mater.* 2004, *69*, 119–125.

[117] Morales-Torres, S.; Maldonado-Hódar, F. J.; Pérez-Cadenas, A. F.; Carrasco-Marín, F. *Phys. Chem. Chem. Phys.* 2010, *12*, 10365–10372.
[118] Maldonado-Hódar, F. J.; Moreno-Castilla, C.; Rivera-Utrilla, J.; Hanzawa, Y.; Yamada, Y. *Langmuir* 2000, *16*, 4367-4373.
[119] Moreno-Castilla, C.; Maldonado-Hódar, F. J. *Carbon* 2005, *43*, 455–465.
[120] Maldonado-Hódar, F.J. *Catal. Today* 2013, *218–219*, 43– 50.
[121] Jirglová, H.; Pérez-Cadenas, A.F.; Maldonado-Hódar, F.J. *Langmuir* 2009, *25*, 2461.
[122] Gallegos-Suárez, E.; Pérez-Cadenas, A.F.; Maldonado-Hódar, F. J.; Carrasco-Marín, F. *Chem. Eng. J.* 2012, *851*, 181–182.
[123] Morales-Torres, S.; Maldonado-Hódar, F. J.; Pérez-Cadenas, A. F.; Carrasco-Marín, F. *Micropor. Mesopor. Mater.* 2012, *153*, 24-29.
[124] Li, W.; Chen, D.; Li, Z.; Shi, Y.; Wan, Y.; Wang, G.; Jiang, Z.; Zhao, D. *Carbon* 2007, *45*, 1757-1763.
[125] Morales-Torres, S.; Maldonado-Hódar, F. J.; Pérez-Cadenas, A. F.; Carrasco-Marín, F. *J. Hazard. Mater.* 2010, *183*, 814-822.
[126] Xu, J.; Wang, A.; Zhang, T. *Carbon* 2012, *50*, 1807 –1816.
[127] Rodríguez-Reinoso, F. *Carbon* 1998, 36, 59-175.
[128] Figueiredo, J. L.; Pereira, M.F. *Catal. Today*, 2010, *150*, 2–7.
[129] Sua, D.S.; Perathoner S.; Centi, G. *Catal. Today,* 2012, *186*, 1–6.
[130] Carbon Materials for Catalysis. P. Serp and J.L. Figueiredo (Eds). John Wiley & Sons, Inc., Hoboken, New Jersey. 2009.
[131] Chang, B.; Fu, J.; Tiana, Y.; Dong, X. *RSC Adv.* 2013, *3*, 1987-1994.
[132] Shu, Q.; Nawaz, Z.; Gao, X.; Liao, Y.; Zhang, Q.; Wang, D.; Wang, J. *Bioresource Tech.* 2010, *101*, 5374–5384.
[133] Demir-Cakan, R.; Makowski, P.; Antonietti, M.; Goettmann, F.; Titirici, M. M. *Catal. Today* 2010, *150*, 115–118.
[134] Matos, I.; Duarte Neves, P.; Castanheiro, J. E.; Pérez-Mayoral, E.; Martin-Aranda, R.; Duran-Valle, C.; Vital, J.; Botelho do Rego, A. M.; Fonseca, I. M. *Appl. Catal. A: Gen.* 2012, *439– 440*, 24– 30.
[135] López-Sanz, J.; Pérez-Mayoral, E.; Soriano, E.; Omenat-Morán, D.; Durán, C. J.; Martín-Aranda, R. M.; Matos, I.; Fonseca, I. *ChemCatChem* 2013, *5*, 3736-3742.
[136] Moreno-Castilla, C.; Maldonado-Hódar, F. J.; Carrasco-Marín, F.; Rodríguez-Castellón, E. *Langmuir* 2002, *18*, 2295-2299.
[137] Maldonado-Hódar, F. J.; Moreno-Castilla, C.; Rivera-Utrilla, J. *Appl. Catal. A: Gen.* 2000, *203*, 151–159.
[138] Pastrana-Martínez, L. M.; Morales-Torres, S.; Papageorgiou, S. K.; Katsaros, F. K.; Romanos, G. E.; Figueiredo, J. L.; Faria, J. L.; Falaras, P.; Silva, A. M. T. *Appl. Catal. B: Env.* 2013, *142– 143*, 101–111.
[139] Machado, B. F.; Morales-Torres, S.; Pérez-Cadenas, A. F.; Maldonado-Hódar, F. J.; Carrasco-Marín, F.; Silva, A. M. T.; Figueiredo, J. L.; Faria, J. L. *Appl. Catal. A: Gen.* 2012, *425-426*,161-169.
[140] Liu, Y.; Fang, Y.; Lu, X.; Wei,Z.; Li, X. *Chem. Eng. J.* 2013, *229*, 105–110.
[141] Kim, Y.; Guo, X.; Kim, G. J. *Catal. Today* 2010, *150*, 91–99.
[142] Silva, A. R.; Figueiredo, J. L.; Freire, C.; de Castro, B. *Micropor. Mesopor. Mater.* 2004, *68*, 83-88.

[143] Raja, P. K.; Chokkalingam, A.; Priya, S. V.; Balasubramanian, V. V.; Benziger, M. R.; Aldeyab, S. S.; Jayavell, R.; Ariga, K.; Vinu, A. *J. Nanosci. Nanotechnol.* 2012, *12*, 4613-20.

[144] Chaijitrsakool,; T, Tonanon, N.; Tanthapanichakoon, W.; Tamon, H.; Prichanont, S. *J. Mol. Catal. B: Enzymatic* 2008, *55*, 137-141.

In: Comprehensive Guide for Mesoporous Materials. Volume 1 ISBN: 978-1-63463-958-3
Editor: Mahmood Aliofkhazraei

Chapter 5

ORDERED MESOPOROUS CARBONS: SYNTHESIS, MODIFICATION AND APPLICATION IN ADSORPTION PROCESSES

Joanna Goscianska* and Robert Pietrzak
Adam Mickiewicz University in Poznań, Faculty of Chemistry,
Poznań, Poland

ABSTRACT

Owing to the unique physicochemical and electric properties the ordered mesoporous carbons have been a subject of intense research for many years. They have a well-developed surface area and great pore volume and have been found suitable for a number of applications including catalysis , energy storage and transformation , nanotechnology and as adsorbents in processes of air, water and industrial gas purification .

This chapter presents a survey of methods for the synthesis of mesoporous carbons with particular emphasis on the hard- and soft-template methods. The factors influencing the ordering as well as textural and structural parameters of ordered mesoporous carbons are identified and discussed such as the physicochemical properties of these materials , the so far applied methods of their modification and their potential use as adsorbents of various compounds from the liquid phase , including amino acid s , vitamins and dyes .

1. INTRODUCTION

Mesoporous carbon materials having pores of diameters from the range 2 - 50 nm have been the subject of great interest at many research centres in the world. In the process of their synthesis it is possible to control precisely the size of pores and their volume, so also their surface area [1-5]. Till recently mesoporous carbons have been synthesised mainly by partial carbonisation and activation of different raw materials of organic origin. Because of the low cost of their production and interesting properties the materials are widely used in the

* e-mail : asiagosc@amu.edu.pl, pietrob@amu.edu.pl

environmental engineering , despite their inhomogeneous structure and a broad pore volume distribution .

Recently, the methods of mesoporous carbons synthesis by hard- and soft-template have become particularly popular. They offer a possibility of obtaining materials of highly ordered structure [1-3]. The hard-template method is based on the use of solid state matrices, usually silica , whose pores are filled with the carbon precursor, then the precursor is subjected to carbonisation and finally the template is dissolved. Besides mesopores , the carbon materials also have micropores localised in the walls. In the method of soft-template synthesis no solid matrices are used. In this method the number of its stages is decreased, so the cost is lower and realisation is easier also in the industrial scale [3].

The ordered carbon materials showing unique physicochemical, electric and chemical properties related to their well-developed surface area , large pore volume and low dispersion of the pore size distribution function , have found a variety of applications in catalysis , energy storage and transformation , nanotechnology and as adsorbents in air, water and industrial gas purification (Table 1) [6-24].

This chapter presents a survey of methods for the synthesis of mesoporous ordered carbons, characterisation of their structures and textures, ways of modifications and potential applications in adsorption processes.

Table 1. Some examples of mesoporous carbon materials application

Application of mesoporous carbons	Reference
SUPERCAPACITORS	[6,7]
ADSORBENTS of:	
biomolecules – amino acids , vitamin B_{12} , chicken egg white albumin, bovine serum albumin	[5,8,9]
gases – CO_2, CH_4, N_2O, NH_3	[10]
dyes – methyl orange, methylene blue, malachite green	[11-13]
dibenzothiophene	[14]
metals - Pb(II), Cd(II), U(VI)	[15,16]
CATALYSTS in the reaction:	
cleavage of allylundecylcarbonate	[17]
dehydrogenation of propane to propylene	[18,19]
alcohol electrooxidation	[20]
formic acid oxidation	[21]
CO and methanol oxidation	[22]
ENERGY STORAGE	[23,24]

2. METHODS FOR THE SYNTHESIS OF MESOPOROUS CARBONS

2.1. Hard-Template Synthesis of Mesoporous Carbon Materials

The first method for the synthesis of mesoporous carbon materials was that of hard template, with different inorganic porous materials (aluminium oxide membranes , zeolites or amorphous silica gels [25-32]) as solid matrices. In the beginning of 1990s, Ryoo et al., used ordered mesoporous silica (OMS) as hard templates [33]. The synthesis with their use runs through a few stages [1-3, 33, 34]. At first the matrix is subjected to impregnation with the carbon precursor that undergoes polymerisation on the matrix surface. Then, the silica-carbon composites are subjected to carbonisation in anaerobic conditions and finally the matrix is removed by washing out with a solution of hydrofluoric acid or sodium hydroxide (Figure 1). The ordered mesoporous carbons (OMC) obtained are the negative replicas of the original molecular sieves OMS.

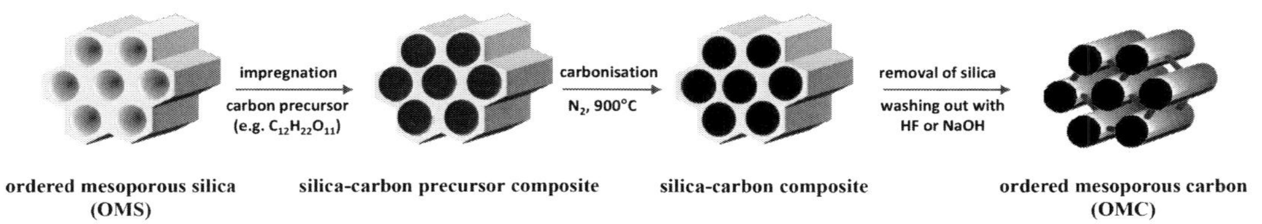

Figure 1. Synthesis of mesoporous carbons by the hard-template method.

So far the templates have been silica matrices of 2-D hexagonal structure (space group p6mm), 3-D cubic structure (regular - Ia3d, Im3m, Pm3n and Fm3m) and 3-D hexagonal structure (P4/mmm and $P4_2$/mnm) [1-3, 33, 34]. As 2-D original molecular sieves only such silica as e.g., SBA-15 is used, in which the hexagonally ordered mesopores are connected by irregular micropores giving a micro-mesoporous 3-D structure, suitable for getting the negative replica [34]. The above-mentioned silica materials are obtained as a result of controlled hydrolysis and condensation of selected silica precursors in the presence of surfactants or triblock copolymers , capable of ordered phase formation [35-38]. As a result of intermolecular interactions , thermal and chemical equilibrium between the inorganic precursor and organic template, the ordered mesoporous silica of well-developed spatial structure is produced. At the last stage of OMS synthesis the surfactant is removed by calcination in the air atmosphere or extraction with the adequately chosen solvents .

The most popular precursors of mesoporous carbons include [33, 34, 39-42]:

- sucrose ,
- furfuryl alcohol,
- acetylene,
- aromatic and polyaromatic hydrocarbons such as benzene , anthracene, naphthalene , pyrene,
- polyacrylonitrile,
- synthetic tars.

Table 2 gives a list of mesoporous carbon materials obtained by hard-template method with different ordered silica and different carbon precursors.

Table 2. Example of the reported mesoporous carbons

Silicate template	Carbon precursor	Ordered mesoporous carbon	Space group	Reference
MCM-41	Sucrose	Hexagonal carbon nanowire arrays	*P6mm*	[43]
	Furfuryl	Self-supported nanowire arrays	Partially ordered	[43]
MCM-48	Sucrose	CMK-1	$I4_132$ or lower	[33]
	Acetylene	CMK-4	$Ia\bar{3}d$	[44]
SBA-15	Sucrose	CMK-3	*P6mm*	[5] [34]
	Furfuryl	CMK-5	*P6mm*	[39][45]
	Acenaphthen	G-CMK-3	*P6mm*	[40]
	Mesophase pitches	Graphitized CMK-3	*P6mm*	[42]
SBA-16	Sucrose	Spherical mesoporous carbon	$Im\bar{3}m$	[46]
	Furfuryl	Spherical mesoporous carbon	$Im\bar{3}m$	[46]
SBA-3	Sucrose	Small pore mesoporous carbon	*P6mm*	[47]
SBA-1	Sucrose	CMK-2	unknown	[47]
FDU-5	Sucrose	Monolithic mesoporous carbon	$Ia\bar{3}d$	[48]
	Furfuryl	Tubular mesoporous carbon	$Ia\bar{3}d$	[49]
FDU-12	Sucrose	Spherical mesoporous carbon	unknown	[50]
KIT-6	Sucrose	CMK-4	$Ia\bar{3}d$	[51][52]
	Mesophase pitches	Graphitized CMK-4	$Ia\bar{3}d$	[42]

The structure and ordering of mesoporous carbons obtained by hard-template method depend on the following [1, 3]:

- type of matrix used,
- type of carbon precursor,
- carbonisation conditions.

The influence of the above mentioned factors have been discussed below.

Type of Matrix Used

Ryoo et al., [33] obtained the mesoporous carbon labelled as CMK-1 with the use of MCM-48 silica (of Ia3d symmetry) as a template [53, 54]. The 3-D structure of this silica was made of two systems of pores [55, 56]. The XRD profiles of the carbon revealed a number of sharp reflexes at small angles, which indicated a uniform structure of the elementary cell of the size 8.0-9.5 nm. The CMK-1 structure was made of micropores of the size 0.5 – 0.8 nm and mesopores of the size of 3.0 nm. The total pore volume was a sum of the volume of micropores of 0.3 cm^3/g and that of mesopores of 1.1 cm^3/g. A well-developed surface area of this carbon (1380 m^2/g) means that it can be successfully used as a catalyst and adsorbent for the removal of organic impurities [33].

Lee et al., [57] obtained mesoporous carbon labelled as SNU-2 with the use of mesoporous hexagonal silica and aluminosilicate as solid matrices. In 2000, Jun et al., [34] proposed the method of CMK-3 synthesis by the negative replica of the mesoporous silica SBA-15. CMK-3 is built of the parallel carbon rods stiffly connected by much thinner carbon fibres. The successful synthesis of CMK-3 has proved the micro-mesoporous 3-D structure of SBA-15 silica [58]. CMK-3 material is characterised by a narrow pore volume distribution with a maximum at 4.5 nm, while the maximum of pore volume distribution of SBA-15 is at 9.2 nm [34].

Other ordered silica materials such as SBA-16, FDU-5, FDU-12, KIT-6, obtained with the use of polymer matrices, have been also applied as hard templates for the synthesis of ordered mesoporous carbon materials [59]. The research groups of Ryoo and Zhao have independently synthesised porous silica of Ia3d structure , characterised by uniform large pores and obtained the highly ordered carbon replicas [40, 51, 60]. KIT-6 was used for production of carbon replicas of the same symmetry Ia3d and rodlike or tubular structure, depending on the synthesis conditions. Carbonisation of sucrose gives a rodlike carbon material which is a faithful negative replica of mesoporous silica of Ia3d structure [51]. In contrast to the ordered carbon materials obtained with MCM-48 as a matrix , the carbon materials obtained with silica matrix of Ia3d structure preserved the structure of the silica, without the structural transformation observed for CMK-1 synthesis [61].

Carbon mesofoams have been obtained with silica foams as hard matrices [62, 63]. Phenol as a carbon precursor was selectively introduced into the pores of the silica matrix under reduced pressure. The pore size distribution function for the carbon foam revealed the presence of two types of pores of sizes 27 and 3.6 nm. The pores of 27 nm came from the silica matrix as they were not filled with phenol under a low pressure, while the pores of 3.6 nm were the replicas of the walls of the mesoporous silica foam [61].

It should be mentioned that the problems with syntheses of negative replicas with pores greater than 6 nm impose certain limitations on the use of ordered mesoporous silica as matrices for the synthesis of mesoporous carbon materials . The problem is that typical ordered silica materials have pores of diameters 1-6 nm so the pores of this size are made in the mesoporous carbons. In order to obtain larger pores, other hard templates have to be used, e.g., colloidal silica of particles of different sizes [3]. By mixing the colloidal dispersed silica nanoparticles with the micrometric particles of carbon precursor, followed by drying , carbonisation and dissolving of silica, it is possible to obtain carbon materials with uniform spherical mesopores of the size corresponding to those of the colloidal silica used. Despite the disordered structure of the carbons, the sizes of their pores and the thicknesses of the pore walls are much greater than those in the carbon materials synthesised by reverse replication of ordered silica obtained with the use of surfactants or block copolymers .

Type of Carbon Precursor

Structural and adsorptive properties of carbon materials depend not only on the type of silica template but also on the properties of the carbon precursor. The use of sucrose , glucose , furfuryl alcohol as carbon precursors leads to obtaining materials not only with mesopores but also with micropores localised in the pore walls [3, 34]. Thanks to the presence of micropores, the carbons are characterised by large surface area and large pore volume. The carbon replicas synthesised with the use of precursors containing linked aromatic rings , e.g., liquid crystal carbon tar , acenaphthene, naphthalene , anthracene, pyrrole or pyrene, show much higher mechanical strength but smaller surface area S_{BET} and smaller pore volume [40, 64]. Carbon can also be obtained from such substances as propylene , acetylene, ethylene , styrene , acetonitrile , as a result of chemical decomposition of vapours [65-68]. Different polymers , e.g., phenol resins , polydivinylbenzene, polyacrylnitrile, polyvinyl chloride forming in the process of carbon synthesis in the matrix pores, are also widely used as carbon precursors [33, 69-72].

The amount and concentration of the carbon precursor used for the carbon synthesis has also great impact on the type and ordering of the forming porous structure . Complete filling up of the pore system of the mesoporous silica SBA-15 with furfuryl alcohol, known as the volume templating, gives CMK-3 carbon of rodlike structure. Coating the SBA-15 matrix with a thin layer of the carbon precursor, known as the surface templating, gives CMK-5 carbon of cylindrical structure. It has been found that when silica matrix is impregnated with a 20% solution of furfuryl alcohol, it is impossible to get any stiff mesostructure, most probably because of the carbon deficiency . In order to obtain CMK-5, it is necessary to use furfuryl alcohol in a concentration of at least 30%. On the other hand, when the concentration of this alcohol exceeds 80%, we get CMK-3 carbon [73]. When an admixture of another element should be introduced to the final product, the impregnation is performed with compounds being sources of carbon and if this element is nitrogen , e.g., aniline . An example is the synthesis of mesoporous carbon modified with nitrogen with the use of KIT-6 silica. The carbon obtained by negative replication has a 3-D cubic structure with an enantiomeric pair of independently interpenetrating 3-D continuous network of mesoporous channels that are mutually intertwined and separated by carbon walls. It has been established that nitrogen is bonded to carbon through the imine and amine group s. Thanks to such a simple method of synthesis it is possible to obtain carbon materials with nitrogen built in their structure, which permits the application of these materials as energy carriers and in electronic devices [74].

Carbonisation Conditions

In the process of carbonisation (or pyrolysis), the materials are subjected high temperature in the atmosphere of neutral gas in order to decompose the precursor. The heat energy supplied causes breaking up of the weakest chemical bonds and stimulates generation of free radicals . Some of the radicals undergo stabilisation as a result of bonding hydrogen . The process of carbonisation involves polymerisation and polycondensation , which permits getting a uniform mass of the carbonisate characterised by a high content of carbon [75].

Carbonisates of tailored physicochemical properties are obtained by a proper choice of conditions of pyrolysis such as temperature , rate of temperature increase, duration of carbonisation and type of atmosphere of this process [75]. The most important is the final temperature of the process that should be high enough for breaking up of weak bonds and distillation of volatile products of the thermal decomposition of the carbon precursor to the surrounding atmosphere as well as high enough to ensure ordering of the compact carbon substance [75].

The rate of temperature increase should also be carefully adjusted. If it is too high, the particular phases of thermal decomposition of carbon and the secondary reactions of the products of pyrolysis overlap , which is unfavourable for getting carbon substance with highly ordered structure . Moreover, upon high rate of heating, a large volume of volatile components is released from the carbonised material in a short time, which leads to formation of large size pores.

The type of atmosphere and pressure at which carbonisation takes place influences the thermal decomposition of the material. The gas being blown through the reactor removes all volatile products, which leads to a significant increase in the carbonisate reactivity . Usually the process of carbonisation is carried out in the atmosphere of a neutral gas, but it can also be performed under vacuum or under elevated pressure.

2.2. Soft-Template Synthesis of Mesoporous Carbon Materials

In the direct synthesis of mesoporous carbon materials , known also as the soft-template method, there is no need to use solid matrices [3]. In this synthesis, the thermally cross-linked polymer (carbon precursor) is obtained in the presence of self-assembled block polymer (matrix) which undergoes decomposition at high temperatures. As a result of the process, an ordered nanocomposite made of two polymers is obtained. One of the polymers undergoes decomposition when subjected to high temperature , while the other easily undergoes carbonisation in a neutral atmosphere , leading to formation of ordered mesoporous carbon. The reduction of stages of synthesis needed to obtain mesoporous carbon in the soft-template method as well as the thermal degradation of the copolymer matrix and the use of environmentally friendly reagents (usually water and ethanol) permit the use of soft-template method on a large scale [3].

The first paper on the synthesis of ordered mesoporous carbons has been published by Liang et al., [76]. Films of mesoporous carbons have been synthesised by soft-template method with diblock polymer styrene and poly(4-vinyl pyridine) (PS-P4VP) as templates and resorcinol together with formaldehyde as carbon precursors. Unfortunately, high cost of PS-P4VP significantly restricts the use of this method for carbon production. Tanaka et al., [77] have obtained for the first time ordered mesoporous carbon using a cheap triblock copolymer , ethylene polyoxide – propylene polyoxide – ethylene polyoxide ($EO_{106}PO_{70}EO_{106}$) of the commercial name Pluronic F127 as the soft-template matrix . These two procedures published by Liang et al., and Tanaka et al., have initiated a dynamic development of investigation related to the synthesis of mesoporous carbon, to which a great contribution has been brought by the groups of S. Dai [3, 78] and D. Zhao [79].

Figure 2 presents a scheme of direct synthesis of mesoporous carbon proposed in [78]. It involves polymerisation of phloroglucine and formaldehyde in the presence of triblock copolymer Pluronic F127 ($EO_{106}PO_{70}EO_{106}$) as a matrix , in an acidic environment . Phloroglucine interacts with hydrophilic blocks of the copolymer and is built into its ordered structure . Formaldehyde molecules penetrate into the hydrophilic part of the copolymer and react with phloroglucine to give phenolic resin. As a result of self-assembly of the block copolymer and polymerisation of phloroglucine and formaldehyde, a polymer nanocomposite is formed and then upon controlled heating and carbonisation it is transformed into mesoporous carbon.

The group headed by Zhao [79] have obtained highly ordered mesoporous polymers and carbon structures in the process based on the interaction of a triblock copolymer capable of ordered phase formation with a soluble phenolic resin of small molecular mass , which was resol, (Figure 3). By controlling the amount of phenol in proportion to the amount of the copolymer, it is possible to obtain carbon materials of 2-D hexagonal (p6m), 3-D continuous (*Ia*3*d*) and body centred cubic (regular) (Im3m) mesostructure. The synthesis of these carbon materials needs the stage of phenolic resin carbonisation preceded by decomposition of the triblock copolymer at 400°C.

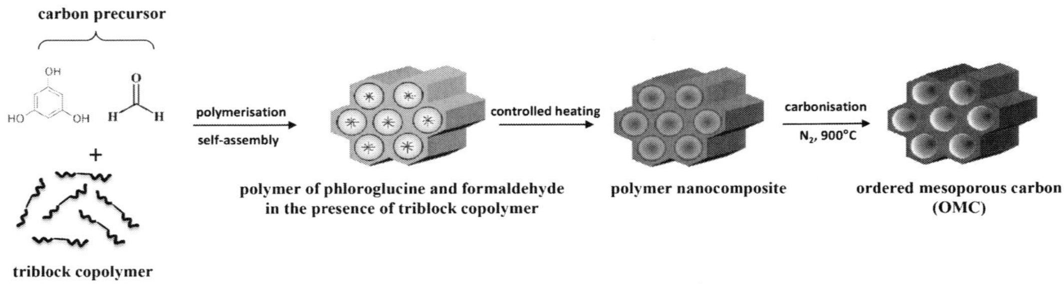

Figure 2. Synthesis of ordered mesoporous carbons by the soft-template method [78].

The regular material FDU-14 which is either polymer resin or carbon of gyroidal mesostructure, has been for the first time synthesised with Pluronic P123 ($EO_{20}PO_{70}EO_{20}$) as the soft-template. The materials obtained, the polymer resin or carbon, have ordered structures, well-developed surface area (670-1490 m^2/g), high pore volume (0.65 – 0.85 cm^3/g), uniform pores of diameters from the range 3.9 – 7.0 nm and thick pore walls (6 – 8 nm). The open porous structure of carbon materials with thick pore walls ensures high thermal stability in anaerobic atmosphere at temperatures even higher than 1400°C. The gas liberated upon calcination has a significant influence on the porosity of these carbon materials.

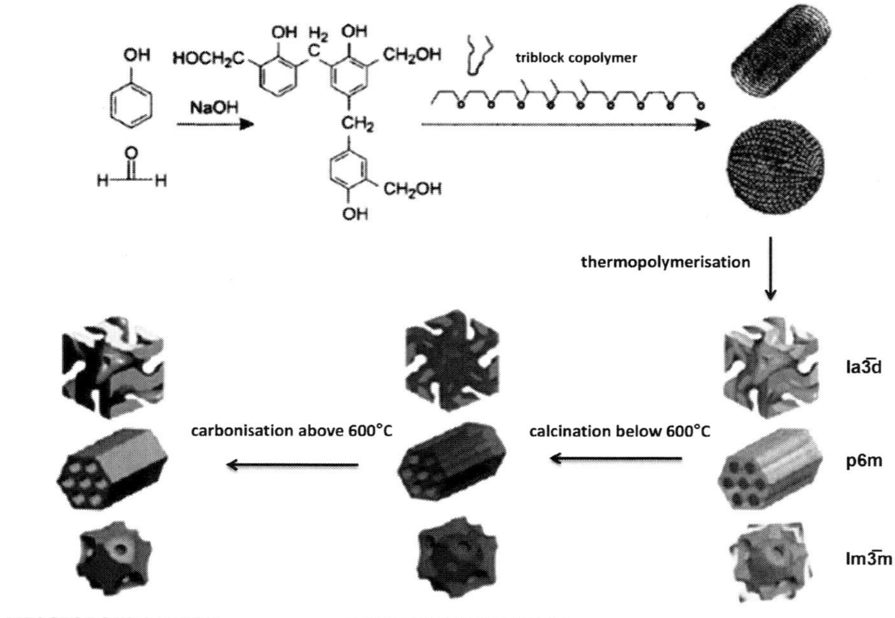

Figure 3. Preparation of ordered mesoporous polymer resins and carbon frameworks with resol as carbon precursor [3, 79].

A relatively simple method for the synthesis of ordered mesoporous carbons based on the use of triblock copolymer as a template and phenol and formaldehyde as carbon precursor, in a water solution , has been proposed in [80]. In a water solution with addition of NaOH, it has been possible to synthesise highly ordered FDU-16 carbons of body centred structure (Im3m), using triblock copolymer Pluronic F127. By applying hexadecane or decane with Pluronic P123, FDU-15 carbons of 2-D hexagonal (p6m) structure have been obtained.

An interesting attempt at getting highly ordered mesoporous polymer -silica and carbon - silica nanocomposites has been reported in [81]. This synthesis is based on the self-assembly of the tree-component mixture of the reagents followed by the removal of all template components. The organic precursor is resol polymer, the inorganic precursor is preliminary hydrolysed solution of tetraethoxysilane , while the template is triblock Pluronic F127. The highly ordered carbon-silica nanocomposite is subjected to oxidation of carbon in air or dissolution of silica in hydrofluoric acid , which gives ordered mesoporous silica (OMS) and ordered mesoporous carbon (OMC), respectively.

3. Modification of Mesoporous Carbon Materials

From the point of view of OMC application in catalytic and adsorption processes, the important parameters are well-developed surface area , pore diameter and pore volume, the presence of heteroatoms, nanoparticles of metals or metal oxides and salts (Table 3) [3, 11, 82-100].

Table 3. Some examples of mesoporous carbons modification

Modification of mesoporous carbons	**Reference**
Metal nanoparticles :	
iron	[82]
nickel	[83-87]
silver	[88,89]
gold	[90-93]
Metal oxide nanoparticles :	
alumina	[94]
silica	[94]
iron(III) oxide	[95]
titania	[96-99]
Metal salt nanoparticles:	
nickel(II) chloride	[100]
lanthanum(III) chloride	[11]

Mesoporous carbon materials containing nanoparticles of metals or metal oxides , can be obtained by hard- and soft-template methods, but for the practical reasons the soft-template method is much more promising, (Figure 4). The nanoparticles of metals, metal oxides or salts can be added in the process of synthesis or in a later treatment .

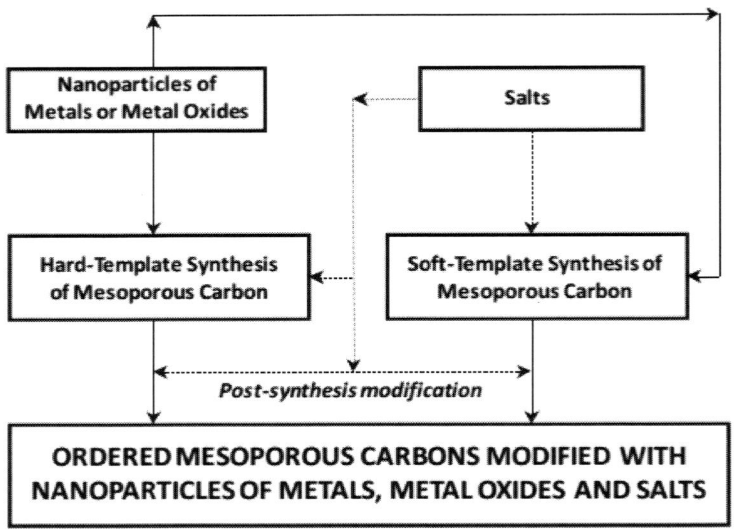

Figure 4. Schematic illustration of the synthesis of ordered mesoporous carbons by hard- and soft-templating methods with the addition of metal, metal oxide and salt nanoparticles .

When applying the soft-template method, inorganic nanoparticles are most often introduced in the process of synthesis . According to one popular procedure, metal or metal oxide nanoparticles of defined size and amount are added in the process of synthesis [94, 101]. According to another also often applied procedure, metal salt nanoparticles are added, which upon gradual heating of the material in a neutral atmosphere are transformed to nanoparticles of metal oxide, metal carbides or other inorganic compounds , depending on the type of metal [83, 84, 95].

Zhao et al., [95] have proposed a procedure for the synthesis of mesoporous carbon materials containing Fe_2O_3 nanoparticles with the use of iron (III) citrate. It has been proved that carbons with a small admixture of iron(III) oxide (close to 9 wt.%) are characterised by ordered hexagonal structure , have uniform mesopores of about 4 nm in diameter, well-developed surface area (590 m^2/g) and large pore volume (0.48 cm^3/g). The group of Wang [82] has obtained mesoporous carbons modified with iron nanoparticles by the soft-template method with hydrated iron(III) nitrate(V) as a precursor. The product containing about 10 wt.% of iron showed a specific catalytic activity in the reaction of phenol oxidation with hydrogen peroxide in water solutions.

Introduction of nickel to carbon has also brought interesting results. Jaroniec et al., [84] have applied soft-template synthesis of mesoporous carbon with resorcinol and formaldehyde as carbon precursors, tetraethoxysilane (TEOS) as a source of silica and nickel(II) nitrate(V) as a source of nickel particles. The removal of silica by an NaOH solution gave mesoporous carbon with nanoparticles of nickel, while the oxidation of carbon in this hybrid material gave mesoporous silica with NiO nanoparticles. The authors of [85] have obtained a CMK-5 type carbon of very thin walls with highly dispersed nickel nanoparticles. The nickel precursor was 6-hydrated nickel(II) nitrate(V) in 2-propanol. The nickel nanoparticles were found to have diameter close to 3 nm and to be relatively uniformly distributed in carbon pores. The authors of [86] have described the synthesis of mesoporous carbon material with nickel nanoparticles which showed magnetic properties , by the soft-template method.

In [87] the authors have proposed a method for the synthesis of graphitised carbon containing nickel nanoparticles involving pyrolysis of polystyrene and with the use of silica colloids as hard-template. Besides nickel(II) nitrate(V), also nickel(II) chloride in the presence of hydrochloric acid , has been used for the synthesis of ordered mesoporous carbons [100]. In the same paper [100] a new method for the synthesis of a binary carbon catalyst containing nickel and platinum nanoparticles has been proposed. The composite obtained was used in the process of methanol electrooxidation. Carbon with nickel nanoparticles has been also produced by the soft-template method with Pluronic F127 copolymer [83]. This carbon is resistant to acids and can be applied in magnetically controlled separation.

Ever since the first synthesis of silver nanoparticles , attempts have been made to use their unique chemical and physical properties . One of the problems has been to find new and more effective materials supporting them. The ordered carbon CMK-3 containing silver nanoparticles was obtained by impregnation of carbon with $AgNO_3$ solution followed by thermal decomposition of this salt [88]. With the help of transmission electron microscopy the silver nanoparticles of about 3 nm in size were proved to be in the tunnel pores of CMK-3. Another interesting method for the synthesis of mesoporous carbon with silver nanoparticles with employing nanocolloid silica as a hard-template has been proposed by Jaroniec et al., [89]. In this method, the colloidal nanoparticles of silica and silver were pressed into a monolith which was then impregnated with phenol resin used as a carbon precursor. Thus obtained composite was subjected to carbonisation and then the silica template was dissolved by hydrofluoric acid . The final product was mesoporous carbon containing silver nanoparticles, characterised by large and uniform pores being replicas of the shape and size of silica particles used (spheres of about 27 nm in diameter), large surface area (close to 830 m^2/g) and large pore volume (close to 2.32 cm^3/g).

Modification of ordered mesoporous carbon composites with gold has been investigated only recently, much later than modification with silver . So far the method of electrodeposition has been successfully applied for introduction of gold nanoparticles to ordered mesoporous carbon [90]. The composite has been used in a sensor for detection of hydrogen peroxide . The authors of [91] have proposed a method for the synthesis of highly dispersed gold nanoparticles of diameters not exceeding 7 nm on the surface of ordered mesoporous carbon containing amine group s (>NH and $-NH_2$). The material was used as a catalyst of catalytic conversion of benzoic aldehyde, piperidine and phenylacetylene into propargyl amine, which is a semiproduct in the synthesis of multifunctional amine derivatives . The selectivity of gold nanoparticles deposited on mesoporous carbon used as a catalyst of linoleic acid isomerisation was studied in [92]. The properties of the carbon-gold composite were compared with those of the composite obtained by gold nanoparticles deposition on silica . The gold nanoparticles had very small size, with diameters near 2 nm. Kim et al., [93] has described the synthesis of a composite made of carbon capsules of mesoporous walls whose pores contained gold nanoparticles. The mesoporous carbon-gold capsules had large surface area BET (over 1000 m^2/g) and large total pore volume (over 1.0 cm^3/g).

In [94] the soft-template method has been applied for direct introduction of aluminium oxide and silica colloids in the amounts of 10 wt.% and 20 wt.% into mesoporous carbon . Irrespective of the type of nanoparticles introduced, when their amount was low, no deterioration in the adsorption properties was noted when compared with those of mesoporous carbon with no nanoparticles. The properties of carbon composites containing relatively large amounts of nanoparticles depended on the chemical properties of the composite surface and the size of nanoparticles.

Nanoporous carbons have been modified with titanium dioxide nanoparticles . Zhao group [96] has synthesised highly ordered mesoporous carbon material containing TiO_2 in concentrations from a wide range 20 wt.% - 80 wt.%, with the use of triblock Pluronic P123, phenol resin as carbon precursor and titanium tetrachloride ($TiCl_4$). The composite showed ordered hexagonal 2-D mesostructure, high thermal stability up to 700°C, large surface area (465 m^2/g) and mesopores of 4.1 nm in diameter. The carbon composites with TiO_2 nanoparticles were found suitable catalysts of rhodamine B decomposition . As an extension of these studies, carbon-titanium composites were obtained with the use of double TiO_2 precursor, that is the acidic titanium tetrachloride ($TiCl_4$) and basic tetrabutoxytitanium ($Ti(OC_4H_9)_4$) [97], containing up to 87 wt.% of TiO_2. The composites can be used for adsorption of proteins and for electrocatalytic reduction of hydrogen peroxide . The authors of [98] have also obtained the carbon-titanium composite with titanium dioxide in the form of anatase as a titanium precursor. This composite was synthesised to be used in ion-lithium cells. Because of the photocatalytic properties of titanium dioxide, the composites with TiO_2 are used in the processes of photocatalytic decomposition of organic compounds . Mesoporous graphitised carbon of the specific surface area of about 300 m^2/g was obtained at 950°C with the use of monodisperse colloidal silica of particles of 26 nm in diameter as the hard-template, styrene as a carbon precursor and nickel as a catalyst . The composite of TiO_2 with graphitised mesoporous carbon was synthesised by depositing TiO_2 onto mesoporous carbon obtained by the sol-gel method. This material was used for the photocatalytic decomposition of rhodamine B and phenol in water solutions [99].

4. Mesoporous Carbons in Adsorption Processes

Thanks to their well-developed surface area , large pore volume and specific mechanical and chemical properties , mesoporous carbon materials have been widely used as adsorbents of biologically active compounds (e.g., amino acid s , drugs , vitamins), and for purification of air, water and industrial gases [5, 8-16, 102-119].

The effectiveness of adsorption of the organic active substances from the liquid phase on mesoporous carbons depends on [5]:

- molecular mass of the adsorbate,
- size and geometric dimensions of the adsorbate particles,
- solubility of the adsorbate,
- polarity of the adsorbate,
- type of functional groups in the adsorbate.

The adsorption can involve intermolecular interaction forces, e.g., van der Waals forces (physical adsorption) or chemical forces (chemical adsorption). Chemical adsorption is realised with the involvement of electrons from the adsorbent and adsorbate. The intermolecular interactions in the physical adsorption depend on the type of adsorbent molecules and the character of adsorbent surface.

The ordered system of mesoporous pores in molecular sieves ensures a uniform distribution of adsorbed biologically active compounds and permits a control of the kinetics of their release. Mesoporous carbon materials have been so far used for adsorption of ibuprophen [104, 105], β-carotene [106, 107], tetracycline chloride [108], vitamin B1 2 [9,109], albumine [9], lysozyme [110, 111], L-histidine [8,112], L-phenylalanine [5,8], and polypeptides [113]. In the most often used method for the adsorption of adsorbate particles into the pores of mesoporous sieves the active substances from organic or buffer solutions are subjected to continuous stirring. Irrespective of the adsorption duration, the amount of adsorbed biologically active substance usually does not exceed 50 wt.%. In some reactions , the beneficial effect of slow evaporation of solvent from the suspension of the adsorbate has been reported to lead to increased effectiveness of adsorption. Upon adsorption of ibuprophen, changes in carbon porosity have been observed [104, 105]. In carbon materials containing micro- and mesopores , a decrease in the volume of micropores was noted as ibuprophen prefers localisation in pores of this size. When the available micropores have been filled, ibuprophen makes a single, uniform thin film on the inner surface of mesopores. In the adsorption processes performed in the liquid phase , the choice of a proper solvent is of great importance. Adsorption of ibuprophen was performed using aprotic solvents DMSO, DMA, DMF or hexane and protolytic solvents (ethanol and methanol). The use of hexane permitted obtaining the highest concentration of ibuprophen inside the pores of mesoporous carbons [104, 105]. The research group of Vinu [112] presented a possibility of the use of mesoporous carbons of CMK-3 type in the process of L-histidine adsorption from the liquid phase. Carbon materials were obtained by hard-template method with sucrose as a carbon precursor and ordered silica SBA-15 as a hard-template. The sorption abilities of CMK-3 towards L-histidine were compared with those of the commercial activated carbon s. The isotherms of L-histidine adsorption show a sharp increase at low final concentrations of solutions, which indicates a high affinity of this amino acid and the adsorbent surface. The amount of L-histidine adsorbed increased with increasing initial concentrations of this amino acid in solutions. When solutions of low L-histidine concentration are used the adsorption is a random process on the surface of mesoporous carbons. When the concentration of the initial L-histidine solution increases, the nonpolar groups from two or more side chain s of L-histidine approach one another to the distance allowed by their van der Waals radii, which leads to close packing of L-histidine on the carbon surface. In this process hydrophobic bonds are generated. The amount of L-histidine adsorbed on mesoporous carbons of CMK-3 type depends on pH of L-histidine solution and increases for pH from 5.0 to 7.5 and then decreases when pH further increases to 9.6. The maximum adsorption is noted for pH 7.5, which is close to the isoelectric point of L-histidine. To confirm the close packing of L-histidine on CMK-3, the nitrogen adsorption isotherms were recorded before and after L-histidine adsorption. The size of pores and surface of mesoporous carbons drastically decreased after adsorption of this amino acid [112]. Similar results have been reported by Goscianska et al., who studied L-histidine and L-phenylalanine adsorption on mesoporous carbons obtained by hard-template method with the silica materials KIT-6, SBA-16 and SBA-15 as the solid matrices and sucrose [5] or furfuryl alcohol [8] as carbon precursors.

Mesoporous carbons are inert, stable, and usually hydrophobic and thus show high affinity to organic pollutants . These materials have been widely applied as adsorbents of dyes , e.g., methylene blue [114, 115], methyl orange [11, 12, 103, 116-118], methylthionine chloride [117], rhodamine B [117], Congo red [117], crystalline violet [119] or malachite green [13]. The presence of dyes in water is a serious problem in many branches of industry , especially in paper industry, dye production, textile and cosmetic industries . Even small amounts of dyes (of ppm order) are undesirable as they colour water and disturb the living processes of organisms living in water environment . They should be carefully removed from water and wastewater . The chemical and biological methods that have been so far used for their removal are often ineffective and much is expected of the adsorption processes. Mesoporous carbons show high sorption capacity towards dyes and high rate of dye adsorption caused by diffusion inside the pores. Mohammadi et al., [116] have applied mesoporous CMK-3 carbons of hexagonal structure for adsorption of methyl orange. The carbon was obtained by hard-template with sucrose as carbon precursor and ordered silica SBA-15 as a matrix . The influence of such parameters as initial concentration of methyl orange, pH of methyl orange solutions, contact time and temperature on the sorption capacity of CMK-3. It has been proved that the time of adsorption has significant influence on the amount of the dye adsorbed. For the first 30 minutes the process is rapid, then the amount of the adsorbed dye gradually increases to reach the equilibrium state after 60 minutes, when the amount of methyl orange adsorbed changes no longer. Another important parameter is pH of methyl orange solutions. With pH increasing from 3 to 9, the amount of methylene orange adsorbed on CMK-3 decreases, but it increases with increasing initial concentration of methylene orange solutions.

Dong et al., [117] have synthesised magnetic mesoporous carbon materials by chemical activation followed by reduction of $Fe(NO_3)_3$. All the materials have large surface area s (975–1321 m^2/g), high pore volumes (0.85–1.08 cm^3/g), and bimodal pores (2.8–2.9 and 3.5–4.1 nm). The adsorption capacity and behaviour of four dyes , methylthionine chloride (MC), methyl orange (MO), rhodamine B (RB), and Congo red (GR), were studied on these magnetic bimodal mesoporous carbon materials. With increasing initial concentrations of methyl orange, methylene blue , rhodamine B and Congo red, the amount of dyes adsorbed increases until saturation , when dye particles have occupied all available free space in the mesoporous carbons. In order to understand the adsorption correlation of the adsorbent and adsorbate, they have divided the dyes into three groups according to their structures: (1) methylene blue and methylene orange which possess short 1-D chain molecular structure s, (2) rhodamine with the 2-D planar molecular structure ; and (3) Congo red which has a long 1-D chain molecular structure. The adsorption capacity increases in the order C-4@Fe > C-3@Fe > C-2@Fe > C-1@Fe for groups 1 and 2. As follows from the above analysis, for the typical monolayer Langmuir adsorption, the surface area is important for adsorption capacity, which is understandable as a larger surface area has more sites available for adsorption. In addition, small size pores (within the feasible size of adsorbate size) show higher capillary effect, so the adsorbate is not able to fall off easily, so the equilibrium amount of adsorbed substance is greater. Therefore, with increasing surface area, C-4@Fe which has more micropores and small mesopores shows increasing adsorption capacity towards methylene blue, methyl orange and rhodamine B. However, the highest adsorption of Congo red is observed for C-1@Fe, which has an ordered mesoporous structure similar to that of CMK-3.

As malachite green shows bactericidal and fungicidal properties, it had been commonly used in water systems, however its use was stopped after discovery of its high toxicity and carcinogenic influence. Tian et al., [13] studied the adsorption of malachite green on the surface of mesoporous carbon synthesised by hard-template with silica SBA-15 as a solid matrix and furfuryl alcohol as carbon precursor. The effect of different volume ratios of furfuryl alcohol (FA) to trimethylbenzene (TMB) of 1:1, 2:1, 3:1 was tested. The carbons with the above ratios of FA to TMB were labelled as C-1, C-2, C-3. They showed well-developed surface area s of 1491 (C-1), 1256 (C-2) and 1177 m^2/g (C-3) and large pore volumes 1.29 (C-1), 1.14 (C-2), 0.83 cm^3/g (C-3). With increasing FA to TMB volume ratios, the surface area and pore volume decrease, leading to deterioration of adsorption capacity towards malachite green.

CONCLUSION

Recently much attention has been devoted to the search for new adsorbents that would permit effective removal of air and water pollutants . Ordered mesoporous carbon materials offer a number of properties attractive from this point of view, including well-developed surface area , large pore volume, specific mechanical and chemical behaviour. The information on the methods for their synthesis and on their physicochemical properties presented in this chapter, can be of interest to those wish to get involved in this area of research. It should be also mentioned that because of their nontoxicity , mesoporous carbons can be applied as adsorbents of biologically active compounds and are considered as possible carriers of therapeutic substances from which the drugs could be released in controlled way, like e.g., for encapsulation of poorly water soluble or insoluble substances.

REFERENCES

[1] Ryoo, R; Joo, S; Kruk, M; Jaroniec, M. *Adv. Mater.* 2001, 9, 13.
[2] Yang, H; Zhao, D. *J. Mater. Chem.* 2005, 15, 1217.
[3] Liang, C; Li, Z; Dao, S. *Angew. Chem. Int. Ed.* 2008, 47, 3696.
[4] Kyotani, T. *Carbon* 2000, 38, 269.
[5] Goscianska, J; Olejnik, A; Pietrzak, R. *J. Taiwan Inst. Chem. Eng.* 2014, 45, 347.
[6] Yuan, DS; Zeng, J; Chen, J; Liu, Y. *Int. J. Electrochem. Sci.* 2009, 4, 562.
[7] Prasad, K; Dhawale, D; Joseph, S; Anand, C; Wahab, M; Mano, A; Sathish, C; Balasubramanian, V; Sivakumar, T; Vinu, A. *Microp. Mesop. Mater.* 2013, 172, 77.
[8] Goscianska, J; Olejnik, A; Pietrzak, R. *Adsorption* 2013, 19, 581.
[9] Xu, D; Yoon, S-H; Mochida, I; Qiao, W; Wang, Y; Ling, L. *Microp. Mesop. Mat.* 2008, 115, 461.
[10] Saha, D; Deng, S. *J. Colloid Interface Sci.* 2010, 345, 402.
[11] Goscianska, J; Marciniak, M; Pietrzak, R. *Chem. Eng. J.* 2014, 247, 258.
[12] Derylo-Marczewska, A; Marczewski, AW; Winter, S; Sternik, D. *Appl. Surf. Sci.* 2010, 256, 5164.
[13] Tian, Y; Liu, P; Wang, X; Lin, H. *Chem. Eng. J.* 2011, 171, 1263.
[14] Nejad, N; Shams, E; Amini, M; Bennett, J. *Microp. Mesop. Mater.* 2013, 168, 239.
[15] Moreno-Tovar, R; Terres, E; Rangel-Mendez, JR. *Appl. Surf. Sci.* 2014, 303, 373.

[16] Wang, Y-Q; Zhang, Z-B; Liu, Y-H; Cao, X-H; Liu, Y-T; Li, Q. *Chem. Eng. J.* 2012, 198-199, 246.
[17] Gokulakrishnan, N; Kania, N; Leger, B; Lancelot, C; Grosso, D; Monflier, E; Ponchel, A. *Carbon* 2011, 49, 1290.
[18] Liu, L; Deng, Q-F; Agula, B; Ren, T-Z; Liu, Y-P; Zhaorigetu, B; Yuan, Z-Y. *Catal. Today* 2012, 186, 35.
[19] Liu, L; Deng, Q-F; Agula, B; Zhao, X; Ren, T-Z; Yuan, Z-Y. *Chem. Commun.* 2011, 47, 8334.
[20] Son, S; Wang, K; Liu, Y; He, C; Liang, Y; Fu, R; Wu, D; Wang, Y. *Int. J. Hydrogen Energy* 2013, 38, 1405.
[21] Maiylagan, T; Nassr, ABA ; Alaje, TO; Bron, M; Scott, K; *J. Powder Sources* 2012, 211, 147.
[22] Lin, D-H; Jiang, Y-X; Chen, S-R; Chen, S-P; Sun , S-G. *Electrochimica Acta* 2012, 67, 127.
[23] Huang, C-C ; Li, Y-H; Wang, Y-W; Chen, C-H. *Int. J. Hydrogen Energy* 2013, 38, 3994.
[24] Wang, G; Kuang, S; Wang, D; Zhuo, S. *Electrochimica Acta* 2013, 113, 346.
[25] Kawashima, D; Aihara, T; Kobayashi, Y; Kyotani, T; Tomita, A. *Chem. Mater.* 2000, 12, 3397.
[26] Han, S; Hyeon, T. *Carbon* 1999, 37, 1645.
[27] Han, S; Hyeon, T. *Chem. Commun.* 1999, 1955.
[28] Bandosz, TJ; Jagiello, J; Putyera, K; Schwarz, JA. *Chem. Mater.* 1996, 8, 2023.
[29] Kyotani, T; Tsai, L-F; Tomita, A. *Chem. Mater.* 1996, 8, 2109.
[30] Kyotani, T; Nagai, T; Inoue, S; Tomita, A. *Chem. Mater.* 1997, 9, 609.
[31] Johnson, SA; Brigham, ES; Ollivier, PJ; Mallouk, TE. *Chem. Mater.* 1997, 9, 2448.
[32] Rodriguez-Mirasol, J; Cordero, T; Radovic, LR; Rodriguez, JJ. *Chem. Mater.* 1998, 10, 550.
[33] Ryoo, R; Joo, SH; Jun, S. *J. Phys. Chem. B* 1999, 103, 7743.
[34] Jun, S; Joo, H; Ryoo, R; Kruk, M; Jaroniec, M; Liu, Z; Ohsuna, T; Terasaki, O. *J. Am. Chem. Soc.* 2000, 122, 10712.
[35] Vartuli, JC; Roth, WJ; Beck, JS; McCullen, SB; Kresge, CT . *Molecular Sieves, Science and Technology, Synthesis*. Springer-Verlag Berlin Heidelberg 1998.
[36] Huo, Q; Margolese, D; Ciesla, U; Demuth, D; Feng, P; Gier, T.; Sieger, P; Firouzi, A; Chmelka, B; Schüth, F; Stucky, GD. *Chem. Mater.* 1994, 6, 1176.
[37] Monnier, A; Schüth, F; Huo, Q; Kumar, D; Margolese, D; Maxwell, RS; Stucky, G; Krishnamurty, M; Petroff, P; Firouzi, A; Janicke, M; Chmelka, B. *Science* 1993, 261, 1299.
[38] Huo, Q; Margolese, D; Ciesla, U; Feng, P; Gier, T; Sieger, P; Leon, R; Petroff, PM ; Ciesla, U; Schüth, F; Stucky, GD. *Nature* 1994, 368, 317.
[39] Joo, H; Choi, J; Oh, I; Kwak, J; Liu, Z; Terasaki, O; Ryoo, R. *Nature* 2001, 412, 169.
[40] Kim, T; Park, I; Ryoo, R. *Angew. Chem., Int. Ed.* 2003, 42, 4375.
[41] Lu, A; Kiefer, A; Schmidt, W; Schüth, F. *Chem. Mater.* 2004, 16, 100.
[42] Yang, H; Yan, Y; Liu, Y; Zhang, F; Zhang, R; Li, M; Meng, Y; Zhao, D. *J. Phys. Chem. B* 2004, 108,17230.
[43] Tian, B; Che, S; Liu, Z; Liu, X; Fan, W; Tatsumi, T; Terasaki, O; Zhao, D. *Chem. Commun.* 2003, 2726.

[44] Ryoo, R; Joo, SH; Jun, S; Tsubakiyama, T; Terasaki, O. *Zeolites and mesoporous materials at the dawn of the 21st century*. Elsevier, Montpellier, France , 2001.
[45] Che, S; Lund, K; Tatsumi, T; Iijima, S; Joo, H; Ryoo, R; Terasaki, O. *Angew. Chem. Int. Ed.* 2003, 42, 2182.
[46] Kim, T-W; Ryoo, R; Gierszal, KP; Jaroniec, M; Soloyyov, LA; Sakamoto, Y; Terasaki, O. *J. Mater. Chem.* 2005, 15, 1560.
[47] Ryoo, R; Joo, SH; Kruk, M; Jaroniec, M. *Adv. Mater.* 2001, 13, 677.
[48] Yang, H; Shi, Q; Liu, X; Xie, S; Jiang, D; Zhang, F; Yu, C; Tu, B; Zhao, D. *Chem. Commun.* 2002, 2842.
[49] Che, S; Garcia-Bennet, AE; Liu, X; Hodgkins, RP; Wright, PA; Zhao, D; Terasaki, O; Tatsumi, T. *Angew. Chem. Int. Ed.* 2003, 42, 3930.
[50] Fan, J; Yu, C; Gao, F; Lei, J; Tian, B; Wang, L; Luo , Q; Tu, B; Zhou, W; Zhao, D. *Angew. Chem. Int. Ed.* 2003, 42, 3146.
[51] Kleitz, F; Choi, SH; Ryoo, R. *Chem. Commun.* 2003, 2136.
[52] Sakamoto, Y; Kim, TW; Ryoo, R; Terasaki, O. *Angew. Chem. Int. Ed.* 2004, 43, 5231.
[53] Beck, JS; Vartuli, JC; Roth, WJ; Leonowicz, ME; Kresge, CT ; Schmitt, KD; Chu, CT; Olson, DH; Shepard, EW; McCullen, SB; Higgins, JB; Shlenker, JL. *J. Am. Chem. Soc.* 1992, 114, 10834.
[54] Kresge, CT ; Leonowicz, ME; Roth, WJ; Vartuli, JC; Beck, JS. *Nature* 1992, 359, 710.
[55] Tanev, PT; Pinnavaia, TJ. Science 1995, 267, 865.
[56] Carlsson, A; Kaneda, M; Sakamoto, Y., Terasaki, O; Ryoo, R; Joo, SH. *J. Electron. Microsc.* 1999, 48, 795.
[57] Lee, J; Yoon, S; Oh, SM; Shin, S-H; Hyeon, T. *Adv. Mater.* 2000, 12, 359.
[58] Ryoo, R; Ko, CH; Kruk, M; Antochshuk, V; Jaroniec, M. *J. Phys. Chem. B* 2000, 104, 11465.
[59] Deng, Y; Yu, T; Wan, Y; Shi, Y; Meng, Y; Gu, D; Zhang, L; Huang, Y; Liu, C; Wu, X; Zhao, D. *J. Am. Chem. Soc.* 2007, 129, 1690.
[60] Liu, R; Shi, Y; Wan, Y; Meng, Y; Zhang, F; Gu, D; Chen, Z; Tu, B; Zhao, D. *J. Am. Chem. Soc.* 2006, 128, 11652.
[61] Lee, J; Han, S; Hyeon, T. *J. Mater. Chem.* 2004, 14, 478.
[62] Lee, J; Sohn, K; Hyeon, T. *J. Am. Chem. Soc.* 2001, 123, 5146.
[63] Han, S; Kim, S; Choi, W; Park, H; Yoon, J; Hyeon, T. *Micropor. Mesopor. Mater.* 2003, 58, 131.
[64] Kim, CH; Lee, D-K; Pinavaia, TJ. *Langmuir* 2004, 20, 5157.
[65] Vix-Gueterl, C; Boulard, S; Parmentier, J; Werckmann, J; Patarin, J. *Chem. Lett.* 2002, 1062.
[66] Zhang, W-H; Liang, C; Sun , H; Shen, Z; Guan, Y; Ying, P; Li, C. *Adv. Mater.* 2002, 14, 1776.
[67] Xia, YD; Mokaya, R. *Adv. Mater.* 2004, 16, 886.
[68] Xia, YD; Mokaya, R. *Adv. Mater.* 2004, 16, 1553.
[69] Yoon, SB; Kim, JY; Yu, JS. *Chem. Commun.* 2001, 559.
[70] Moriguchi, I; Koga, Y; Matsukura, R; Teraoka, Y; Kodama, M. *Chem. Commun.* 2002, 1844.
[71] Lu, A; Kiefer, A; Schmidt, W; Schüth, F. *Chem. Mater.* 2004, 16, 100.
[72] Fuertes, AB; Alvarez, S. *Carbon* 2004, 42, 3049.
[73] Lu, A; Li, W; Schmidt, W; Kieler, W; Schüth, F. *Carbon* 2004, 42, 2939.

[74] Vinu, A; Anandan, S; Anand, C; Srinivasu, P; Ariga, K; Mori, T. *Microp. Mesop. Mater.* 2008, 109, 398.
[75] Nowicki, P; Wachowska, H. *Wiadomości Chemiczne* 2008, 62, 9.
[76] Liang, C; Hong, KL; Gouichon, GA; Mays, JW.; Dai, S. *Angew. Chem. Int. Ed.* 2004, 43, 5785.
[77] Tanaka, S; Nishiyama, N; Egashira, Y; Ueyama, K; *Chem. Commun.* 2005, 2125.
[78] Liang, C; Dai, S. *J. Am. Chem. Soc.* 2006, 128, 5316.
[79] Meng, Y; Gu, D; Zhang, F; Shi, Y; Cheng, L; Feng, D; Wu, Z; Chen, Z; Wan, Y; Stein, A; Zhao, D. *Chem. Mater.* 2006, 18, 4447.
[80] Zhang, F; Meng, Y; Gu, D; Yan, Y; Chen, Z; Tu, B; Zhao, D. *Chem. Mater.* 2006, 18, 5279.
[81] Liu, R; Shi, Y; Wan, Y; Meng, Y; Zhang, F; Gu, D; Chen, Z; Tu, B; Zhao, D. *J. Am. Chem. Soc.* 2006, 128, 11652.
[82] Li, J; Gu, J; Li, H; Liang, Y; Hao, Y; Sun , X; Wang, L. *Microp. Mesop. Mater.* 2010, 128, 144.
[83] Yao, J; Li, L; Song, H; Liu, C; Chen, X. *Carbon* 2009, 47, 436.
[84] Sterk, L; Górka, J; Jaroniec, M. *Colloids Surf A* 2010, 362, 20.
[85] Fulvio, PF; Liang, C; Dai, S; Jaroniec, M. *Europ. J. Inorg. Chem.* 2009, 605.
[86] Wang, X; Dai, S. *Adsorption* 2009, 15, 138.
[87] Lei, Z; Xiao, Y; Dang, L; Bai, S; An, L. *Microp. Mesop. Mater.* 2008, 109, 109.
[88] Cao, JM; Cao, YL; Chang, X; Zheng, MB ; Liu, JS; Ji, HM . *Stud. Surf. Sci. Catal.* 2005, 156, 423.
[89] Jaroniec, M; Choma, J; Górka, J; Zawiślak, A. *Chem. Mater.* 2008, 20, 1069.
[90] Wang, L; Bo, X; Bai, J; Zhu, L; Guo, L. *Electroanal.* 2010, 22, 2536.
[91] Datta, KK; Subareddy, BV; Ariga, K; Vinu, A. *Angew. Chem. Int. Ed.* 2010, 49, 5961.
[92] Simakova, OA; Leino, A-R; Campo, B; Arvela, PM ; Kordas, K; Mikkola, J-P; Murzin, DY. *Catal. Today* 2010, 150, 32.
[93] Kim, M; Sohn, K; Na, HB; Hyeon, T. *Nano Lett.* 2002, 2, 1383.
[94] Górka, J; Jaroniec, M. *J. Phys. Chem. C* 2008, 112, 11657.
[95] Zhai, Y; Dou, Y; Liu, X; Tu, B; Zhao, D. *J. Mater. Chem.* 2009, 19, 3292.
[96] Liu, R; Ren, Y; Shi, Y; Zhang, F; Zhang, L; Tu, B; Zhao, DY. *Chem. Mat.* 2008, 20, 1140.
[97] Qian, X; Wan, Y; Wen, Y, Jia, N; Li, H; Zhao, DY. *J. Colloid Interface Sci.* 2008, 328, 367.
[98] Das, SK; Darmakolla, S; Bhattacharyya, AJ. *J Mater. Chem.* 2010, 20, 1600.
[99] Xiao, Y; Dang, L; An, L; Bai, S; Lei, Z. *Chinese J. Catal.* 2008, 29, 31.
[100] Zhou, J; He, J; Wang, T; Sun , D; Zhao, G; Chen, X. *J. Mater. Chem.* 2008, 18, 5776.
[101] Jaroniec, M; Górka, J; Choma, J; Zawiślak, A. *Carbon* 2009, 47, 3034.
[102] Florent, M; Tocci, M; Bandosz, J. *Carbon* 2013, 63, 283.
[103] Tian, Y; Wang, X; Pan, Y. *J Hazardous Mater* 2012, 213-214, 361.
[104] Wang, X; Liu, P; Tian, Y. *Microp. Mesop. Mater.* 2011, 142, 334.
[105] Wang, X; Liu, P; Tian, Y; Zhang, L. *Microp. Mesop. Mater.* 2011, 145, 98.
[106] Muhammad, M; Choong, TSY; Chuah, TG; Yunus, R; Yap, YHT. *Chem. Eng. J.* 2010, 164, 178.
[107] Muhammad, M; Khan, MA; Choong, TSY. *J. Chem.* 2013, http://dx.doi.org/10.1155/2013/235836.

[108] Yuan, X; Xing, W; Zhou, S-P; Han, Z; Wang, G; Gao, Z; Yan, Z-F. *Microp. Mesop. Mater.* 2009, 117, 678.
[109] Guo, Z; Zhu, G; Gao, B; Zhang, D; Tian, G; Chen, Y; Zhang, W; Qui, S. *Carbon* 2005, 43, 2344.
[110] Lei, Z; Cao, Y; Dang, L; Lo, A-Y; Yo, N; Liu, S. *J. Colloids Interface Sci.* 2009, 339, 439.
[111] Vinu, A; Miyahara, M; Hossain, KZ; Nakanishi, T; Ariga, S. *Stud. Surf. Sci. Catal.* 2005, 156, 637.
[112] Vinu, A; Hossain, KZ; Kumar, GS; Ariga, K. *Carbon* 2006, 44, 530.
[113] Chen, R-H; Zhong, G-Y; Wang, X-F. *Asian J. Chem.* 2013, 25, 5099.
[114] Yan, C; Wang, C; Yao, J; Zhang, L; Liu, X. *Colloids Surf A* 2009, 333, 115.
[115] Marczewski, AW. *Appl. Surf. Sci.* 2010, 256, 5145.
[116] Mohammadi, N; Khani, H; Gupta, VK; Amereh, E; Agarwal, S. *J. Colloid Interface Sci.* 2011, 362, 457.
[117] Dong, Y; Lin, H; Qu, F. *Chem. Eng. J.* 2013, 193-194, 169.
[118] Zhang, P; An, Q; Guo, J; Wang, C-C . *J. Colloid Interface Sci.* 2013, 389, 10.
[119] Zhuang, X; Wan, Y; Feng, C; Shen, Y; Zhao, D. *Chem. Mat.* 2009, 21, 706.

In: Comprehensive Guide for Mesoporous Materials. Volume 1 ISBN: 978-1-63463-958-3
Editor: Mahmood Aliofkhazraei

Chapter 6

PROPENE SYNTHESIS FROM ETHENE OR ETHANOL: CATALYSIS AND CHARACTERIZATION OF NICKEL ION-LOADED MESOPOROUS SILICA

Masakazu Iwamoto* and Masashi Tanaka
Research and Development Initiative, Chuo University,
Kasuga, Bunkyo-ku, Tokyo, Japan

ABSTRACT

Nickel ion-loaded mesoporous silica (Ni-M41) was found to show good activity for one-step conversion of ethene or ethanol to propene by us. The conversion of ethene was 68% and the propene selectivity was 48% in a continuous gas-flow reaction at 673 K. The following reaction scheme was suggested based on the partial pressure dependences and the reactions of the intermediate compounds : dimerization of ethene to 1-butene, isomerization of 1-butene to 2-butene, and metathesis of 2-butene and ethene to propene. Various physicochemical techniques were applied to study the nickel species loaded on M41 and elucidated the formation of 2:1 nickel phyllosilicate-like species and the existence of three-coordinated Ni^{2+} ions on five-membered Si-O rings , which were suggested as active site s for the dimerization reaction of ethene. The conversion of ethanol on Ni-M41 was finally introduced and the reaction mechanism s were discussed. Two routes from ethanol to propene were suggested; the dehydration route and the dehydrogenation route. All of the results indicate that the formation of propene from ethene or ethanol could be achieved without using the shape selectivity well known in zeolite catalysis .

1. INTRODUCTION

New types of catalytic technologies for the production of propene ($C_3^=$) are desired to realize reduction of the emission of carbon dioxide . In addition, although the mainstay of

* Phone: +81-3-3817-1612, Fax: +81-3-3817-7371, E-mail: iwamotom@tamacc.chuo-u.ac.jp.

petrochemical industries in the world is still ethene ($C_2^=$), the need for $C_3^=$ rapidly increases due to the increasing demand of polypropene, propene oxide, etc. [1]. This trend also increases the necessity of environmentally-benign production of $C_3^=$. Three kinds of measures are applied or suggested as summarized in Table 1. First is the so-called mild-cracking or the catalytic cracking of naphtha. The increment of $C_3^=$, however, is limited due to narrow range of applicable reaction conditions for the mild-cracking. The catalytic cracking has been studied in the world but the deactivation of the catalysts is still the severe problem [2].

Table 1. Catalytic technologies for the production of propene

Reaction	Technology	Catalyst	Reference
Cracking of naphtha	Mild cracking		
	Catalytic cracking	Zeolites	2
Conversion of ethene to propene	Metathesis of ethene and butenes	WO_3/Al_2O_3	3
	Conversion of ethene to propene	$Mo(CO)_6$-Al_2O_3	4
		WO_3-TiO_2	5
		Zeolites	6
		Ni-MCM-41	19
		$W(H)_3/Al_2O_3$	20
Conversion of bio-ethanol to propene	Dehydration	Zeolites	6b, 7-9
	Dehydrogenation	Ni-MCM-41	21
	Ketonization of acetic acid	Sc/In_2O_3	23
		Y/CeO_2	24

Second is the conversion of $C_2^=$ to $C_3^=$ where the source of $C_2^=$ would be the natural gas including the shale gas. One of answers is metathesis of $C_2^=$ and butenes ($C_4^=$), for instance, the ABB Lummus process [3]. Its disadvantage is necessity of equimolar amounts of $C_2^=$ and $C_4^=$. Another solution is direct conversion of $C_2^=$ to $C_3^=$ (ETP) without any addition of other hydrocarbons . This would be the most desirable but no good catalyst for the reaction has been found. Supported molybdenum [4] and tungsten oxide [5] have been reported as the catalysts, but their activity was so low as to be observed only in a closed recirculation system. On the other hand, various zeolites have been employed as the catalysts for this reaction [6]. The reaction includes oligomerization and polymerization of lower olefins , subsequent decomposition to yield $C_3^=$ or the others on strong acid sites of zeolites, and selective evolution of $C_3^=$ due to the shape -selectivity of zeolite pore windows . This process has the limitations of selectivity due to the shape selectivity and the disadvantage of the life time owing to the coke formation . The objective of this chapter is the selective formation of $C_3^=$ without the shape-selectivity.

The third idea is the conversion of bio-ethanol (bEtOH) to $C_3^=$. The use of bEtOH as an additive for automobile fuels has increased rapidly all over the world. This is one way of using renewable resources to suppress carbon dioxide emissions, while another challenge is the conversion of bEtOH to various olefins and their use for production of chemicals and polymers [1, 6b, 7-11]. The latter would be very significant for the long-term fixation of carbon dioxide. Many efforts have therefore been devoted to the development of systems for converting bEtOH to $C_2^=$ and other lower olefins. In particular conversion to $C_3^=$ is desirable due to the greater demand for $C_3^=$ derivatives as has been already introduced [1].

The catalytic activity of Ni ion for the dimerization or oligomerization of olefins was found 60 years ago and widely studied [12]. In the case of heterogeneous catalysis , Ozaki et al. [13] reported high catalytic activity of Ni/SiO_2 for the dimerization, though severe deactivation during the reaction prevented it from being applied in the practical process. They also found that acidic supports to be effective for enhancement of the catalytic activity of nickel . Similar catalytic activity was also confirmed on various Ni-zeolites [14] or on Ni-supported MCM-41 [15] in a closed recirculation system. During our study on the catalytic activity of silica MCM-41 [16-18], we tried the dimerization reaction of $C_2^=$ to $C_4^=$ on Ni ion-loaded mesoporous silica (Ni-MCM-41, abbreviated as Ni-M41). The study revealed that the produced $C_4^=$ reacted with remaining $C_2^=$ to yield $C_3^=$. This is the first result for the finding that Ni-M41 was active and had long life-time for the ETP reaction in the gas-phase flow reaction [19]. Around the same time, Basset et al. [20] independently reported the catalytic activity of tungsten hydride-supported alumina for the same reaction although the conversion of $C_2^=$ was low.

Catalytic conversions of EtOH on zeolites [6b, 7-9] and metal oxides [10, 11] have widely been studied. On zeolites, the activity and selectivity in many studies reported so far were insufficient. The major weakness is catalyst deactivation [6b, 7-9]. EtOH can also react on metal oxide surfaces, to give various chemicals . Acid sites are recognized to lead to dehydration of EtOH, giving $C_2^=$, while basic sites lead to dehydrogenation to yield acetaldehyde (AcH) [10, 11]. As a result, many kinds of products, for example aldehydes , ketones , $C_2^=$, and $C_4^=$, were observed on oxide catalysts. In this catalysis $C_4^=$ and other higher olefins were produced by oligomerization of $C_2^=$, but as far as we are aware significant $C_3^=$ production on oxide catalysts has not been reported. The results on the ETP reaction on Ni-M41 led us to apply the catalyst for the conversion of EtOH to $C_3^=$ since M41 is active for the dehydration of EtOH to yield $C_2^=$ [21a-c]. This was first confirmed by us [21] and subsequently by Sugiyama et al. [22]. The pore diameters of M41 are usually 1.5-5.0 nm, and therefore the product distribution on the catalysts is not controlled by the shape selectivity. The reaction mechanism /pathways are of interest, and would be suggested in this chapter.

2. Conversion of Ethene to Propene on Ni-M41 Catalysts

2.1. Catalytic Activity

The ETP reaction was examined on Ni-M41 as a function of reaction temperature , in which M41 was prepared in the reported procedure [25] and nickel ion was loaded by the template ion exchange (TIE) method [26]. The dimerization of $C_2^=$ to $C_4^=$ mainly proceeded at 573 K. When 0.5 g of Ni(15)-M41 (the numerical value in parentheses represents the Si/Ni ratio) was used, the degrees of conversion of $C_2^=$ and selectivity to $C_4^=$ were 43 and 93%, respectively. The production ratio of 1-, trans-2-, and cis-2-butene was 0.5:1.0:0.3. At 673-723 K the major products were $C_3^=$ and $C_4^=$. Hexenes, the product of $C_2^=$ trimerization, were observed at the wide temperature range but the yields were always less than 5%. No $C_4^=$ was produced on M41 alone, indicating the necessity of nickel ion for the reaction. Small change in the catalytic activity was observed for the $C_3^=$ formation in the initial stage at 673 K, but the activity became stable within 2 h and no deactivation was found during the 10 h experiment. The XRD patterns and the surface area s of Ni-M41 remained unchanged after the catalytic runs. Thus the stability of Ni-M41 under the present reaction conditions could be confirmed.

The correlations between the product distribution and the reaction conditions were investigated. In the range P_{C2H4}=10-50%, the conversion levels of $C_2^=$ and to $C_3^=$ and $C_4^=$ increased monotonously with increasing P_{C2H4}. At P_{C2H4}=49.7%, the respective conversions to C_3, C_4, and C_6 olefins were 33, 29, and 6% on 0.3 g of Ni(20)-M41. The carbon balance was 99.8%, which indicates almost no production of "unknown products". The degree of conversion to $C_3^=$, 33%, appears to rather small but it should be noted that the concentration of unreacted $C_2^=$ was about 34% under the present conditions and the ratio of $C_3^=/C_2^=$=33/34 in carbon basis would be sufficiently great.

Figure 1 shows the change in product distribution as a function of the weight of Ni-M41 employed, i.e., the contact time dependence. Clearly, longer contact time resulted in greater conversion of $C_2^=$ and better selectivity of $C_3^=$, while the selectivity of $C_4^=$ decreased and that of hexenes was almost constant. Propene is indeed the secondary product in the consecutive reaction of $C_2^=$ on Ni-M41. At 0.5 g of Ni(43)-M41, the degrees of $C_2^=$ conversion and $C_3^=$ selectivity were 55 and 54%, respectively. At last it should be noted that catalytic activity of nickel ion loaded on mesoporous silica for the ETP reaction was recently confirmed by several groups [27], though the deactivation of the catalysis was observed.

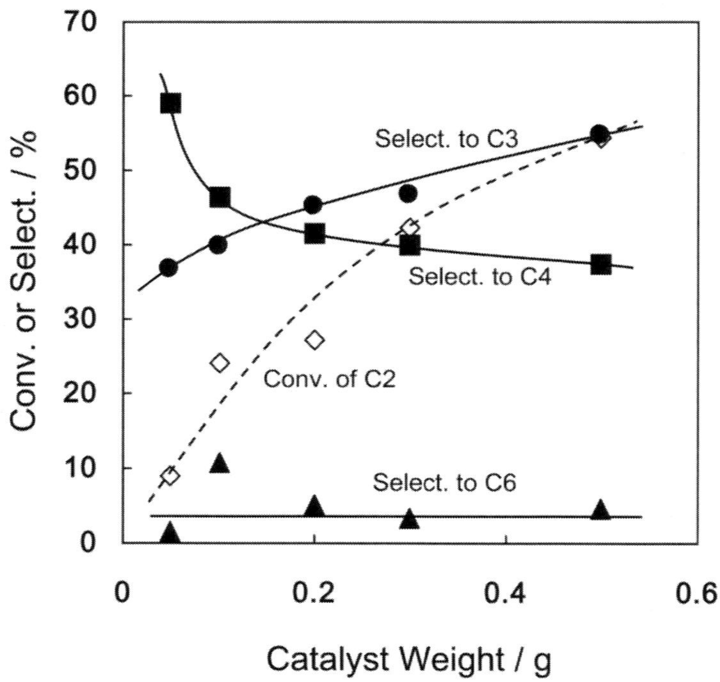

Figure 1. Changes in ethene conversion and product distribution at 673 K with weight of Ni(43)-M41. The codes of C2-C6 mean ethene, propene , butenes, and hexenes.

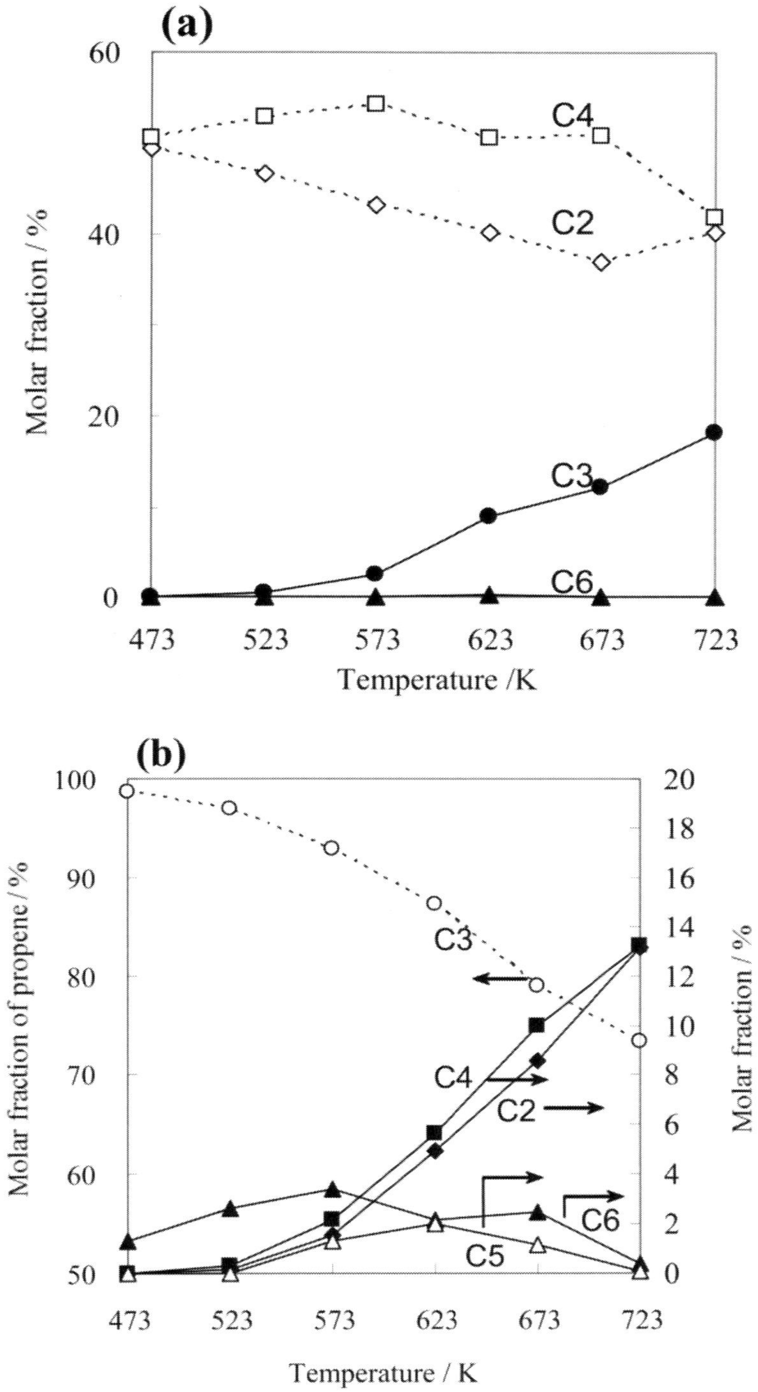

Figure 2. Metathesis reaction of ethene and 1-butene (a, $P_{C2H4}=P_{C4H8}=5\%$) or propene (b, $P_{C3H6}=10\%$) on 0.3 g of Ni(15)-M41. The codes of C2-C6 mean ethene, propene, butenes, pentenes, and hexenes. In Figure 2b, the left vertical axis is the amount of propene and the right those of the products.

2.2. Reaction Pathways

The reaction of $C_2^=$ and 1-butene was studied to clarify the mechanism of the $C_3^=$ formation . As shown in Figure 2a, one can recognize the selective formation of $C_3^=$ on Ni-M41 at 623 K and higher temperature . The increment in the amount of $C_4^=$ at 523-573 K is due to the dimerization of $C_2^=$. The selective production of $C_3^=$ would indicate the progress of metathesis reaction on this catalyst . To confirm the reaction pathway in more detail, we examined two kinds of reactions . The first was the reaction of 1-hexene. When 1-hexene was introduced onto the Ni-M41 catalyst, methane , $C_2^=$, $C_4^=$, and pentenes besides $C_3^=$ were produced, indicating the random scission of carbon -carbon bonds of 1-hexene. This indicates little possibility that $C_2^=$ and 1-butene first afford hexenes and the resulting hexenes homolytically decompose to give $C_3^=$ selectively. The second reaction examined was the retro-metathesis reaction. Namely the reaction of $C_3^=$ was investigated on Ni-M41 and really proceeded to yield equimolar $C_2^=$ and $C_4^=$ as shown in Figure 2b. The amounts of by-products were always small. It was further confirmed in separate experiments that parent M41 was not active for the reaction of $C_2^=$ and $C_4^=$. All of the results therefore strongly suggest the metathesis reaction on Ni-M41.

Although at the present we cannot preclude the possibility of decomposition mechanism of higher olefins because the other types of reaction mechanism s have been suggested on Cr [28] or Zr [29], we believe that the metathesis mechanism is the most plausible for the $C_3^=$ formation on Ni-M41. At first two $C_2^=$ molecules dimerize to give 1-$C_4^=$ on Ni (Reaction 1 in Figure 3), resulting 1-$C_4^=$ then isomerizes to 2-$C_4^=$ on acid sites of M41 (Reaction 2), and finally the metathesis of produced 2-$C_4^=$ with unreacted $C_2^=$ proceeds to form $C_3^=$ on Ni (Reaction 3). The acidic property of silica M41 has already been reported by us [16, 17] and the other groups [15, 18] and the isomerization of 1-$C_4^=$ to 2-$C_4^=$, a typical acid-catalyzed reaction, was indeed confirmed on silica M41 [15].

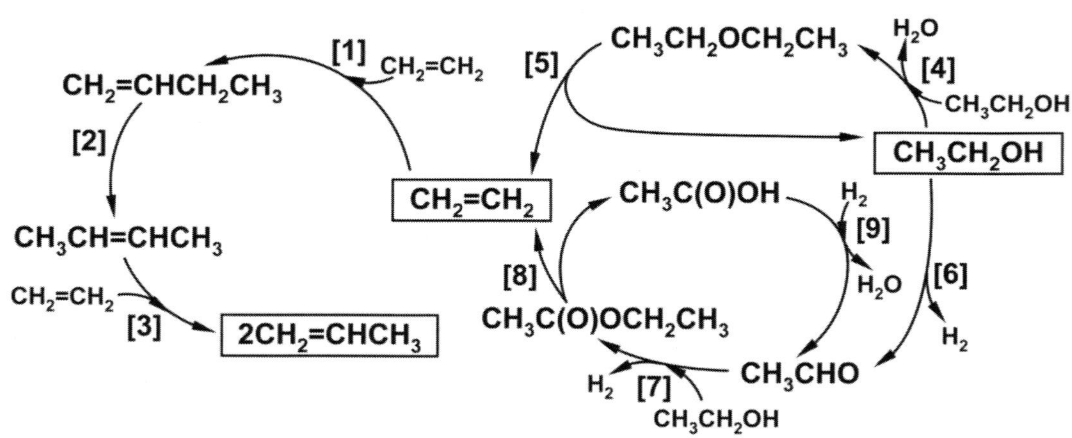

Figure 3. Proposed reaction mechanism for the conversion of ethene or ethanol to propene on nickel ion-loaded MCM-41.

As has been summarized by Grubbs [30] and Arpe [1], the metathesis reaction is one of the important organic reactions . It is well known that the catalytically active species for the reaction are confined to Mo, W, Ru, and Re etc. irrespective of the world-wide study. The results in this section indicated that nickel ion-loaded mesoporous silica is active for the metathesis of $C_4^=$ and $C_2^=$ to yield $C_3^=$ at 573-723 K in the gas-phase flow reaction. Mori et al. [31] suggested the possibility of metathesis on a Ni(0) complex, while Baker et al. [32] concluded no progress of a metathesis reaction on Ni complexes. At the moment, no report claims nickel ion as an active species for the metathesis. It would be worth to note that the surface density of Ni is approximately 0.5 Ni/nm^2 in the case of Ni(20)-M41 on the assumption of the even distribution of nickel on the surface, suggesting the high dispersion of the nickel ions loaded by the TIE method. The valences of nickel ion in the mesoporous silica were not studied here. There are two possibilities for the redox cycles of nickel species, Ni(I)-Ni(III) and Ni(0)-Ni(II). The TPR experiments [19b] indicated the difficult reduction of nickel species loaded on M41 to Ni(0), which would be one important factor for generation of the catalytic activity . Therefore we speculated that the Ni(I)-Ni(III) system would be the possible redox cycle for the metathesis reaction. The stability of Ni(I) in the zeolites [33a] and mesoporous materials [33b] might support the speculation that Ni(I) is an active center and a Ni(III) carbene is produced as intermediates.

2.3. Addition Effect of Aluminum Ions on Ni-M41

The catalytic activity of various metal ion -loaded MCM-41s for this reaction was examined. The conversion levels on $C_2^=$ on Al(Si/Al=22)-, Ti(30)-, V(22)-, Cr(43)-, Mn(20)-, Fe(25)-, Co(16)-, Cu(37)-, Zn(28)-, Zr(23)-, Mo(30)-, or W(30)-M41 were all 5% or less at 673 K, and most of the products were "unknown products". Ru or Re loaded MCM-41, prepared by the conventional impregnation method, showed no activity for the reaction. This would be due to the lack of activity of Ru or Re for the dimerization of $C_2^=$. Clearly only Ni-M41 shows the unique activity for the ETP reaction at 673 K in the gas-phase.

Addition effect of various metal ion s onto Ni-M41 was then examined [34]. The metal ions were loaded by the TIE method using not-calcined Ni-M41 (Si/Ni=ca. 28) and the Si/M (M: the second metal ion, 12 kinds) ratios were 19-183 in which the amount loaded strongly depended on the kinds of metal ions. Unfortunately the addition of the second metal ions did not result in improvement of the $C_3^=$ yields. The catalytic activity of aluminum ion-loaded Ni-M41 is shown in Table 2 as a typical example. The $C_3^=$ selectivity was the same as that of Ni-M41 but the $C_2^=$ conversion degree somewhat decreased. At the present, no additive was found to improve the catalytic activity of Ni-M41.

Next, the effect of pre-introduction of aluminum ions into the silica wall of M41 was studied based on the following two possibilities: the change in the properties of the nickel ion due to the some interactions with an aluminum ion [15] and the promotion of the isomerization of butenes on a newly introduced Brønsted acid site [15, 35, 36]. Ni-M41 samples with the Si/Al ratios of 1212 (the parent), 531, and 175 (Ni-AlM41) were prepared using tetraethyl orthosilicate (TEOS) and sodium aluminate, where the amounts of nickel loaded on the AlM41 were the Si/Ni ratios of 20, 21, and 26, respectively. The catalytic activity for the ETP reaction is summarized in Table 2 [34]. Please note that the catalytic activity of Ni-AlM41 was measured without the water addition because the pore structure of Ni-AlM41 prepared from TEOS was easily destructed by the hydrothermal reaction, indicating that the comparison of the absolute values between the upper and the lower rows is meaningless. The data of the lower row of Table 2 showed the increase in the $C_2^=$ conversion and the $C_3^=$ selectivity with the increasing amount of aluminum in M41. The similar change in the $C_3^=$ selectivity were reported [27b], though the $C_3^=$ selectivity reported was very low. Aluminum ions loaded appropriately in the silica network might lead the interaction of Ni-O-Al, which could work as an active site for the ETP reaction. The results indicated the strong dependency of the catalytic activity on the supporting method, which is the problem to be solved in the near future.

Table 2. Addition effect of aluminum ions on the catalytic activity of Ni-M41 for the ETP reaction at 673 K

Additive	Si/Al ratio	Si/Ni ratio	Conversion/% C2	Selectivity/% C3	C4	C5	C6
Ni-M41 in which Al ions were added as second metal ions onto Ni-M41							
none[a,b]	191	23	37.1	43.5	38.4	7.4	4.1
Al[a,b,c]	26	26	33.0	43.4	38.6	6.7	1.5
Ni-M41 which possessed Al ions pre-introduced in the silica wall							
none[d,e]	1212	20	2.4	11.7	49.8	0.0	0.0
Al[d,f,g]	531	21	7.6	48.8	47.1	3.9	0.0
Al[d,f,g]	175	26	21.7	50.4	41.9	6.3	1.6

The codes of C2-C6 mean ethene , propene , butenes, pentenes, and hexenes. [a] M41 was prepared using colloidal silica as silicon source. [b] Catalyst wt. 0.3 g, total flow rate 11 ml min^{-1}, P_{C2H4} 9.3%, P_{H2O} 1.4% (N_2 balance). The yields of products obtained after 200 min were stated. [c] Aluminum ions were loaded by the TIE method on not-calcined Ni-M41. [d] M41 was prepared using TEOS as silicon source. [e] Catalyst wt. 0.3 g, total flow rate 11 ml min^{-1}, P_{C2H4} 9.3%, P_{H2O} 0% (N_2 balance). The yields of products obtained after 120 min were stated. [f] The aluminum ions were added into the silicon source in the preparation of M41. [g] Catalyst wt. 0.1 g, total flow rate 11 ml min^{-1}, P_{C2H4} 9.3%, P_{H2O} 0% (N_2 balance). The yields of products obtained after 120 min were stated.

3. Characterization of Silica Wall and Nickel Species of Ni-M41

3.1. Characterization of Ni-M41 Prepared from Colloidal Silica

Ni-loaded M41 and SiO_2 samples were prepared by the TIE and the conventional impregnation (IMP) methods, which were named as Ni-M41, Ni/M41, and Ni/SiO_2. The following results and discussion will be described on the premise of no essential difference in the pore structures among the M41 samples employed here. XRD measurements confirmed no nickel -related crystalline phases on the TIE-catalysts, but showed formation of NiO particles on the IMP catalysts. The activity of the IMP catalysts (Ni/M41 and /SiO_2) was compared with that of the TIE catalyst (Ni-M41) [19b]. In contrast to the high and stable activity of Ni-M41, the activity of Ni/M41 or Ni/SiO_2 was very low and decreased with the reaction time .

The Ni K-edge XAFS spectra were measured and compared with the literature [37-43]. The distance (2.08 Å) and the coordination number (6.9) of the first shells (oxygen backscatterer) on Ni-M41 indicated the presence of hexacoordinated Ni^{2+} ions [38] in the TIE sample. The distances of the second shell (Ni and Si backscatterer) of Ni-M41, 3.05 and

3.36 Å, were longer than those of NiO and Ni/M41, and similar to those of layered nickel silicates. In contrast, the Ni/M41 and Ni/SiO_2 catalysts gave the spectra assignable to NiO species . It follows that the preparation methods have the essential effect for the appearance of the catalytic activity through the change in loading states of nickel ion on the supports.

The detailed properties of layered nickel silicates on M41 were further characterized by the IR measurement [44] and the TPR (temperature programmed reduction) technique [19b]. The IR spectra of Ni-M41 exhibited the bands at 3626, 710, and 667 cm^{-1} assignable to 2:1 nickel phyllosilicate [45] which consists of a NiO_6 layer sandwiched by two silica layers [39]. The same IR spectra were reported by Lehmann et al. [46]. Based on the many TPR experiments reported so far [42, 43, 47, 48], we could summarize the TPR peak regions of nickel on silica as follows: Ni-oxide, the cationic form of nickel on the silica surface, and the nickel ion forming some surface composite compounds could be reduced at approximately 600–800, 800–900, and 900–1000 K, respectively. The reduction temperature of Ni-M41, 931 K, clearly fell within the region of the reduction of composite compounds. Later, the SEM-EDS analysis of the calcined sample revealed the uniform loading of the nickel species on the pore wall of Ni-M41 [49]. The results further supported the above conclusion that the nickel ion in the TIE sample formed the 2:1 nickel phyllosilicate-like structure on the surface.

The surface layer ed nickel silicate is reported to be produced by loading of nickel ion as ammine-complexes onto silica in basic aqueous solution , an equilibrium adsorption (EA) method of $[Ni(NH_3)_x]^{2+}$ [38, 40, 45, 48]. We have applied the method reported by Hadjiivanov et al. [48] to prepare the samples containing the surface layered nickel silicate to evaluate its role for the catalysis . The EXAFS spectra of the EA catalysts (named as Ni-NH_3/M41 and Ni-NH_3/SiO_2) were very similar to those of layered nickel silicate, indicating the formation of layered nickel silicate on silica surface by the EA method. The catalytic activity of the Ni-NH_3/M41 and Ni-NH_3/SiO_2 samples was lower than that of the TIE sample, but much greater than those of the IMP samples. In our report [19b] we wrote that more detailed investigation of the EA catalysts possibly leads to raising their catalytic activity to the same levels as that of the TIE catalyst . This estimation was indeed realized partly by Lehmann et al. [27a]. All results presented here showed that the TIE method is the most effective to prepare the active nickel ion-loaded catalysts and the high catalytic activity would result from the effective formation of surface layered nickel silicate-like structure .

3.2. Detailed Characterization of Nickel Species and Silica Wall

The reaction pathway of the present ETP reaction was composed of three steps. The first step, the ethene dimerization , was here investigated as a function of the pore size and the amount of nickel . The parent M41 was prepared using a fumed silica and sodium silicate solution . These two managements were to avoid the complications resulting from the consecutive reaction and the Si-O-Al sites. The surfactants $C_nH_{2n+1}N(CH_3)_3Br$ (n=10, 14, 18) were used as templates and the samples obtained were abbreviated as Ni-M41-10, -14, and -18, respectively. The pore diameters were approximately 1.7, 2.5, and 3.5 nm. The changes in the ethene partial pressure in a conventional volumetric apparatus were measured at 298 K as a function of reaction time [44], and the reaction rate constants obtained were plotted in

Figure 4a. Ni-M41 having smaller pores and higher Si/Ni ratio showed greater catalytic activity per nickel.

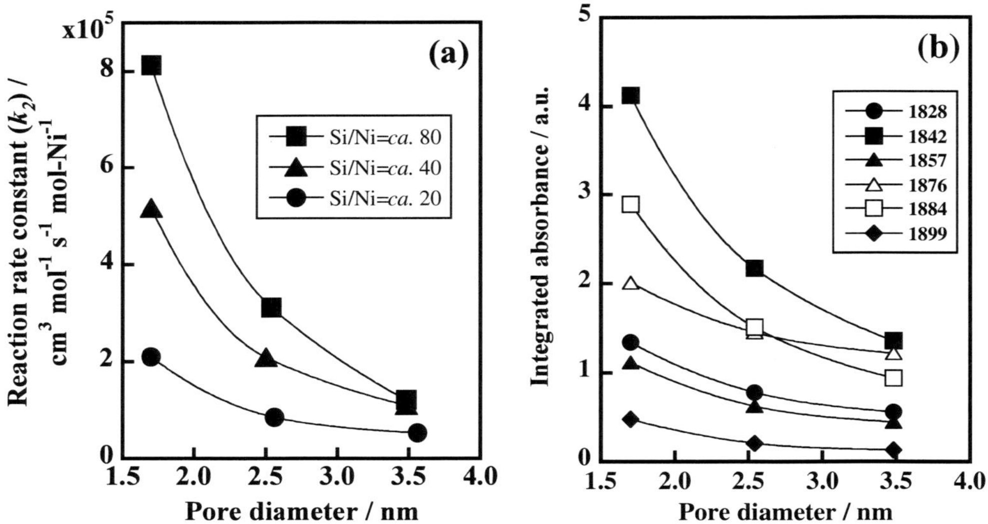

Figure 4. (a) Dependence of the catalytic activity of Ni-M41 for ethene dimerization on the pore size and the Si/Ni ratio. The reaction rate constants were calculated based on ethene pressure changes as a function of the reaction time . (b) Correlation between peak areas of the IR absorption bands of NO species and the pore diameter of Ni-M41 with a Si/Ni ratio of 73-85. The numerical values in Figure (b) correspond to the wave number s (cm^{-1}) of the respective absorption bands.

In the IR spectra of Ni-M41 a clear band appeared at 570 cm^{-1}, which was attributable to five-membered rings consisting of Si-O [50]. Although theoretical calculations suggested that there might be four- or five-membered rings in the framework of M41 [51] besides the six-membered rings [52], only a few experimental results supported the existence of such small rings in mesoporous silica [50b, 53]. Interestingly, the intensity of the 570 cm^{-1} band decreased with loading of nickel [44], suggesting that the five-membered rings interacted with nickel ions or provided ion-exchange sites for nickel ions.

The Ni K-edge XAFS spectra of Ni-M41 with different pore size s were measured. The coordination numbers of silicon on Ni(73)-M41-10, Ni(78)-M41-14, and Ni(85)-M41-18 were 2.8, 3.2, and 3.1, respectively. When nickel ion is loaded on the five-membered ring, the number of silicon atoms adjacent to nickel ion is three as referred in report [44]. The number on the six-membered ring is four. The above data and the others suggested easier formation of the five-membered ring on the smaller pore sample and preferential loading of nickel ion on it. This was further supported by the observations of UV -Vis-NIR DR spectra assignable to Ni^{2+} ions with distorted octahedral, distorted tetrahedral, and trigonal structures [44]. It was suggested that trigonal Ni^{2+} ions were located on the five-membered rings and distorted octahedral and tetrahedral Ni^{2+} ions on the six-membered rings.

Ni^{+} ions on silica and M41 were reported to be active for ethene dimerization [15, 54], but the nickel species loaded on M41 by the TIE method could only be reduced above 773 K [19b]. High stability of Ni^{2+} on Ni-M41 was further confirmed by ESR measurements [44]. Thus the nickel ions active for the ethene dimerization on Ni-M41 should be not Ni^{+} but other nickel species. The introduction of ethene on Ni(73)-M41-10 exhibited the decrement of the intensities of UV -Vis-NIR DR bands due to Ni^{2+} ions with a trigonal or distorted tetrahedral coordination environment that is coordinatively unsaturated. In addition, an isolated Ni^{2+} ion on zeolite was suggested as an active site for the ethene dimerization [55]. Either or both of the three- and four-coordinated Ni^{2+} ions would act as adsorption sites for ethene.

To investigate the active site s in more detail, NO was employed as a probe molecule. The FT-IR spectra of NO adsorbed on Ni-M41 evacuated at 673 K could be deconvoluted to six components at 1899, 1884, 1876, 1857, 1842, and 1828 cm^{-1}. The 1884 and 1842 cm^{-1} bands were attributed to the symmetric and asymmetric stretching vibrations of dinitrosyl species on Ni^{2+}/SiO_2 [56]. The other bands were all assigned to mononitrosyl species. The bands at 1876 and 1857 cm^{-1} were attributed to two types of NO adsorbates on Ni^{2+}/SiO_2 [56] and those at 1899 and 1828 cm^{-1} to those on Ni^{2+} lying at zeolite -like structure and on NiO, respectively [57]. The amounts of di- and mono-nitrosyl species increased as the pore size decreased as shown in Figure 4b or as the Si/Ni ratio increased as shown in the report [44]. The comparison of the results with the catalytic activity (Figure 4a) strongly suggested that the changes in the amounts of dinitrosyl species on Ni^{2+}/SiO_2 closely followed the trend in the catalytic activity. The dinitrosyl species would be formed on the three-coordinated Ni^{2+} ions , because the ions were coordinatively unsaturated and were approached by two molecules . We propose that the three-coordinated Ni^{2+} ions situated on the five-membered Si-O rings are the active sites for ethene dimerization .

3.3. Hypothetical Structure of Nickel Ions on Mesoporous Silica

The following two problems will be discussed here: why the five-membered Si-O rings were preferentially produced in smaller pore M41s and how it was formed on the layered nickel silicate structure . The surface structure of the 2:1 nickel phyllosilicate is shown in Figure 5a, which is composed of a NiO_6 layer and two silica layers of six-membered Si-O rings [39]. The six-membered Si-O rings in the first layer are located slightly diagonally on those in the second layer. Since the layer of nickel silicate prepared by the EA method was looked like to be developed parallel to the surface [40], the direction of the nickel silicate layers in Ni-M41 was assumed to be along the pore surface. The surface structure of Ni-M41 with the pore diameter of 1.8 nm was estimated here. The distance between the silica layers is 0.4 nm (Figure 5a) and therefore the diameter of the second layer in the M41 structure should be 2.6 nm as shown in Figure 5b. In the estimation of the surface structure of Ni-M41, preservation of the structure of six-membered rings in the second layer was postulated. The first layer without any distortion or shrinkage could not be put on the second layer because of its smaller diameter than that of the second one. The best fit of the first layer was achieved when the first layer was composed of the sequence of the five-membered Si-O rings. The ring size of a typical six-membered ring is 0.525 nm and that of a five-membered one 0.357 nm. The length of the circumference of the second layer is 8.16 nm, which can be covered by 15.6 six-membered rings. On the other hand, the length of the first layer is 5.65 nm and is covered by 15.8 five-membered rings. The good agreement of the numbers of the building units suggested unstrained overlapping the five-membered rings of the first layer to the six-membered rings of the second layer.

The proposed structure is summarized in Figure 5b. The 2:1 nickel phyllosilicate structure was rolled up, in which the six-membered Si-O rings in the second layer was somewhat distorted but the fundamental structure was approximately maintained. On the other hand the building unit of the first layer was replaced with the five-membered one. Each ring in the first layer was overlapped on the ring in the second layer in a similar manner to the original layered structure (Figure 5a). This model implies that the larger pore size would result in the increase in the rate of a six-membered ring in the first layer, coinciding with the experimental finding. The five-membered rings would be efficient for the formation three-coordinated Ni ions as suggested in the report [44]. We know that more experimental evidences should be obtained to confirm reliability of the suggested model.

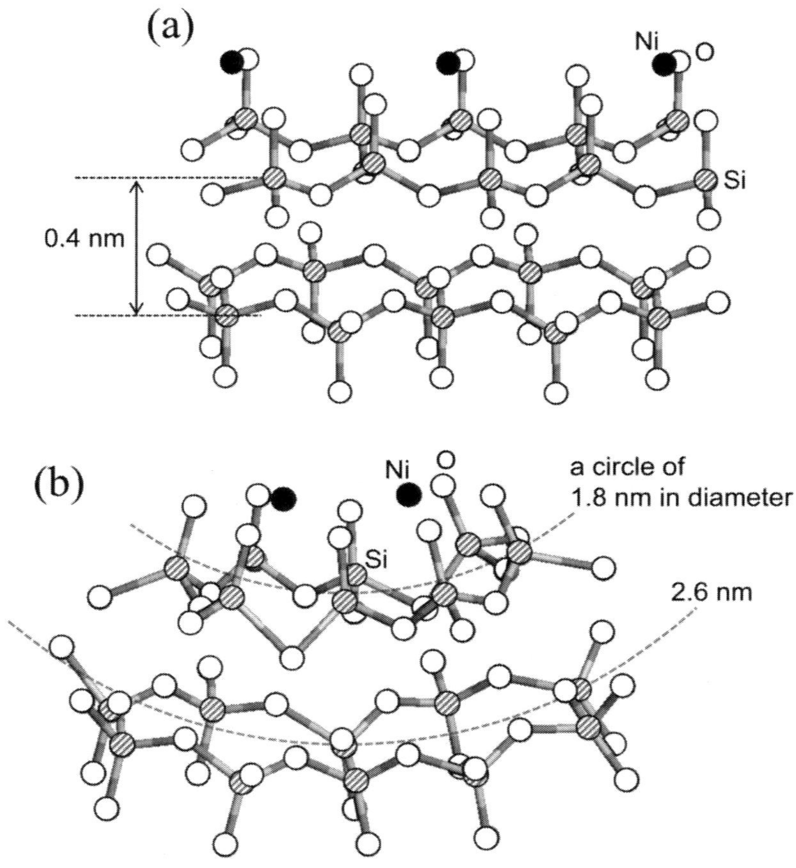

Figure 5. (a) Oblique view of the layers of 2:1 nickel phyllosilicate, in which nickel is located on the top and the hydrogen atoms in OH groups are omitted [39]. (b) Oblique view of rolled layers of 2:1 nickel phyllosilicate consisting of five-membered (first layer) and six-membered rings (second layer).

4. Reaction of Ethanol on Ni-M41

4.1. Catalytic Activity

The temperature dependence of EtOH conversion over Ni-M41 is shown in Figure 6. Diethyl ether (DEE) was mainly obtained at around 523 K. DEE was an intermediate compound in the dehydration , decomposing to yield EtOH and $C_2^=$ at higher temperatures [21c]. The $C_2^=$ yield sharply increased at 573 K, and reached approximately 70% at 623 K or above. The $C_4^=$ yield reached a maximum at 623 K, while maxima in $C_3^=$ yield occurred at 673 and 723 K. Notably, AcH was formed at 573-723 K, although not in large amounts, which will be discussed later. The stability of Ni-M41 was examined at 673 K. The catalytic activity did not change during 20 h of continuous time on stream, and the carbon -based mass balances were always approximately 100%, within the experimental errors [21]. It should be noted, however, that the stable catalytic activity was limited only at the low concentration of ethanol . For example, gradual decrease in the conversion to $C_3^=$ during the reaction was observed in the experiments at the partial pressure of EtOH of 30%. The deposition of carbon should be weakened under the high partial pressure of EtOH.

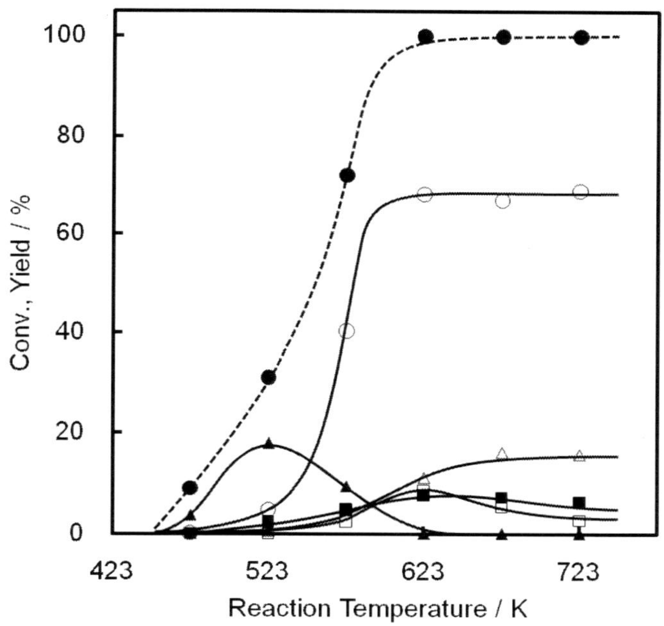

Figure 6. Reaction temperature dependence of conversion of EtOH on a Ni(23)-M41 catalyst . Catalyst wt. 0.2 g, total flow rate 10 ml min^{-1}, P_{EtOH} 5.6 kPa (N_2 balance). Conversion of EtOH (closed circle), yield of $C_2^=$ (open circle), $C_3^=$ (open triangle), $C_4^=$ (open square), DEE (closed triangle), and AcH (closed square).

The yields of $C_2^=$, $C_3^=$, $C_4^=$, and AcH in the experiments of Figure 6 were 67, 16, 5, and 7%, respectively, at 673 K. The distribution should be compared with those of the reaction of $C_2^=$ on the same catalyst . At 673 K and P_{C2H4}=10 vol %, the $C_2^=$ conversion and selectivity to $C_3^=$ and $C_4^=$ were 42, 47, and 40%, respectively [21]. Clearly, the product distribution for the EtOH reaction was different from that of the $C_2^=$ reaction. This difference might result from a change in active site s through adsorption of EtOH, AcH, or intermediates, or from a difference in reaction mechanism . The dependence on the partial pressure of EtOH was also studied at P_{EtOH}=3.0-13.2 kPa and a space velocity (SV) of 1,000 h^{-1}. Little dependence of $C_3^=$ formation on P_{EtOH} was observed on the present conditions, probably indicating strong adsorption of EtOH on the active sites.

A similar reaction of EtOH on proton-exchanged ZSM-5 zeolites was reported to be retarded in the presence of water vapor [7-9]. The effect of water addition on the catalytic activity of Ni-M41 was studied here. When EtOH/water ratios were varied in the range 100/0-75/25 (w/w), the EtOH conversion levels and product distribution changed only a little. This result is very significant for the application of the present system to bEtOH conversion, because coarsely distilled bEtOH usually contains 5-10 vol % water.

The product distribution was studied as a function of space velocity at 673 K. The dependence is summarized in Figure 7. At SV=70,000 h^{-1} the conversion level of EtOH was approximately 50%, but at 20,000 h^{-1} or below it increased to 95% or more. It follows that Ni-M41 is very active for the catalytic conversion of EtOH. The product distribution depended strongly on the space velocity. At higher SVs (in the region 10,000-100,000 h^{-1}) AcH was produced in large amounts, and the amount decreases monotonically with decreasing SV. DEE was also produced and its formation showed a behavior similar to that of AcH, although the amount formed was very small. On the other hand the yields of $C_3^=$ and $C_4^=$ gradually increased with decreasing SV. $C_2^=$ was always a major product in the reaction, and its yield reached a maximum at 2,000-3,000 h^{-1}. Since at lower SV $C_3^=$ sequentially produced and at higher SV AcH also produced as discussed below, the selectivity to $C_2^=$ would diminish at both SV regions.

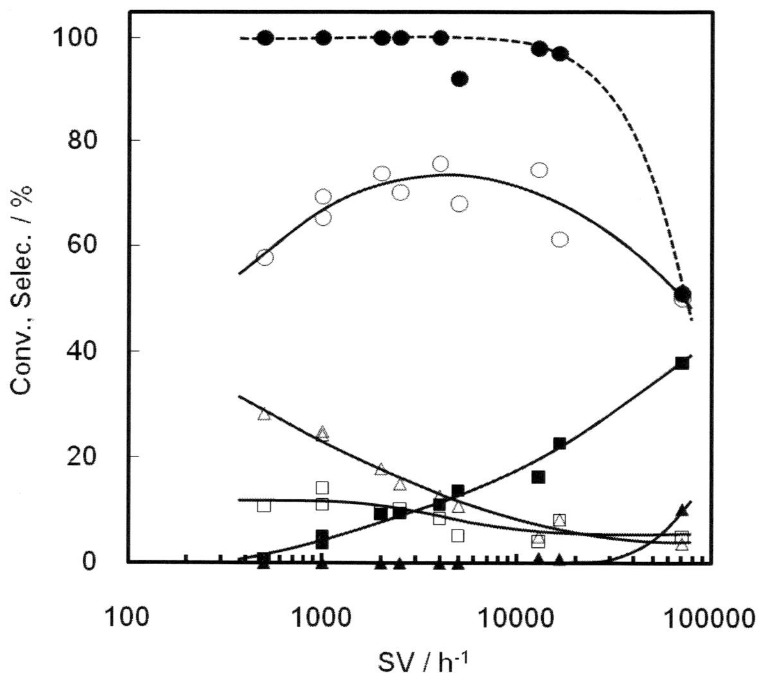

Figure 7. Conversion of EtOH on Ni-M41 with Si/Ni=23-28 as a function of space velocity . Catalyst wt. 0.05-0.4 g, total flow rate 10-300 ml min^{-1}, P_{EtOH} 5.6 kPa (N_2 balance). Conversion of EtOH (closed circle), selectivity of $C_2^=$ (open circle), $C_3^=$ (open triangle), $C_4^=$ (open square), DEE (closed triangle), and AcH (closed square).

4.2. Reaction Pathways

Figures 6 and 7 indicated the presence of two reaction pathways , DEE and AcH routes . We will first discuss conversion of EtOH to DEE, $C_2^=$, and then $C_3^=$. According to the SV dependence below 5,000 h^{-1} in Figure 7, the yield of $C_2^=$ considerably decreased with decreasing the SV and those of $C_4^=$ and $C_3^=$ increased, especially $C_3^=$. Even at lower SV the formation of hydrocarbon equal to or higher than C5 was limited, though undetectable products by GC was produced with several percentage. Therefore, there are little possibility of the C5-C6 hydrocarbons as the intermediate for $C_3^=$ production.

DEE was employed as a substrate and the product distribution was examined as a function of reaction temperature . At 573 and 673 K the conversions of DEE were almost 100% whereas the product distributions were greatly changed. At 573 K, the yields of $C_2^=$, $C_3^=$, and $C_4^=$ were approximately 80, 4, and 16%, respectively. On the other hand, those at 623 K were approximately 63, 22, and 7%. In the latter the $C_2^=$ and $C_4^=$ clearly decreased and $C_3^=$ increased. The product distributions exhibited that ethene was dimerized to 1-butene (Reaction 1, Figure 3), it was isomerized to 2-butene (Reaction 2), and finally the propene production proceeded via metathesis of 2-butene and ethene (Reaction 3), in the same way as the reactions in Section 2.

The reaction of DEE at 523 K on Ni-M41 afforded almost equimolar yields of EtOH and $C_2^=$, which was similar to those on an M41 catalyst [21c]. The little production of DEE at SV 70,000 h^{-1} (Figure 7) indicated the followings. First, ethanol was converted to DEE by the intermolecular dehydration (Reaction 4). DEE did not change to two ethene molecules via intramolecular dehydration but decomposed to $C_2^=$ and EtOH (Reaction 5). $C_2^=$ produced was toward dimerization and EtOH again reacted on the catalyst.

On the other hand, it was necessary to explain the phenomena of AcH production at higher SV and then the decrease in AcH and the increase in $C_2^=$ with decreasing SV, shown in Figure 7. The dehydrogenation of EtOH to yield AcH is widely accepted to proceed on various catalysts. The possibility that a mixture of AcH and hydrogen gives $C_2^=$ was examined here. An equimolar mixture of AcH and H_2 (6 kPa, respectively, N_2 balance) was introduced onto the Ni-M41 catalyst at 573-773 K and a SV of 4,600 h^{-1}, but no reaction was observed except for condensation of AcH and low levels of methane and $C_2^=$ formation (2-3%, respectively). The possibility that the hydrogenation of AcH and subsequent dehydration of resultant EtOH might lead to the formation of $C_2^=$ is therefore ruled out.

Next, an equimolar mixture of EtOH and AcH (5 kPa, respectively, N_2 balance) was introduced onto Ni-M41 at 673 K and SV of 990 h^{-1}. The conversion levels of EtOH and AcH were 100 and 65% and the selectivity to $C_2^=$, $C_3^=$, and $C_4^=$ were 61, 15, and 12%, respectively. The selectivity was very similar to that observed in Figures 6 and 7. It follows that AcH could be converted into $C_2^=$ in the presence of EtOH. Since 0.5-1% of ethyl acetate (ETA) was observed in the experiments of Figure 7, the possibility of ETA as intermediate was discussed. ETA is considered to be produced via Reaction 7 in Figure 3. Although the Tishchenko reaction (Reaction 7') could be postulated for the ETA formation , it was already confirmed in our previous experiment that the Tishchenko reaction did not proceed on Ni-M41.

$$2CH_3CHO \rightarrow CH_3COOCH_2CH_3 \text{ (ETA)} \qquad \text{(Reaction 7')}$$

The decomposition of ETA (Reaction 8) was proposed to yield $C_2^=$ and acetic acid (AcOH). Reaction 8 is not popular but the reverse reaction of Reaction 8 was already confirmed to catalytically proceed and was put to practical use by Showa Denko K.K., Japan [58]. To postulate Reaction 8 is therefore legitimate. The sequence of Reactions 6-7-8 would result in the formation of $C_2^=$ from AcH and EtOH through ETA as the intermediate. AcOH produced would return to AcH according to Reaction 9 and thus was not detected as a product. Finally it would be worthy to note that the reaction pathways on scandium -modified In_2O_3 [23] and yttrium -modified CeO_2 [24] are entirely different from those in Figure 3. In addition, these catalysts showed long-term stability even at partial pressure of EtOH of 30%.

Conclusion

This chapter has for the first time claimed the gas-phase metathesis of lower olefins on nickel -containing catalysts at around 673 K. The reaction mechanism of the conversion of $C_2^=$ to $C_3^=$ is suggested to be the dimerization of $C_2^=$, the isomerization of produced 1-$C_4^=$ to 2-$C_4^=$, and the metathesis of 2-$C_4^=$ and $C_2^=$ to yield $C_3^=$. The 2:1 nickel phyllosilicate-like species was proposed to be produced on the M41 pore wall and work as an active site . The three- and four-coordinated Ni^{2+} ions were suggested to produce on the five- and six-membered Si-O rings , the former of which was suggested to be the active sites for the reaction. The Ni-M41 catalyst was also active for the conversion of EtOH to $C_3^=$. EtOH was first converted to $C_2^=$ on Ni-M41 through DEE or AcH as an intermediate and then to $C_3^=$. The present findings indicated that the formation of $C_3^=$ from $C_2^=$ and EtOH could be achieved not using the shape selectivity like zeolites but by developing the reaction mechanism-controlled catalysis .

Acknowledgements

The authors are deeply grateful to Professor Emeritus Atsumu Ozaki of Tokyo Institute of Technology for his fruitful discussions and also to Dr. Tetsuo Suzuki and Mrs. Osamu Takahashi, Hiroshi Ohashi, and Takahiro Kakinuma of the NEDO research group for their helpful discussions. The work mentioned in this chapter was financially supported by three Grants-in-Aids (JSPS, NEDO, and ALCA) from the ministries (MEXT and METI) of Japan .

References

[1] Arpe, H.J. *Industrial Organic Chemistry*, 4th ed.; Wiley-VCH Verlag BmbH & Co., Weinheim, Germany , 2003. Olah, G.A.; Molnar, A. *Hydrocarbon Chemistry*, 2nd ed.; John Wiley & Sons, Inc., Hoboken, U.S.A., 2003, 30. Morschbacker, A. *Polym. Rev.* 2009, *49*, 79-84.

[2] For example; (a) Corma, A.; Mengual, J.; Miguel, P.J. *Appl. Catal. A: Gen.* 2013, *460-461*, 106-115. (b) Zhang, J.; Shan, H.; Chen, X.; Li, C.; Yang, C. *Ind. Eng. Chem. Res.* 2013, *52*, 6366-6376.
[3] Banks, R.L.; Kukes, S.G. *J. Mol. Catal.* 1985, *28*, 117-131.
[4] O'Nill, P.P.; Rooney, J.J. *J. Am. Chem. Soc.* 1972, *94*, 4383-4384.
[5] Yamaguchi, T.; Tanaka, Y.; Tanabe, K. *J. Catal.* 1980, *65*, 442-447.
[6] (a) Lu, J.; Zhao, Z.; Xu, C.; Duan, A.; Zhang, P. *Catal. Lett.* 2006, *109*, 65-70 and the references therein. (b) Oikawa, H.; Shibata, Y.; Baba, T. *Appl. Catal. A: Gen.* 2006, *312*, 181-185. (c) Koyama, T.; Hayashi, Y.; Horie, H.; Kawauchi, S.; Matsumoto, A.; Iwase, Y.; Sakamoto, Y.; Miyaji, A.; Motokura K.; Baba, T. *Phys. Chem. Chem. Phys.* 2010, *12*, 2541-2554.
[7] For example; Phillips, C.B.; Datta, R. *Ind. Eng. Chem. Res.* 1997, *36*, 4466-4475. Brando, P.; Philippou, A.; Rocha, J.; Anderson, M.W. *Catal. Lett.* 2002, *80*, 99-102. Aguayo, A.T.; Gayubo, A.G.; Atutxa, A.; Olazar, M.; Bilbao, J. *Ind. Eng. Chem. Res.* 2002, *41*, 4216-4224. Gayubo, A.G.; Alonso, A.; Valle, B.; Aguayo, A.T.; Olazar, M.; Bilbao, J. *Chem. Eng. J.* 2011, *167*, 262-277.
[8] For example; Takahara, I.; Saito, M.; Inaba, M.; Murata, K. *Catal. Lett.* 2005, *105*, 249-252. Song, Z.; Takahashi, A.; Nakamura, I.; Fujitani, T. *Appl. Catal. A: Gen.* 2010, *384*, 201-205.
[9] Arias, D.; Colmenares, A.; Cubeiro, M.L.; Goldwasser, J.; Lopez, C.M.; Machado, F.J.; Sazo, V. *Catal. Lett.* 1997, *45*, 51-58.
[10] For example; Golay, S.; Doepper, R.; Renken, A. *Chem. Eng. Sci.* 1999, *54*, 4469-4474. Bakoyannakis, D.N.; Zamboulis, D.; Stalidis, G.A.; Deliyanni, E.A. *J. Chem. Technol. Biotechnol.* 2001, *76*, 1159-1164. Varisli, D.; Dogu, T.; Dogu, G. *Ind. Eng. Chem. Res.* 2009, *48*, 9394-9401. Kamiguchi, S.; Chihara, T. *Catal. Lett.* 2003, *85*, 97-100.
[11] Kamimura, Y.; Sato, S.; Takahashi, R.; Sodesawa, T.; Akashi, T. *Appl. Catal. A: Gen.* 2003, *252*, 399-410. Nagashima, O.; Sato, S.; Takahashi, R.; Sodesawa, T. *J. Mol. Catal. A: Chem.* 2005, *227*, 231-239. Tsuchida, T.; Kubo, J.; Yoshioka, T.; Sakuma, S.; Takeguchi, T.; Ueda, W. *J. Catal.* 2008, *259*, 183-189.
[12] Speiser, F.; Braunstein, P.; Saussine, L. *Acc. Chem. Res.* 2005, *38*, 784-793 and the references therein.
[13] Shiba, T.; Ozaki, A. *Nippon Kagaku Zasshi* 1953, *74*, 295-297. Ozaki, A. *Nippon Kagaku Zasshi* 1956, *77*, 888-892. Kimura, K.; Ai, H.; Ozaki, A. *J. Catal.* 1970, *18*, 271-280. Sohn, J.R.; Ozaki, A. *J. Catal.* 1980, *61*, 29-38.
[14] Ghosh, A.K.; Kevan, L. *J. Phys. Chem.* 1990, *94*, 3117-3121. Zheng, L.; Wang, G.; Bai, X. *Stud. Surf. Sci. Catal.* 1986, *28*, 965-972.
[15] Hertmann, M.; Pöppl, A.; Kevan, L. *J. Phys. Chem.* 1996, *100*, 9906-9910.
[16] Iwamoto, M.; Tanaka, Y.; Sawamura, N.; Namba, S. *J. Am. Chem. Soc.* 2003, *125*, 13032-13033.
[17] Tanaka, Y.; Sawamura, N.; Iwamoto, M. *Tetrahedron Lett.* 1998, *39*, 9457-9460. Ishitani, H.; Iwamoto, M. *Tetrahedron Lett.* 2003, *44*, 299-301. Murata, H.; Ishitani, H.; Iwamoto, M.; *Org. Biomol. Chem.* 2010, *8*, 1202-1211.
[18] Trong, D.; Joshi, P.N.; Lemay, G.; Kaliaguine, S. *Stud. Surf. Sci. Catal.* 1995, *97*, 543-549. Yamamoto, T.; Tanaka, T.; Funabiki, T.; Yoshida, S. *J. Phys. Chem. B* 1998, *102*,

5830-5839. Ito, A.; Kodama, T.; Maeda, S.; Masaki, Y. *Tetrahedron Lett.* 1998, *39*, 9461-9464.

[19] (a) Iwamoto, M.; Kosugi, Y. *J. Phys. Chem. C* 2007, *111*, 13-15. (b) Ikeda, K.; Kawamura, Y.; Yamamoto, T.; Iwamoto, M. *Catal. Commun.* 2008, *9*, 106-110.

[20] Taoufik, M; Le Roux, E.; Thivolle-Cazat, J.; Basset, J.M. *Angew. Chem. Int. Ed.* 2007, *46*, 7202-7205.

[21] (a) Kasai, K.; Haishi, T.; Iwamoto, M. *Shokubai* 2007, *49*, 126-128. (b) Iwamoto, M.; Kasai, K. Jp. Patent P2008-255104 A. (c) Haishi, T.; Kasai, K.; Iwamoto, M. *Chem. Lett.* 2011, *40*, 614-616. (d) Iwamoto, M.; Kasai, K.; Haishi, T. *ChemSusChem* 2011, *4*, 1055-1058.

[22] Sugiyama, S.; Kato, Y.; Wada, T.; Ogawa, S.; Nakagawa, K.; Sotowa, K. *Top. Catal.* 2010, *53*, 550-554.

[23] Mizuno, S.; Kurosawa, M.; Tanaka, M.; Iwamoto, M. *Chem. Lett.* 2012, *41*, 892-894. Iwamoto, M.; Mizuno, S.; Tanaka, M. *Chem. Eur. J.* 2013, *19*, 7214-7220.

[24] Hayashi, F.; Iwamoto, M. *ACS Catal.* 2013, *3*, 14-17. Hayashi, F.; Tanaka, M.; Lin, D.; Iwamoto, M. *J. Catal.* 2014, *316*, 112-120.

[25] Abe, T.; Tachibana, Y.; Uematsu, T.; Iwamoto, M. *J. Chem. Soc. Chem. Commun.* 1995, 1617-1618.

[26] Iwamoto, M.; Tanaka, Y. *Catal. Surv. Jpn.* 2001, *5*, 25-36.

[27] (a) Lehmann, T.; Wolff, T.; Zahn, V.M.; Veit, P.; Hamel, C.; Seidel-Morgenstern, A. *Catal. Commun.* 2011, *12*, 368-374. (b) Alvarado Perea, L.; Wolff, T.; Veit, P.; Hilfert, L.; Edelmann, F.T.; Hamel, C.; Seidel-Morgenstern, A. *J. Catal.* 2013, *305*, 154-168. (c) Frey, A.S.; Hinrichsen, O. *Microporous Mesoporous Mater.* 2012, *164*, 164-171.

[28] Liu, B.; Nakatani, H.; Terano, M. *J. Mol. Catal. A: Chem.* 2003, *201*, 189-197.

[29] Negishi, E.; Takahashi, T. *Acc. Chem. Res.* 1994, *27*, 124-130.

[30] Grubbs, R.H. Ed.; *Handbook of metathesis*, Wiley-VCH Verlag BmbH & Co., Weinheim, Germany , 2003. Grubbs, R.H. *Tetrahedron* 2004, *60*, 7117-7140.

[31] Sato, Y.; Saito, N.; Mori, M. *J. Org. Chem.* 2002, *67*, 9310-9317.

[32] Baker, M.V.; Brown, D.H.; Skelton, B.W.; White, A.H. *Aust. J. Chem.* 2002, *55*, 655-660.

[33] (a) Xiao, S.; Meng, Z. *J. Chem. Soc. Faraday Trans.* 1994, *90*, 2591-2595. Schoonheydt, R.A.; Roodhooft, D. *J. Phys. Chem.* 1986, *90*, 6319-6323. Elev, I.V.; Shelimov, B.N.; Kazanskii, V.B. *J. Catal.* 1984, *89*, 470-477. (b) Prakash, A.M.; Wasowicz, T.; Kevan, L. *J. Phys. Chem.* 1996, *100*, 15947-15953. Hartmann, M.; Pöppl, A.; Kevan, L. *J. Phys. Chem.* 1995, *99*, 17494-17496.

[34] Ashizawa, H.; Yamamoto, T.; Iwamoto, M. unpublished results.

[35] Uchida, H.; Imai, H. *Bull. Chem. Soc. Jpn.* 1962, *35*, 989-995.

[36] Hulea, V.; Fajula, F. *J. Catal.* 2004, *225*, 213-222.

[37] Clause, O.; Kermarec, M.; Bonneviot, L.; Villain, F.; Che, M. *J. Am. Chem. Soc.* 1992, *114*, 4709-4717.

[38] Carriat, J.Y.; Che, M.; Kermarec, M.; Verdaguer, M.; Michalowicz, A. *J. Am. Chem. Soc.* 1998, *120*, 2059-2070.

[39] Martin, G.A.; Renoupre, A.; Dalmaiim, G.; Imelik, B. *J. Chim. Phys.* 1970, *67*, 1147-1160.

[40] Burattin, P.; Che, M.; Louis, C. *J. Phys. Chem. B* 1997, *101*, 7060-7074.

[41] Yang, J.C.; Shul, Y.G.; Louis, C., Che, M. *Catal. Today* 1998, *44*, 315-325.

[42] Yang, Y.; Lim, S.; Du, G.; Chen, Y.; Ciuparu, D.; Haller, G.L. *J. Phys. Chem. B* 2005, *109*, 13237-13246.

[43] Clause, O.; Bonneviot, L.; Che, M.; Dexpert, H. *J. Catal.* 1991, *130*, 21-28.

[44] Tanaka, M.; Itadani, A.; Kuroda, Y.; Iwamoto, M. *J. Phys. Chem. C* 2012, *116*, 5664-5672.

[45] Kermarec, M.; Carriat, J.Y.; Burattin, P.; Che, M.; Decarreau, A. *J. Phys. Chem.* 1994, *98*, 12008-12017.

[46] Lehmann, T.; Wolff, T.; Hamel, C.; Veit, P.; Garke, B.; Seidel-Morgenstern, A. *Microporous Mesoporous Mater.* 2012, *151*, 113-125.

[47] Wojcieszak, R.; Moteverdi, S.; Mercy, M.; Nowak, I.; Ziolek, M.; Bettahar, M.M. *Appl. Catal. A: Gen.* 2004, *268*, 241-253.

[48] Hadjiivanov, K.; Mihaylov, M.; Klissurski, D.; Stefanov, P.; Abadjieva, N.; Vassileva, E.; Mintchev, L. *J. Catal.* 1999, *185*, 314-323.

[49] Tanaka, M.; Kuroda, Y.; Iwamoto, M. *J. Phys. Chem. C* 2012, *116*, 22649-22651.

[50] (a) Jansen, J.C.; van der Gaag, F.J.; van Bekkum, H. *Zeolites* 1984, *4*, 369-372. (b) Wang, H.; Liu, Y.; Pinnavaia, T.J. *J. Phys. Chem. B 2006*, 110, 4524-4526. (c) Sidhpuria, K.B.; Parikh, P.A.; Bahadur, P.; Jasra, R.V. *J. Porous Mater.* 2008, *15*, 481-489.

[51] Feuston, B.P.; Higgins, J.B. *J. Phys. Chem.* 1994, *98*, 4459-4462.

[52] Wong, J.; Angel, C.A. *Glass Structure by Spectroscopy*, Marcel Dekker, New York, 1976.

[53] Nachtigall, P.; Frolich, K.; Drobná, H.; Bludský, O.; Nachtigallová, D.; Bulánek, R. *J. Phys. Chem. C* 2007, *111*, 11353-11362.

[54] Kazansky, V.B.; Elev, I.V.; Shelimov, B.N. *J. Mol. Catal.* 1983, *21*, 265-274. Bonneviot, L.; Olivier, D.; Che, M. *J. Mol. Catal.* 1983, *21*, 415-430.

[55] Mlinar, A.N.; Baur, G.B.; Bong, G.G.; Getsoian, A.; Bell, A.T. *J. Catal.* 2012, *296*, 156-164.

[56] Morrow, B.A.; Moran, L.E. *J. Catal.* 1980, *62*, 294-303. Mihaylov, M.; Hadjiivanov, K. *Langmuir* 2002, *18*, 4376-4383.

[57] Niu, G.; Huang, Y.; Cao, Z.; Huang, Y.; Li, G. *Appl. Surf. Sci.* 1999, *141*, 35-42.

[58] Tsuji, K.; Uchida, H.; Nakajou, T.; Sano, K. *Chem. Chem. Ind.* 2008, *61*, 206-208.

In: Comprehensive Guide for Mesoporous Materials. Volume 1 ISBN: 978-1-63463-958-3
Editor: Mahmood Aliofkhazraei

Chapter 7

ENGINEERING NANOPOROUS POLYMERIC MATERIALS FROM NANOSTRUCTURED MACROMOLECULAR FRAMEWORKS

Daniel Grande **[*]** ***and Benjamin Le Droumaguet***
Institut de Chimie et des Matériaux Paris-Est,
CNRS-Université Paris-Est Créteil, Thiais, France

ABSTRACT

Over the last decades, nanoporous polymeric materials have been the subject of widespread interest and intense research as they can arguably be considered for nanotechnological applications, e.g., advanced nanofiltration membranes, interlayer dielectrics, high surface area catalytic supports, as well as size/shape-selective nanoreactors. Organic porous materials have unique properties that distinguish them from their inorganic nanostructured analogues, such as tunable mechanical properties, ease to be functionalized, and especially lower production cost. Designing nanoporous polymeric materials with controlled morphology and functionality thus constitutes an active research area. In this context, original approaches using suitable porogen templates have emerged: such templates are able to induce specific structural pores within the residual frameworks. These template-oriented routes are very interesting as a wide array of nanoporous polymers with a well-defined porosity can be derived from miscellaneous macromolecular architectures. Smart methodologies include molecular imprinting, selective removal of a porogen homogeneously blended within a thermostable matrix, selective removal of one partner from (semi-)interpenetrating polymer networks, selective destruction of one block in self-organized block copolymers, as well as removal of self-assembled molecules from supramolecular architectures. This chapter affords a critically selected appraisal of these synthetic strategies developed to create functional nanoporous polymers and investigates the scope and limitations associated with the utilization of such routes.

[*] Phone: +33 (0)1 49 78 11 77, Fax: +33 (0)1 49 78 12 08 Email: grande@icmpe.cnrs.fr

List of Abbreviations

18NDM	1,8-Naphthalene-dimethanol
32HNA	3-hydroxy-2-naphthoic acid
62HNA	6-hydroxyl -2-naphthoic acid
ACN	acetonitrile
AM	acrylamide
BADMA	bisphenol A dimethacrylate
BCP	block copolymer
BHPB	3,5-bis(5-hexylcarbamoylpentyloxy)benzoic acid decyl ester
DBSA	dodecylbenzene sulfonic acid
DCBE	dicyanate of bisphenol E
DCM	dichloromethane
DEAEMA	diethylaminoethyl methacrylate
DHN	1,5-dihydroxynaphtalene
DMF	dimethylformamide
DTT	dithiothreitol
DUDMA	diurethane dimethacrylate
DVB	divinyl benzene
EDC	*N*-(3-dimethylaminopropyl)-*N'*-ethylcarbodiimide hydrochloride
EGD(M)A	ethylene glycol di(meth)acrylate
hbPTE	hyperbranched poly(triazene ester)
HE (M)A	2-hydroxyethyl (meth)acrylate
HF	fluorhydric acid
HOBt	1-hydroxybenzotriazole
HPLC	high-performance liquid chromatography
IPN	interpenetrating polymer network
MAA	methacrylic acid
MAM	methacrylamide
MDA	4,4'-methylenedianiline
MIP	molecularly imprinted polymer
M(M)A	methyl (meth)acrylate
M_n	number-average molar mass
MOCA	4,4'-methylenebis(2-chloroaniline)
N	norbornene
N-HSS	sulfo-*N*-hydroxysuccinimide
P2VP	poly(2-vinylpyridine)
P4VP	poly(4-vinylpyridine)
PAA	poly(allylamine)
PB	poly(butadiene)
PBrS	poly(4-bromostyrene)
PCHE	poly(cyclohexylethylene)
PCL	poly(ε-caprolactone)
PCN	polycyanurate
PDMA	poly(*N*,*N*-dimethyl acrylamide)

PDMS	poly(dimethylsiloxane)
PDP	pentadecylphenol
PE	polyethylene
PEGDA	poly(ethylene glycol) diacrylate
PEO	poly(ethylene oxide)
PFS	poly(4-fluorostyrene)
PHEMA	poly(2-hydroxyethyl methacrylate)
PI	poly(isoprene)
PIM	polyimide
PLA	poly(D,L-lactide)
PLLA	poly(L-lactide)
PM (M)A	poly(methyl (meth)acrylate)
PPG	poly(propylene glycol)
P(PMDSS)	poly(pentamethyldisilylstyrene)
PPO	poly(propylene oxide)
PS	polystyrene
PSTPA	poly(styrene triphenylamine)
P*t*BA	poly(*tertio*-butyl methacrylate)
S	styrene
TBAF	tetrabutylammonium fluoride
TFA	trifluoroacetic acid
UV	ultra-violet
VP	2-vinyl pyridine
Đ	polydispersity index
f	volume fraction

INTRODUCTION

In recent years, nanomaterials have been the focus of nanoscience and nanotechnology , which is an ever-growing multidisciplinary field of research and development around the world. Nanomaterials display unique surface, structural, and bulk properties which justifies their applications in many areas, including ion-exchange , separation, catalysis , purification , energy storage , etc. Atoms, ions , or molecules can be entrapped on their surfaces and in the nanometer -sized pore space.

They offer new opportunities in areas of inclusion chemistry, host -guest synthesis and molecular reactors at the nanoscale for engineering nanoparticles , nanostructures , and more generally nanomaterials . Nanoscale is particularly important, because at this scale –by adjusting the size in the range of 1-100 nm–, molecular interactions, processes, and assembly of building blocks can be controlled and lead to the formation of tailored geometries of the materials with desirable properties. Porous materials exhibit porosity ratios (volume ratios of pore space to the total volume of the material) up to 0.95. Pores can be classified into two types, i.e., open pores which connect to the surface of the material and closed pores which are isolated from the outside. Pores have various shapes and morphologies (cylindrical, spherical or more complex shapes). Acccording to IUPAC [1], *micropores* are smaller than 2 nm in diameter, *mesopores* are 2 to 50 nm large and *macropores* are larger than 50 nm. The term of *nanoporous* is often used to refer to materials having pore diameters between 1 and 100 nm. Nanoporous materials may have different pore size s and surface properties , and therefore they are convenient for various applications which require different sets of performance criteria [2]. Among the major areas of uses, one can mention advanced microelectronics , energy conversion and storage , biomolecule immobilization , diagnosis, controlled drug delivery , tissue engineering , separation and purification techniques , gas or ion transport , as well as supported catalysis [3-8].

Organic porous materials have unique properties compared with their inorganic counterparts, such as flexibility , elasticity, easy functionalization , and high (bio)compatibility , which are important features for specific applications, and they are also characterized by a low-cost production. In the last two decades, much attention has been paid to the design of polymer -based nanoporous materials .

The basic idea associated with the design of (nano) porous polymers relies on the introduction of various types of porogens (solvents , gases, small molecules or macromolecules) within polymer structure s, followed by their selective removal. Solvents and gases used as porogens create pores by physically- or chemically-induced phase separation, but the microstructures of the resulting materials are generally characterized by a broad pore size distribution .

In contrast, the utilization of suitable porogen templates enables the generation of nanoporous polymers with a defined porosity . Various types of simple organic molecules and specific macromolecules have thus been used as effective porogen templates for creating a wealth of nanoporous materials with desired characteristics (Table 1). These elegant template-oriented strategies mainly include molecular imprinting [9], selective removal of a porogen homogeneously blended within a thermostable matrix [10], selective removal of one partner from (semi-)Interpenetrating Polymer Networks (IPNs) [11], selective degradation of block copolymers (BCPs) [12], and selective destruction of supramolecular assemblies [13].

In the case of direct templating, the templated material is a copy of the structure - directing agent. During the synthetic process, no changes in order and scale length structure occur as the resulting polymer contains pores with the shape and size of the template. In the case of indirect templating, the template reorganization due to supramolecular self-assembly or further chemical reaction with a co-template occurs during the synthesis , and removal of one such modified template leads to the formation of a porous material with pores having a shape and size different from that of the starting template. The compatibility of the polymer matrix with the template also plays a key role in the morphology and the final properties of the resulting porous polymer.

As a matter of fact, the properties of porous polymers mainly depend on the topology and connectivity of constituting pores. Their morphology is important to be fine-tuned in order to target specific applications. This chapter focuses on the possibilities offered by the main synthetic strategies for engineering nanoporous polymeric materials from miscellaneous nanostructured macromolecular frameworks. The general principles of these pore generation approaches will be illustrated through several examples based on the use of molecular and macromolecular templates, and major applications of corresponding porous polymers will be addressed.

Table 1. Major template-oriented strategies for the production of porous polymers and their main characteristics

Methodology	Porogenic agent	Porous structure	Pore size range	Targeted applications
Molecular imprinting	small molecules or macromolecules, even cells	cross-linked poly(meth) acrylates, polystyrenes, polyurethanes, etc.	1 – 50 nm	separation, detection , solid phase extraction , biosensors , biomedical and pharmaceutical systems
Selective removal of a porogen homogeneously blended within thermostable matrices	photo- or thermo-labile thermoplastics, oligomers	epoxy- and cyanurate-based thermosets, phenolic resol resins	≤ 1 μm	interlayer dielectrics, membranes or adsorbents working under severe conditions
Selective removal of one partner from (semi-) interpenetrating Polymer Networks (IPNs)	un-cross-linked oligomers or polymers, polymeric sub-network	cross-linked poly(meth) acrylates, polystyrenes, polyethylenes, etc.	10 nm to a few μm	chromatography, ultrafiltration
Selective degradation of block copolymers (BCPs)	sacrificial polymeric block or sequence	polystyrenes, poly(meth) acrylates, etc. (cross-linked or not)	≥ 10 nm	nanofiltration, confined chemistry (catalysis, nanomaterials design, ...)
Selective destruction of supramolecular assemblies	self-assembled (macro)molecules	polystyrene- *b*-poly(4-vinyl pyridine), polyaniline, polyisoprene, phenolic resins, etc.	1 nm – 10 μm	smart membranes, confined chemistry (catalysis, nanomaterials design, ...)

1. Molecular Imprinting

Molecular imprinting is a powerful and versatile methodology for designing mesoporous polymeric materials with pores structured on a molecular scale [9, 14]. The preparation of such polymeric scaffolds relies on the copolymerization of functional and cross-linking monomers in the presence of a molecular template and additional solvents as porogenic agents. The first step of this methodology, the so-called imprinting process, involves the homogeneous mixing of functional co-monomers with the target analyte along with their self-assembly in the presence of an additional porogenic solvent, thus leading to covalent complexes (Figure 1b) [15], non-covalent ones (Figure 1a) [16] or hybrid ones [17] between the template molecule and the functional (co-)monomer(s). It is thus crucial for the functional co-monomers to interact tightly with the template molecules . After cross-linking, the template molecules are removed from the materials by solvent extraction or bond cleavage , thus giving rise to molecularly imprinted cavities within the resulting networks, depending on the shape and chemical structure of the templates. One should bear in mind that highly specific molecular recognition is a crucial requirement for potential applications of MIPs, such as waste water purifications, chromatography , and use as sensors .

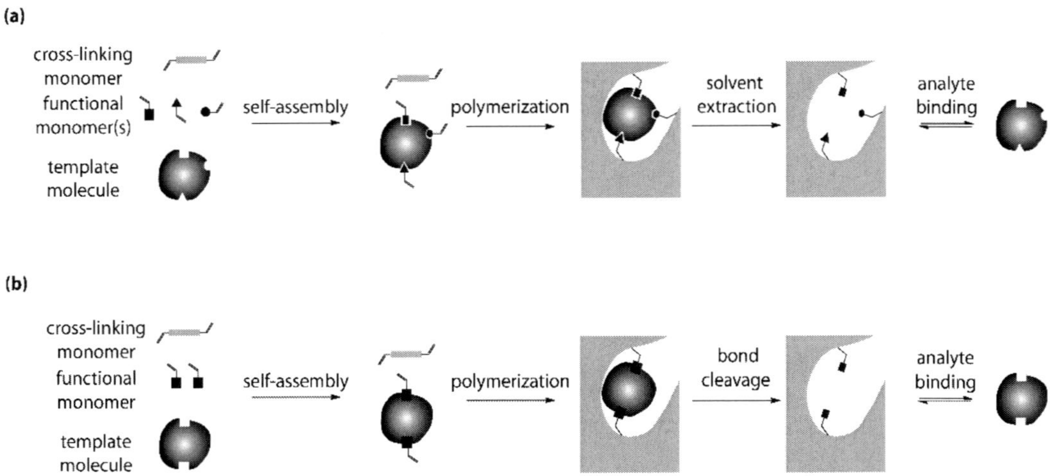

Figure 1. Schematic illustration of molecular imprinting *via* (a) non-covalent and (b) covalent templating.

The molecularly imprinted cavities are in fact binding sites that should normally have high affinity and specificity for a target analyte, thus approaching those of antibody -antigen systems. The solvent, inert towards the molecular template and the monomers , not only allows for the solubilization of all the reactants but also affords further porosity within the porous structures that is necessary for the subsequent elution of the template molecule and for the analyte to access to the imprinted cavities. In order to obtain an optimized MIP for a given target analyte, it is crucial to tune some parameters, such as the type and molar ratio of the functional co-monomers. One of the many attractive features of the molecular imprinting methodology relies on the diverse range of templates that can be employed (Table 2). Small organic molecules or analogues, such as pharmaceuticals [18-20], pesticides [21, 22], fungicides [23], nucleotide bases [24,25], amino -acids [26-29], sugars [30] or steroids [31] have been used as templates. MIPs have originally been employed as stationary phases in HPLC, notably for chiral separation, but their use has then been extended to other analytical techniques , including thin layer chromatography [32], capillary electrochromatography [33], solid phase extraction [34], immunoassay-type binding assays [31], and chemical sensors . Larger molecules, i.e., peptides , have been successfully imprinted *via* similar protocols [35]. Finally, much larger templates have also been used to prepare MIPs, such as proteins [36], cells [37] or even mineral crystals [38]. In this case, the MIP synthesis required specially adapted protocols.

Most of the MIPs are based on the free-radical copolymerization of vinyl or acrylic functional and cross-linking monomers , because of the broad choice of monomers available (Table 2).

Table 2. Summary of model templates and suitable functional and cross-linking monomers used for preparing MIPs

Template molecule	**Organic monomer or (co)polymer**	**Cross-linking agent**	**Porogenic solvent**	**References**
pharmaceuticals	MAA, MAM, VP, HEMA, DEAEMA	EGDMA	DMF, DCM, ACN	[18–20]
pesticides	MAA	EGDMA	DCM, toluene	[21,22]
fungicides	MMA, AM	DVB, EGDMA	ACN	[23]
nucleotide bases	MAA	EGDMA	ACN	[24]
amino- acids	VP, MAA	EGDMA, hybrid cross-linkers	ACN, toluene-dodecanol	[26–29]
sugars	PAA HCl	epichlorohydrin	water	[30]
steroids	MAA	EGDMA	ACN	[31]
mineral crystals	6-methacrylamido-hexanoic acid	DVB	-	[38]
peptides	sugar modified 4-vinylphenylboronates	EGDMA	-	[35]
cells	PAA	adipoyl chloride	-	[37]

They can for instance be hydrogen -bonding , hydrophilic, hydrophobic, negatively or positively charged but also metal-coordinating, thus allowing for tight interactions with suitably chosen templates. Some key parameters have been shown to be crucial regarding the pore size distribution of MIPs and their specific surface area.

Thus, the strength of the template functional monomer complex (which depends on degree of complementarity of the chemical functionalities in the template molecule with those in the functional monomers) and the solvent nature (e.g., dielectric constant) play important roles on the porosity features [23]. MIPs have generally been prepared in the form of thin films ; nonetheless, in some cases they can be obtained as monoliths, but also as nanoparticles inserted in polymeric nanofibers *via* the electrospinning technique for molecular recognition purposes [39].

2. SELECTIVE REMOVAL OF A POROGEN TEMPLATE IN THERMOSTABLE POLYMERS

An elegant strategy for generating nanopores within thermally stable polymer matrices relies on the incorporation of thermo- or photo-labile hbPTE porogens into epoxy resin precursors. In this way, closed pore size s below 10 nm could be obtained after complete decomposition of the labile polymers into volatiles, thus allowing for the engineering of materials that are well-suited for interlayer device applications (Figure 2). Polymers bearing triazene units decompose rapidly upon laser irradiation or heating up to 160–200 °C without leaving solid residues [40, 41].

The polymer matrix has thus to be high-temperature stable with no softening while the degradation of the globular molecules takes place. The preparation of hbPTE consists in a low-temperature polycondensation of triazene containing AB_2 monomers (e.g., 1-carboxyphenyl-4,4-di(2-hydroxyethyl)triazene or 1-carboxyphenyl-4-hydroxyethyl-4-methyltriazene) at neutral pH to avoid undesired early degradation of the labile units [42, 43]. For the different hbPTE prepared, the degree of branching was generally close to 50 % while M_n up to 8000 g.mol^{-1} could be obtained. To attain a homogeneous blending of hbPTE within the polymer matrix , it was necessary to modify the polarity of hbPTE by turning one of the hydroxyl termini of the triazene monomer into the corresponding *O-tert*-butyldimethylsilyl moiety before or after formation of the hbPTE globular molecule.

Nanoporous epoxy thermosets can be used as insulating materials due to the lowering of the dielectric constant by substitution of a portion of the matrix material with air. This particular property made them suitable candidates as structural materials in manufacturing microelectronic devices and components, such as computer chip packaging and circuit boards, and they have seen their sizes really decreased in the last decades. While many of the porous epoxy thermosets were achieved with pore size s larger than the submicron scale, more recent studies have investigated the possibility to produce nanoporosity within epoxy resins by using amphiphilic diblock copolymers , e.g., PE-*b*-PEO [44] or PEO-*b*-PDMS [45], triblock copolymers, e.g., PCL-*b*-PB-*b*-PCL [46] or PEO-*b*-PPO-*b*-PEO [47], as well as PPO-*b*-PEO-based star-shaped [48] porogenic BCPs, after curing PEO or PCL with MDA or MOCA, respectively.

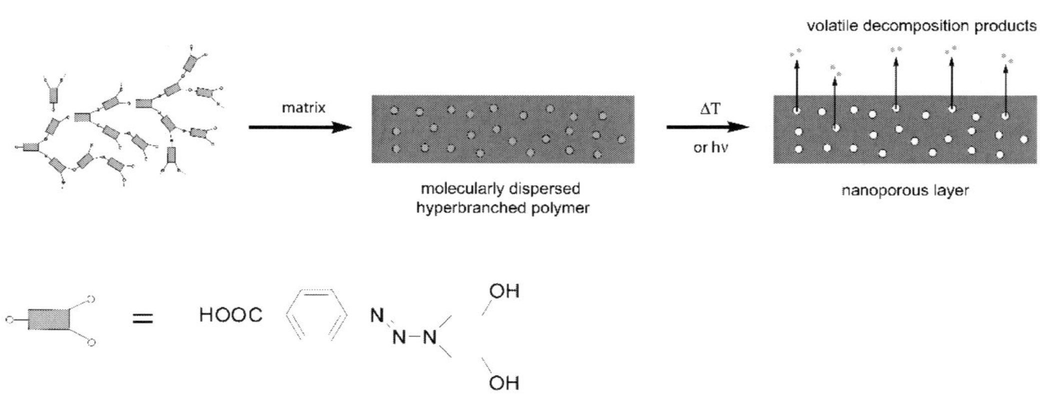

Figure 2. Template approach towards nanoporous thermostable polymers using hbPTE as a porogen.

Alternatively, porous phenolic resol resins could be produced *via* blending of an epoxy thermoset with PEO-*b*-PPO-*b*-PEO after its selective thermal curing [49]. Nanostructured thermosets based on a bisphenol A-type epoxy resin and an amphiphilic reactive diblock copolymer , i.e., PI-*b*-P4VP, were also produced by Guo et al. [50], and after degradation of the PI block by ozonolysis , they allowed for the generation of nanoporous epoxy frameworks with average pore diameter of about 60 nm uniformly distributed in the bulk material.

Numerous advanced technologies require a combination of properties for polymer films , including high thermal stability , chemical inertness , excellent resistance towards solvents and aggressive media , etc. Polycyanurates (PCNs) represent a family of thermosetting polymers that meet such requirements , in addition to their attractive intrinsic features, i.e., excellent dimensional stability , high purity , low dielectric constant s (2.6–3.2), inherent flame -retardancy, and high adhesion to conductor metals and composites [51, 52]. They thus constitute promising materials for producing membranes , adsorbents , and filters working under severe conditions (high temperatures, corrosive media, etc.). Kiefer et al. [53] synthesized such porous systems from DCBE using cyclohexane as a porogen. An alternative route to nanoporous PCN networks without using high temperatures was developed through a smart methodology involving the polycyclotrimerization of DCBE in the presence of an oligomer reactive toward cyanate groups, i.e., dihydroxy-telechelic poly(ε-caprolactone) (PCL) [54, 55]. Indeed, under mild hydrolytic conditions (70 °C in ethanol /phosphate buffer mixture) or by solvent extraction , PCL segments can be selectively removed, thus playing the role of porogens. A full characterization of these PCN-based porous matrices indicated pore size distributions ranging from 10 to almost 150 nm and increasing pore volumes, depending on the PCN/PCL mass ratio (from 5/95 to 50/50). Alternatively, another approach recently reported relies on the preparation of nanoporous PCN networks through the quantitative Soxhlet extraction of inert high-boiling temperature plasticizers initially incorporated [56].

3. Nanoporous Networks Derived from (Semi-)Interpenetrating Polymer Networks

An alternative and original strategy for engineering nanoporous cross-linked materials has been forward put through the utilization of (semi-)interpenetrating polymer networks (IPNs) as nanostructured precursors. IPNs represent an intimate combination of two independently cross-linked polymers , at least one being obtained in the immediate presence of the other. The network structure may result from either chemical cross-linking of polymer chain s by covalent bond ing or physical cross-linking of segments of multiple polymer chains containing crystalline or glassy phase domains [57]. On the other hand, semi-IPNs are prepared by cross-linking polymer chains in the presence of another un-cross-linked polymer that will afford the porosity to the material, upon its solvent extraction or selective degradation . The most important types of IPNs are defined as follows: sequential IPNs are created by forming the two constituting sub-networks *via* two successive and independent cross-linking processes, while simultaneous IPNs are derived from two cross-linking processes carried out simultaneously. Both sub-networks must be generated by non-interfering reaction routes , namely different polymerization mechanism s and/or distinct experimental condition s. Due to their interlocking framework, macrophase segregation is prevented in IPNs, and the spatial scale of phase separation may be restricted to domain sizes of a few hundreds or even tens of nanometers . It is noteworthy that IPN morphology is mainly governed by the interplay between two types of competitive factors: thermodynamic ones, i.e., intermolecular interactions between both partners, *versus* kinetic ones, i.e., competition between cross-linking and phase separation. Hence, the microphase-separated morphologies developed in IPNs can be synthetically controlled by the proper selection of component nature and composition , as well as polymerization kinetics and mechanisms. By resorting to selective degradation methods, such polymeric architectures are particularly interesting precursors to nanoporous networks when they arise from the combination of two partners that exhibit a contrasted degradability under specific conditions.

Since pioneering studies developed by Widmaier et al. [58, 59] on the preparation of IPNs based on labile or stable P*n*BA and PS sub-networks cross-linked by either acrylic acid anhydride (hydrolyzable) or DVB (stable), respectively, several reports have flourished in the following years on the synthesis of nanoporous materials through the IPN strategy. Du Prez et al. [60] prepared sequential IPNs based on the successive polymerization of an acrylate -terminated poly(1,3-dioxolane) bis-macromonomer with M(M)A, followed by the copolymerization of M(M)A with tetraethylene glycol dimethacrylate . In this particular case, poly(1,3-dixolane) segments were degraded using trifluoromethane sulfonic acid or strong Lewis acid vapors, affording the corresponding PM (M)A porous networks. Biocompatible PHEMA-based porous hydrogels could also be obtained from semi-IPNs composed of PHEMA and hydrolyzable poly[(D,L-lactide)-*co*-(ε-caprolactam)] as the un-cross-linked partner that was totally degraded under mild alkaline conditions (pH = 7.4, T = 37 °C) [61]. Electron beam irradiation was also successfully used to generate the porosity from IPNs composed of polyethylene and poly(methyl methacrylate -*co*-butyl methacrylate) [62-64]. An irradiation dose of 800 KGy not only allowed the partial decomposition of PMMA sub-network but also permitted further PE cross-linking in the resulting porous material. This selective degradation notably allowed for the production of open nanoporous PE structures that could find a potential application in the field of polymer membranes for ultrafiltration. IPN precursors based on PE and poly(dodecyl methacrylate-*co*-ethyl methacrylate) [63, 64] prepared by Pionteck's group afforded nanoporous materials with smaller pore size , as polymethacrylates possessing long aliphatic side-chains degraded less than their counterparts with short pendant chains.

PLA/PMMA -based (semi-)IPNs have thoroughly been investigated notably by Rohman et al. [65, 66] in order to engineer mesoporous networks with defined porosity , and they have particularly studied the influence of the PMMA cross-link density and the chemical nature of the cross-linker (DUDMA *vs.* BADMA) on the porosity of the resulting PMMA networks. The first approach applying the extraction of un-cross-linked oligoesters from PLA/PMMA-based semi-IPNs constitutes a straightforward and effective methodology (Figure 3a), provided the extent of miscibility between both partners is high, i.e., when using 1 mol % dimethacrylate -containing samples based on DUDMA preferentially [65]. Alternately, the second strategy implying the quantitative hydrolysis of the polyester sub-network from PLA/PMMA-based IPNs offers more versatility (Figure 3b), since mesoporous polymers can be obtained with a wide range of dimethacrylate contents (nearly up to 10 mol %), regardless of their nature [66]. In the latter case, the mesoporosity can be attributed to the good degree of chain interpenetration associated with both sub-networks in IPN precursors, due to their peculiar interlocking framework.

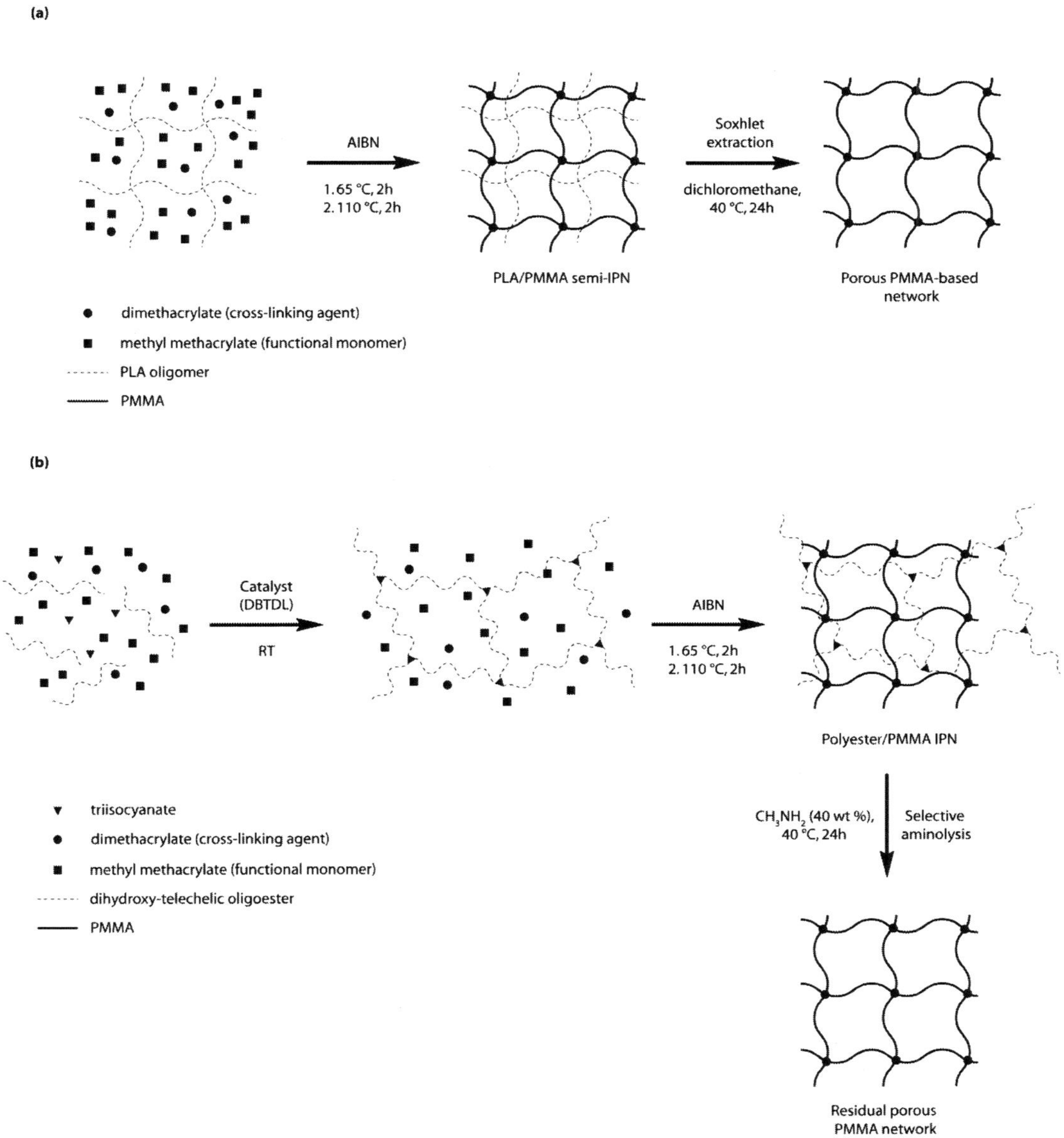

Figure 3. Design of nanoporous networks (a) by oligoester extraction from PLA/PMMA semi-IPNs or (b) from partially hydrolyzable IPNs.

Recently, PCL/PS semi-IPNs have also been designed for the development of porous monoliths meant for capillary electrochromatography. Extraction of the un-cross-linked PCL oligomers from the PS matrix allows for the generation of the corresponding porous materials that display both macropores and nanopores in the 30 nm range, and good separation efficiency when analyzing a series of alkyl benzenes [67].

4. Etching of a Sacrificial Block from Block Copolymer Precursors

4.1. Principle and General Considerations

Pioneering studies of this approach have been reported by Nakahama and co-workers [68]. Nevertheless, Hillmyer's group have probably contributed most to the development of highly ordered porous polymeric materials , in the form of thin films or monoliths, from di- or multi-BCP precursors that are subjected to selective etching of one of the constituent blocks [69, 70]. So far, two main methodologies have been used to generate ordered BCP precursors. The first one relies on the initial casting of a BCP solution onto a suitable substrate to form a thin film, followed by thermal or solvent annealing to reduce defects and improve pattern ordering in the self-organized structure [71]. The second strategy entails the microstructure alignment of BCPs upon the shear flow induced by a channel die processing to afford highly oriented structures [69, 70]. Once the material ordered by one of these two methodologies, the subsequent step of the process consists in the selective removal of the sacrificial block under selective conditions, depending on the inherent nature of both the etched and remaining block (Figure 4). One should bear in mind that the morphology (cylinders, gyroids, etc) observed after degradation of the expendable segment is mainly dependent on the volume fraction of each block in the BCP precursor. It seems also essential for each block of the BCP to be obtained with rather low polydispersity index, typically with *Đ* lower than 1.2, if one expects highly ordered materials after orientation. Other parameters are also very important regarding the self-organization process and the desired morphology. Indeed, BCP molar mass and molecular architecture, along with the Flory-Huggins interaction parameter between alien blocks, should be finely tuned in order to tailor the morphology of the as-obtained materials. When preparing porous polymers from BCP precursors, optimization of the self-organization process is required in order to improve microphase separation and microdomain orientation. In this regard, experimental condition s, such as solvent used for casting, annealing conditions, and shear alignment, should be carefully taken into consideration. The etching process should also be as mild as possible regarding the polymeric frameworks as it should not alter/destroy the nanostructures stemming from the previous nanostructuration step. Finally, careful attention should be focused onto the thermal and mechanical properties as well as solvent resistance of the resulting porous frameworks, as they should be stable enough to keep their original structure. Well-defined porous morphologies with pore diameters ranging from 10 to 100 nm are typically produced.

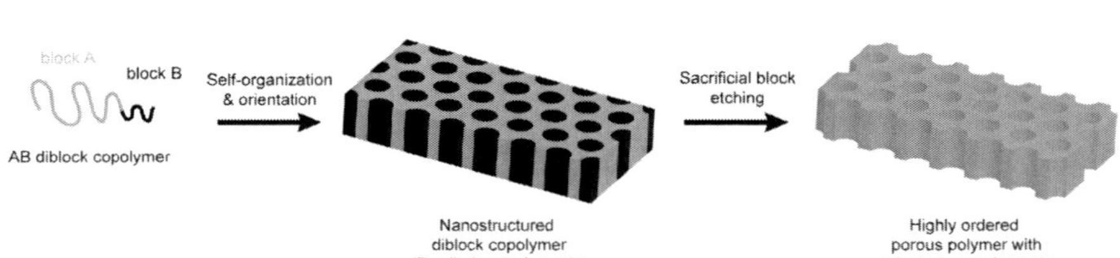

Figure 4. Schematic illustration of the design of nanoporous material from a diblock copolymer precursor displaying a hexagonally close-packed morphology .

4.2. Selective Degradation of Sacrificial Block

Two main strategies have hitherto been developed for the selective etching of the sacrificial component from the self-organized BCP precursor. The first one relies on the degradation of the expendable segment itself. In this case, etching conditions are clearly dictated by the nature of the concerned block. The sacrificial components used to generate nanopores in BCP precursors are diversified and include PLA [69-74], PDMS [75-77], PPG [78], PI [79] or PMMA [80-82] blocks that can be degraded by alkaline hydrolysis , reaction with HF or TBAF, oxygen plasma treatment (or degradation in the presence of fuming nitric acid), ozonolysis or UV treatment followed by acetic acid extraction , respectively (Table 3).

Alternatively, ongoing research concerns the development of more efficient and environmentally benign etching strategies. In this context, the insertion of an easily cleavable linker group at the junction between the microphase-separated blocks seems to be a promising route, since it allows for site-specific scission between both blocks. This elegant strategy does not require degradation of the entire backbone of the polymer segment in the sacrificial minor phase. The chemical nature of the linker should thus be suitably chosen so as to allow for its subsequent cleavage under experimental condition s that do not alter/destroy the morphology obtained during the self-organization process. "Smart" methodologies involving the cleavable of the chemical junction between incompatible blocks have been successfully put forward in the literature for the generation of nanoporous polymeric frameworks. For instance, hetero-Diels-Alder adduct [83], orthonitrobenzyl [84, 85], disulphide [86] or trityl ether [87] functional groups, can be inserted at the junction between both blocks and allowed for further specific cleavage in rather mild conditions (Table 4).

Table 3. Experimental conditions required for selective degradation of sacrificial block from semi-degradable BCPs and main porous characteristics of the as-obtained frameworks

BCP nature	Sacrificial block	Etching conditions	Porous morphology	Pore size (nm)	References
PS-*b*-PLA	PLA	alkaline hydrolysis	lamellae, gyroids, cylinders	10–80	[69-74]
PI-*b*-PDMS	PDMS	reaction with HF or TBAF	cylinders	14-50	[75-77]
PIM-*b*-PPG	PPG	oxygen plasma treatment or degradation with fuming nitric acid	not determined	not determined	[78]
PS-*b*-PI	PI	ozonolysis	lamellae, cylinders, bicontinuous, spheres	20–30	[79]
PB-*b*-PMMA and PS-*b*-PMMA	PMMA	UV treatment followed by acetic acid extraction	cylinders	10–50	[80-82]

Table 4. Experimental conditions required for cleavage of chemical junction of BCPs and main porous characteristics of the as-obtained frameworks

BCP nature	Nature of chemical j unction	Junction cleavage conditions	Porous morphology	Pore size (nm)	References
PS-*b*-PEO	hetero-Diels-Alder adduct	heating (90 °C)	not determined	40–200	[83]
PS-*b*-PEO and PFP(M)A-*b*-PEO	orthonitrobenzyl	UV irradiation (350 nm)	nanochannels	~35	[84,85]
PS-*b*-PEO	disulphide	reaction with DTT	nanochannels	~20	[86]
PS-*b*-PEO	trityl ether	reaction with TFA	nanochannels	~12	[87]

Alternatively, a ruthenium (II)-terpyridine bis-complex [88] or a $-SO_3^-$@$-NH_3^+$ ionic pair [89] have been introduced into "pseudo" BCPs, allowing for further facile degradation of the interfacial junction moiety. Such design strategies for the incorporation of a functional junction essentially do not alter the self-organization behaviour of the BCPs because of the negligible volume fraction of the linking entities. Moreover, they pave the way for decoration of pore walls with reactive functional groups and expand the choice of components for the BCP precursors.

4.3. Morphological and Chemical Structure of Nanoporous Materials

Diblock copolymers constituted by one PLA segment have been widely used for the preparation of nanoporous materials in conjunction with a PS [69-74], PSTPA [90], PFS [91], PCHE [92], Poly(S-*co*-HE (M)A) [93] or Poly(N-*co*-S) block [94]. Different morphologies for porous PS or modified PS structures were obtained after extensive studies that reported the possibility to obtain spherical or gyroid cavities [95], oriented nanochannels, i.e., hollow cylinders [69, 70, 73], and alternating lamellae (Figure 5) [73], depending on the PLA volume fraction and the polydispersity index of the BCP precursors.

Different groups took advantage of the PMMA degradability under UV or γ irradiation exposure to generate PS- [82, 96], PFS- [97], PBrS- [98] or 1,2-PB-based [80] porous materials . The morphology obtained from these BCP precursors was generally a hexagonally-close packed structure with average pore diameters spread from 8 to 50 nm. Upon HF or TBAF degradation of PDMS sacrificial block in self-organized PS-*b*-PDMS [95], PI-*b*-PDMS [75-77] or PB-*b*-PDMS [99] BCP precursors, various porous templates with gyroid or hexagonally close-packed morphology and average pore diameters in the 14–50 nm range were generated.

Although the use of BCPs allows for the preparation of porous polymeric materials characterized by highly ordered morphologies, their synthesis and nanostructuration may be quite costly and time-consuming. Moreover, the utilization of BCPs for the generation of nanoporous polymer structure s with suitable mechanical and morphological stability generally requires cross-linking of the remaining macromolecular chains in the obtained nanoporous materials .

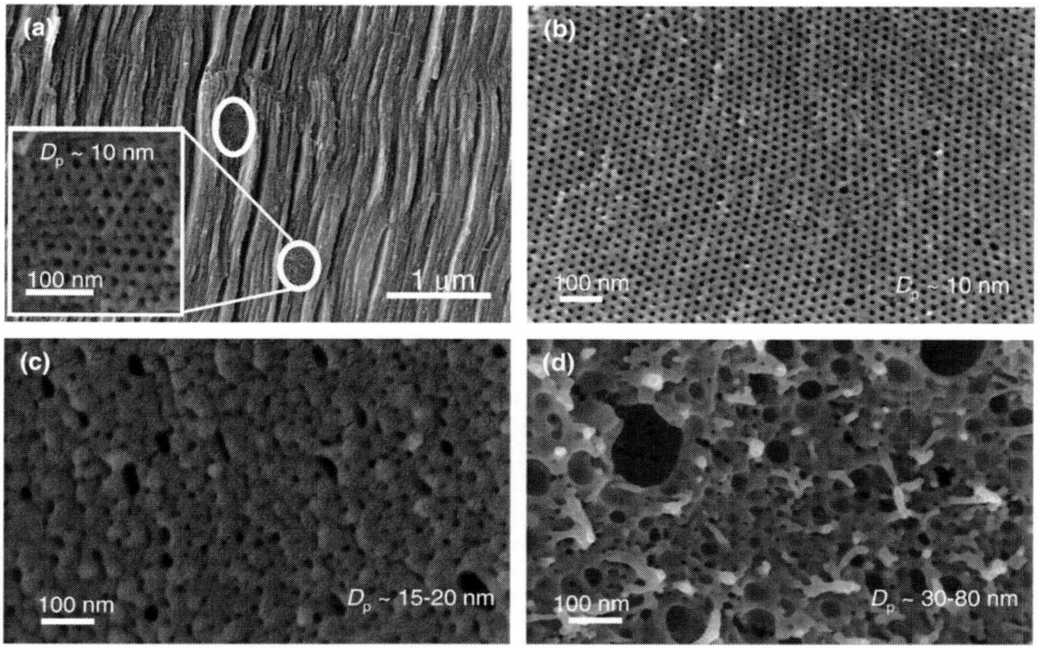

Figure 5. Cross-sectional SEM micrographs of porous materials derived from PS-*b*-PLA copolymers displaying f_{PLA} ~ 0.35: (a) PS_{105}-*b*-PLA_{50}, $Đ = 1.5$; (b) PS_{100}-*b*-PLA_{50}, $Đ = 1.1$; (c) PS_{200}-*b*-PLA_{90}-*b*-PLA_{167}, $Đ = 1.2$; (d) PS_{370}-*b*-PLA_{167}, $Đ = 1.3$. An estimation of the pore diameter (D_p) range is indicated. Reprinted with permission [73]. Copyright 2011, Elsevier.

4.4. Pore Surface Functionalization

Etching one block from self-organized BCP precursors enables further functionalization of the pore surface with several model compounds . This has notably been demonstrated by functionalizing carboxy-decorated nanochannels of porous PS frameworks with amino - containing molecules *via* amide bond formation (Figure 6a) in the presence of a catalytic coupling reagent such as HOBt/EDC [72, 73] or EDC/*N*-HSS [81]. Alternatively, Majdoub et al. [74] also demonstrated the possibility to turn the sulfonic acid functionalities decorating the pore surface of nanoporous PS into the corresponding sulfonamide ones by simple reaction with an amine -containing compound after preliminary activation of the sulfonyl functionality in the presence of thionyl chloride (Figure 6b).

Thus, the possibility to generate pores with various chemical functionalities should allow for further facile derivatization of the pore surface with desired moieties, depending on the targeted application.

Furthermore, multiblock copolymers possessing an expendable block have been successfully synthesized and used as precursors for the preparation of a broad range of functional nanoporous materials with pore size s in the 10–65 nm range. Miscellaneous original studies involve ABC triblock copolymers, e.g., PS-*b*-PMMA -*b*-PEO [100], PS-*b*-

PDMA-*b*-PLA [101], PS-*b*-PI-*b*-PLA [102], PI-*b*-P(PMDSS)-*b*-PI [103], PDMS-*b*-P*t*BA-*b*-PS [104], PB-*b*-P2VP-*b*-P*t*BA or PS-*b*-P2VP-*b*-P*t*BA [105], among many others. In conclusion, the use of ABC triblock terpolymers possessing a terminal etchable block and a central functionalized one as nanostructured precursors to nanoporous frameworks constitutes an elegant means to tailor pore surface chemistry .

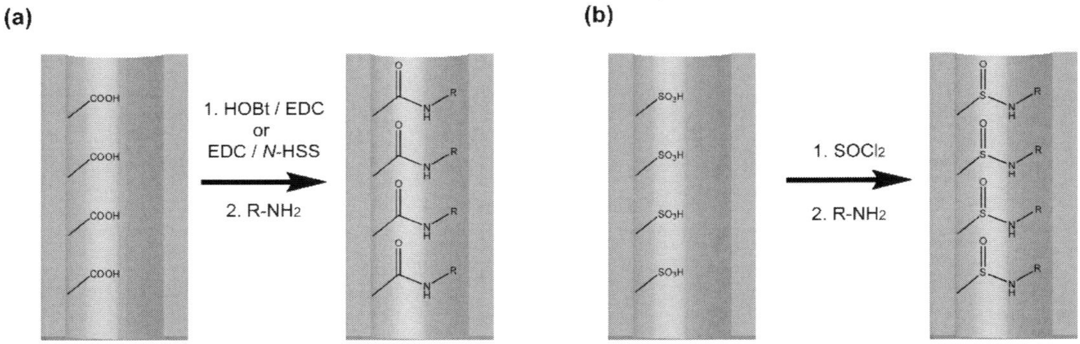

Figure 6. Pore surface functionalization of carboxy- (a) or sulfonyl- (b) functionalized porous PS *via* activation and amide bond formation in the presence of model amine compounds .

5. Self-Assembly of Copolymers with Organic (Macro)Molecules

More recently, more complex (co)polymeric systems have also been developed for porous material synthesis purposes. They rely on the self-assembly of copolymers with homopolymers , copolymers of different nature or organic molecules [13], respectively. Self-assembly of structures into complex supramolecular architectures allow for the preparation of nanoporous scaffolds upon selective extraction of the template homopolymer, copolymer or organic molecules from the principal copolymer matrix . Thus, different systems include PS-*b*-P4VP in the presence of 18NDM, 32HNA or 62HNA [106], PS-*b*-P4VP [107] or PTBS-*b*-PS-*b*-P4VP [108] in the presence of PDP, PS-*b*-P4VP in the presence of DHN [109] or PS-*b*-PEO in the presence of DBSA [110].

Alternatively, different self-assembled polymer blend systems have permitted the generation of miscellaneous nanoporous frameworks upon removal of PI by ozonolysis [111], PS and PVP by pyrolysis (Figure 7) [112], PEO by alkaline degradation [113] or PMMA by UV irradiation [114].

Another interesting strategy was developed by Mésini's group; it involved the self-assembly of an organogelator, i.e., BHPB, known to self-assemble into nanotubes, homogeneously suspended in EGDA difunctional monomer solution [115-117]. Upon polymerization of the diacrylate and subsequent leaching of the BHPB-based nanotubes, nanoporous PEGDA materials could be obtained with pore diameters lower than 100 nm.

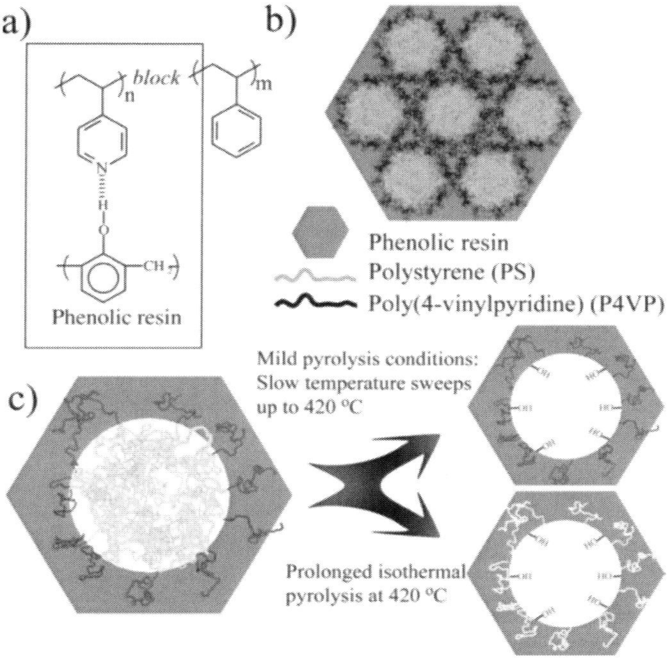

Figure 7. Schematic illustration of the methodology employed to engineer nanoporous phenolic resins from self-assembled PS-*b*-P4VP and phenolic resins : (a) structures of PS-*b*-P4VP and phenolic resin and corresponding intermolecular hydrogen bonds , (b) self-assembled polymers , and (c) etching methodology used to generate the nanoporosity. Reprinted with permission [112]. Copyright 2007, Wiley-VCH.

CONCLUSION

In summary, polymeric nanoporous materials can be successfully engineered from nanostructured macromolecular frameworks by five major template-oriented methodologies in order to obtain tailorable porous features, depending on the application targeted. Upon varying the template structure , the synthetic strategy and pore generation method, a wide array of functional polymeric materials with various pore size s and connectivities can be obtained. Although the last decades have seen the blooming of such smart polymer -based nanostructures , some necessary requirements , notably regarding their mechanical properties , still need to be fulfilled for their possible everyday use as advanced materials in important technological areas, including separation and purification techniques , controlled drug delivery , and nanoconfined chemistry among many others.

REFERENCES

[1] Sing, K. S. W.; Everett, D. H.; Haul, R. A. W.; Moscou, L.; Pierotti, J.; Rouquerol, J.; Siemieniewska, T., Reporting Physisorption Data for Gas/Solid Systems, with Special Reference to the Determination of Surface Area and Porosity. *Pure Appl. Chem.,* 1985, 57, 603-619.

[2] Wu, D. C.; Xu, F.; Sun , B.; Fu, R. W.; He, H. K.; Matyjaszewski, K., Design and Preparation of Porous Polymers. *Chem. Rev.,* 2012, 112, 3959-4015.

[3] Koros, W. J.; Fleming, G. K.; Jordan , S. M.; Kim, T. H.; Hoehn, H. H., Polymeric membrane materials for solution -diffusion based permeation separations. *Prog. Polym. Sci.,* 1988, 13, 339-401.

[4] Thomas, J. D. R., Selective-membrane electrodes for analysis. *Analyst,* 1994, 119, 203-208.

[5] Dunleavy, M., Polymeric membranes . A review of applications. *Med. Device Technol.,* 1996, 7, 14-21.

[6] Tavolaro, A.; Drioli, E., Zeolite membranes . *Adv. Mater.,* 1999, 11, 975-996.

[7] Svec, F.; Peters, E. C.; Sykora, D.; Yu, G.; Frechet, J. M. J., Monolithic stationary phases for capillary electrochromatography based on synthetic polymers : Designs and applications. *J. High Resolut. Chromatogr.,* 2000, 23, 3-18.

[8] Hentze, H. P.; Antonietti, M., Porous polymers and Resins for biotechnological and biomedical applications . *Rev. Mol. Biotechnol.,* 2002, 90, 27-53.

[9] Wulff, G., Enzyme-like Catalysis by Molecularly Imprinted Polymers. *Chem. Rev.,* 2001, 102, 1-28.

[10] Hedrick, J. L.; Carter, K. R.; Labadie, J. W.; Miller, R. D.; Volksen, W.; Hawker, C. J.; Yoon, D. Y.; Russell, T. P.; McGrath, J. E.; Briber, R. M., Nanoporous polyimides . *Adv. Polym. Sci.,* 1999, 141, 1-43.

[11] Balaji, R.; Boileau, S.; Guérin, P.; Grande, D., Design of porous polymeric materials from miscellaneous macromolecular architectures: an overview. *Polym. News,* 2004, 29, 205-212.

[12] Hillmyer, M., Nanoporous Materials from Block Copolymer Precursors. *Adv. Polym. Sci.,* 2005, 190, 137-181.

[13] Ikkala, O.; ten Brinke, G., Functional materials based on self-assembly of polymeric supramolecules. *Science,* 2002, 295, 2407-2409.

[14] Haupt, K.; Mosbach, K., Molecularly imprinted polymers and their use in biomimetic sensors . *Chem. Rev.,* 2000, 100, 2495-2504.

[15] Wulff, G., Molecular imprinting in cross-linked materials with the aid of molecular templates - a way towards artificial antibodies. *Angew. Chem., Int. Ed.,* 1995, 34, 1812-1832.

[16] Mosbach, K.; Ramstrom, O., The emerging technique of molecular imprinting and its future impact on biotechnology . *Bio-Technology,* 1996, 14, 163-170.

[17] Klein, J. U.; Whitcombe, M. J.; Mulholland, F.; Vulfson, E. N., Template-mediated synthesis of a polymeric receptor specific to amino acid sequences. *Angew. Chem.-Int. Ed.,* 1999, 38, 2057-2060.

[18] Jafari, M. T.; Rezaei, B.; Zaker, B., Ion Mobility Spectrometry as a Detector for Molecular Imprinted Polymer Separation and Metronidazole Determination in Pharmaceutical and Human Serum Samples. *Anal. Chem.,* 2009, 81, 3585-3591.

[19] Chassaing, C.; Stokes, J.; Venn, R. F.; Lanza, F.; Sellergren, B.; Holmberg, A.; Berggren, C., Molecularly imprinted polymers for the determination of a pharmaceutical development compound in plasma using 96-well MISPE technology . *J. Chromatogr. B,* 2004, 804, 71-81.

[20] Suarez-Rodriguez, J. L.; Diaz-Garcia, M. E., Fluorescent competitive flow-through assay for chloramphenicol using molecularly imprinted polymers . *Biosens. Bioelectron.,* 2001, 16, 955-961.

[21] Ferrer, I.; Lanza, F.; Tolokan, A.; Horvath, V.; Sellergren, B.; Horvai, G.; Barcelo, D., Selective trace enrichment of chlorotriazine pesticides from natural waters and sediment samples using terbuthylazine molecularly imprinted polymers . *Anal. Chem.,* 2000, 72, 3934-3941.

[22] Zhu, X. L.; Yang, J.; Su, Q. D.; Cai, J. B.; Gao, Y., Selective solid-phase extraction using molecularly imprinted polymer for the analysis of polar organophosphorus pesticides in water and soil samples. *J. Chromatogr. A,* 2005, 1092, 161-169.

[23] Al Kobaisi, M.; Tate, M.; Rix, C.; Jakubov, T. S.; Mainwaring, D. E., The effect of molecular imprinting on the pore size distribution of polymers . *J. Int. Adsorpt. Soc.,* 2007, 13, 315-321.

[24] Huang, Y. C.; Lin, C. C.; Liu, C. Y., Preparation and evaluation of molecularly imprinted polymers based on 9-ethyladenine for the recognition of nucleotide bases in capillary electrochromatography. *Electrophoresis,* 2004, 25, 554-561.

[25] Li, Z.; Ding, J. F.; Day, M.; Tao, Y., Molecularly imprinted polymeric nanospheres by diblock copolymer self-assembly . *Macromolecules,* 2006, 39, 2629-2636.

[26] Sibrian-Vazquez, M.; Spivak, D. A., Enhanced enantioselectivity of molecularly imprinted polymers formulated with novel cross-linking monomers . *Macromolecules,* 2003, 36, 5105-5113.

[27] Wei, X. L.; Li, X.; Husson, S. M., Surface molecular imprinting by atom transfer radical polymerization . *Biomacromolecules,* 2005, 6, 1113-1121.

[28] Li, X.; Husson, S. M., Adsorption of dansylated amino acid s on molecularly imprinted surfaces: A surface plasmon resonance study. *Biosens. Bioelectron.,* 2006, 22, 336-348.

[29] Li, M.; Lin, X. C.; Xie, Z. H., Investigation of enantiomer recognition of molecularly imprinted polymeric monoliths in pressurized capillary electrochromatography screening the amino acid s and their derivatives . *J. Chromatogr. A,* 2009, 1216, 5320-5326.

[30] Parmpi, P.; Kofinas, P., Biomimetic glucose recognition using molecularly imprinted polymer hydrogels . *Biomaterials,* 2004, 25, 1969-1973.

[31] Jiang, T. H.; Zhao, L. X.; Chu, B. L.; Feng, Q. H.; Yan, W.; Lin, J. M., Molecularly imprinted solid-phase extraction for the selective determination of 17 beta-estradiol in fishery samples with high performance liquid chromatography . *Talanta,* 2009, 78, 442-447.

[32] Wulff, G.; Haarer, J., Enzyme-analog built polymers .29. The preparation of defined chiral cavities for the race mic-resolution of free sugars. *Makromol. Chem.,* 1991, 192, 1329-1338.

[33] Fischer, L.; Muller, R.; Ekberg, B.; Mosbach, K., Direct enantioseparation of beta-adrenergic blockers using a chiral stationnary phase prepared by molecular imprinting . *J. Am. Chem. Soc.,* 1991, 113, 9358-9360.

[34] Sellergren, B.; Shea, K. J., Influence of polymer morphology on the ability of imprinted network polymers to resolve enantiomers . *J. Chromatogr.,* 1993, 635, 31-49.

[35] Kotrotsiou, O.; Chaitidou, S.; Kiparissides, C., On the synthesis of peptide imprinted polymers by a combined suspension-Epitope polymerization method. *Mater. Sci. Eng. B-Adv. Funct. Solid-State Mater.,* 2009, 165, 256-260.

[36] Shi, H. Q.; Tsai, W. B.; Garrison, M. D.; Ferrari, S.; Ratner, B. D., Template-imprinted nanostructured surfaces for protein recognition . *Nature,* 1999, 398, 593-597.

[37] Aherne, A.; Alexander, C.; Payne, M. J.; Perez, N.; Vulfson, E. N., Bacteria-mediated lithography of polymer surfaces. *J. Am. Chem. Soc.,* 1996, 118, 8771-8772.

[38] D'Souza, S. M.; Alexander, C.; Carr, S. W.; Waller, A. M.; Whitcombe, M. J.; Vulfson, E. N., Directed nucleation of calcite at a crystal-imprinted polymer surface. *Nature,* 1999, 398, 312-316.

[39] Chronakis, I. S.; Jakob, A.; Hagström, B.; Ye, L., Encapsulation and Selective Recognition of Molecularly Imprinted Theophylline and 17β-Estradiol Nanoparticles within Electrospun Polymer Nanofibers. *Langmuir,* 2006, 22, 8960-8965.

[40] Lippert, T.; Stebani, J.; Ihlemann, J.; Nuyken, O.; Wokaun, A., Excimer-laser ablation of novel triazene polymers - influence of structural parameters on the ablation characteristics. *J. Phys. Chem.,* 1993, 97, 12296-12301.

[41] Lippert, T.; Wokaun, A.; Stebani, J.; Nuyken, O.; Ihlemann, J., Triazene polymers designed for excimer laser ablation . *Angew. Makromol. Chem.,* 1993, 206, 97-110.

[42] Eigner, M.; Komber, H.; Voit, B., Hyperbranched Poly(triazene ester)s as Novel Globular Photolabile and Thermolabile Polymers. *Macromol. Chem. Phys.,* 2001, 202, 245-256.

[43] Eigner, M.; Voit, B.; Estel, K.; Bartha, J. W., Labile hyperbranched poly(triazene ester)s - decomposition behavior and their use as porogens in thermally stable matrix polymers . *e-Polymers,* 2002, 2, 386-403.

[44] Guo, Q.; Thomann, R.; Gronski, W.; Staneva, R.; Ivanova, R.; Stühn, B., Nanostructures , Semicrytalline Morphology, and Nanoscale Confinement Effect on the Crystallization Kinetics in Self-Organized Block Copolymer/Thermoset Blends. *Macromolecules,* 2003, 36, 3635-3645.

[45] Guo, Q.; Chen, F.; Wang, K.; Chen, L., Nanostructured thermoset epoxy resin templated by an amphiphilic poly(ethylene oxide)-block-poly(dimethylsiloxane) diblock copolymer . *J. Polym. Sci., Part B: Polym. Phys.,* 2006, 44, 3042-3052.

[46] Meng, F.; Zheng, S.; Zhang, W.; Li, H.; Liang, Q., Nanostructured Thermosetting Blends of Epoxy Resin and Amphiphilic Poly(ε-caprolactone)-block-polybutadiene - block-poly(ε-caprolactone) Triblock Copolymer. *Macromolecules,* 2006, 39, 711-719.

[47] Guo, Q.; Thomann, R.; Gronski, W.; Thurn-Albrecht, T., Phase Behavior, Crystallization, and Hierarchical Nanostructures in Self-Organized Thermoset Blends of Epoxy Resin and Amphiphilic Poly(ethylene oxide)-block-poly(propylene oxide)-block-poly(ethylene oxide) Triblock Copolymers. *Macromolecules,* 2002, 35, 3133-3144.

[48] Guo, Q.; Wang, K.; Chen, L.; Zheng, S.; Halley, P. J., Phase behavior, crystallization , and nanostructures in thermoset blends of epoxy resin and amphiphilic star-shaped block copolymers . *J. Polym. Sci., Part B: Polym. Phys.,* 2006, 44, 975-985.

[49] Kosonen, H.; Ruokolainen, J.; Torkkeli, M.; Serimaa, R.; Nyholm, P.; Ikkala, O., Micro- and Macrophase Separation in Phenolic Resol Resin/PEO-PPO-PEO Block Copolymer Blends: Effect of Hydrogen-Bonded PEO Length. *Macromol. Chem. Phys.,* 2002, 203, 388-392.

[50] Guo, Q.; Liu, J.; Chen, L.; Wang, K., Nanostructures and nanoporosity in thermoset epoxy blends with an amphiphilic polyisoprene -block-poly (4-vinyl pyridine) reactive diblock copolymer . *Polymer,* 2008, 49, 1737-1742.

[51] Nair, C. P. R.; Mathew, D.; Ninan, K. N., Cyanate ester resins , recent developments. *Adv. Polym. Sci.,* 2001, 155, 1-99.

[52] Fainleb, A. Thermostable polycyanurates: Synthesis, Modification, Structure and Properties; Nova Science Publishers: New York, 2010.

[53] Kiefer, J.; Hilborn, J. G.; Hedrick, J. L.; Cha, H. J.; Yoon, D. Y.; Hedrick, J. C., Microporous cyanurate networks via chemically induced phase separation. *Macromolecules,* 1996, 29, 8546-8548.

[54] Grande, D.; Grigoryeva, O.; Fainleib, A.; Gusakova, K.; Lorthioir, C., Porous thermosets via hydrolytic degradation of poly(epsilon-caprolactone) fragments in cyanurate-based hybrid networks. *Eur. Polym. J.,* 2008, 44, 3588-3598.

[55] Grigoryeva, O.; Gusakova, K.; Fainleib, A.; Grande, D., Nanopore generation in hybrid polycyanurate/poly(ε-caprolactone) thermostable networks. *Eur. Polym. J.,* 2011, 47, 1736-1745.

[56] Grande, D.; Grigoryeva, O.; Gusakova, K.; Fainleib, A., Novel mesoporous high-performance films derived from polycyanurate networks containing high-boiling temperature liquids . *Eur. Polym. J.,* 2013, 49, 2162-2171.

[57] Kim, S. C.; Sperling, L. H. IPNs around the world: science and engineering ; Wiley-VCH: Chichester, UK , 1997.

[58] Widmaier, J. M.; Sperling, L. H., Phase continuity in sequential poly(n-butyl acrylate)/polystyrene interpenetrating polymer networks . *Macromolecules,* 1982, 15, 625-631.

[59] Widmaier, J.-M.; Sperling, L. H., The use of labile crosslinker for morphological studies in interpenetrating polymer networks . *Br. Polymer J.,* 1984, 16, 46-48.

[60] Du Prez, F.; Goethals, E. J., "Compatibilizing effect" in interpenetrating polymer networks . *Macromol. Chem. Phys.,* 1995, 196, 903-914.

[61] Kayaman-Apohan, N.; Baysal, B. M., Semi-Interpenetrating Hydrogel Networks of Poly(2-hydroxyethyl methacrylate) with Poly[(D,L-lactic acid)-co-(ε-caprolactam)]. *Macromol. Chem. Phys.,* 2001, 202, 1182-1188.

[62] Hu, J.; Pompe, G.; Schulze, U.; Pionteck, J., Synthesis, electron irradiation modification and characterization of polyethylene/poly(butyl methacrylate -co-methyl methacrylate) interpenetrating polymer network. *Polym. Adv. Technol.,* 1998, 9, 746-751.

[63] Hu, J.; Schulze, U.; Pionteck, J., Degradation of interpenetrating polymer networks based on PE and polymethacrylates by electron beam irradiation . *Polymer,* 1999, 40, 5279-5284.

[64] Pionteck, J.; Hu, J.; Schulze, U., Membrane properties of microporous structures prepared from polyethylene/polymethacrylate IPN. *J. Appl. Polym. Sci.,* 2003, 89, 1976-1982.

[65] Rohman, G.; Grande, D.; Laupretre, F.; Boileau, S.; Guerin, P., Design of porous polymeric materials from interpenetrating polymer networks (IPNs): Poly(DL-lactide)/poly(methyl methacrylate) -base d semi-IPN systems. *Macromolecules,* 2005, 38, 7274-7285.

[66] Rohman, G.; Laupretre, F.; Boileau, S.; Guerin, P.; Grande, D., Poly(D,L-lactide)/poly(methyl methacrylate) interpenetrating polymer networks : Synthesis, characterization, and use as precursors to porous polymeric materials . *Polymer,* 2007, 48, 7017-7028.

[67] Lav, T.-X.; Carbonnier, B.; Guerrouache, M.; Grande, D., Porous polystyrene -based monolithic materials templated by semi-interpenetrating polymer networks for capillary electrochromatography. *Polymer,* 2010, 51, 5890-5894.

[68] Lee, J. S.; Hirao, A.; Nakahama, S., Polymerization of monomers containing functional silyl groups. 5. Synthesis of new porous membranes with functional groups. *Macromolecules,* 1988, 21, 274-276.

[69] Zalusky, A. S.; Olayo-Valles, R.; Taylor, C. J.; Hillmyer, M. A., Mesoporous Polystyrene Monoliths. *J. Am. Chem. Soc.,* 2001, 123, 1519-1520.

[70] Zalusky, A. S.; Olayo-Valles, R.; Wolf, J. H.; Hillmyer, M. A., Ordered Nanoporous Polymers from Polystyrene−Polylactide Block Copolymers. *J. Am. Chem. Soc.,* 2002, 124, 12761-12773.

[71] Vayer, M.; Hillmyer, M. A.; Dirany, M.; Thevenin, G.; Erre, R.; Sinturel, C., Perpendicular orientation of cylindrical domains upon solvent annealing thin films of polystyrene -b-polylactide. *Thin Solid Films,* 2010, 518, 3710-3715.

[72] Gorzolnik, B.; Davidson, P.; Beurroies, I.; Denoyel, R.; Grande, D., Novel Functional Mesoporous Materials Obtained from Nanostructured Diblock Copolymers. *Macromol. Symp.,* 2010, 287, 127-134.

[73] Grande, D.; Penelle, J.; Davidson, P.; Beurroies, I.; Denoyel, R., Functionalized ordered nanoporous polymeric materials : From the synthesis of diblock copolymers to their nanostructuration and their selective degradation . *Microporous Mesoporous Mater.,* 2011, 140, 34-39.

[74] Majdoub, R.; Antoun, T.; Le Droumaguet, B.; Benzina, M.; Grande, D., Original route to polylactide–polystyrene diblock copolymers containing a sulfonyl group at the junction between both blocks as precursors to functional nanoporous materials . *React. Funct. Polym.,* 2012, 72, 495-502.

[75] Cavicchi, K. A.; Zalusky, A. S.; Hillmyer, M. A.; Lodge, T. P., An Ordered Nanoporous Monolith from an Elastomeric Crosslinked Block Copolymer Precursor. *Macromol. Rapid Commun.,* 2004, 25, 704-709.

[76] Hansen, M. S.; Vigild, M. E.; Berg, R. H.; Ndoni, S., Nanoporous Crosslinked Polyisoprene from Polyisoprene-Polydimethylsiloxane Block Copolymer. *Polym. Bull.,* 2004, 51, 403-409.

[77] Szewczykowski, P. P.; Andersen, K.; Schulte, L.; Mortensen, K.; Vigild, M. E.; Ndoni, S., Elastomers with Reversible Nanoporosity. *Macromolecules,* 2009, *42*, 5636-5641.

[78] Hedrick, J. L.; Miller, R. D.; Hawker, C. J.; Carter, K. R.; Volksen, W.; Yoon, D. Y.; Trollsås, M., Templating Nanoporosity in Thin-Film Dielectric Insulators. *Adv. Mater.,* 1998, 10, 1049-1053.

[79] Chen, S.-Y.; Huang, Y.; Tsiang, R. C.-C., Ozonolysis efficiency of PS-b-PI block copolymers for forming nanoporous polystyrene . *J. Polym. Sci., Part A: Polym. Chem.,* 2008, 46, 1964-1973.

[80] Sekine, R.; Sato, N.; Matsuyama, T.; Akasaka, S.; Hasegawa, H., Radiation -induced fabrication of polymer nanoporous materials from microphase-separated structure of

diblock copolymers as a template. *J. Polym. Sci., Part A: Polym. Chem.,* 2007, 45, 5916-5922.

[81] Li, Y.; Ito, T., Surface Chemical Functionalization of Cylindrical Nanopores Derived from a Polystyrene–Poly(methylmethacrylate) Diblock Copolymer via Amidation. *Langmuir,* 2008, 24, 8959-8963.

[82] Xu, T.; Kim, H.-C.; DeRouchey, J.; Seney, C.; Levesque, C.; Martin, P.; Stafford, C. M.; Russell, T. P., The influence of molecular weight on nanoporous polymer films . *Polymer,* 2001, 42, 9091-9095.

[83] Glassner, M.; Blinco, J. P.; Barner-Kowollik, C., Formation of nanoporous materials via mild retro-Diels-Alder chemistry. *Polym. Chem.,* 2011, 2, 83-87.

[84] Kang, M.; Moon , B., Synthesis of Photocleavable Poly(styrene -block-ethylene oxide) and Its Self-Assembly into Nanoporous Thin Films. *Macromolecules,* 2009, 42, 455-458.

[85] Zhao, H.; Gu, W.; Thielke, M. W.; Sterner, E.; Tsai, T.; Russell, T. P.; Coughlin, E. B.; Theato, P., Functionalized Nanoporous Thin Films and Fibers from Photocleavable Block Copolymers Featuring Activated Esters. *Macromolecules,* 2013, 46, 5195-5201.

[86] Ryu, J.-H.; Park, S.; Kim, B.; Klaikherd, A.; Russell, T. P.; Thayumanavan, S., Highly Ordered Gold Nanotubes Using Thiols at a Cleavable Block Copolymer Interface. *J. Am. Chem. Soc.,* 2009, 131, 9870-9871.

[87] Zhang, M.; Yang, L.; Yurt, S.; Misner, M. J.; Chen, J. T.; Coughlin, E. B.; Venkataraman, D.; Russell, T. P., Highly Ordered Nanoporous Thin Films from Cleavable Polystyrene-block-poly(ethylene oxide). *Adv. Mater.,* 2007, 19, 1571-1576.

[88] Fustin, C. A.; Lohmeijer, B. G. G.; Duwez, A. S.; Jonas, A. M.; Schubert, U. S.; Gohy, J. F., Nanoporous Thin Films from Self-Assembled Metallo- Supramolecular Block Copolymers. *Adv. Mater.,* 2005, 17, 1162-1165.

[89] Takahashi, A.; Rho, Y.; Higashihara, T.; Ahn, B.; Ree, M.; Ueda, M., Preparation of Nanoporous Poly(3-hexylthiophene) Films Based on a Template System of Block Copolymers via Ionic Interaction. *Macromolecules,* 2010, 43, 4843-4852.

[90] Crossland, E. J. W.; Cunha, P.; Scroggins, S.; Moratti, S.; Yurchenko, O.; Steiner, U.; Hillmyer, M. A.; Ludwigs, S., Soft-Etch Mesoporous Hole-Conducting Block Copolymer Templates. *ACS Nano,* 2010, 4, 962-966.

[91] Crossland, E. J. W.; Cunha, P.; Ludwigs, S.; Hillmyer, M. A.; Steiner, U., In situ Electrochemical Monitoring of Selective Etching in Ordered Mesoporous Block-Copolymer Templates. *ACS Appl. Mater. Interfaces,* 2011, 3, 1375-1379.

[92] Wolf, J. H.; Hillmyer, M. A., Ordered Nanoporous Poly(cyclohexylethylene). *Langmuir,* 2003, 19, 6553-6560.

[93] Seo, M.; Amendt, M. A.; Hillmyer, M. A., Cross-Linked Nanoporous Materials from Reactive and Multifunctional Block Polymers. *Macromolecules,* 2011, 44, 9310-9318.

[94] Phillip, W. A.; Amendt, M.; O'Neill, B.; Chen, L.; Hillmyer, M. A.; Cussler, E. L., Diffusion and Flow Across Nanoporous Polydicyclopentadiene-Based Membranes. *ACS Appl. Mater. Interfaces,* 2009, 1, 472-480.

[95] Ndoni, S.; Vigild, M. E.; Berg, R. H., Nanoporous Materials with Spherical and Gyroid Cavities Created by Quantitative Etching of Polydimethylsiloxane in Polystyrene–Polydimethylsiloxane Block Copolymers. *J. Am. Chem. Soc.,* 2003, 125, 13366-13367.

[96] Jeong, U.; Ryu, D. Y.; Kim, J. K.; Kim, D. H.; Russell, T. P.; Hawker, C. J., Volume Contractions Induced by Crosslinking: A Novel Route to Nanoporous Polymer Films. *Adv. Mater.,* 2003, 15, 1247-1250.

[97] Fu, G. D.; Yuan, Z.; Kang, E. T.; Neoh, K. G.; Lai, D. M.; Huan, A. C. H., Nanoporous Ultra-Low-Dielectric-Constant Fluoropolymer Films via Selective UV Decomposition of Poly(pentafluorostyrene)-block-Poly(methyl methacrylate) Copolymers Prepared Using Atom Transfer Radical Polymerization. *Adv. Funct. Mater.,* 2005, 15, 315-322.

[98] Chen, W.; Park, S.; Wang, J.-Y.; Russell, T. P., Fabrication of Nanoporous Block Copolymer Thin Films through Mediation of Interfacial Interactions with UV Cross-Linked Polystyrene. *Macromolecules,* 2009, 42, 7213-7216.

[99] Ndoni, S.; Li, L.; Schulte, L.; Szewczykowski, P. P.; Hansen, T. W.; Guo, F.; Berg, R. H.; Vigild, M. E., Controlled Photooxidation of Nanoporous Polymers. *Macromolecules,* 2009, 42, 3877-3880.

[100] Bang, J.; Kim, S. H.; Drockenmuller, E.; Misner, M. J.; Russell, T. P.; Hawker, C. J., Defect-Free Nanoporous Thin Films from ABC Triblock Copolymers. *J. Am. Chem. Soc.,* 2006, 128, 7622-7629.

[101] Rzayev, J.; Hillmyer, M. A., Nanochannel Array Plastics with Tailored Surface Chemistry. *J. Am. Chem. Soc.,* 2005, 127, 13373-13379.

[102] Bailey, T. S.; Rzayev, J.; Hillmyer, M. A., Routes to Alkene and Epoxide Functionalized Nanoporous Materials from Poly(styrene -b-isoprene -b-lactide) Triblock Copolymers. *Macromolecules,* 2006, 39, 8772-8781.

[103] Avgeropoulos, A.; Z-H Chan, V.; Lee, V. Y.; Ngo, D.; Miller, R. D.; Hadjichristidis, N.; Thomas, E. L., Synthesis and Morphological Behavior of Silicon-Containing Triblock Copolymers for Nanostructure Applications. *Chem. Mater.,* 1998, 10, 2109-2115.

[104] Guo, F.; Jankova, K.; Schulte, L.; Vigild, M. E.; Ndoni, S., One-Step Routes from Di- and Triblock Copolymer Precursors to Hydrophilic Nanoporous Poly(acrylic acid)-b-polystyrene . *Macromolecules,* 2008, 41, 1486-1493.

[105] Sperschneider, A.; Schacher, F.; Gawenda, M.; Tsarkova, L.; Müller, A. H. E.; Ulbricht, M.; Krausch, G.; Köhler, J., Towards Nanoporous Membranes based on ABC Triblock Terpolymers. *Small,* 2007, 3, 1056-1063.

[106] Chua, K. S.; Koh, A. P.; Lam, Y. M., The influence of polarity of additive molecules on micelle structures of polystyrene -block-poly(4-vinylpyridine) in the fabrication of nano-porous templates. *J. Colloid Interface Sci.,* 2010, 351, 69-76.

[107] Fahmi, A. W.; Gutmann, J. S.; Vogel, R.; Gindy, N.; Stamm, M., Rheology Pathway to Macroscale Ordered Nanostructures of Polymeric Nanotemplates: Nanopores, Nanosheets and Nanofibers. *Macromol. Mater. Eng.,* 2006, 291, 1061-1073.

[108] Gao, J.; Zhang, P.; Fu, J.; Li, B.; Han, Y.; Yu, X.; Pan, C., Surface morphology evolution of poly(styrene -block-4-vinylpyridine) (PS-b-P4VP)(H^+) and poly(methyl methacrylate) -dibenzo-18-crown -6-poly(methyl methacrylate) (PMCMA) supramolecular film. *Polymer,* 2007, 48, 2425-2433.

[109] Laforgue, A.; Bazuin, C. G.; Prud'homme, R. E., A Study of the Supramolecular Approach in Controlling Diblock Copolymer Nanopatterning and Nanoporosity on Surfaces. *Macromolecules,* 2006, 39, 6473-6482.

[110] Lee, W.; Zhang, X.; Briber, R. M., A simple method for creating nanoporous block-copolymer thin films . *Polymer,* 2010, 51, 2376-2382.

[111] Adachi, M.; Okumura, A.; Sivaniah, E.; Hashimoto, T., Incorporation of Metal Nanoparticles into a Double Gyroid Network Texture. *Macromolecules,* 2006, 39, 7352-7357.

[112] Valkama, S.; Nykänen, A.; Kosonen, H.; Ramani, R.; Tuomisto, F.; Engelhardt, P.; ten Brinke, G.; Ikkala, O.; Ruokolainen, J., Hierarchical Porosity in Self-Assembled Polymers: Post-Modification of Block Copolymer–Phenolic Resin Complexes by Pyrolysis Allows the Control of Micro- and Mesoporosity. *Adv. Funct. Mater.,* 2007, 17, 183-190.

[113] Gamys, C. G.; Vlad, A.; Bertrand, O.; Gohy, J.-F., Functionalized Nanoporous Thin Films From Blends of Block Copolymers and Homopolymers Interacting via Hydrogen Bonding. *Macromol. Chem. Phys.,* 2012, 213, 2075-2080.

[114] Li, X.; Zhao, S.; Zhang, S.; Kim, D. H.; Knoll, W., Fabrication of Metallized Nanoporous Films from the Self-Assembly of a Block Copolymer and Homopolymer Mixture. *Langmuir,* 2007, 23, 6883-6888.

[115] Díaz, N.; Simon, F.-X.; Schmutz, M.; Rawiso, M.; Decher, G.; Jestin, J.; Mésini, P. J., Self-Assembled Diamide Nanotubes in Organic Solvents. *Angew. Chem., Int. Ed.,* 2005, *44, 3260*-3264.

[116] Simon, F.-X.; Khelfallah, N. S.; Schmutz, M.; Díaz, N.; Mésini, P. J., Formation of Helical Mesopores in Organic Polymer Matrices. *J. Am. Chem. Soc.,* 2007, 129, 3788-3789.

[117] Nguyen, T.-T.-T.; Simon, F.-X.; Khelfallah, N. S.; Schmutz, M.; Mesini, P. J., Mesoporous polymeric catalysts synthesized from self-assembled organic nanotubes as templates. *J. Mater. Chem.,* 2010, 20, 3831-3833.

In: Comprehensive Guide for Mesoporous Materials. Volume 1 ISBN: 978-1-63463-958-3
Editor: Mahmood Aliofkhazraei

Chapter 8

UNLOCKING THE POTENTIAL OF UV LIGHT FOR THE SYNTHESIS OF MESOPOROUS INORGANIC FILMS

Abraham Chemtob[1], Héloïse De Paz-Simon[1], Céline Croutxé-Barghorn[1], Séverinne Rigolet[2], Laure Michelin[2], Loïc Vidal[2] and Bénédicte Lebeau[2]

[1]Laboratory of Macromolecular Photochemistry and Engineering, University of Haute-Alsace, ENSCMu, Mulhouse Cedex, France

[2]Institut de Science des Matériaux de Mulhouse, University of Haute-Alsace, Mulhouse Cedex, France

ABSTRACT

In the synthesis of mesoporous materials, there is a high interest to find simple, efficient and rapid processing methods. Over the last decade, UV irradiation has made a valuable contribution in this field especially for mesoporous silica films. Initially implemented to eliminate the organic template by photodegradation, this technology has proven its utility for patterning, driving sol-gel polymerization and controlling self-assembly. Most photo-mediated processes rely on in situ photogeneration of Brönsted photoacids to promote hydrolysis and condensation of siloxane species. Many advantages can result from a light-induced polymerization as regard to efficiency (yields, reactivity), environmental impact (solvent-free, low energy consumption) and improved control (temporal and spatial control). This chapter provides for the first time a state-of-the-art and comprehensive review of domains in which light has enabled technical innovations in the synthesis of mesoporous inorganic films.

INTRODUCTION

Most industrial applications arising from photoinitiated polymerization are dominated by cross-linked polymer films, mainly for temperature-sensitive substrates like wood, paper and plastics [1]. The salient features of this technology area are a 100 % solid content formulation (no solvents or volatiles), a fast reticulation (in less than a few seconds) and a high energy saving process [2]. Despite these advantages, photopolymerization is still a niche process that has poorly penetrated other fields of materials chemistry. For example, there have been very few studies addressing the use of UV light in the preparation of inorganic or hybrid mesoporous materials . The reasons for this limited development are twofold: most UV-mediated polymerizations rely on organic monomers and resins , while mesoporous materials result from an inorganic polymerization, also referred to as sol-gel process [3]. Additionally, the relative randomness during photopolymerization precludes any extended spatial correlation , yielding generally amorphous structures devoid of spatial order at the mesoscopic scale [4].

Nonetheless, it is useful and relevant to initiate a discussion on the possible added value provided by UV light in the synthesis of mesoporous inorganic materials . In the literature, several examples have indeed demonstrated the benefits of a successful marriage of UV irradiation with the production of mesoporous films in three distinct fields: (i) to eliminate surfactant template; (ii) to prepare micro- or nanopatterned structures; and more recently, (iii) to make micrometer thick mesoporous silica film s through a light induced self-assembly approach. The present chapter attempts, for the first time, to provide a rational and comprehensive description of these three distinct application fields in which UV was successfully involved for the synthesis of mesoporous inorganic films. Before that, it seems relevant to give an overview of the general concepts and principles underlying the preparation of mesoporous films.

1. Processing of Mesoporous Films

1.1. Surfactant Templated Approach

The discovery of surfactant -templated silica mesophases in 1992 has led to a great deal of attention on the synthesis of ordered mesoporous inorganic oxides to extend the applications of microporous materials (with average pore diameters less than 2 nm) [5]. According to IUPAC notation [6], mesoporous materials contain larger pores with diameters ranging from 2 to 50 nm, the upper limit marking the demarcation with macroporous materials . They are characterized by high surface area s comprised between 500 and 1000 m^2 g^{-1} and uniform pore size s [7]. Another essential feature is the presence of spatial order on mesoscale leading to well-defined pore arrangement with a geometrical structure . (It is essential to acknowledge the existence of other mesoporous solids consisting of a spatially disordered and non-crystalline arrangement of pores, although they exhibit only short-range order.) Generally, mesoporous material may exhibit no atomic ordering found in crystalline solid phase , so that the pore walls are generally vitreous on the atomic scale without solvothermal treatment .

Before the onset of porosity , the first synthesis step is to create solid organic-inorganic hybrid mesophases. To this end, the sol-gel polymerization of inorganic precursors is combined with the formation lyotropic mesophases. These latter are liquid crystalline phases formed by the addition of solvents , and consisting of amphiphilic organic molecules ’ or macromolecules ' aggregates acting as template [5]. Depending on the mechanism, polymerization can either promote ordered mesostructures or simply fix the preformed mesophases, so that the solid hybrid structures cannot be distorted and disrupted by physical or chemical forces. Subsequently, mesoporosity is induced by removing the supramolecular arrays of surfactant molecules. *Self-assembly* and *sol-gel polymerization* are thus two crucial events underlying the synthesis of mesoporous materials , as sketched in Scheme 1:

- The simplest *self-assembly* mechanism relies on a hydrophilic continuous phase (inorganic precursor and solvent) selectively hydrating the hydrophilic moieties of each surfactant molecule. This "schizophrenic" relationship drives the surfactant molecules to self-assemble in lyotropic mesophases, with the precursor locally confined in the ordered hydrophilic regions thus minimizing its exposure to hydrophobic regions. Ideally, such "intermediate" hybrid mesostructures formed will show a direct correlation between their surfactant array size and shape and the final pore size and arrangement.
- *Sol-gel chemistry* is the mineral polymerization process converting the molecular inorganic precursor into a hydrophilic oxo-polymer cross-linked network. In the case of alkoxysilane as molecular silica source, the versatility of the sol-gel process is particularly relevant to work under conditions where the condensation rate of the hydrophilic silica precursors can be minimized [8]. The key idea is not to disturb the self-assembly process by premature gelation that would favor amorphous kinetically-trapped products. The concept of mesophase retention during condensation of inorganic species is of high significance as liquid crystal phase s inherently lack mechanical stability .

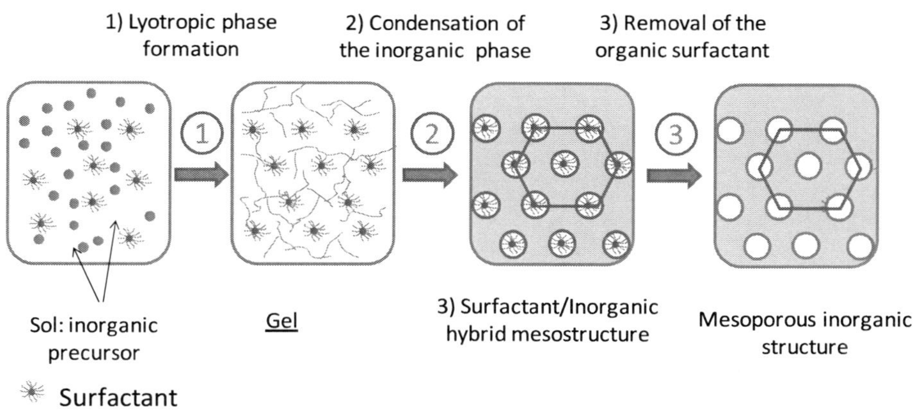

Scheme 1. Schematic representation of the surfactant -templated approach to mesoporous inorganic film.

1.2. Mesoporous Inorganic Film

A majority of works on mesoporous materials has concentrated on silica [9], transition metal oxide [10-11] and organic–inorganic hybrid structures [12-13] with an emphasis on powder and monolith morphologies suitable to catalysis [14-16] and adsorption [17] applications. Since 1996, alternative pathways have been designed to synthesize mesoporous materials shaped as thin films in order to further widen their application spectrum. On this issue, the opinion of Jeffrey Brinker (Sandia National Laboratories) is particularly evocative: [18] "*When I first became aware of the Mobil work on mesoporous silica, I thought it would be an important thing to make these materials as films , because after calcination to remove the surfactant , they would be useful as membranes , sensors , and low dielectric constant insulators* ".

First reported by the groups of Ozin [19] and Brinker [20], the Evaporation Induced Self-Assembly (EISA) process is the most common method for the preparation of mesostructured thin films . Thanks to its versatility , the EISA process has rapidly gained in popularity and earned recognition as the reference methodology .

Practically [21-23], *(i) a homogeneous isotropic solution is first prepared*, including surfactant , prehydrolyzed low-polymerized inorganic precursor (sol), catalyst , water and large excess of a volatile solvent usually ethanol . Surfactant aggregation is hindered by the initial surfactant concentration in solution far below the CMC and the presence of a good solvent. Moreover, suppression of inorganic condensation is essential before film deposition , and the pH solution needs to be adjusted in order to favor fast hydrolysis of the inorganic precursor and slow condensation. *(ii) The latter isotropic solution is cast by spin or dip coating* resulting in the evaporation of the volatile components and the increase of the surfactant concentration above the CMC. This causes two major events forming the cornerstones of the EISA approach: formation of lyotropic mesophases and inorganic network through condensation reactions . Control of RH and volatile compound evaporation rate during film deposition is essential to self-assembly . *(iii) Prior to porosity liberation , the inorganic network is generally stabilized by post-treatment* (NH_3 or thermal induced condensation). *(iv) The template is eliminated* by thermal decomposition , solvent or aqueous solution selective extraction to yield ordered mesoporous films . Figure 1 summarizes the different steps of the EISA process.

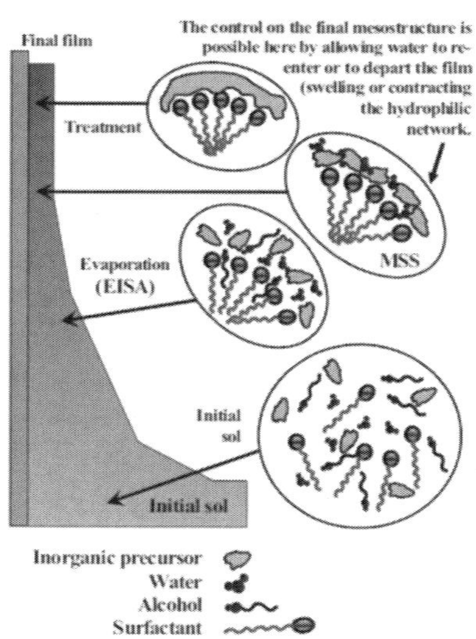

Figure 1. Schematic representation of the different steps of the EISA process to form mesostructured organic/inorganic film (copyright permission granted) [23].

1.3. Optimizing Mesoporous Film Preparation

The shaping of mesoporous materials as thin films is a key feature for applications as interlayer dielectrics in microelectronics [24], sensors [25], membranes [26] or photonic devices [27]. The EISA process has the advantage of allowing a fast mesophase formation . In addition, surface area , pore dimension and geometry are efficiently tailored through the adjustment of numerous chemical and processing parameters such as initial sol composition , pH , aging time, RH , evaporation rate, deposition technique, etc. [23] Its success is reflected today in the increasing number of publications and its impact on many technologically important fields.

Although the EISA process has reached maturity and created a considerable wealth of films , there is a continuous pressure to find easier, more affordable and eco-efficient preparation conditions. Several alternatives based on vapor phase synthesis [28-31], growth at air-water or oil -water interface [32], and electro-assisted deposition [33-34] have brought some improvements as regard process intensification. By contrast, there have been very few studies reporting the use of UV light at a specified stage of the synthesis procedure. Yet, this technology is widely implemented industrially for the preparation of organic polymer films , and well-known as an eco-efficient process [35] The following parts shall give a comprehensive overview of the UV technology's potential for improving the synthesis of mesoporous inorganic films.

2. UV -Driven Silica Condensation for Photopatterning Mesoporous Silica Films

The possibility to pattern mesoporous silica film s under UV exposure was first demonstrated by Brinker et al. in 2000 [36-37] Subsequently, other patterning techniques were implemented like dip-pen nanolithography , ink-jet printing and selective dewetting to create spatially-controlled silica mesoporous patterns [38] However, the first choice was a simple UV patterning technique. Their synthesis procedure differed from the EISA protocol only by the addition of a photoacid generator (PAG) in the sol. To facilitate its incorporation in the initial precursor/surfactant solution (TEOS /EtOH/water /HCl/$C_{16}H_{33}(OCH_2CH_3)_{10}OH$: 1/20/3.1/0.0065/0.063 to 0.127 moles ratio), an amphiphilic PAG (diaryliodonium salt bearing a long hydrocarbon chain) was employed (0.0156 moles ratio). As expected from the evaporation induced self-assembly route, the film deposition enabled the formation of poorly condensed but well-ordered hexagonal hybrid silica-surfactant mesophases. The resultant mesostructured film was then exposed to UV light through a mask. The spatially-controlled irradiation caused the local decomposition of the PAG in the exposed areas, which promoted the selective siloxane condensation resulting in patterned regions of more or less highly condensed silica.

As illustrated in the different schemes of Figure 2, three types of patterns were generated. In scheme 1, the non-exposed areas are much less condensed and can be readily etched with an aqueous base solution to remove totally the hybrid film. The result is a hierarchically structured film in which organization resides on multiple length scales: mesoporosity (2-10 nm), film thickness (50-500 nm), and pattern size (0.5-2 μm). A second aspect of the approach revealed in Scheme 2 is the optical definition of film mesostructure. For this, the patterned surfactant -silica hybrid film was heated to create adjoining regions of different refractive index ($\Delta n = n_{irradiated} - n_{unirradiated} > 0$). Remarkably, Δn was controlled by the UV dose which determines the siloxane condensation of the exposed parts. The ability to pattern the refractive index was exploited to create optical diffraction gratings . At higher surfactant concentration, a third application displayed in Scheme 3 shows a light -induced mesophase transformation from hexagonal to tetragonal structure related to silica-surfactant reorganization due to an increase of siloxane condensation.

The idea of UV patterning was taken up and elaborated on several years later by Watkins et al. at Cornell University [40-41]. For these authors, the main issue was the contradiction between diffusivity of photoacid species (to unexposed areas) in a mobile poorly condensed film and spatial restriction imposed by the lithographic approaches for pattern generation. Hence, the authors proposed a method decoupling mesophase formation and inorganic network formation. In their procedure, template mesophases were first generated from bulk amphiphilic block copolymer film (Pluronic F108: $(PEO)_{135}(PPO)_{52}(PEO)_{135}$) including a triphenylsulfonium triflate as PAG. The organic films were then exposed to UV radiation through a photomask. The triflic acid photogenerated in the illuminated regions tended to segregate into hydrophilic domains of the block copolymer. Upon infusion of silicon alkoxides, patterned silicate network were thus formed. Removal of the template by calcination (400 °C) provided a patterned mesoporous silica film. Figure 3 shows schematically the different steps of the process and the two levels of structuration using optical and electronic transmission microscopy . Using a variant procedure, patterned mesoporous films exhibiting ordered mesochannels parallel to the substrate were also achieved [42].

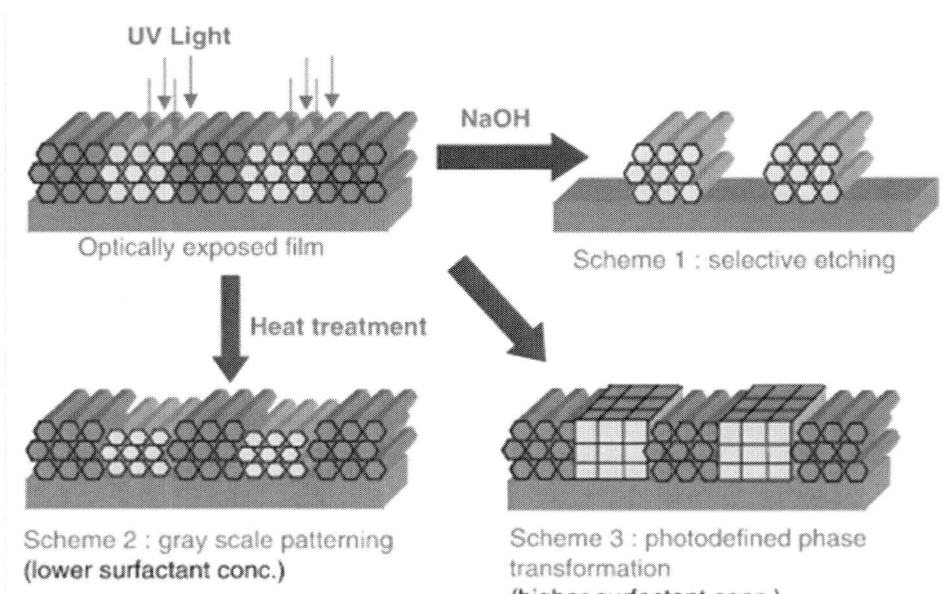

Figure 2. Summary of the different applications of patterned polymerization induced by UV irradiation (copyright permission granted) [39].

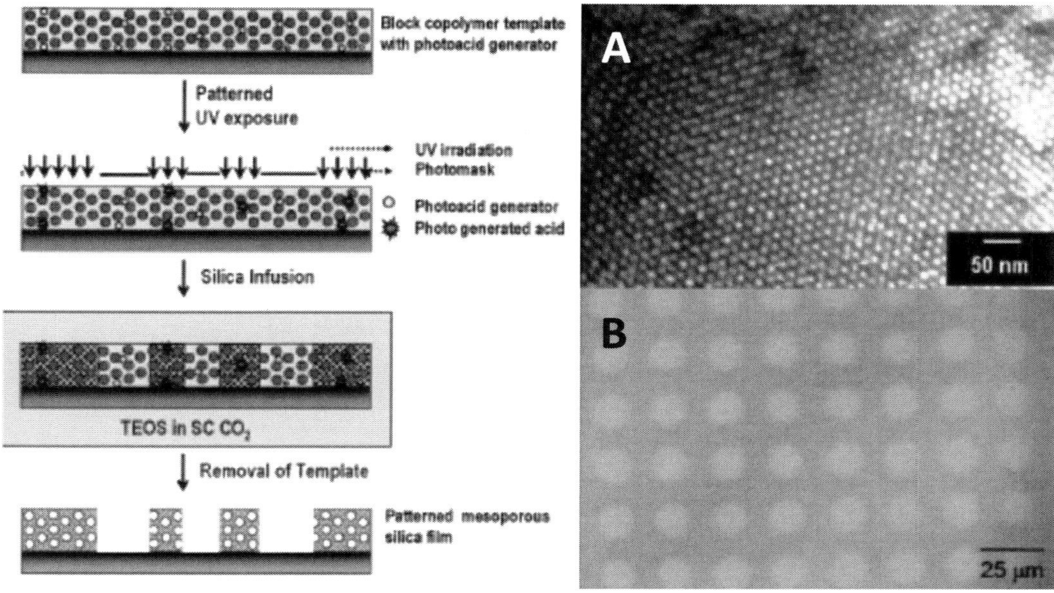

Figure 3. (Left) Three different steps for the preparation of micro-patterned mesoporous silica film s . (Right) Transmission electron (A) and optical (B) images show respectively the mesostructures and device level microstructure present in the mesoporous silica film formed in presence of F108 copolymer template (copyright permission granted) [40].

However, facilitated diffusion of photoacids in this system led to modest pattern resolution with poorly defined boundaries. To overcome this problem, the same authors used chemically amplifiable block copolymer template such as poly(styrene -*b*-tert-butoxycarbonyyloxy styrene) (PS-*b*-PtbocSt). In addition to condensing TEOS , photoacids change the PtbocSt block in a more hydrophilic poly(4-hydroxystyrene) block (PHOSt). This establishes a higher polarity and glass transition temperature gap between the exposed and non-exposed areas, which restricts the diffusion of photogenerated acids in non-exposed areas, thereby providing sharper boundaries. Well-defined patterned domains were obtained but the mesostructured structures were not ordered due to low phase separation between the two blocks of the copolymer. More recently, they proposed in another example, patterned mesoporous silicate films using positive and negative tone strategies using poly-(styrene-*b*-tert-butyl acrylate) (PS-*b*-PtbA) templates [43] Template exposure through the mask triggered area selective generation of acid from the PAG, which in turn deprotected the PtbA block to yield poly(acrylic acid) block. Heating above 120 °C enabled its cross-linking into insoluble poly(acrylic anhydride).

Song et al. [44] reported in 2008 a similar methodology to patternable mesoporous silica thin films using a Brij78/Poly(hydrostyrene) (PHOSt) mixture, respectively amphiphilic block copolymer and homopolymer.

The photoacid-catalyzed TEOS condensation caused both a selective phase segregation of the homopolymer and selective templating of the Brij mesostructure. As a result, calcination afforded not only mesopores (from the Brij phase) but also macropores (from the PHOSt phase) as shown in Figure 4.

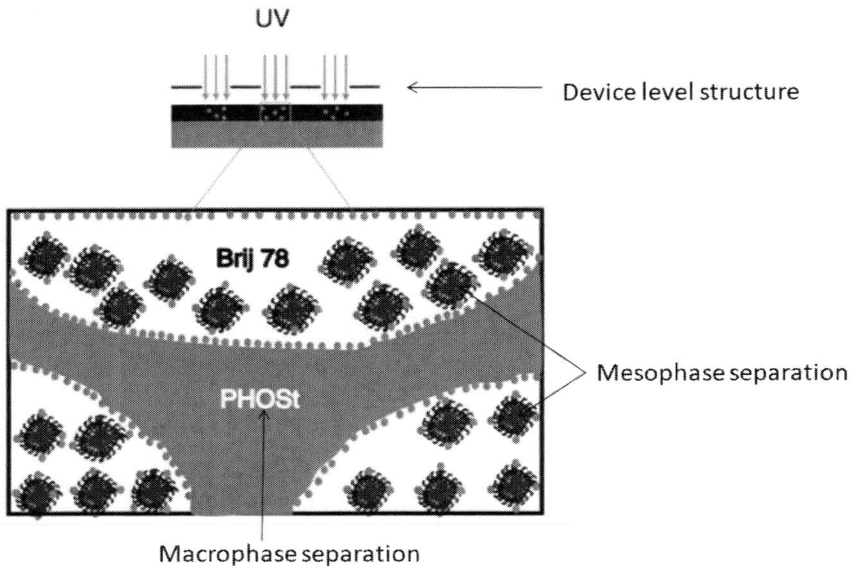

Figure 4. Hierarchically organized porous silica films produced by polymerization -induced phase separation of PHOSt (macropores) and self-assembly of Brij 78 (mesopores) (copyright permission granted) [44].

3. Photodegradation of Organic Template

A second topic in which UV light and mesoporous materials have overlapped is the elimination of the organic template. Such photoablation or photocalcination approach was reported independently in 2000 by two research groups: Hozumi et al. (Nagoya University, Japan) and Parikh et al. (University of California, USA). Their motivation was the selective removal of the organic template while preserving not only the integrity of the silica mesostructure (size, geometry) but also that of the substrate which can be solvent or thermally sensitive.

Indeed, thermal calcination is known to cause film distortion or mesostructure collapse owing to the considerable amount of internal stress generated by film shrinkage during the heating at high temperature . UV ablation is anticipated to bring up substantial advantages to thermally unstable silica /surfactant mesostructured materials which tend to collapse owing to an excessive thermal instability. As a first difference, the group of Hozumi et al. focused on photoablation of ionic surfactant (cetyltrimethylammonium chloride) *versus* block copolymer template (Brij, Pluronic) for the Parikh group. The second difference resides in the type of UV irradiation device. Hozumi et al. [45-46] employed highly energetic photons (deep UV) arising from a monochromatic excimer lamp (λ = 172 nm, 10 mW cm^{-2}) requiring harmful and expensive vacuum chamber. By contrast, Parikh et al. implemented a more affordable low-pressure mercury (LPM) lamp [47-49] Preferred applications of this lamp include conventionally disinfection of water , air and surfaces in the food and beverage industry and in water works. Even if 85 % of the emission occurs near 254 nm, the LPM lamp has also lower levels of radiation at wavelengths about 185 nm.

Although a complete destruction of the template was reported in both cases, the mechanism is not completely elucidated. Short wavelengths (172 nm) can facilitate the dissociation of oxygen from the ambient atmosphere to produce *in situ* ozone and atomic oxygen. For the longer wavelengths of the LPM lamp (254 nm emission line), a direct cleavage of C-C and C-H bonds can occur to destroy the surfactant molecules into ions , free radicals and excited polymer fragments . Finally, the activated organic species can be further decomposed via a complex set of oxidation processes involving oxygen and ozone to form simpler volatile products such CO_2 , H_2O and N_2 escaping from the film sample. According to Hozumi et al., the elimination rate of the organic species is dependent on O_2 concentration and may vary with the pressure of the chamber [46] The same authors found that photocalcination caused much less damage in the film distortion and stress as well as the conservation of most silanol moieties. This later point is very interesting from the perspective of pore modification since the superficial silanol moieties can act as anchoring point for silane coupling reagents . Parikh et al. used FTIR spectroscopy to assess the progressive elimination of the copolymer surfactant during irradiation time, using in particular the disappearance of CH stretching modes as a probe (Figure 5A). A complete removal of the template was demonstrated in 60 – 120 min, as well as a concomitant densification of the silica skeleton [49]. Finally, both groups exploited UV spatial control to form micropatterned silica films alternating mesoporous inorganic domains (silica) and mesostructured hybrid domains (silica/surfactant) as shown in Figure 5B.

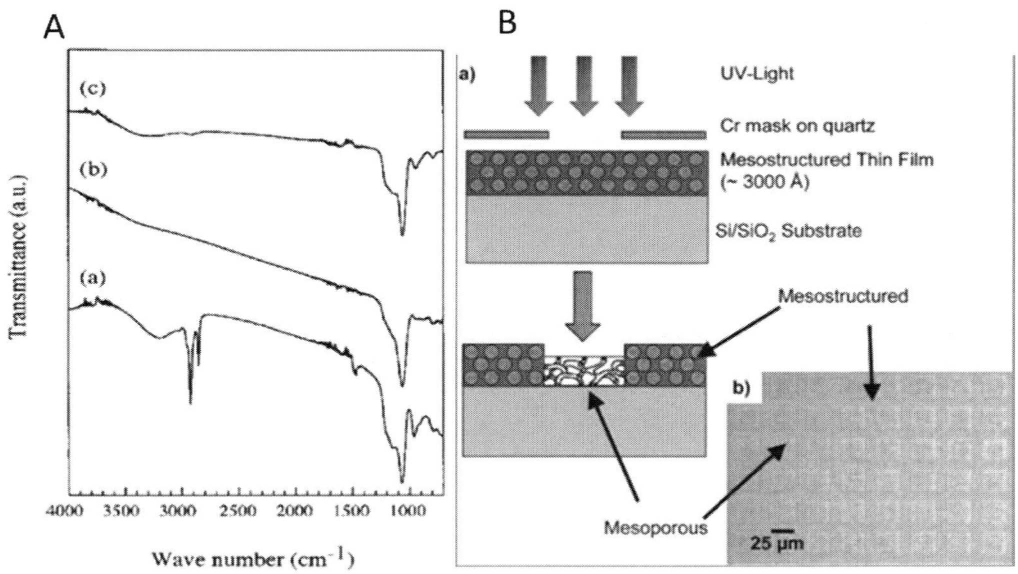

Figure 5. A) IR spectra of as-synthesized mesostructured silica /surfactant films (a) and the same films after thermocalcination at 573 K for 1 h (b) and photocalcination for 3h (c) [45]. B) Preparation of patterned mesostructured/mesoporous silica thin films using a photomask [50].

4. UV INDUCED SELF-ASSEMBLY

4.1. Concept of Light Induced Micellization

The third application places light as a critical feature in the preparation of mesoporous inorganic films since its role is now to control the self-assembly process. Such concept of UV -mediated self-assembly was first proved in organic solution , at surfactant concentrations sufficiently low to prevent the formation of lyotropic mesophases. The main idea was first a UV-induced micellization of specific light-responsive block copolymers [51]. Practically, a specific UV-sensitive block is photochemically modified in a way to change its affinity for the solvent, thereby causing the copolymer micellization. This concept of light-induced copolymer self-assembly may appear out of the scope , but the reader will see that it has many similarities with self-assembly in silica /surfactant system. In this paragraph, we will only review micellization induced through photolysis [52], but recent studies revealed also several cases of micellization trigerred by photoelectron transfer [44] and photoClaisen rearrangement [16].

The first study of Yoshida et al. in 2007 showed how a UV light could trigger on demand micellization from a dichloromethane solution including a soluble poly(4-tert-butoxystyrene)-*b*-polystyrene copolymer (PBSt-*b*-PSt) [53]. First, bis(alkylphenyl)iodonium hexafluorophosphate acting as PAG was added to the organic solution. Upon UV irradiation provided by a high-pressure mercury lamp, rapid micellization was confirmed by electron microscopy showing the presence of spherical particles. As depicted in Figure 6, micellization was attributed to the photoinduced hydrolysis of the hydrophobic PBSt block into hydrophilic and insoluble poly(4-vinylphenol)(PVPh), thus creating in situ an amphiphilic structure .

Rieger et al. [54] reported also the photoinduced micellization of PBSt-*b*-PSt block copolymer in CH_2Cl_2, but the main difference was the addition of siloxane cross-linkers. These latter species confined in the hydrophilic core locked in the micelle structure by sol-gel condensation ; the result was the formation of hybrid organic/inorganic nanoparticles . In 2011, Bertrand et al. reported the synthesis and self-assembly of a novel photocleavable block copolymers (poly(dimethoxynitrobenzyl acrylate)-*b*-polystyrene (PDMNBA-*b*-PS) [55-56].

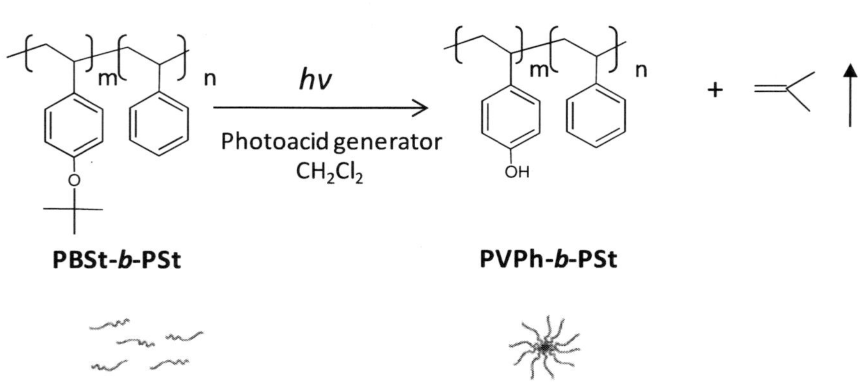

Figure 6. Micellization induced by photolyisis of PBSt-b-PSt block copolymer in CH_2Cl_2 [51].

Upon UV irradiation, the hydrophobic PDMNBA block was converted into a fully hydrophilic poly(acrylic acid) (PAA) block insoluble in the solvent (chloroform). Thus, self-assembly occurred, forming micelles exhibiting PAA core and PS corona.

4.2. Mesoporous Silica Films through Light Induced Self-Assembly

The next step was thus a light induced self-assembly process applicable to the synthesis of mesostructured surfactant /silica films. In 2012, Chemtob et al. released a simple approach towards mesoporous silica film s based on a fast photoinduced sol-gel process utilizing photoacid generators [57]. In contrast to EISA, the procedure began with nonhydrolyzed alkoxy silicate precursors in absence of solvent and even water. For the first time, self–assembly became photochemically induced, and was thus no longer dependent on film deposition conditions, solvent evaporation and the complex chemistry associated with a silica sol. The procedure relies only on three components: i. poly(dimethoxysiloxane) (PDMOS) is a fully alkoxylated oligomeric precursor, serving both as non-volatile silica source and solvent for the template; ii. diphenyliodonium salt ($\Phi_2I^+ PF_6^-$) used as PAG; and iii. an amphiphilic block copolymer (P123, $(PEO)_{19}$-$(PPO)_{69}$-$(PEO)_{19}$) behaving as structure directing agent. Their mixture forms a homogeneous solution cast as micrometer thick film. Prior to irradiation, the XRD pattern showed no peak diffraction confirming the absence of preformed mesophases (Figure 7). By contrast, a single broad diffraction peak appeared after UV irradiation and calcination. The absence of higher angle reflections ruled out a long-range order (no space group), suggesting rather a disordered mesoporous structure with a uniform pore motif [58]. Such XRD patterns could be reconciled with the wormlike pore structure of MSU-X59-60 or HMS61-62 powders disclosed in the mid-90s. Short-ranger ordering was also supported by TEM images. Indeed, Figure 7 revealed disordered wormlike channels, and the *d*-spacing of ca. 8 nm was thus related to pore–pore distance.

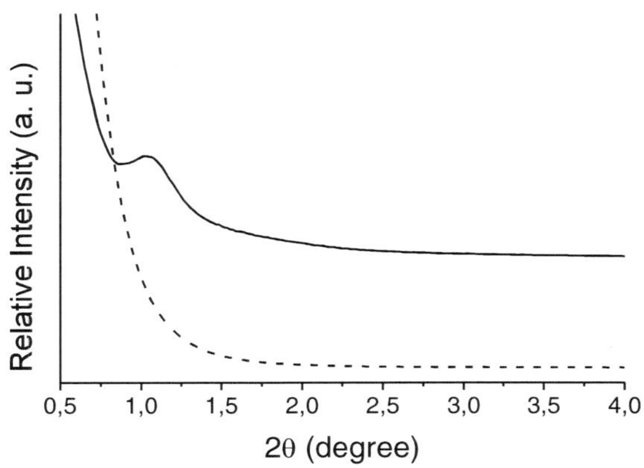

Figure 7. XRD pattern (left) of a P123/PDMOS = 0.50 (weight ratio) liquid film before UV irradiation (dotted line). The pattern of the solid silica film after UV irradiation and calcination is also shown for comparison (solid line). UV irradiation was performed under a UV conveyor using a microwave-powered mercury vapour lamp (H bulb, 10 passes, emitted light dose =1.46 J/cm²). Calcination was performed without preliminary hydrothermal treatment at 480 °C for 4 h in air. The TEM image (right) of the worm-like mesoporous silica film.

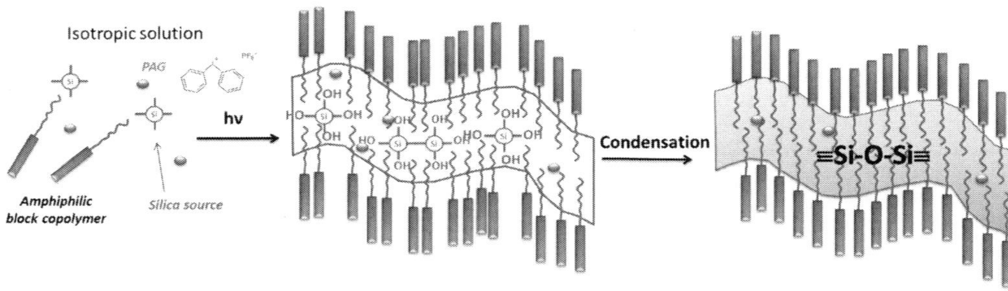

Figure 8. Processing pathway for the preparation of mesoporous silica films via LISA process.

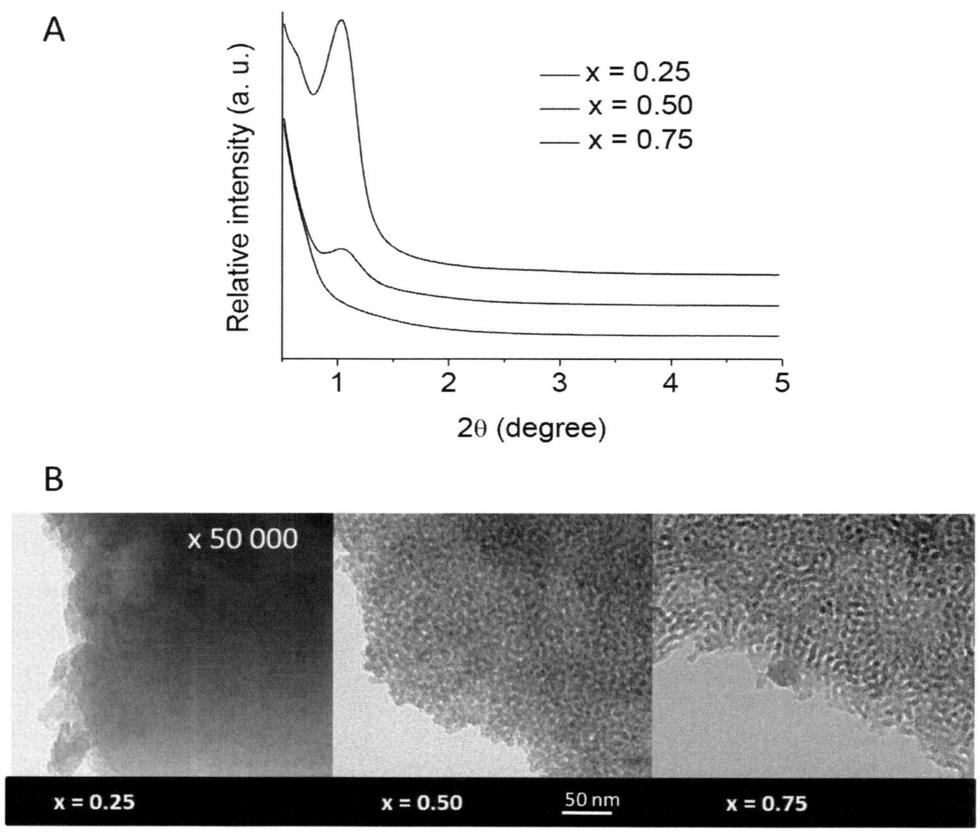

Figure 9. (A) XRD patterns obtained for calcined samples at various P123/PDMOS weight ratios: x = 0.25 (——), 0.50 (——) and 0.75 (——). (B) TEM images of the corresponding calcined films.

The previous results enabled to propose an original self-assembly mechanism [63] sketched in Figure 8 and referred to as light induced self-assembly (LISA). Upon UV exposure, the PAG photolysis caused the generation of a series of photoproducts (cations, radicals, radical cationic species) resulting from the homolytic and heterolytic Φ-I bond cleavage. The resulting species interacted with hydrogen donors (i.e., from precursor molecules or impurities) to create Brönsted acid of the structure $H^+PF_6^-$. These protonic superacids are thought to play a prominent role in the catalysis of the sol-gel process [64].

Hydrolysis of PDMOS induced a polarity change in the media from a hydrophobic to a hydrophilic environment, thus fostering the formation of lyotropic mesophases in which segregated PEO blocks were in contact with silica while the PPO blocks were in the core of the micelles. By contrast with previous studies on micellization, the driving force is not to change the copolymer structure to make it amphiphilic, but the modification of the solvation properties (the change from hydrophobic alkoxy precursor to hydrophilic silanol species). The third step involved the condensation and densification of the spatially confined inorganic network to fix the hybrid mesostructure. Finally, the last step permitted the advent of porosity by eliminating the template.

As can be seen graphically in Figure 9, increasing the P123/PDMOS (x) ratio did not affect the *d*-spacing value, but the single XRD peak became more intense and narrow indicating enhanced phase segregation. By contrast, the sample possessing the lowest concentration in P123 ($x = 0.25$) lacked of any discernible mesoscopic feature, reflecting an amorphous structure. Consequently, a threshold concentration in template ($x \geq 0.50$) was required to induce mesostructuration. In a recent contribution, Chemtob et al. established that controlling the relative humidity and the irradiance provided a rational basis to promote mesophase ordering. A range of periodically ordered silica /surfactant mesophases (hexagonal, lamellar, cubic) were thus achieved by this means [65].

CONCLUSION

As early as 2000, the potential of UV light in the preparation of mesoporous inorganic films has been demonstrated [36]. The first attempts simply introduced UV irradiation at a specified stage of a multi-step procedure. The main feature was the incorporation of photogenerated Brönsted acids to catalyze silica condensation. Such species were widely recognized for the initiation of cationic photopolymerization, but their catalytic role in sol-gel chemistry was not at that time widespread. The spatially controlled generation of photoacids upon UV exposure enabled to create patterned mesoporous films. Before this breakthrough, specific patterning of mesoporous films produced via conventional methods was not easily compatible with dip-coating or spin -coating of the precursor solution onto a substrate without further processing.

A second successful application field consisted in using UV light as a less aggressive way to eliminate the organic template from the silica /surfactant mesostructured film. Compared to calcinations process conducted at high temperature, such photochemical method was suitable to temperature-sensitive substrates and decreased the risk of film cracking and mesostructure damages. Very recently, UV has emerged also as an efficient vector to trigger both surfactant self-assembly and sol-gel polymerization.

The consequence was a simple, efficient, and solvent-free route to mesoporous silica film s. It is remarkable to observe, more than 20 years after the initial investigations of Kresge and co-workers, how UV irradiation can provide improvements as regards to process conditions [66-67].

REFERENCES

[1] Fouassier, J. P.; Lalevée, J., Photoinitiators for Polymer Synthesis: Scope, Reactivy and Efficiency. Wiley-VCH: Weinheim, Germany , 2012.

[2] Schwalm, R., UV Coatings Basics, Recent Developments and New Applications. First ed.; Elsevier: Amsterdam, The Netherlands , 2007; p 223.

[3] Sanchez, C.; Rozes, L.; Ribot, F.; Laberty-Robert, C.; Grosso, D.; Sassoye, C.; Boissiere, C.; Nicole, L., "Chimie douce": A land of opportunities for the designed construction of functional inorganic and hybrid organic-inorganic nanomaterials . *C. R. Chim.,* 2010, 13, 3-39.

[4] Forney, B. S.; Guymon, C. A., Nanostructure Evolution during Photopolymerization in Lyotropic Liquid Crystal Templates. *Macromolecules,* 2010, 43, 8502-8510.

[5] Xu, R.; Pang, W.; Yu, J.; Huo, Q.; Chen, J., Synthesis, structure , and characterization of mesoporous materials . In Chemistry of Zeolites and Related Porous Materials: Synthesis and Structure, John Wiley & Sons ed.; 2007; pp 467-601.

[6] McCusker, L. B., IUPAC nomenclature for ordered microporous and mesoporous materials and its application to non-zeolite microporous mineral phases. *Reviews in mineralogy and geochemistry,* 2005, 57, 1-16.

[7] Zhao, D.; Wan, Y.; Zhou, W., *Ordered Mesoporous Materials.* Wiley-VCH: Weinheim, 2013.

[8] Dunphy, D. R.; Smarsly, B.; Brinker, C. J., Control of morphology in mesoporous and mesostructured hybrid materials In Supramolecular Chemistry of Organic-Inorganic Hybrid Materials Rurack, K.; Martinez-Manez, R., Eds. 2010; pp 531-545.

[9] Wan, Y.; Zhao, On the Controllable Soft-Templating Approach to Mesoporous Silicates. *Chem. Rev.,* 2007, 107, 2821-2860.

[10] Wang, J.; Zhou, J.; Li, Z.; He, Y.; Lin, S.; Liu, Q.; Zhang, M.; Jiang, Z., Mesoporous mixed metal oxides derived from P123-templated Mg-Al layered double hydroxides . *J. Solid State Chem.,* 183, 2511-2515.

[11] Crepaldi, E. L.; Soler-Illia, G. J. d. A. A.; Grosso, D.; Cagnol, F.; Ribot, F.; Sanchez, C., Controlled Formation of Highly Organized Mesoporous Titania Thin Films: From Mesostructured Hybrids to Mesoporous Nanoanatase TiO2. *J. Am. Chem. Soc.,* 2003, 125, 9770-9786.

[12] Nicole, L.; Boissiere, C.; Grosso, D.; Quach, A.; Sanchez, C., Mesostructured hybrid organic-inorganic thin films . *J. Mater. Chem.,* 2005, 15, 3598-3627.

[13] Sanchez, C.; Boissière, C.; Grosso, D.; Laberty, C.; Nicole, L., Design, Synthesis, and Properties of Inorganic and Hybrid Thin Films Having Periodically Organized Nanoporosity. *Chem. Mater.,* 2008, 20, 682-737.

[14] Corma, A., From Microporous to Mesoporous Molecular Sieve Materials and Their Use in Catalysis. *Chem. Rev.,* 1997, 97, 2373-2420.

[15] Galarneau, A.; Desplantier-Giscard, D.; Di Renzo, F.; Fajula, F., Thermal and mechanical stability of micelle-templated silica supports for catalysis . *Catal. Today,* 2001, 68, 191-200.

[16] Taguchi, A.; Schüth, F., Ordered mesoporous materials in catalysis . *Micropor. Mesopor. Mat.,* 2005, 77, 1-45.

[17] Liu, A.; Hidajat, K.; Kawi, S.; Zhao, D., A new class of hybrid mesoporous materials with functionalized organic monolayers for selective adsorption of heavy metal ion s . *Chem. Commun.,* 2000, 1145-1146.

[18] Brinker, C. J., Evaporation-Induced Self-Assembly: Functional Nanostructures Made Easy. *MRS Bull.,* 2004, 29, 631-640.

[19] Yang, H.; Kuperman, A.; Coombs, N.; Mamiche-Afara, S.; Ozin, G. A., Synthesis of oriented films of mesoporous silica on mica. *Nature,* 1996, 379, 703-705.

[20] Brinker, C. J.; Lu, Y.; Sellinger, A.; Fan, H., Evaporation-Induced Self-Assembly: Nanostructures Made Easy. *Adv. Mater.,* 1999, 11, 579-585.

[21] Cagnol, F.; Grosso, D.; Soler-Illia, G.; Crepaldi, E. L.; Babonneau, F.; Amenitsch, H.; Sanchez, C., Humidity-Controlled Mesostructuration in CTAB-templated Silica Thin Film Processing. The Existence of a Modulable Steady State. *J. Mater. Chem.,* 2003, 13, 61-66.

[22] Gibaud, A.; Grosso, D.; Smarsly, B.; Baptiste, A.; Bardeau, J. F.; Babonneau, F.; Doshi, D. A.; Chen, Z.; Brinker, C. J.; Sanchez, C., Evaporation-Controlled Self-Assembly of Silica Surfactant Mesophases. *J. Phys. Chem. B,* 2003, 107, 6114-6118.

[23] Grosso, D.; Cagnol, F.; Soler-Illia, G. J. d. A. A.; Crepaldi, E. L.; Amenitsch, H.; Brunet-Bruneau, A.; Bourgeois, A.; Sanchez, C., Fundamentals of Mesostructuring Through Evaporation-Induced Self-Assembly. *Adv. Funct. Mater.,* 2004, 14, 309-322.

[24] Hatton, B. D.; Landskron, K.; Whitnall, W.; Perovic, D. D.; Ozin, G. A., Spin-Coated Periodic Mesoporous Organosilica Thin Films - Towards a New Generation of Low-Dielectric-Constant Materials. *Adv. Funct. Mater.,* 2005, 15, 823-829.

[25] Bertolo, J. M.; Bearzotti, A.; Falcaro, P.; Traversa, E.; Innocenzi, P., Sensoristic Applications of Self-Assembled Mesostructured Silica Films. *Sens. Lett.,* 2003, 1, 64-70.

[26] Kumar, P.; Guliants, V. V., Periodic Mesoporous Organic-Inorganic Hybrid Materials: Applications in Membrane Separations and Adsorption. *Micropor. Mesopor. Mat.,* 2010, 132, 1-14.

[27] Fuertes, M. C.; Lopez-Alcaraz, F. J.; Marchi, M. C.; Troiani, H. E.; Luca, V.; Miguez, H.; Soler-Illia, G. J. A. A., Photonic Crystals from Ordered Mesoporous Thin-Film Functional Building Blocks. *Adv. Funct. Mater.,* 2007, 17, 1247-1254.

[28] Nishiyama, N.; Tanaka, S.; Egashira, Y.; Oku, Y.; Ueyama, K., Enhancement of Structural Stability of Mesoporous Silica Thin Films Prepared by Spin-Coating. *Chem. Mater.,* 2002, 14, 4229-4234.

[29] Nishiyama, N.; Tanaka, S.; Egashira, Y.; Oku, Y.; Ueyama, K., Vapor-Phase Synthesis of Mesoporous Silica Thin Films. *Chem. Mater.,* 2003, 15, 1006-1011.

[30] Tanaka, S.; Nishiyama, N.; Oku, Y.; Egashira, Y.; Ueyama, K., Nano-Architectural Silica Thin Films with Two-Dimensionally Connected Cagelike Pores Synthesized from Vapor Phase. *J. Am. Chem. Soc.,* 2004, 126, 4854-4858.

[31] Maruo, T.; Tanaka, S.; Hillhouse, H. W.; Nishiyama, N.; Egashira, Y.; Ueyama, K., Disordered mesoporous silica low-k thin films prepared by vapor deposition into a triblock copolymer template film. *Thin Solid Films,* 2008, 516, 4771-4776.

[32] Yang, H.; Coombs, N.; Dag, O.; Sokolov, I.; Ozin, G. A., Free-standing mesoporous silica film s ; morphogenesis of channel and surface patterns. *J. Mater. Chem.,* 1997, 7, 1755-1761.

[33] Goux, A. l.; Etienne, M.; Aubert, E.; Lecomte, C.; Ghanbaja, J.; Walcarius, A., Oriented Mesoporous Silica Films Obtained by Electro-Assisted Self-Assembly (EASA). *Chem. Mater.,* 2009, 21, 731-741.

[34] Walcarius, A., Mesoporous materials and electrochemistry . *Chem. Soc. Rev.* 2013, *42*, 4098-4140.

[35] Schnabel, W., Polymers and Light. Wiley-VCH: Weinheim, 2007.

[36] Doshi, D. A.; Huesing, N. K.; Lu, M. C.; Fan, H. Y.; Lu, Y. F.; Simmons-Potter, K.; Potter, B. G.; Hurd, A. J.; Brinker, C. J., Optically, Defined Multifunctional Patterning of Photosensitive Thin-Film Silica Mesophases. *Science,* 2000, 290, 107-111.

[37] Doshi, D. A.; Fan, H.; Huesing, N.; Hurd, A.; Brinker, C. J., Photo-definable self-assembled materials . US Patent 7332264: 2004.

[38] Fan, H.; Lu, Y.; Stump, A.; Reed, S. T.; Baer, T.; Schunk, R.; Perez-Luna, V.; Lopez, G. P.; Brinker, C. J., Rapid prototyping of patterned functional nanostructures . *Nature,* 2000, 405, 56-60.

[39] Dhaval A. Doshi; Nicola K. Huesing; Mengcheng Lu; Hongyou Fan; Yunfeng Lu; Kelly Simmons-Potter; B. G. Potter, J.; Alan J. Hurd; Brinker, a. C. J., *Science,* 2000, 290, 107-111.

[40] Nagarajan, S.; Bosworth, J. K.; Ober, C. K.; Russell, T. P.; Watkins, J. J., Simple Fabrication of Micropatterned Mesoporous Silica Films Using Photoacid Generators in Block Copolymers. *Chem. Mater.,* 2007, 20, 604-606.

[41] Yao, L.; Woll, A. R.; Watkins, J. J., Directed Assembly of Block Copolymer Templates for the Fabrication of Mesoporous Silica Films with Controlled Architectures via 3-D Replication. *Macromolecules,* 2013, 46, 6132-6144.

[42] Chen, H.-T.; Crosby, T. A.; Park, M.-H.; Nagarajan, S.; Rotello, V. M.; Watkins, J. J., Accessibility of Cylindrical Channels within Patterned Mesoporous Silica Films using Nanoparticle Diffusion. *J. Mater. Chem.,* 2009, 19, 70-74.

[43] Nagarajan, S.; Russell, T. P.; Watkins, J. J., Dual-Tone Patterned Mesoporous Silicate Films Templated From Chemically Amplified Block Copolymers. *Adv. Funct. Mater.,* 2009, 19, 2728-2734.

[44] Song, L.; Wu, Q.; Vogt, B. D., Templating Multiple Length Scales by Combining Phase Separation, Self-Assembly and Photopatterning in Porous Films. *J. Colloid Interf. Sci.,* 2008, 328, 374-384.

[45] Hozumi, A.; Yokogawa, Y.; Kameyama, T.; Hiraku, K.; Sugimura, H.; Takai, O.; Okido, M., Photocalcination of Mesoporous Silica Films Using Vacuum Ultraviolet Light. *Adv. Mater.,* 2000, 12, 985-987.

[46] Hozumi, A.; Sugimura, H.; Hiraku, K.; Kameyama, T.; Takai, O., Low-Temperature Elimination of Organic Components from Mesostructured Organic−Inorganic Composite Films Using Vacuum Ultraviolet Light. *Chem. Mater.,* 2000, 12, 3842-3847.

[47] Clark, T.; Ruiz, J. D.; Fan, H.; Brinker, C. J.; Swanson, B. I.; Parikh, A. N., A new application of UV -ozone treatment in the preparation of substrate -supported, mesoporous thin films . *Chem. Mater.,* 2000, 12, 3879-3884.

[48] Parikh, A. N.; Navrotsky, A.; Li, Q.; Yee, C. K.; Amweg, M. L.; Corma, A., Non-thermal calcination by ultraviolet irradiation in the synthesis of microporous materials . *Micropor. Mesopor. Mat.,* 2004, 76, 17-22.

[49] Dattelbaum, A. M.; Amweg, M. L.; Ruiz, J. D.; Ecke, L. E.; Shreve, A. P.; Parikh, A. N., Surfactant removal and silica condensation during the photochemical calcination of thin film silica mesophases. *J. Phys. Chem. B,* 2005, 109, 14551-14556.

[50] Dattelbaum, A. M.; Amweg, M. L.; Ecke, L. E.; Yee, C. K.; Shreve, A. P.; Parikh, A. N., Photochemical Pattern Transfer and Enhancement of Thin Film Silica Mesophases. *Nano Letters,* 2003, 3, 719-722.

[51] Yoshida, E.; Kuwayama, S., Photo-Induced Micellization of Block Copolymers. *Polymers,* 2010, 2, 623-648.

[52] Yoshida, E.; Kuwayama, S., Photolysis-induced micellization of a poly (4-tert-butoxystyrene)-block-polystyrene diblock copolymer . *Colloid Polym. Sci.,* 2008, 286, 1621-1627.

[53] Yoshida, E.; Kuwayama, S., Micelle formation induced by photolysis of a poly(tert-butoxystyrene)-block-polystyrene diblock copolymer . *Colloid Polym. Sci.,* 2007, 285, 1287-1291.

[54] Anger, C.; Deubel, F.; Salzinger, S.; Stohrer, J.; Halbach, T.; Jordan , R.; Veinot, J. G. C.; Rieger, B., Organic–Inorganic Hybrid Nanoparticles via Photoinduced Micellation and Siloxane Core Cross-Linking of Stimuli-Responsive Copolymers. *ACS Macro Lett.,* 2013, 2, 121-124.

[55] Bertrand, O.; Schumers, J.-M.; Kuppan, C.; Marchand-Brynaert, J.; Fustin, C.-A.; Gohy, J.-F., Photo-induced micellization of block copolymers bearing 4,5-dimethoxy-2-nitrobenzyl side groups. *Soft Matter,* 2011, 7, 6891-6896.

[56] Gohy, J.-F.; Zhao, Y., Photo-responsive block copolymer micelles : design and behavior. *Chem. Soc. Rev.,* 2013.

[57] De Paz, H.; Chemtob, A.; Crest, F.; Croutxe-Barghorn, C.; Michelin, L.; Rigolet, S.; Vidal, L.; Lebeau, B., Thick Mesostructured Films via Light Induced Self-Assembly. *RSC Adv.,* 2012.

[58] Park, I.; Pinnavaia, T. J., Large-pore mesoporous silica with three-dimensional wormhole framework structures. *Micropor. Mesopor. Mat.,* 2009, 118, 239-244.

[59] Kim, S.-S.; Pauly, T. R.; Pinnavaia, T. J., Non-ionic surfactant assembly of wormhole silica molecular sieves from water soluble silicates. *Chem. Commun.,* 2000, 835-836.

[60] Bagshaw, S. A.; Prouzet, E.; Pinnavaia, T. J., Templating of Mesoporous Molecular Sieves by Nonionic Polyethylene Oxide Surfactants. *Science,* 1995, 269, 1242.

[61] Tanev, P. T.; Pinnavaia, T. J., A Neutral Templating Route to Mesoporous Molecular Sieves. *Science,* 1995, 267, 865-867.

[62] Tanev, P. T.; Pinnavaia, T. J., Biomimetic Templating of Porous Lamellar Silicas by Vesicular Surfactant Assemblies. *Science,* 1996, 271, 1267-1269.

[63] De Paz-Simon, H.; Chemtob, A.; Crest, F.; Croutxe-Barghorn, C.; Michelin, L.; Vidal, L.; Rigolet, S.; Lebeau, B., Thick Mesostructured Films via Light Induced Self-Assembly. *RSC Adv.,* 2012, 2, 11944-11952.

[64] De Paz, H.; Chemtob, A.; Croutxé-Barghorn, C.; Le Nouen, D.; Rigolet, S., Insights into Photoinduced Sol–Gel Polymerization: An in Situ Infrared Spectroscopy Study. *J. Phys. Chem. B,* 2012, 116, 5260-5268.

[65] De Paz-Simon, H.; Chemtob, A.; Croutxe-Barghorn, C.; Rigolet, S.; Michelin, L.; Vidal, L., Periodic Mesostructured Silica Films Made Simple Using UV Light. *J. Phys. Chem. C,* 2014, 118, 4959–4966.

[66] Kresge, C. T.; Roth, W. J., The discovery of mesoporous molecular sieve s from the twenty year perspective. *Chem. Soc. Rev.,* 2013, 42, 3663-3670.

[67] Lebeau, B.; Galarneau, A.; Linden, M., Introduction for 20 years of research on ordered mesoporous materials . *Chem. Soc. Rev.,* 2013, 42, 3661-3662.

In: Comprehensive Guide for Mesoporous Materials. Volume 1 ISBN: 978-1-63463-958-3
Editor: Mahmood Aliofkhazraei

Chapter 9

MESOPOROUS CARBON AEROGELS: SYNTHESIS, CHARACTERISATION AND ENVIRONMENTAL APPLICATIONS

Liviu Cosmin Coteţ, Andrada Măicăneanu and Virginia Danciu
Faculty of Chemistry and Chemical Engineering,
Department of Chemical Engineering,
Babeş-Bolyai University, Cluj-Napoca, Romania

ABSTRACT

Mesoporous undoped and metal (Fe or Cu) doped carbon aerogels obtained using a sol-gel process, followed by low temperature supercritical drying and various pyrolysis conditions were synthesized. The resulting materials were characterized by means of elemental analysis, nitrogen adsorption -desorption , TEM , SEM, AFM and XRD methods. The presence of graphene-like structures was evidenced in the case of Fe-doped carbon aerogels . Due to their special morpho-structural properties: high surface area and pore volume, narrow mesopore size distribution , low density, etc., these materials were tested in different applications such as: phenol chemical wet air oxidation and phenol and pesticide adsorption from aqueous solutions .

I. INTRODUCTION

Nowadays, the main problems regarding the industrial development are related to the widely distributed toxic substances , such as heavy metals and organic compounds , in surface and ground waters, which in some concentration values are harmful for human health and the surrounding environment [1].

In agriculture , organic pesticides including insecticides and herbicides used for production increase have become a real environmental problem since parts of them were identified in waters [2], fruit, vegetables [3] and even in animal bodies [4]. Scientific studies related to the presence of these non-biodegradable compounds in drinking water and foods have confirmed the negative impact on the metabolism of live beings [2, 5-7]. All of these problems are reflected in life level and are associated with the lack of access to clean and fresh water for billions of human beings [1]. Beside these, the demand for fresh water is expected to increase along with the population growth , further stressing traditional sources. In this context, low-cost and more efficient methods are needed to disinfect and decontaminate the waters from various sources, without further stressing the environment or endangering human health by the treatment itself.

Several processes for organic compounds abatement from wastewaters, which can be included in two large categories, recuperative methods (separation) and destructive methods, were studied [8, 9], Tables 1 and 2.

Table 1. Recuperative methods for organic compounds removal from wastewaters

Methods			Adsorbents	
precipitation	[1,2]		activated carbon	[10-15]
coagulation–flocculation			multi-walled carbon nanotubes	[10, 11, 16]
sedimentation			C_{18} cartridges	[10]
flotation			shale ash	[17]
filtration			carbon fibres	[18]
membrane processes			graphene-based materials	[9, 19, 20]
electrochemical techniques				
biological process				
solvent extraction				
reverse osmosis				
evaporation				
adsorption				
ion exchange				

Table 2. Destructive methods for organic compounds removal from wastewaters

total oxidation by air or oxygen – WAO, SCWO, CWAO *	[21-26]
wet oxidation with chemical o xidants (ozone and H_2O_2 – WPO, CWPO**, and oxidative polymerization in the presence of peroxidases)	
oxidation with chlorine, chlorine dioxide, potassium permanganate, ferrate (VI) ion	[21]
electrochemical treatment (indirect electro-oxidation, direct anodic oxidation)	[21, 27, 28]
photocatalytic oxidation	[21, 29, 30]
supercritical water g asification	[21]
electrical discharge (electro-hydraulic, pulsed corona, glow discharge electrolysis)	[21]
sonochemical processes	[21, 31]
biodegradation (microbial or fungi species)	[21,32,33]
Catalysts	
Ru, Rh, Pt, Ir, Ni, Ag supported on TiO_2, CeO_2, Al_2O_3	[34, 35]
metal oxides CuO, CoO, Cr_2O_3, NiO, MnO_2, Fe_2O_3, ZnO, CeO_2	[34, 36-38]
activated carbon	[39]
iron and copper immobilized on synthetic zeolites	[40, 41]
pillared clays	[42, 43]

*WAO – non catalytic wet air oxidation , SCWO – supercritical water oxidation, CWAO – catalytic wet air oxidation.

**WPO – wet peroxide oxidation , CWPO – catalytic wet peroxide oxidation.

Chemical oxidation with all its alternatives (Table 2) is widely used for the treatment of wastewaters in order to remove organic pollutants, and has as its final objective total oxidation (mineralization) of the organic contaminants to CO_2, H_2O and inorganics or, at least, their transformation into harmless products [23]. Also, from among the oxidation processes, the catalytic ones become useful alternatives due to the working conditions, which are milder [34]. Some catalysts used to destroy organic pollutants are presented in Table 2.

Among the organic pollutants, phenol and phenolic compounds have gained increased attention in the last two decades due to their toxicity and frequency of their presence in industrial wastewaters [44]. Moreover, phenol is considered to be an intermediate in the oxidation route of higher molecular weight aromatics and therefore is usually taken as a model compound in research studies [21, 45]. The choice of treatment, in the case of wastewaters polluted with phenol and phenolic compounds, depends on the concentration, which can vary from 0.1 to 6,800 mg/dm^3 depending on the wastewater source (pulp and paper industry, refineries, coking operations, coal processing) [21], economics, efficiency, easy control and reliability [36].

In the last decades, new forms of carbon nanomaterial – carbon aerogel (CA) and carbon xerogel (CX) – have been promoted as mesoporous and micro-mesoporous adsorbents, respectively. Their efficiency is owed to their unique properties, such as controllable porosity, thermal stability, large pore volume and high specific surface area [46]. These materials were successfully applied in adsorption of both organic compounds, e.g., dye [47-52], antibiotics [53], toluene [54], phenol [8], methane [55], and inorganic ions [56-58].

Due to their morpho-structural properties carbon aerogels and xerogels are widely used as support matrices for various catalytic active centres, represented especially by metal species [59]. These are embedded in the three-dimensional framework of the carbon nanoporous material either by an adsorption process or by *in situ* synthesis process either by using metal species as catalytic precursors [59] or by an ion-exchange process between the wet organic gel and metal doping solution (see Chapter II.2). In addition, carbon aerogels and xerogels can be obtained in the form of monoliths, beads, powders or thin films. As metal species, Ce-, Cr-, Mo-, W-, Fe-, Ru-, Co-, Ni-, Pd-, Pt-, Cu-, Ag-containing carbon aerogels were synthesized and tested as electrodes for electrical double-layer capacitors and as catalysts [59-61]. In order to increase the adsorptive and catalytic properties of carbon xerogel by increasing the mesoporous fraction, two alternatives can be used: either by using surfactants in the synthesis stage or by an optimization of the drying conditions [61].

This chapter will present a study about the synthesis, morpho-structural characterization and adsorptive and catalytic performances of undoped and doped (iron or copper) carbon aerogels, and the results will be compared to other carbon forms, such as carbon xerogels.

II. Sample Synthesis of Undoped and Metal Doped Carbon Aerogels

II.1. Synthesis of Undoped Carbon Aerogels

Carbon aerogels are monolithic nanostructured materials formed by coupling, as a 3D structure , of carbon nanoparticles , which delimit a nanoporous net [63]. The synthesis of these materials is based on a sol-gel process in which a polycondensation reaction of resorcinol (R) with formaldehyde (F) in the catalytic presence of Na_2CO_3 (C) takes place (Scheme 1). This catalyst activates a small part of R to act as positions for the growth of the monomer particles, which after a while will connect to each other forming an organic tridimensional structure that fills the original volume of the aqueous solution [63]. The sol-gel process takes place in a sealed glass bottle and is favoured by a temperature program (e.g., the bottle was kept 1 day at room temperature , 1 day at 50°C and then 3 days at 70°C) [46]. The resulted resorcinol-formaldehyde (RF) wet gel is then washed with acetone or ethanol , dried in supercritical conditions of liquid CO_2 . Practically, solvent molecules from the gel placed in the autoclave dryer are replaced with liquid CO_2 by a CO_2 purging process for 2-4 hours (0-4°C and 50 atm). Then the autoclave is heated to reach CO_2 supercritical conditions (37-40°C and 90 atm). After one hour, the autoclave is depressurized slowly to atmospheric pressure . Finally, a pyrolytic (>500°C/2-4h) process realized in an oven in inert atmosphere (Ar or N_2 flow) is performed. The final product obtained at the end of this process sequence is carbon aerogel [46, 63].

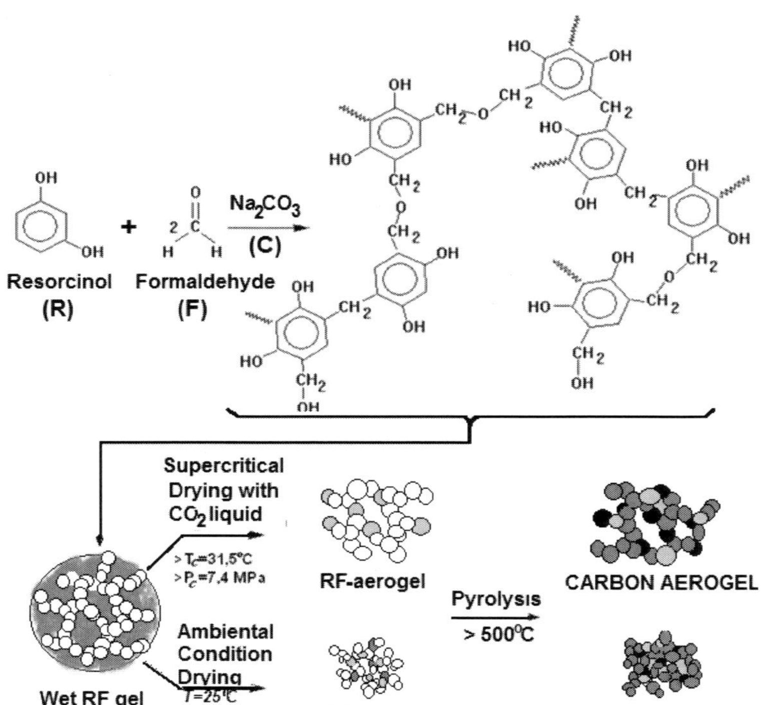

Scheme 1. Carbon aerogel and xerogel synthesis pathways .

In the case of carbon xerogel synthesis , the same sol-gel process is used but the drying step is performed under ambient conditions [64]. Detailed synthesis pathways of carbon aerogel and xerogel are presented in Scheme 1.

An important feature of the sol-gel process is that, by changing the ratio between precursors (R, F) and the catalyst (C) or the precursor and solvent, which are named synthesis parameters (expressed in mol/mol or g/cm^3, see Figure 1 and Table 3 from Chapter III.1), the morpho-structural characteristics of the final materials can be controlled.

Figure 1. Real images of organic aerogels (a,b) and xerogel (b), and carbon aerogels (c) and xerogels (d).

Figure 1 presents real images of monolithic organic aerogels (a, b) and xerogel (b), and carbon aerogels (c) and xerogels (d) obtained using various synthesis parameters and having various geometry . Differences in colour and drying shrinkage of organic aerogel pieces synthesized in identical bottles are visualized in Figure 1a. Figure 1b presents the shrinkage difference between an organic gel dried under ambient conditions (i.e., RF-xerogel, smaller) and the same gel dried in supercritical condition of CO_2 liquid (i.e., RF-aerogel). After the pyrolysis step, the sample colour changes into black and the geometry feature of the precursor organic materials is kept even if a smaller shrinkage is noted (Figure 1c, d).

According to the target application, the geometry of the final monolithic sample is cast by the geometry of the vessel where the sol-gel process is performed. Figure 1 presents rod, tube and plate shaped materials .

II.2. Synthesis of Metal Doped Carbon Aerogels

In order to improve the catalytic [59] or electrochemical [65, 66] properties, metallic species were embedded in carbon aerogel framework, when metal doped carbon aerogels were obtained. Generally, metal doped monolithic carbon aerogels (DCA -Me) can be prepared by the following three main strategies: by addition of the metal precursor to the initial mixture, by ion-exchange or by deposition of the metal precursor on the organic or the carbon aerogel by one of various methods [59].

In this chapter the synthesis of iron and copper doped carbon aerogels performed by an ion-exchange process is described in detail. For some application, homologue carbon xerogels were prepared and used for comparison. Various materials were obtained and used to remove organics from wastewater , and tested as catalysts in wet air oxidation process, (see Chapter IV).

Scheme 2 presents the synthesis pathway of metal doped carbon aerogel and xerogel is presented. The aerogel/xerogel samples prepared using Cu^{2+}, Fe^{2+} and Fe^{3+} as doping ions are named DOA-/DOX-Cu, -$Fe^{(2+)}$ and -$Fe^{(3+)}$, respectively, for "doped organic aerogel/xerogel", and DCA -/DCX-Cu, -$Fe^{(2+)}$ and -$Fe^{(3+)}$, respectively, for "doped carbon aerogel/xerogel" (see Scheme 2).

The synthesis of metal doped carbon aerogels includes the same general process steps (polycondensation , drying and pyrolysis) as those for undoped carbon aerogels (see Chapter II.1), with the following changes (see Scheme 2) [67]:

I. in the sol-gel synthesis , potassium salt of 2,4-dihydroxybenzoic acid is used instead of resorcinol . The aqueous precursor salt solution was obtained by mixing under stirring, 2.4-dihydroxybenzoic acid (A) and K_2CO_3 (K1), A/K1 = 0.5;
II. K_2CO_3 (K2) was used as catalyst instead of Na_2CO_3 at A/K2 = 100, A/H_2O = 0.03 g/cm^3 (Table 4);
III. resulted K^+ doped wet gel was immersed in a 0.1 M $Cu(NO_3)_2$, $Fe(NO_3)_3{\cdot}9H_2O$ or $Fe(OAc)_2$, solution to perform K^+ ion exchange with the corresponding metal ion to obtain the metal doped polymeric gel.

After repeated washing in acetone or ethanol (3 times), the metal doped gels were dried in supercritical conditions and pyrolysed in inert atmosphere (see Scheme 2). In the case of metal doped carbon xerogels the drying step was performed under ambient conditions.

A second copper containing carbon aerogel catalyst was prepared by contacting 0.5 g undoped carbon aerogel (for this preparation , sample CA-"aA" was used, see Table 3) with a Cu^{2+} solution (50 mL 1 g Cu^{2+}/L solution obtained from $CuSO_4{\cdot}5H_2O$ dissolved in distilled water) for 48 hours under batch conditions without stirring. The sample was then filtered and dried in the oven at 105°C for 2 hours (catalyst named Cu/CA-"aA", see Chapter IV.2.2). Half of the obtained sample was sent to pyrolysis at 750°C for 2 hours when a second form of the catalyst was obtained (Cu/CA-"aA"-750).

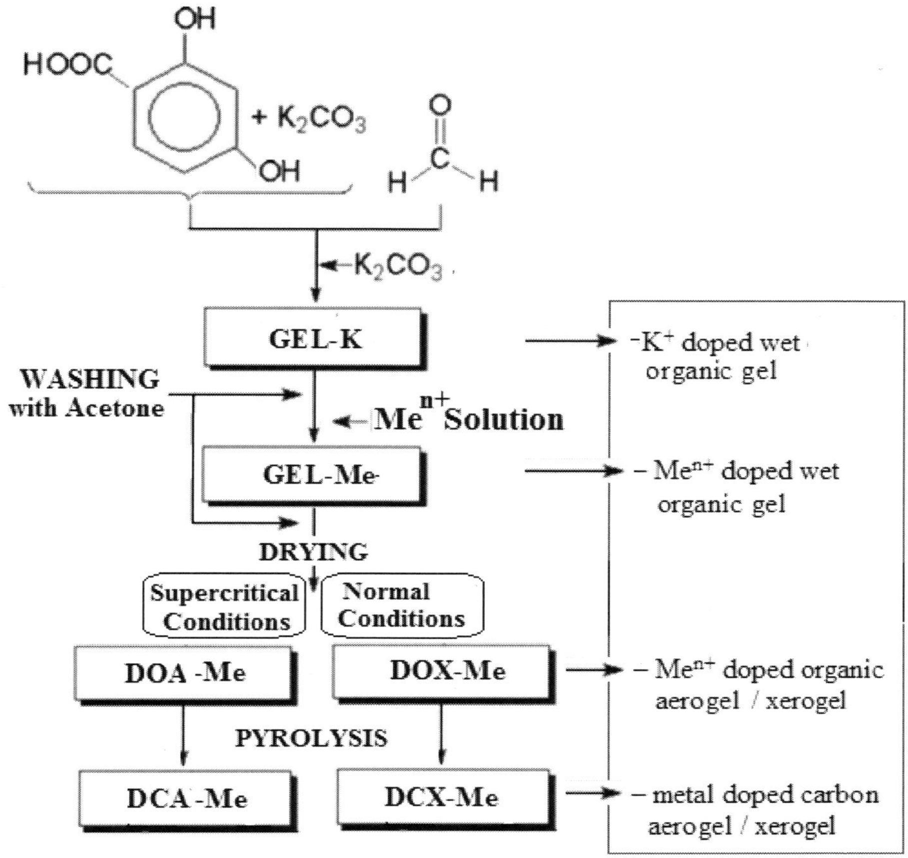

Scheme 2. Synthesis of metal doped carbon aerogel and xerogel obtained by ion-exchange doping process.

III. Morpho-Structural Characterization of Undoped and Iron or Copper Doped Carbon Aerogels

In order to obtain morpho-structural characterization of undoped and metal doped carbon aerogels , nitrogen adsorption -desorption analysis (BET, BJH and HK methods), density measurements , elemental analysis, microscopy investigations (TEM , SEM and AFM) and X-ray diffraction (XRD and EDX) were performed.

III.1. Nitrogen Adsorption-Desorption Investigation, Density Measurements and Elemental Analysis of Undoped and Iron or Copper Doped Carbon Aerogels

BET specific surface area, pore size distribution , volume of mesopores (BJH) and micropores (HK) of undoped and metal doped samples before (RFAs/RFX, DOA-Me/DOX-

Me) and after (CAs/CXs, DCA -Me/DCX-Me) pyrolysis were performed by the Brunauer–Emmett–Teller (BET), Barrett–Joyner–Halenda (BJH) and Horvath-Kawazoe (HK) methods using an ASAP 2000 surface area analyzer (Micrometrics Instruments Corporation) or a Sorptomatic SO-1990 device (Termo Electron Corporation), (Tables 3, 4). Samples of approximately 0.03-0.1 g were heated to 110°C (for OAs/OXs) and 130°C (for CAs/CXs) under vacuum (10^{-5} Torr) for at least 18 h to remove all adsorbed species .

By nitrogen adsorption -desorption investigation, hundreds of square meter per gram BET specific surface areas, specific for mesoporous materials , are obtained for undoped (Table 3) and metal doped (Table 4) organic and carbon aerogels . The materials doped with K (DOA-K and DCA -K) are considered the blank samples for metal doped aerogels (Table 4). The highest surface area (1126 m^2/g) was obtained for CA-"cA" synthesized at R/C = 76.3 and pyrolysed at 850°C for 4h (Table 3). After pyrolysis , an increase in BET surface area is observed; except for the sample doped with a Fe^{2+} salt solution (Table 4) in which case, because of higher iron percent that generates formation of a high amount of graphene-like structures (see Chapter III.2) [66-68], DCA-$Fe^{(2+)}$ presents a lower BET surface area as DOA-$Fe^{(2+)}$.

Table 3. Synthesis parameters (R/C, R/H_2O ratios and pyrolysis temperature), bulk density, porosity (*P*), BET surface area and volume of mesopores (BJH) and micropores (HK) for resorcinol -formaldehyde (RF) and carbon (C) aerogels (a-hA) and xerogels (aX)

Sample	R/C [mol/mol]	R/H_2O [g/cm³]	Pyrolysis [°C]	Density [g/cm³]		S_{BET} [m²/g]		V_{BJH} [cm³/g]	V_{HK} [cm³/g]	Porosity (*P*)** [%]
				RF	C	RF	C	C	C	
				(± 0.01)		(± 30)		(± 0.01)	(± 0.01)	
aA	50	0.200	750	n.a.*	0.73	n.a.	637	0.37	0.28	71.48
aX	50	0.200	750	n.a.	1.37	n.a.	79	-	0.05	46.48
bA	50	0.032	850	0.16	0.27	600	1096	3.76	0.19	88.75
cA	76.3	0.032	850	0.10	0.21	520	1126	3.80	0.17	91.20
dA	96.6	0.032	850	0.08	0.17	617	820	3.86	0.14	93.25
eA	200	0.032	1050	0.03	0.08	520	849	0.90	0.12	96.60
fA	400	0.032	1050	0.05	0.06	309	782	0.76	0.13	97.50
gA	800	0.032	1050	0.13	0.08	140	697	0.38	0.13	96.50
hA	1000	0.077	1050	0.11	0.12	128	682	0.52	0.25	94.70
iA	1500	0.200	1050	0.26	0.24	84	160	0.02	0.21	89.60

* n.a. - not analysed; ** Porosity was calculated:

$$P(\%) = \frac{\rho_G - \rho_S}{\rho_G};$$

Where: ρ_G = 2.56 g/cm³- graphite density and ρ_S– sample density.

Table 4. Synthesis parameters (A/K2, A/H_2O and pyrolysis temperature), BET surface area , bulk density (ρ) and metal % (C_{Me}) of metal doped organic (DOA-Me) and carbon (DCA -Me) aerogels

METAL			DOA-Me		DCA- Me			
	A/K2* mol/mol	A/H_2O* [g/cm^3]	ρ [g/cm^3] (± 0.05)	S_{BET} [m^2/g] (± 30)	Pyrolysis (°C)	ρ [g/cm^3] (± 0.05)	S_{BET} [m^2/g] (± 30)	C_{Me} [wt %] (± 0.5)
K	100	0.03	0.42	233	1050	0.67	545	8.7
Cu			0.47	477		0.58	728	20.8
$Fe^{(2+)}$			0.29	498		0.31	368	17.7
$Fe^{(3+)}$			0.54	523		0.45	745	9.6

* - A = 2,4-dihydroxybenzoic acid ; K2 = K_2CO_3 used as catalyst , see Chapter II.2.

In the case of copper doped samples synthesized under the same sol-gel parameter conditions but pyrolysed at 1050°C for 3h (DCA -Cu-1050, Table 4) and 750°C for 2h (DCA-Cu-750), and dried under ambient conditions (DCX-Cu-750) the following BET surface area s were obtained: 728, 570 and 311 m^2/g respectively. For xerogel samples a lower BET specific surface area than homologue aerogels was observed.

The metal presence into precursor gel structure decreases the shrinkage phenomenon during drying process performed under ambient conditions [69].

The bulk densities (Tables 3, 4) of the organic and carbon samples are estimated by measuring the dimensions and mass of each monolithic sample. Low density values specific for very porous materials are measured. As a general trend, the decreasing of the catalyst content gives rise to lighter aerogels .

A theoretical porosity (*P*) expressed as a function of the graphite density (Table 3) was calculated for undoped carbon samples.

Carbon aerogels are very porous with the CA-"eA" (97.5%) as the most porous sample, and xerogel as the less porous one (46.48%), (Table 3). The gel drying conditions strongly influence the porosity values.

In the case of metal doped carbon aerogels (Table 4) elemental analyses are performed with an inductively coupled plasma-mass spectroscope (ICP-MS). High percentage of metals was obtained in carbon aerogel frameworks. In case of iron doped aerogels the metal percentage is influenced by the iron salt used in the doping process. When a Fe^{2+} salt, $Fe(OAc)_2$, was used as doping salt, a higher concentration of iron was obtained in comparison to the case when a Fe^{3+} salt, $Fe(NO_3)_3{\cdot}9H_2O$, was used. Embedding Fe^{3+} in the gel structure requires, unlike Fe^{2+}, the presence of three optimal neighbouring carboxyl groups, which determines a steric hindrance [67]. The higher metal contents in the DCA -$Fe^{(2+)}$ (see Table 4) probably determines a partial mesopores blocking and a decreasing in the aerogel shrinkage during pyrolysis . These phenomena could also explain the diminishing of the DCA-$Fe^{(2+)}$ bulk density [67].

The presence of a higher amount of good electric conductor graphene-like structures (nanoribbons , nanosheets, nanotubes, see Chapter III.2) into carbon aerogels with a higher iron percentage makes them proper for electrochemical applications [65, 66]. The carbon aerogels doped using a Fe^{3+} salt, which present a lower iron contents, are suitable for catalytic applications (see Chapter IV.2.1) [70].

In Figure 2, pore size distributions (PSD) of CA-"aA", CX-"aA" (a) and CA-"gA" (b), and adsorption -desorption isotherms (insets) of CA-"aA" and CA-"hA" (b) are presented.

The samples prepared with high amount of catalyst (R/C < 100; Figure 1a) reveal a Gaussian PSD and isotherms with hysteretic loops of type IV typically associated to "inkbottle" shape mesopores (2-50 nm diameter).

In this case, during the adsorption process, capillarity condensation is occurring [46]. The xerogel sample is a microporous material, which presents a small mesopore volume and does not present an adsorption-desorption isotherm.

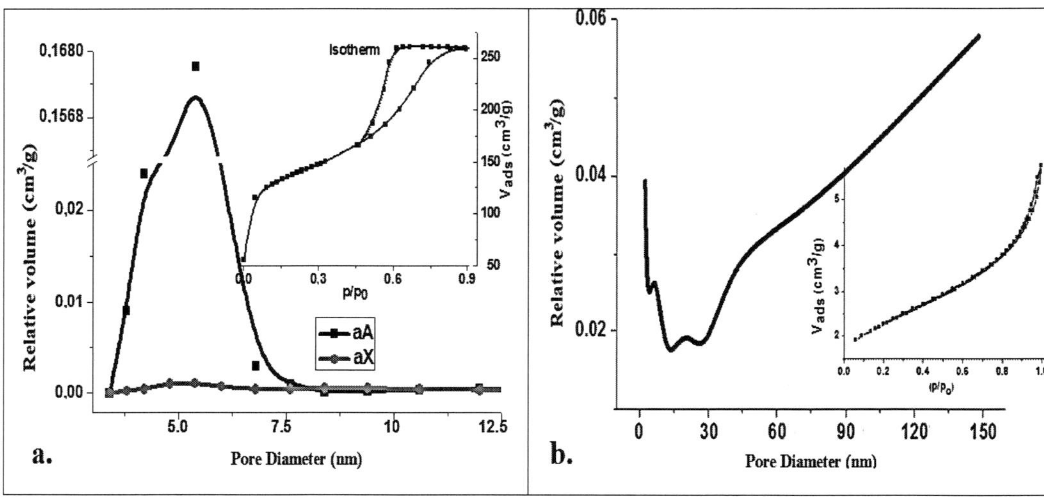

Figure 2. Pore size distributions of "aA" carbon aerogel, "aX" carbon xerogel (a) and "gA" carbon aerogel (b), and adsorption -desorption isotherms of "aA" (inset a) and "hA" (inset b) carbon aerogel, (see Table 3).

In the case of the samples prepared with a smaller amount of catalyst (R/C > 100; Figure 1b) a non-Gaussian PSD and small loop isotherms are observed. These characteristics are specific for macroporous materials with larger and loosely connected particles [46].

In Figure 3, PSD and adsorption -desorption isotherms (insets) of DCA -K (a), -Cu (b), -$Fe^{(2+)}$ (c) and -$Fe^{(3+)}$ (d) are presented. The sample DCA-Cu presents the same feature of PSD (Figure 3b) as the blank sample K-DCA (Figure 3a) with a maximum around 30 nm (± 20 nm) specifics for mesoporous materials .

A similar PSD profile is also observed for both DCA -$Fe^{(2+)}$ (Figure 3c) and -$Fe^{(3+)}$ (Figure 3d). But in these cases the PSD curves shift to the micro-meso size distribution in the case of DCA-$Fe^{(2+)}$ and to the mezo-macro size distribution for DCA-$Fe^{(3+)}$ [67]. All the isotherms present a pronounced loop specific for mesoporous materials (Figure 3, insets).

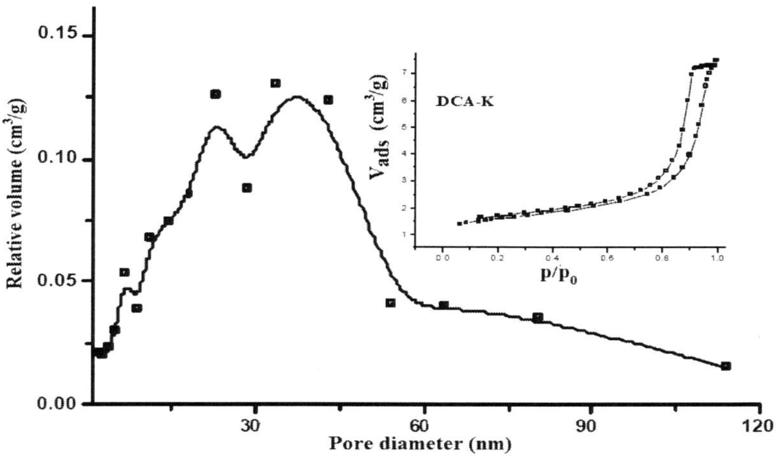

a

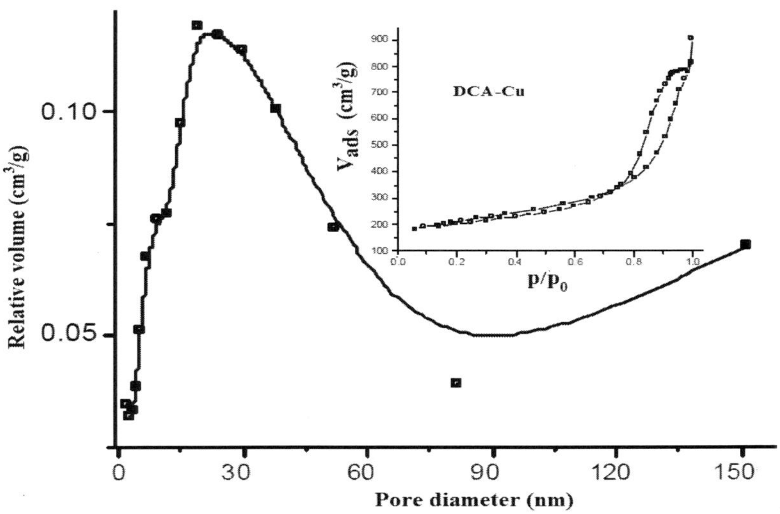

b

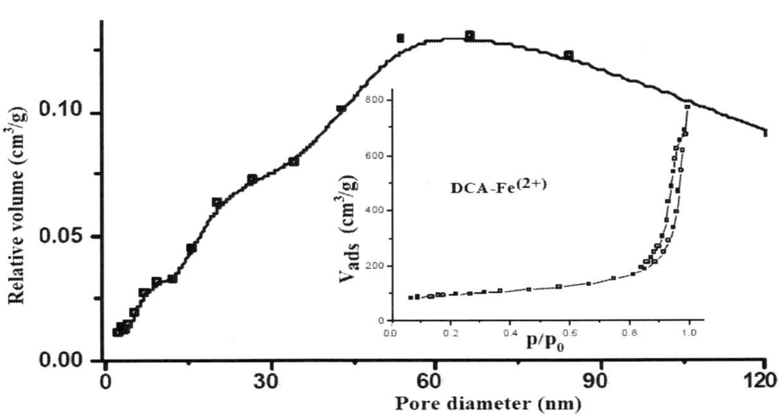

c

Figure 3. (Continued)

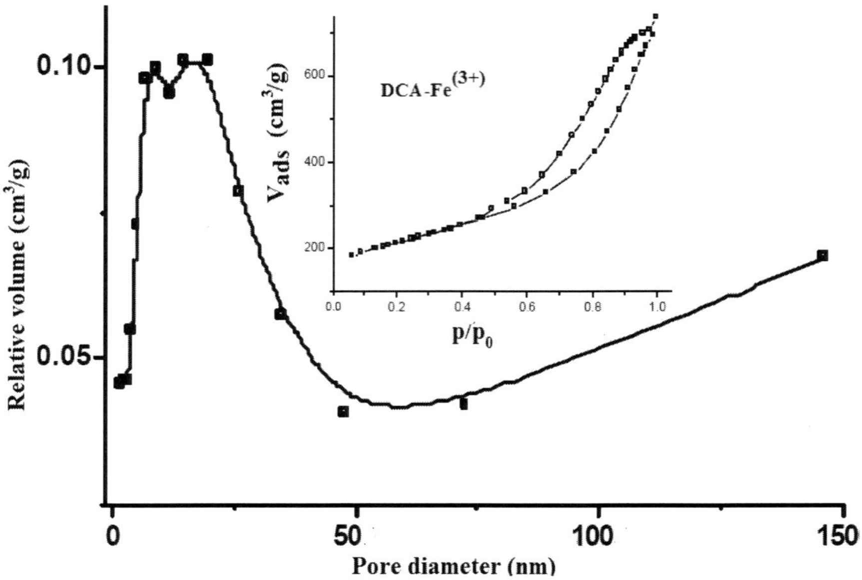

d

Figure 3. Pore size distributions and adsorption -desorption isotherms (inset) of metal doped carbon aerogels pyrolysed at 1050°C for 3h, (see Table 4).

III.2. SEM, TEM and AFM Investigation of Undoped and Iron or Copper Doped Carbon Aerogels

Scanning electron microscopy (SEM) was performed with a Phillips SEM 515 microscope . Transmission electron microscopy (TEM) investigation on the carbon aerogels was performed with a Hitachi H-7000 microscope operating at 125 keV. Atomic force microscopy (AFM) images were performed with PicoSPM from Molecular Imaging Company.

In Figure 4, SEM, TEM and AFM images of three representative undoped carbon aerogels (CA-"dA" -"hA" and -"iA", see Table 3) are presented.

At low R/C ratios, small carbon nanoparticles (~3-10 nm) that are interconnected with large necks making a fibrous tridimensional mesoporous structure are visualized by SEM, TEM and AFM (Figure 4, 1st column).

In contrast, at high R/C ratios, large carbon nanoparticles (tens to thousands nm) that have a fractal shape (smaller particles building larger ones) and are connected by narrow necks making a "string of pearls" meso-microporous structure are visualized (Figure 4, 2nd and 3rd columns) [63, 67]. These two types of aerogels result from supercritical drying of two RF wet gels, which are commonly described as "polymeric" and "colloidal", respectively [63].

Figures 5-8 present TEM images of K, Cu and Fe doped carbon aerogels . In the case of DCA -K (i.e., blank sample) a porous carbon matrix without metal nanoparticles is observed (Figure 5). In the case of DCA-Cu metal nanoparticles well dispersed in porous carbon aerogel are visualized (Figure 6).

In the case of DCA -$Fe^{(2+)}$ (Figure 7) and -$Fe^{(3+)}$ (Figure 8), besides regions where iron nanoparticles are well dispersed in a porous carbon framework (Figures 7a, 8a), regions with isolated or/and conglomerated iron nanoparticles dispersed in graphene-like structures (Figure 7b-d, 11b) and isolated graphene-like structures (Figure 8c-f) were identified [67]. In the case of DCA-$Fe^{(2+)}$ a higher amount and diversity of graphene-like structures – nanosheets (Figure 7b-d), nanoribbons (Figure 7d,e), nanoclew (Figure 7e) and even carbon nanotubes (Figure 7f) were observed [68].

The growth of these graphene-like structures was favoured by the iron presence and some of them tended to cover the iron nanoparticles (see Figure 7d). This behaviour explains the lower catalytic activity (see Chapter IV.2.1) [70] and better electrochemical activity of DCA -$Fe^{(2+)}$ [65, 66].

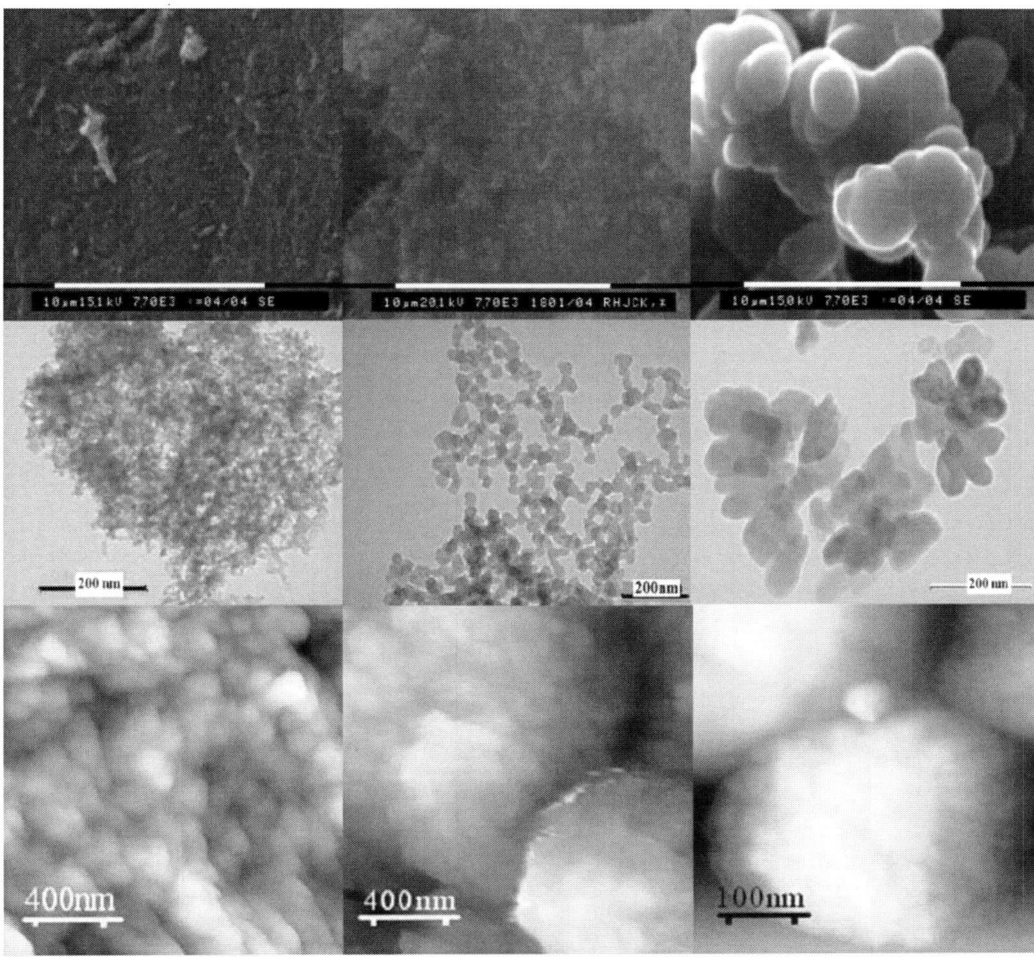

Figure 4. SEM (1st row), TEM (2nd row) and AFM (3rd row) images of "dA" (1st column), "hA" (2nd column) and "iA" (3rd column), (see Table 3).

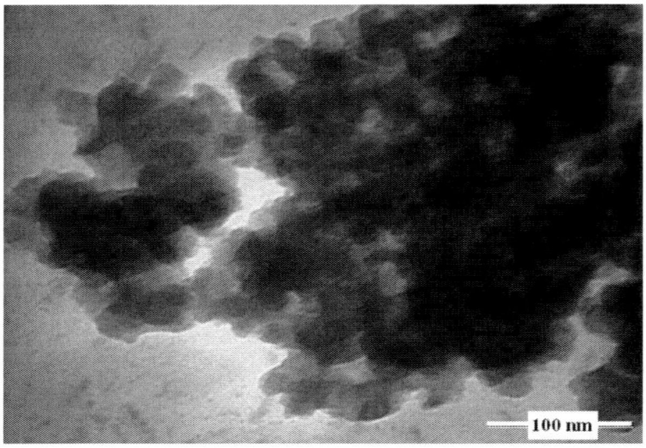

Figure 5. TEM image of DCA -K pyrolysed at 1050°C for 3h, (see Table 4).

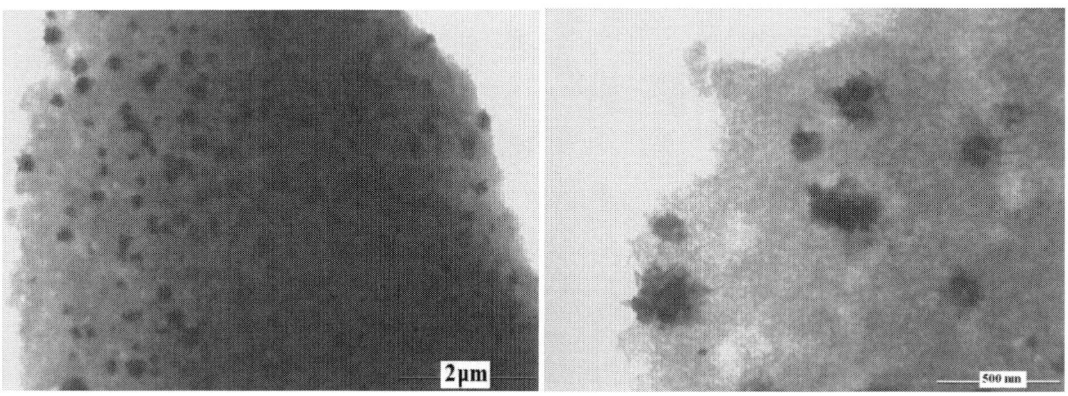

a b

Figure 6. TEM images of DCA -Cu pyrolysed at 1050°C for 3h, (see Table 4).

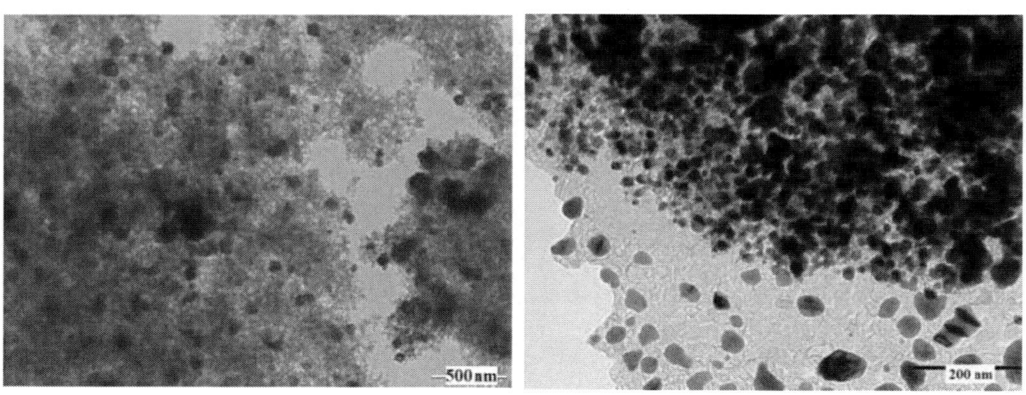

a b

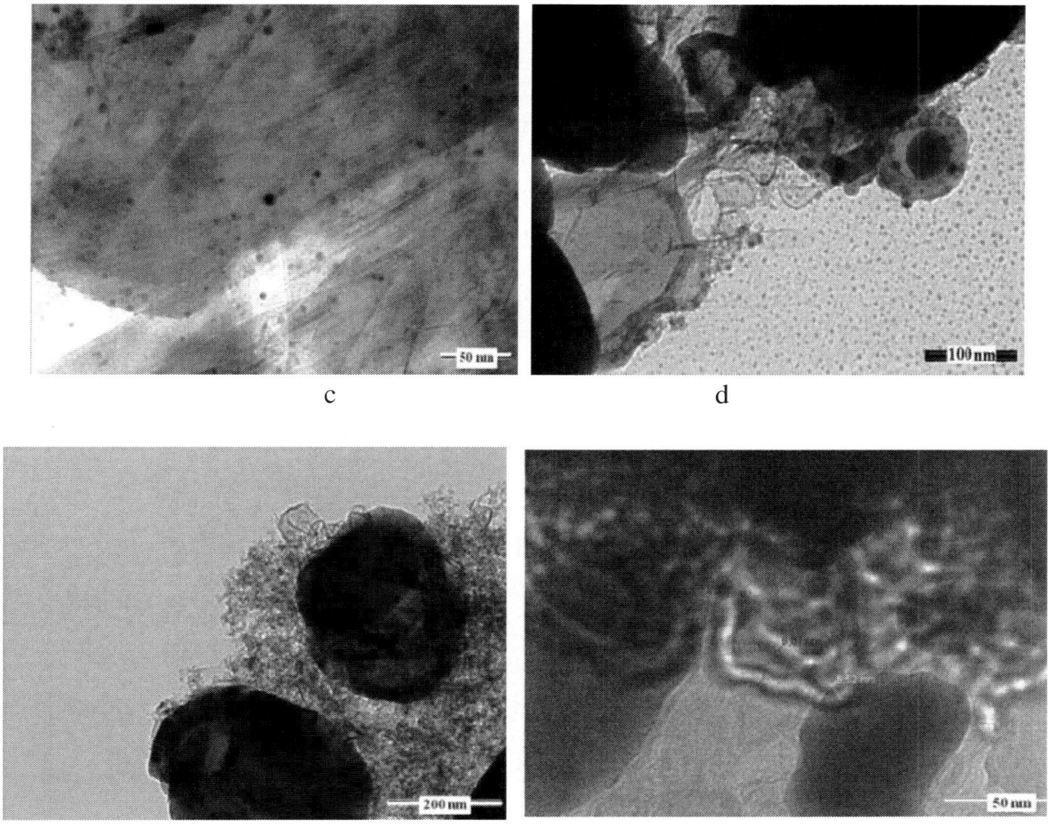

c d

e f

Figure 7. TEM images of various zones of DCA -$Fe^{(2+)}$ pyrolysed at 1050°C for 3h, (see Table 4).

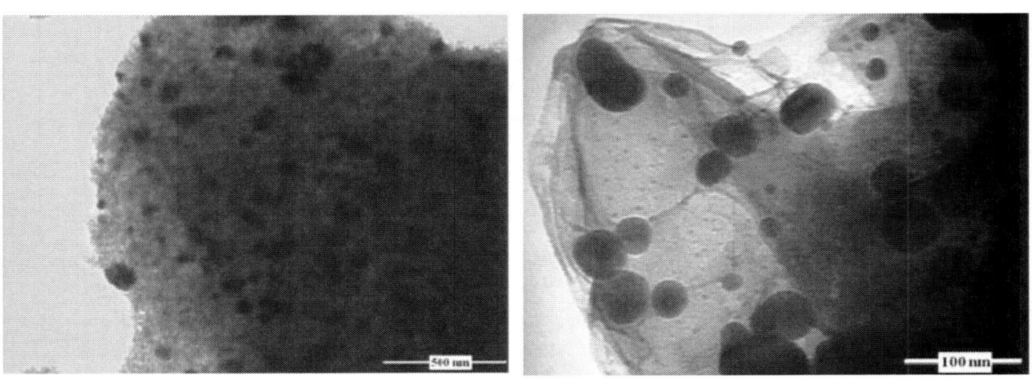

a b

Figure 8. TEM images of DCA -$Fe^{(3+)}$ pyrolysed at 1050°C for 3h, (see Table 4).

SEM investigations for DCA -K (Figure 9), DCA-Cu-750 (Figure 10), DCX-Cu-750 (Figure 11) and Cu/CA-"aA"-750 (Figure 12) were performed. In the case of DCA-K (i.e., blank sample) a porous carbon matrix without metal nanoparticles was observed (Figure 9). In the case of DCA-Cu-750 and DCX-Cu-750 metal nanoparticles well dispersed in porous carbon aerogel (Figure 10, right) and xerogel (Figure 11, right) were observed. The nanoparticle diameter distribution investigation (performed by "ImageJ" Program) reveals that carbon aerogel material presents a smaller value (about 33 nm, Figure 10, left) as corresponding xerogel material (about 64 nm, Figure 11, left). Also, by EDX investigation of Figures 10 and 11 (Table 5), a smaller metal percentage was obtained for Cu doped carbon aerogel (about 5%) than homologue xerogel (about 7%). This difference could be due to the drying conditions of precursor metal doped wet gels: performing the drying process under supercritical conditions (see Chapter II 2) an additional washing process, which takes out the unlinked metallic ions , occurred. Even if DCX-Cu-750 presents the higher particle diameter and metal catalytic amount, the highest catalytic performances are obtained in the case of DCA-Cu-750, (see Chapter IV.2.2).

Figure 9. SEM image of DCA -K pyrolysed at 1050°C for 3h, (see Table 4).

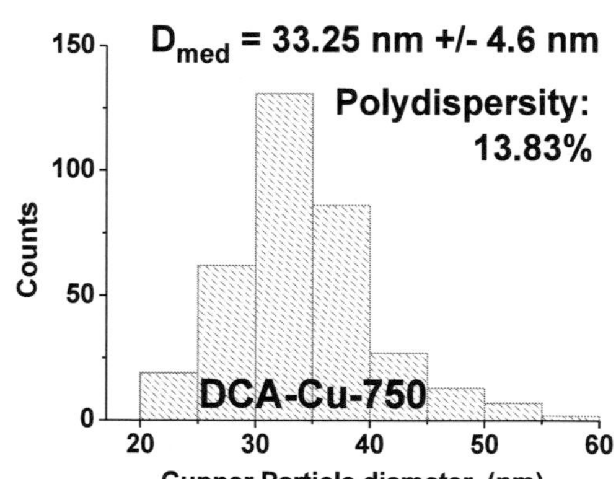

Figure 10. SEM image (left) and Cu particle diameter distribution (right) of DCA -Cu-750 (pyrolysed at 750°C for 2h).

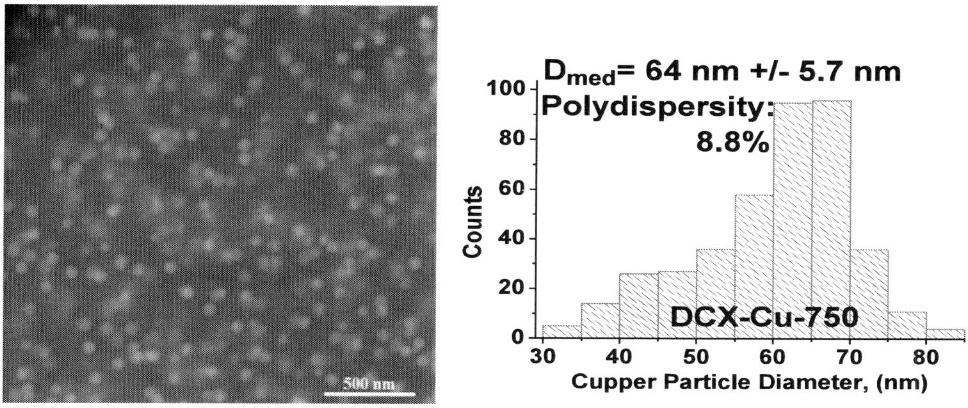

Figure 11. SEM image (left) and Cu particle diameter distribution (right) of DCX-Cu-750 (pyrolysed at 750°C for 2h).

Table 5. EDX investigation of the zones of Figures 10 and 11

Sample	C (wt %)	O (wt %)	Cu (wt %)
DCA- Cu-750	92.5	2.8	4.7
DCX-Cu-750	90.1	2.7	7.2

Figure 12 presents SEM image of copper adsorbed on undoped carbon aerogel and thermally treated in inert atmosphere at 750°C (Cu/CA-“aA”-750). Nonhomogeneous distribution of Cu aggregates (which cannot be considered as particle) on carbon matrix are visible.

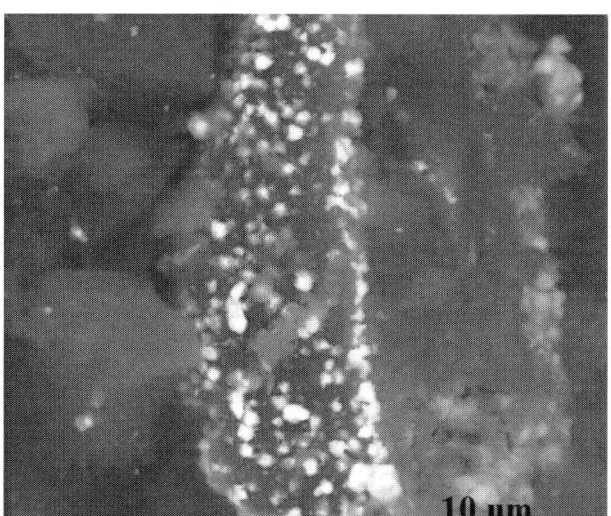

Figure 12. SEM image of Cu/CA-“aA”-750.

III.3. X-Ray Investigation of Undoped and Iron or Copper Doped Carbon Aerogels

The X-ray diffraction patterns (Figure 13) of undoped (CA-“aA”) and K, Fe, Cu doped by ion-exchange process, and Cu doped by adsorption process (Cu/CA-“aA”-750) were recorded in a θ–2θ Bragg–Bretano geometry with a Siemens D5000 powder diffractometer with CuK_ incident radiation (λ = 1.5406Å) and a graphite monochromator. The microstructure information (Table 6) was obtained by a single X-ray profile Warren–Averbach Fourier analysis , with a method developed by Aldea and Indrea using the XRLINE computer program [71-73].

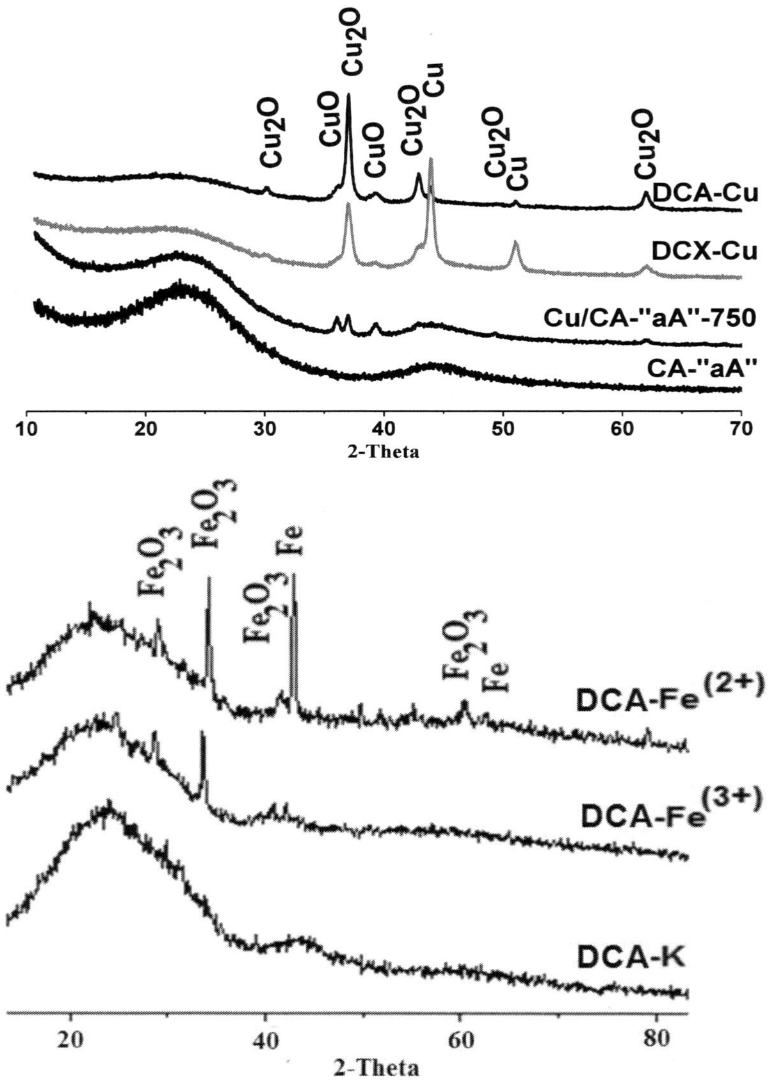

Figure 13. XRD investigation of undoped CA-“aA”, Cu/CA-“aA”-750, DCX-Cu and DCA -Cu (left), and DCA-K, -$Fe^{(3+)}$ and -$Fe^{(2+)}$ (right).

The XRD patterns (Figure 13) of all investigated samples show two broad bands around $2\theta = 22^\circ$ and 43°, indicating an amorphous structure of the carbon matrix [46, 67]. Beside these, in the case of DCA -Me, the presence of a mixture of metal and metal oxide phases is highlighted by the appearance of intense and sharp Bragg peaks [66, 67]. Smaller sharp peaks for Cu/CA-"aA"-750 and no sharp peaks for CA-"aA" and DCA-K are observed.

The relative content of crystalline phases and their corresponding crystallite sizes have been determined by Rietveld refinement analysis of those patterns. In order to determine the micro structural parameters of the metal clusters , the (1 1 1) and (2 0 0) metal peaks from the X-ray diffraction profiles were analysed. The X-ray diffraction line broadening is caused by the small size of the crystallites , the lattice strains, lattice faults, and the experimental diffraction geometry [73, 74].

The effective crystallite mean size (D_{eff}) of the metal nanoparticles together with the root mean square of the microstrains averaged along [h k l] the direction ($\langle\varepsilon^2\rangle^{1/2}_{h\,k\,l}$) are shown in Table 6. By comparing the crystallite sizes presented in Table 6 and taking into account that the metal/metal oxides ratio can be correlated with the electronegativity of the doping element one can assume that the metallic nanoparticles tend to form agglomerates, as can also be seen in the TEM images (Figures 6-8), [67,74].

Table 6. Microstructural parameters of crystalline forms from metal doped carbon aerogels (Table 4)

Probe	D_{eff} **(nm)**	$\langle\varepsilon^2\rangle^{1/2}_{h\,k\,l} \times 10^{-2}$
DCA- K	-	-
DCA- $Fe^{(3+)}$	43	0.3072
DCA- $Fe^{(2+)}$	22	0.2929
DCA- Cu	61	0.2251

IV. Environmental Applications

IV.1. Adsorption

Organic (phenol , alpha-cypermethrin) removal (adsorption) was realized on CA-"aA" and CX-"aX" samples obtained using the procedures described above (see Chapters II.1, III.1). Phenol solution was obtained from analytical purity reagent. Fastac (alpha-cypermethrin) working solution was obtained from pesticide vials FASTAC 10EC available on the market (Bayer). Stock solutions were obtained by dissolving corresponding reagents in distilled water . Solutions used in adsorption experiments were obtained by diluting the stock solution to the desired concentrations (Table 7).

Table 7. Experimental conditions for organics adsorption onto CA-"aA" and CX-"aX"

Sample	**Working conditions**	**Concentration**	**Organics determination procedure**
Phenol adsorption			
CA-"aA"	Batch immobile phases – IP Batch magnetic stirring – MS	5 – 100 mg phenol/ L	COD-Mn
Alpha-cypermethrin adsorption			
CA-"aA" CX-"aX"	Batch magnetic stirring – MS	10 – 70 mg alpha-cypermethrin /L	UV absorption, $\lambda = 276$ nm

Phenol adsorption was realized in a batch reactor with adsorbent and phenolic solution as immobile phases (IP) and with adsorbent and phenolic solution under magnetic stirring (MS), 500 rpm, at room temperature . A quantity of 0.1 g carbon aerogel, brought at a grain size of d < 250 μm using a mortar and pestle and an appropriate sieve, were contacted with 100 mL aqueous solution containing phenol – 5, 10, 25, 50 and 100 mg phenol/L. During the experiment, determination of the organic compounds in solution was carried out every 15 minutes (5, 10, 25 mg phenol/L) and every 1 or 3 hours (50, 100 mg phenol/L) in case of IP, and every 5 minutes (5, 10, 25 mg phenol/L) and 15 minutes (50, 100 mg phenol/L) in case of MS until equilibrium was reached. Phenol concentration in water was determined using $KMnO_4$ as chemical oxygen demand, COD-Mn. This determination is currently used in environmental laboratories for wastewaters characterization (STAS 3002/85, SR ISO 6060/96) according to Romanian legislation (Law 458/2002, 311/2004, 107/1996, 311/2004, 112/2006, HG 188/2002, HG 352/2005).

Alpha-cypermethrin adsorption was realised in a batch reactor with adsorbent and aqueous solution under magnetic stirring (MS), 500 rpm, at various temperatures (25, 35, 45°C). 0.01-0.05 g carbon aerogel or carbon xerogel , brought at a grain size of d < 250 μm, using a mortar and pestle and an appropriate sieve, were contacted with 10 mL solution, containing 10-70 mg alpha-cypermethrin/L. During the experiment, determination of the organic compounds in solution was carried out every 10 min for one hour and then every 30 min until equilibrium /balance was reached. Concentration of alpha-cypermethrin in solution was determined using an UV /VIS spectrophotometer T70+ (PG Instruments Ltd.), $\lambda_{max} = 276$ nm.

The evolution of the organics adsorption process was followed by means of adsorption efficiency (percentage removal), eq. (1).

$$E, \% = \frac{C_i - C_{t(e)}}{C_i} \cdot 100 \qquad (1)$$

where,

C_i is the initial organics concentration, in mg $KMnO_4$/L (phenol), mg alpha-cypermethrin/L

$C_{t(e)}$ is time t or equilibrium /balance organics concentration, in mg $KMnO_4$/L (phenol), mg alpha-cypermethrin/L.

IV.1.1. Phenol Adsorption Results

The results obtained in the case of CA-"aA" adsorbent and phenolic solution as IP in a batch reactor are presented in Figures 14 and 15.

Adsorption efficiency evolutions in time for the three more diluted phenolic solutions (5, 10 and 25 mg phenol /L) are presented in Figure 14. A slow increase in the adsorption efficiency was observed in the first 30 minutes reaching to equilibrium . Equilibrium was reached after 30 minutes for all the three solutions, while the maximum adsorption efficiency was calculated to be 72% for 25 mg/L solution . In the case of the two more concentrated phenolic solutions, 50 and 100 mg/L, equilibrium was reached in 8 hours, while the adsorption efficiency increased by around 11% for 100 mg/L solution.

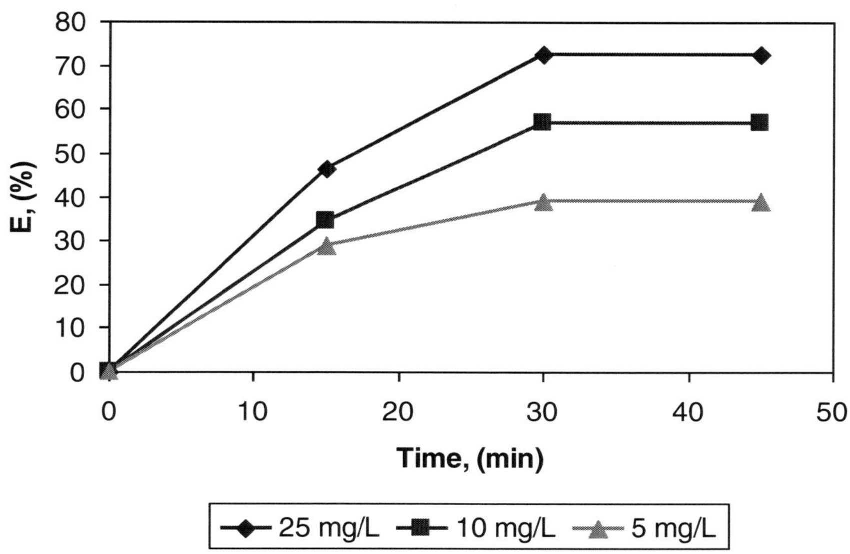

Figure 14. Adsorption efficiency evolution in time during the phenol removal process on CA-"aA" adsorbent (0.1 g), IP, for C_i = 5, 10 and 25 mg phenol/L.

The influence of the initial phenol concentration over the maximum adsorption efficiency, during the phenol removal process on 0.1 g CA-"aA" adsorbent, IP, is presented in Figure 15. It can be observed that with an increase in the initial concentration of phenol, adsorption efficiency increased up to 83%, which is the maximum value, calculated for 100 mg/L solution .

The results obtained in the case of CA-"aA" adsorbent and phenolic solution as MS are presented in Figures 16, 17 and 18. The adsorption efficiency evolution in time for the two sets of solutions is presented in Figures 16 and 17. In the case of the more diluted phenolic solution, maximum adsorption efficiency was reached after 10 minutes (equilibrium), with values very close to those obtained after 5 minutes (Figure 16). Maximum adsorption efficiency was calculated to be 88% for 25 mg/L solution. In the case of 50 and 100 mg/L solutions, Figure 17, equilibrium was reached slower, after 60 minutes, while adsorption efficiency increased by around 7% for 100 mg/L solution.

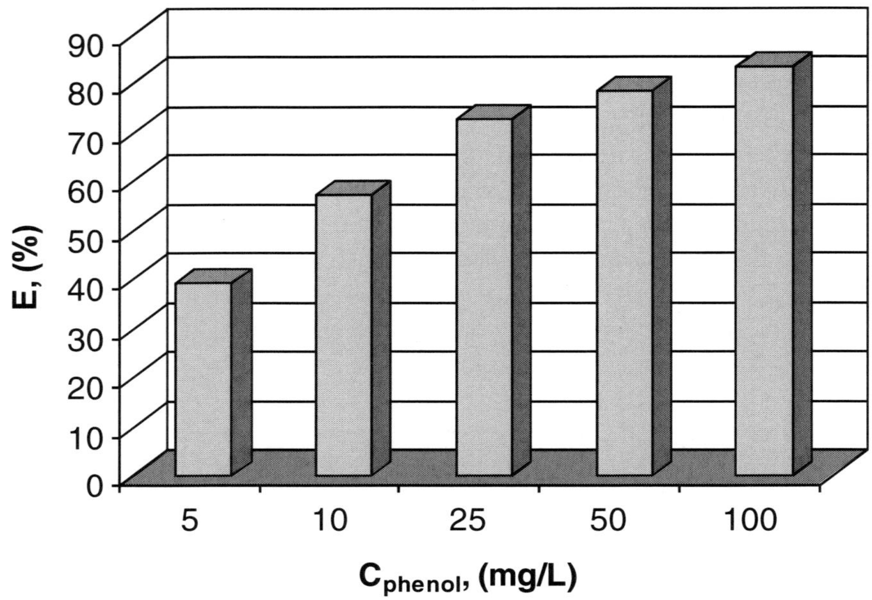

Figure 15. Influence of the initial phenol concentration over the maximum adsorption efficiency, during the phenol removal process on 0.1 g CA-"aA" adsorbent, IP.

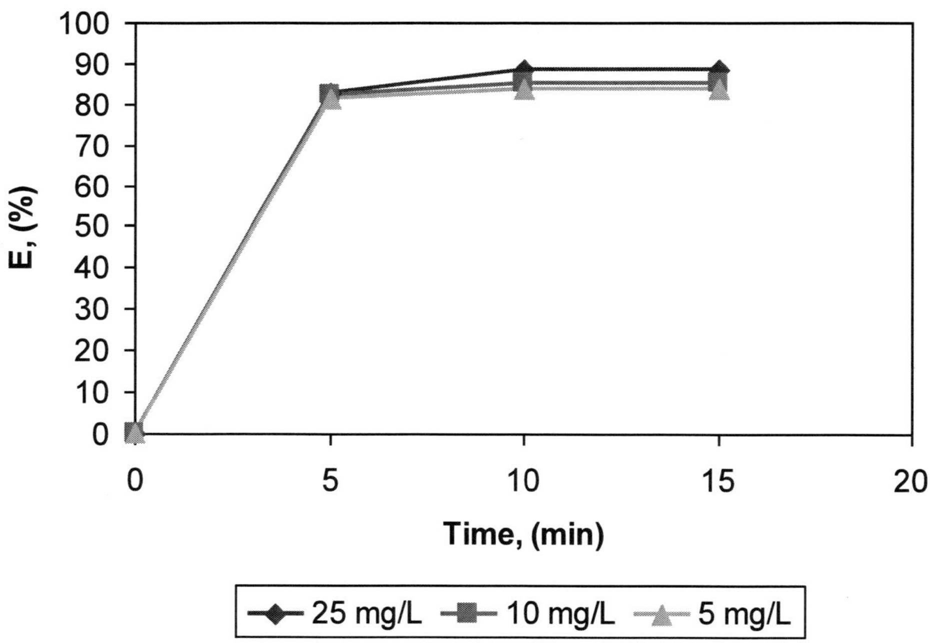

Figure 16. Adsorption efficiency evolution in time during the phenol removal process on CA-"aA" adsorbent (0.1 g), MS, for C_i = 5, 10 and 25 mg phenol/L.

If we compare the maximum adsorption efficiencies calculated for all the five initial concentrations, the same evolution as in the case of IP was observed, Figure 18. A closer inspection of the adsorption efficiency values leads to the conclusion that MS conditions are more efficient for the adsorption of phenol from aqueous solution , Figures 15 and 18, due to the elimination of diffusion limitations.

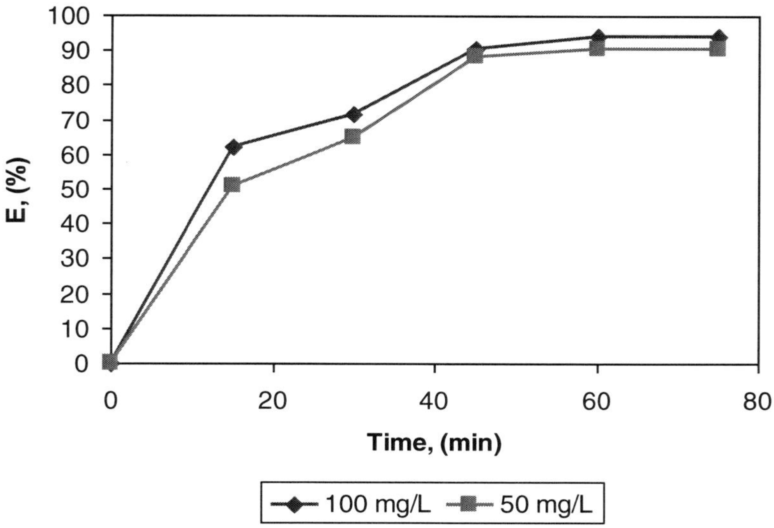

Figure 17. Adsorption efficiency evolution in time during the phenol removal process on CA-"aA" adsorbent (0.1 g), MS, for C_i = 50 and 100 mg phenol/L.

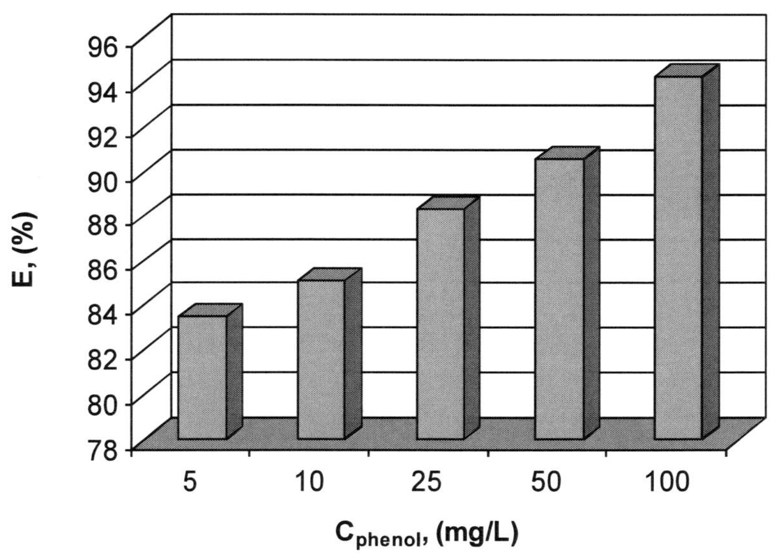

Figure 18. Influence of the initial phenol concentration over the maximum adsorption efficiency, during the phenol removal process on 0.1 g CA-"aA" adsorbent, MS.

The influence of the working conditions (IP and MS) and phenol concentration in aqueous solutions were studied in order to establish their influence over the phenol adsorption process on carbon aerogel.

An increase in the concentration and the utilization of the magnetic stirring led to an increase in the adsorption efficiency. Prepared carbon aerogel proved to be an efficient adsorbent for phenol from wastewaters. Adsorption efficiencies up to 94% were reached [8].

IV.1.2. Alpha-Cypermethrin Adsorption Results

The influence of the adsorbent quantity over the alpha-cypermethrin removal presented in Figure 19 indicated that as the adsorbent quantity increased, removal efficiencies increased too.

This might be due to the fact that for the same initial concentration a higher surface (increased number of adsorption sites) would be available for adsorption. In the case of CA-"aA", removal efficiency increased from 69 to 78%, while in the case of CX-"aX" a smaller increase from 66 to 71% was observed.

The influence of the initial concentration of alpha-cypermethrin on the evolution of the adsorption process was also studied. It was concluded that the process evolved rapidly, with the removal efficiency increasing abruptly in the first 10 minutes, Figure 20. Further on, removal efficiency values increased slowly until equilibrium was reached (60-90 minutes depending on the initial concentration value). As the initial concentration increased, maximum removal efficiency slowly decreased. For example, in the case of CA-"aA" the calculated values were 86 and 77% for 10 and 70 mg alpha-cypermethrin/L, respectively.

The alpha-cypermethrin removal process on CA-"aA" and CX-"aX" was also studied at different temperatures. It was observed that with an increase in temperature , maximum removal efficiency values were increasing for both adsorbents . In the case of CA-"aA" an increase from 78 to 89% was observed when the temperature was increased from 25 to 45°C, C_i = 55 mg/L. In the case of CX-"aX" an increase from 76 to 79% was observed under the same conditions. This suggests that the considered adsorption process is endothermic .

A set of experiments were conducted in order to investigate the adsorption capacity of CA-"aA" subjected to regeneration (pyrolysis 550°C for 2 h), 0.01 g, C_i = 55 mg/L, 25°C. It was observed that in comparison to the fresh sample (88%), the maximum removal efficiency decreased to 80% after the 1st regeneration and to 71% after the 2nd regeneration.

The difference between adsorption capacities for CA-"aA" and CX-"aX", which was in favour of CA was also observed for all the studied parameters. This difference is illustrated in Figure 21 for different initial concentrations.

This difference between CA-"aA" and CX-"aX", was due to the distinction that arouse from their porous structure – in the case of CA-"aA" where its structure was mainly formed by mesopores , an increase in grain size would lead to an increase in pore length, and therefore a longer path for molecules to reach adsorption sites. In the case of CX-"aX" the presence of micropores would lead from the beginning to a lower adsorption capacity whereas an increase in grain size would further increase diffusion limitations [75].

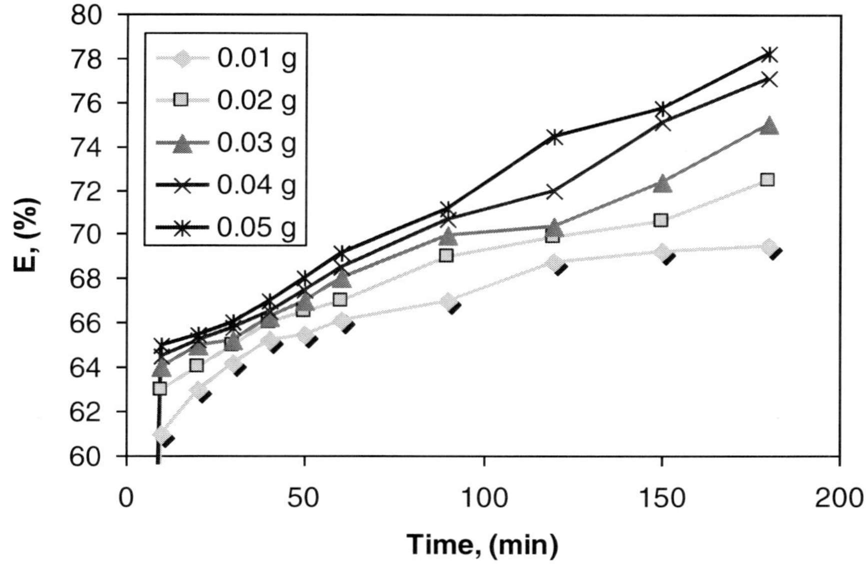

Figure 19. Adsorption efficiency evolution in time during the alpha-cypermethrin removal process on different quantities of CA-"aA" adsorbent, MS, for C_i = 40 mg alpha-cypermethrin/L.

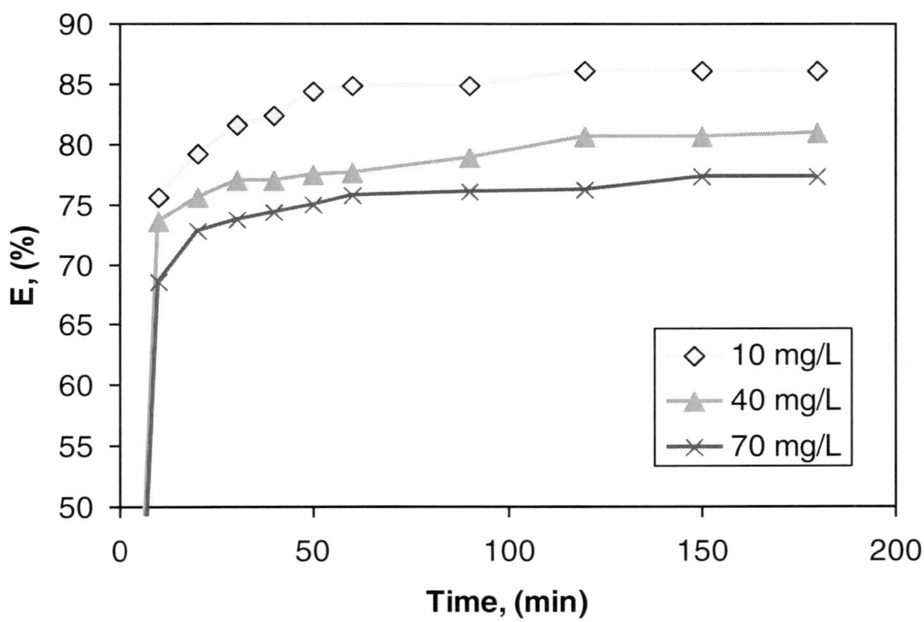

Figure 20. Adsorption efficiency evolution in time during the alpha-cypermethrin removal process on CA-"aA" adsorbent, 0.01 g, MS, C_i = 10-70 mg/L.

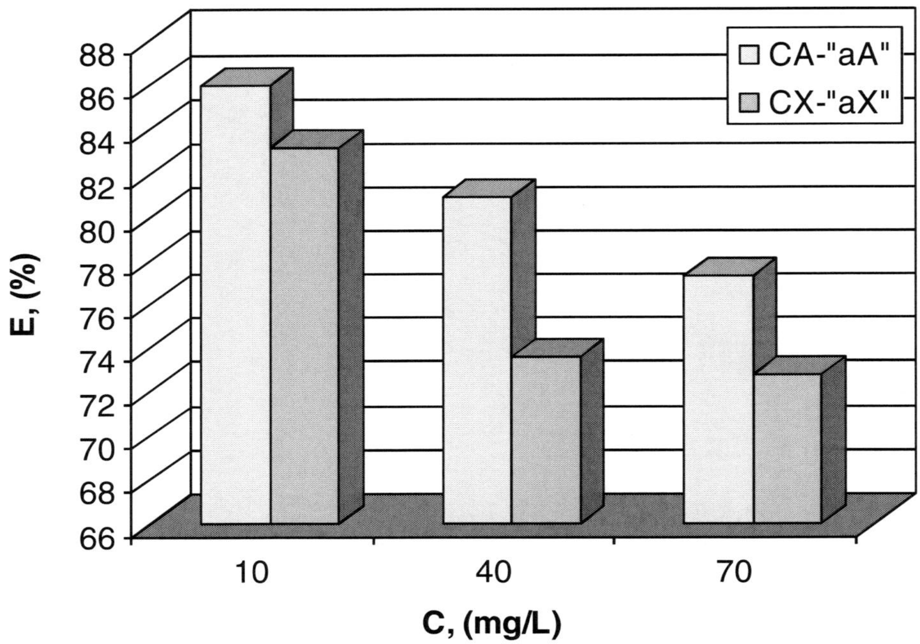

Figure 21. Influence of the initial alpha-cypermethrin concentration over the maximum adsorption efficiency, during the alpha-cypermethrin removal process on 0.01 g CA-“aA” and CX-“aX” adsorbents , MS, C_i = 10-70 mg/L.

IV.2. Catalytic Wet Air Oxidation

The phenol total oxidation or catalytic wet air oxidation (CWAO) was realised using iron catalysts (iron doped carbon aerogel) and copper catalysts (copper doped carbon aerogel and xerogel), Table 8, prepared using the procedures described above (Chapter II.2). Catalysts were brought at a grain size of d < 250 μm using a mortar and pestle and an appropriate sieve. The phenol stock solution was obtained by dissolving the analytical purity reagent in distilled water . Solutions used further on in the experiments were obtained by diluting the stock solution to the desired concentrations. All the experiments were realised by contacting the desired catalysts quantity with 100 mL phenol solution.

The CWAO process was conducted in a thermostated stirred batch reactor (magnetic stirring, 500 rpm) with a continuous feed of air from an air compressor at the desired flow rate and atmospheric pressure .

Various reaction parameters were tested:

- air flow rate: 20, 40 and 60 L/h;
- temperature : 20-80°C;
- catalyst quantity: 0.1, 0.2, 0.3 g;
- phenol concentration: 50, 100, 150, 1000 mg/L;
- solution pH : 4, 6.3, 7.

During the experiments, determination of the organic compounds in solution was carried out every 15 minutes, (first two determinations), and then every 30 minutes. The experiments were carried out for 120 minutes (iron catalysts), until no modifications were observed in the COD-Mn values. Taking into consideration previous results, in the case of copper catalysts, the experiment was conducted for 180 minutes (evolution in time was not considered at this point).

Taking into account the fact that in this stage of the research we were interested to see how this type of material acted as catalyst for the total oxidation of organic compounds we used $KMnO_4$ chemical oxygen demand, COD-Mn (see Chapter IV.1), and total organic carbon (TOC) methods in order to establish the final concentration of the organics in solution .

The TOC analysis was realised using Hach Lange equipment , which included a TOC-X5 stirrer, a LT 200 thermostat, a DR 2800 spectrophotometer and TOC vials (LCK 385, LCK 386) for various concentrations.

In the case of copper catalysts, solutions collected during the experiments were also analysed using an atomic absorption spectrometer (SensAA Dual GBS Scientific Equipment, Australia) in order to establish the copper loss from the catalyst surface.

Table 8. Experimental conditions for phenol CWAO

Catalyst	**Working conditions**	**Concentration**	**Organics determination procedure**
Iron catalysts			
DCA -$Fe^{(3+)}$ DCA -$Fe^{(2+)}$ DCA- K (blank)	Batch reactor Magnetic stirring	100, 1000 mg/L	COD-Mn
Copper catalysts			
DCA -Cu-750 DCX-Cu-750 Cu/CA-“aA” Cu/CA-“aA”-750 CA-“aA” (blank) CX-“aX” (blank)	Batch reactor Magnetic stirring	50 – 150 mg/L	TOC

The evolution of the phenol oxidation process was followed by means of overall efficiency, eq. (2).

$$X, \% = \frac{C_i - C_{t(f)}}{C_i} \cdot 100 \tag{2}$$

where,

C_i is the initial organics concentration, in mg $KMnO_4$/L, mg C/L;
$C_{t(f)}$ is time t or final organics concentration, in mg $KMnO_4$/L, mg C/L.

IV.2.1. Iron Catalysts

The influence of the operating temperature over the evolution of the overall efficiency, air flow 60 L/h, 0.1 g DCA -$Fe^{(3+)}$ catalyst and C_i = 1000 mg/L, is presented in Figure 22. As expected, the overall efficiency increased with the increasing of the temperature. In the first 30 minutes, the efficiency increased up to 51, 56 and 67% for an operating temperature of 20, 60 and 80°C, respectively. The maximum values for overall efficiency are presented in Figure 23. The highest value obtained was 69% at 80°C.

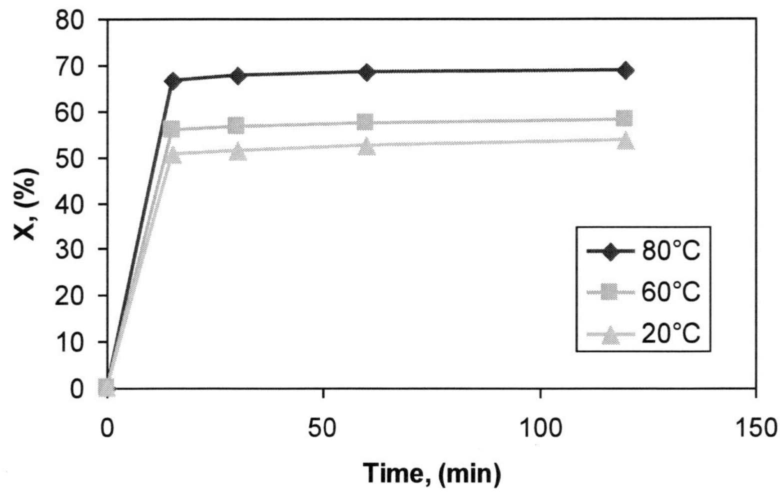

Figure 22. Variation of the overall efficiency in the phenol CWAO process, DCA -$Fe^{(3+)}$ catalyst , at different temperatures and constant air flow, (60 L/h, 0.1 g catalyst and C_i = 1000 mg/L).

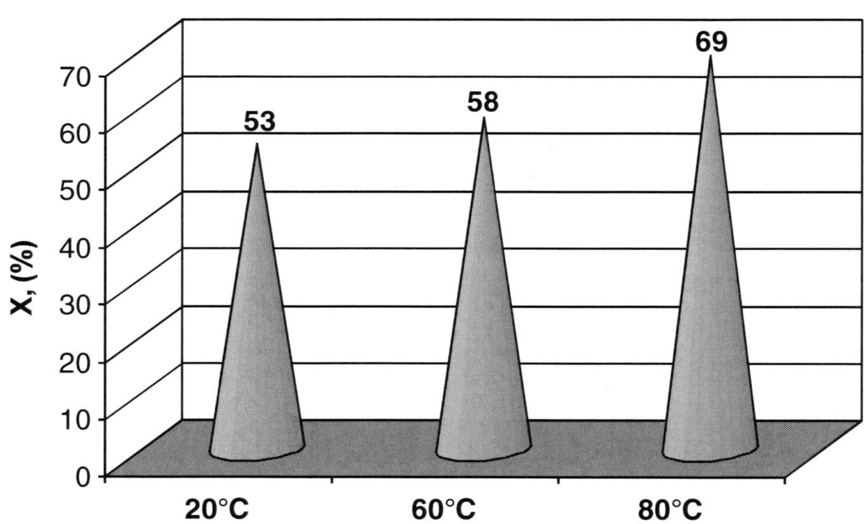

Figure 23. Influence of the operating temperature over the maximum values of overall efficiency in the phenol CWAO process, DCA -$Fe^{(3+)}$ catalyst , at constant air flow, (60 L/h, 0.1 g catalyst and C_i = 1000 mg/L).

The influence of the air flow over the overall efficiency variation in time, for DCA -$Fe^{(3+)}$ catalyst , at constant temperature 80°C, 0.1 g catalyst and C_i = 1000 mg phenol /L, is presented in Figure 24. Maximum increase in the overall efficiency was observed in the first 15 minutes and varied from 36 to 67% for 20 and 60 L/h, respectively. We also observed that the increase in overall efficiency was not as steep as in the case of temperature variation, suggesting that the oxygen diffusion process could be a rate-determining step in the catalytic wet air oxidation process. The maximum values obtained for all air flows are presented in Figure 25.

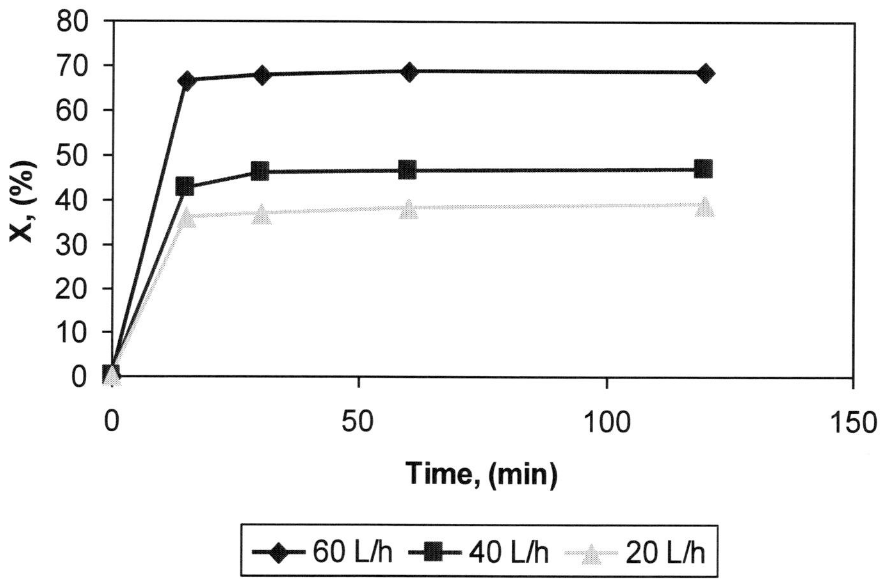

Figure 24. Variation of the overall efficiency in the phenol CWAO process, DCA -$Fe^{(3+)}$ catalyst , at constant temperature and different air flows, (80°C, 0.1 g catalyst and C_i = 1000 mg/L).

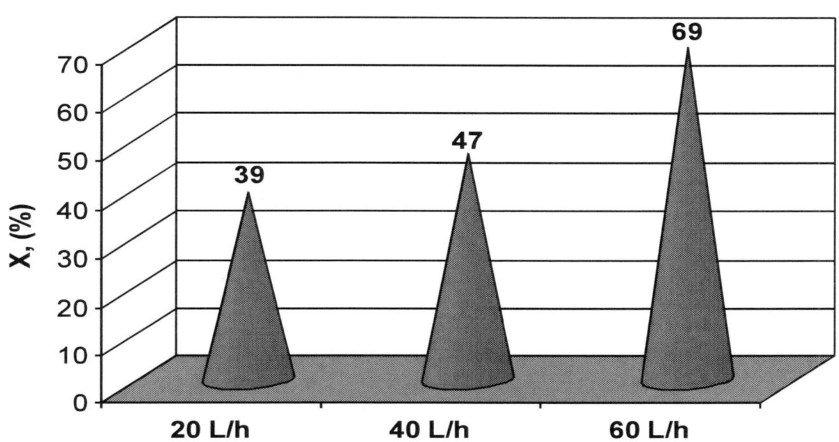

Figure 25. Influence of the operating temperature over the maximum values of overall efficiency in the phenol CWAO process, DCA -$Fe^{(3+)}$ catalyst , at constant temperature, (80°C, 0.1 g catalyst and C_i = 1000 mg/L).

In addition, the influence of the catalyst quantity and that of the phenol initial concentration over the overall efficiency in the CWAO process were studied. The results obtained in these cases are presented in Figures 26 and 27. For an increase at 0.2 g in catalyst quantity an increase by only 11% of the overall efficiency was observed, Figure 26. A 10 times decrease in the phenol solution concentration led to an increase by around 28% in the overall efficiency value (Figure 27) [70].

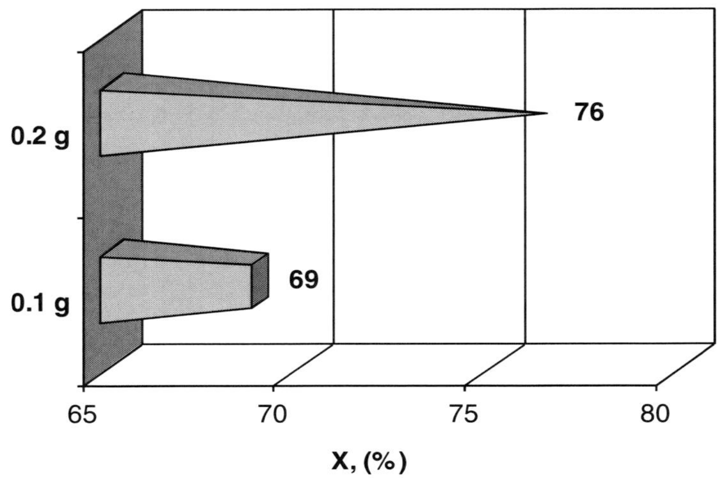

Figure 26. The catalyst quantity influence over the maximum overall efficiencies obtained in the phenol CWAO process, DCA -$Fe^{(3+)}$ catalyst, at constant air flow and constant temperature , (60 L/h, 80°C, C_i = 1000 mg/L).

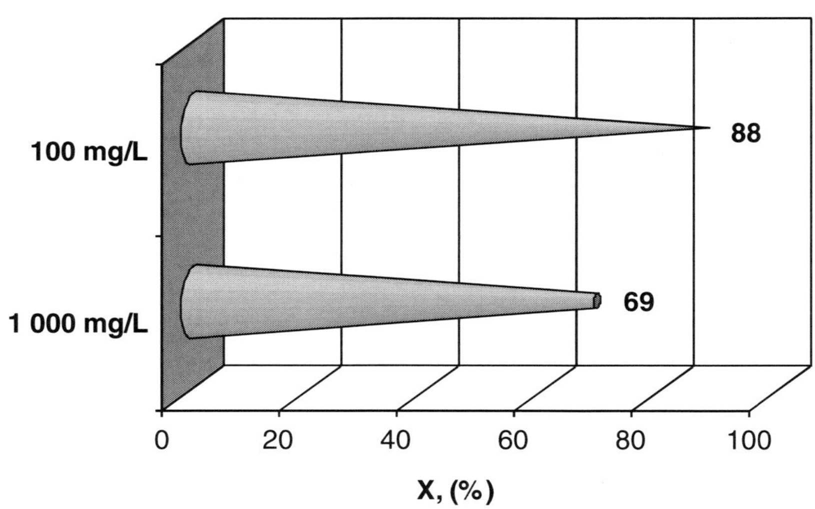

Figure 27. The phenol initial concentration influence over the overall efficiencies in the phenol CWAO process, DCA -$Fe^{(3+)}$ catalyst , at constant air flow and constant temperature , (60 L/h, 80°C, 0.1 g catalyst).

Under the same working conditions , on the DCA -$Fe^{(2+)}$ catalyst we obtained overall efficiency values by 20% smaller than in the case of DCA-$Fe^{(3+)}$, which can be explained by its smaller surface area and partial coverage of iron /iron oxide particles by graphitic clouds. The blank sample, DCA-K, had no activity in the phenol oxidation process.

IV.2.2. Copper Catalysts

The experiments were started considering the influence of air flow over the TOC values, overall efficiency values and Cu loss at 25°C. The results showed that as the air flow increases, the TOC value decreases. Consequently, overall efficiency values increase, Figure 28. A higher difference was observed when the air flow increased from 20 to 40 L/h, while a further increase to 60 L/h had as effect a smaller influence. Therefore, an increase in flow rate, hence in oxygen quantity, favoured the oxidation process. Mass transfer was also intensified. Probably the most important observation was the fact that a copper loss from the catalyst structure was identified. Copper loss increased consistently with an increase in flow rate, Figure 28c.

An increase in the catalyst quantity from 0.1 to 0.2 g led to an increase in overall efficiency from 85 to 96%, Figure 29a. For a further increase to 0.3 g no notable increase was observed. On the other hand, for 0.3 g catalyst an important Cu loss was established, Figure 29b. Therefore, an increase in the catalyst quantity, > 0.2 g, will not have positive effects upon the catalytic process.

The results obtained when the initial concentration of phenol in solution was modified are presented in Figure 30. A decrease in the overall efficiency was observed, from 89 to 77% for 50 and 150 mg/L, respectively.

After the first utilisation, the DCA -Cu-750 catalyst was dried, regenerated (pyrolysis 550°C/Ar for 2 h) and tested again in the CWAO process under the same conditions. Due to the copper loss during the first process, the overall efficiency drops from 85 to 75%, Figure 31.

In order to compare the CWAO overall efficiencies of the DCA -Cu-750 catalyst , several other materials were considered as catalysts and tested under the same conditions. The preparation of all the catalysts is described in Chapter II.2.

The first comparison was realized between DCA -Cu-750 and DCX-Cu-750, Figure 32a. It was observed that in the case of the doped xerogel catalyst , the CWAO efficiency was smaller (62%) in comparison to the doped aerogel catalyst (85%). This difference could be attributed to the difference in their porous structure .

The next comparison was realized between the aerogel doped catalyst and two catalysts obtained using copper adsorption on carbon aerogel (dried and pyrolysed at 750°C, see Chapter II.2, Figure 12), Figure 32b. Out of the three, Cu/CA-“aA”-750, obtained by copper adsorption and pyrolysed, was the most efficient (89%) followed by DCA -Cu-750 (85%) and Cu/CA-“aA” (79%). The difference between DCA-Cu-750 and Cu/CA-“aA”-750 could be attributed to the difference in copper distribution over the catalyst surface, in favour of the catalyst obtained by adsorption, where copper might have been placed towards the surface. In the case of the catalyst obtained by adsorption but without pyrolysis (Cu/CA-“aA”), copper loss during the CWAO process could have the effect of a small value for overall efficiency.

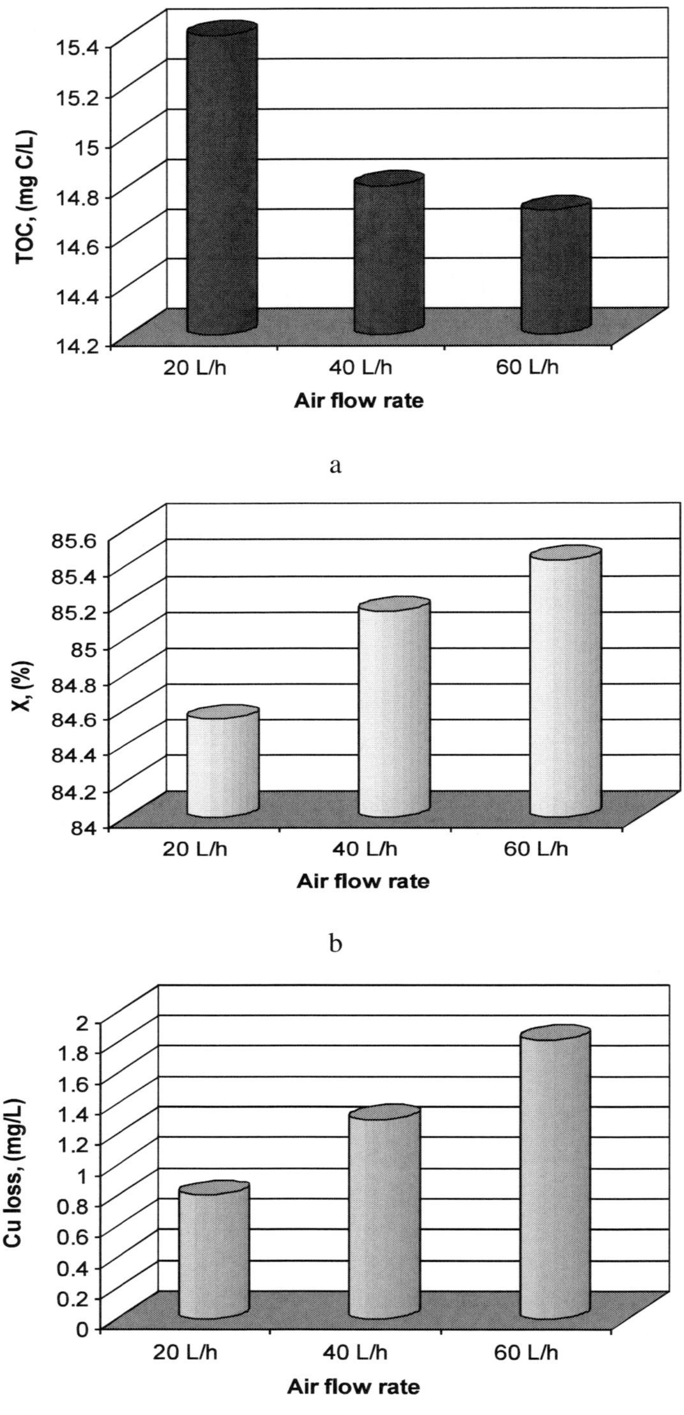

Figure 28. The air flow rate influence over the (a) TOC value, (b) CWAO efficiency, (c) Cu loss on DCA -Cu-750, (0.1g, 100 mg/L, 25°C).

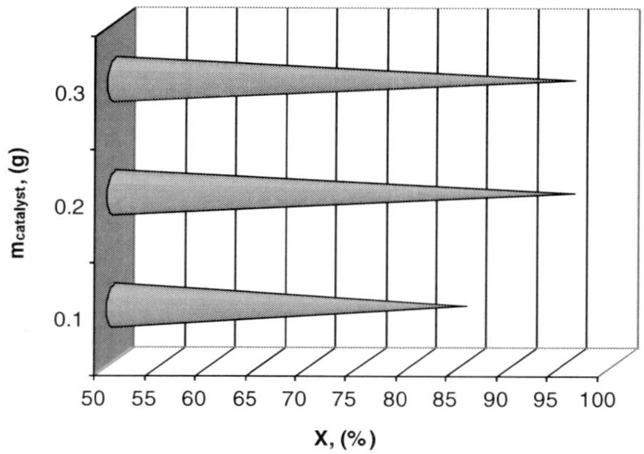

a

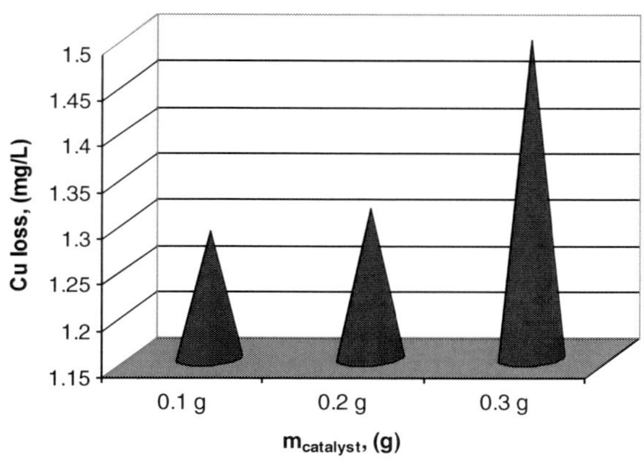

b

Figure 29. DCA -Cu-750 catalyst mass influence over the CWAO efficiency (a) and Cu loss (b), (100 mg/L, 25°C, 40 L/h).

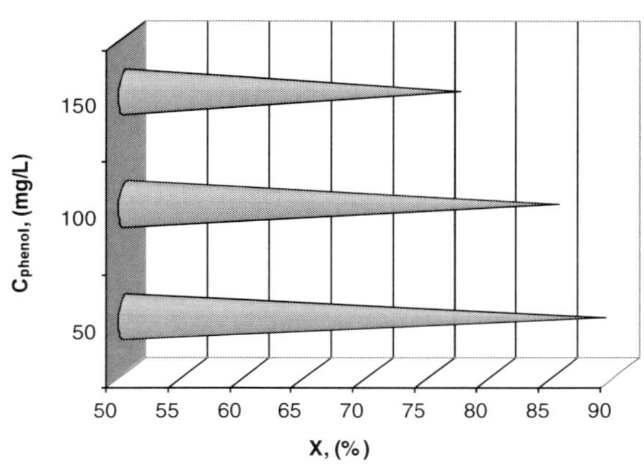

Figure 30. The phenol concentration influence over the CWAO efficiency on the DCA -Cu-750 catalyst , (0.1 g, 25°C, 40 L/h).

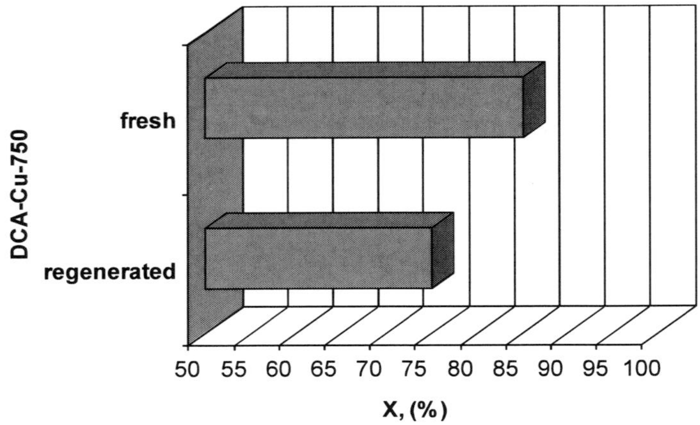

Figure 31. Maximum overall efficiencies for the CWAO process on fresh and regenerated DCA -Cu-750 catalyst , (0.1 g, 25°C, 40 L/h).

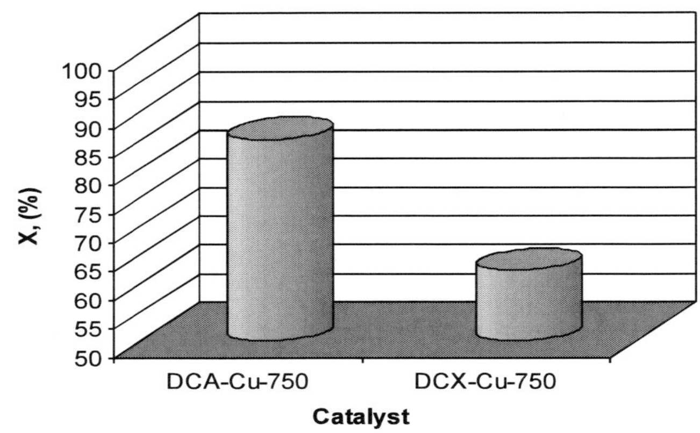

a

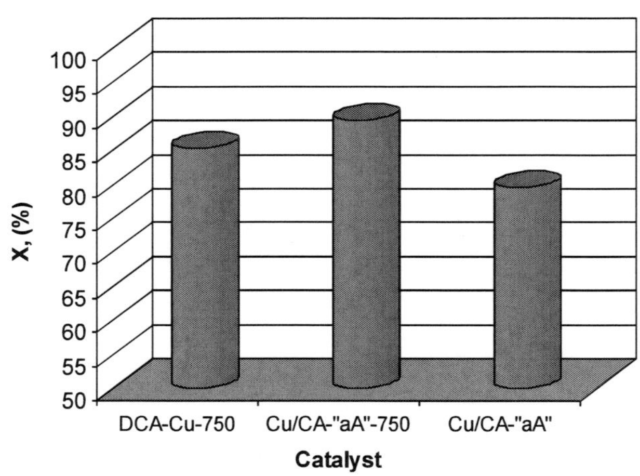

b

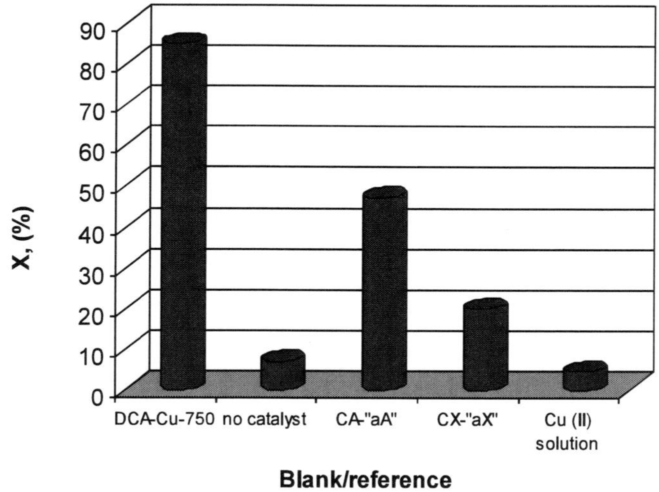

c

Figure 32. Comparative CWAO maximum overall efficiencies for (a) DCA -Cu-750 and DCX-Cu-750, (b) DCA-Cu-750, Cu/CA-“aA”-750 and Cu/CA-“aA”, (c) DCA-Cu-750, no catalyst , Cu-“aA”, CX-“aX” and Cu(II) solution , (0.1 g, 25°C, 40 L/h).

The final comparison was realised between the doped aerogel catalyst and various blank/reference materials (no catalyst, carbon aerogel, carbon xerogel , Cu(II) dissolved in solution in a quantity equivalent to that of 0.1 g doped aerogel), Figure 32c. When no catalyst was present in the solution or cooper was present as Cu(II) ions , the values of efficiency were under 10%. When carbon aerogel or xerogel were present, efficiency values were of 47 and 20%, respectively, but these values were due to the removal of phenol by adsorption .

CONCLUSION

Various samples of undoped and metal (Fe or Cu) doped carbon aerogels and xerogels were synthesized and characterized (N_2 adsorption -desorption investigation, TEM , SEM, AFM , X-ray diffraction and elemental analysis).

The prepared carbon aerogel proved to be an efficient adsorbent for phenol from wastewaters. Adsorption efficiencies up to 94% were reached. In the case of alpha-cypermethrin adsorption , efficiencies up to 89% were reached, with carbon aerogel having better performances than carbon xerogel .

Iron and copper doped carbon aerogel have good performances in the phenol chemical wet air oxidation process. High efficiencies (>75%) were realized under mild conditions (temperature 20-80°C, atmospheric pressure).

REFERENCES

[1] M.A. Shannon, P.W. Bohn, M. Elimelech, J.G. Georgiadis, B.J. Marinas, A.M. Mayes, *Nature* 452 (2008) 301.

[2] K.Y. Foo, B.H. Hameed, *J. Hazard. Mater.*, 175 (2010) 1.

[3] X. Cheng, Z. Zhang, S. Tian, *Spectrochim. Acta, Part A*, 67 (2007) 1270.

[4] A. Ojha, S.K. Yaduvanshi, N. Srivastava, *Pestic. Biochem. Physiol.*, 99 (2011) 148.

[5] M. Cioroiu, D. Tarcau, R. Mocanu, S. Cucu-Man, B. Nechita, M. Luca, *Sci. Total Environ.*, 408 (2010) 2639.

[6] M. Clementia, G.M. Tibonib, R. Causinc, C. La Roccad, F. Maranghid, F. Raffagnatoa, R. Tenconia, *Reprod. Toxicol.*, 26 (2008) 13.

[7] P. Zhou, Y. Wu, S. Yin, J. Li, Y. Zhao, L. Zhang, H. Chen, Y. Liu, X. Yang, X. Li, *Environ. Pollut.*, 159 (2011) 524.

[8] A. Maicaneanu, L.C. Cotet, V. Danciu, M. Stanca, *Studia UBB Chemia*, 54 (2009) 33.

[9] S. Varma, D. Sarode, S. Wakale, B.A. Bhanvase, M.P. Deosarkar, *Int. J. Chem. Phy. Sci.*, 2 (2013) 132.

[10] A.H.; El-Sheikh, J.A.; Sweileh, Y.S.; Al-Degs, A.A.; Insisi, N. Al-Rabady, *Talanta*, 74 (2008) 1675.

[11] Y.S. Al-Degs, M.A. Al-Ghouti, A.H. El-Sheikha, *J. Hazard. Mater.*, 169 (2009) 128.

[12] K. Ignatowicz, Select*J. Hazard. Mater.*, 169 (2009) 953.

[13] O.A. Ioannidou, A.A. Zabaniotou, G.G. Stavropoulos, M.A. Islamb, T.A. Albanis, *Chemosphere*, 80 (2010) 1328.

[14] A. Dhaouadi, L. Monser, N. Adhoum, *J. Hazard. Mater.*, 181 (2010) 692.

[15] I.P.P. Cansado, P.A.M. Mourão, A.I. Falcão, M.M.L. Ribeiro Carrott, P.J.M. Carrott, *Fuel Process. Technol.* 103 (2012) 64.

[16] M.F.L. De Volder, S.H. Tawfick, R.H. Baughman, A.J. Hart, *Science* (2013) 535.

[17] Z. Al-Qodah, A.T. Shawaqfeh, W.K. Lafi, *Adsorption*, 131 (2007) 73.

[18] C.; Faur, H. Metivier-Pignon, P. Le Cloirec, *Adsorption*, 11 (2005) 479.

[19] L.L. Luo , X.X. Gu, J. Wu, S.X. Zhong, J.R. Chen, *Adv. Mater. Res.* 550 – 553 (2012) 2121.

[20] H. Wang, X. Yuan, Y. Wu, H. Huang, X. Peng, G. Zeng, H. Zhong, J. Liang, M.M. Ren, *Adv. Coll. Inter. Sci.*, 195 (2013) 19.

[21] G. Busca, S. Berardinelli, C. Resini, L. Arrighi, *J. Hazard. Mater.*, 160 (2008) 265.

[22] F. Luck, *Catalysis Today*, 53 (1999) 81.

[23] R. Andreozzi, V. Caprio, A. Insola, R. Marrota, *Catalysis Today*, 53 (1999) 51.

[24] B. Legube, N. Karpel Vel Leitner, *Catalysis Today*, 53 (1999) 61.

[25] S. Zrncevic, Z. Gomzi, *Ind. Eng. Chem. Research*, 44 (2005) 6110.

[26] I. V. Perez, S. Rogak, R. Branion, *J. Supercr. Fluids*, 30 (2004) 71.

[27] D. Fino, C. Carlasi Jara, G. Saracco, V. Specchia, P. Spinelli, *J Appl. Electroch.*, 35 (2005) 405.

[28] A. M. Polcaro, A. Vacca, S. Palmas, M. Marcia, *J. Appl. Electroch.*, 33 (2003) 885.

[29] J-M. Herrmann, *Catalysis Today*, 53 (1999) 115.

[30] U. Bali, E. C. Catalkaya, F. Sengul, *J. Envir. Sci. Health, A38*, (2003) 2259.

[31] Y. G. Adewuyi, *Ind. Eng. Chem. Res.*, 40 (2001) 4681.

[32] G. Collins, C. Foy, S. McHugh, T. Mahony, V. O'Flaherty, *Water Reserach*, 39 (2005) 1614.

[33] A. Nuhoglu, B. Yalciu, *Processes in Biochem.*, 40 (2005) 1233.

[34] Y.I. Matattov-Meytal, M. Sheintuch, *Ind. Eng. Chem. Res.*, 37 (1999) 309.

[35] A. Cybulski, J. Trazynski, *Appl. Cat. B: Env.*, 47 (2004) 1.

[36] A. Alejandre, F. Medina, A. Fortuny, P. Salagre, J. E. Sueiras, *Appl. Cat. B: Env.*, 16 (1998) 53.

[37] L. Chang, I-P. Chen, S-S. Lin, *Chemosphere*, 58 (2005) 485.

[38] P. M. Alvarez, D. Mclurgh, P. Plucinski, *Ind. Eng. Chem. Res.*, 41 (2001) 2153.

[39] A. Santos, P. Yustos, S. Gomis, G. Ruiz, F. Garcia-Ochoa, *Ind. Eng. Chem. Res.*, 44 (2005) 3869.

[40] M. Noorjahan, V.D. Kumari, M. Subrahmanyam, L. Panda, *Appl. Cat. B: Env.*, 57 (2005) 291.

[41] N.H. Phu, T.T.K. Hoa, N.V. Tan, H.V. Thang, P.L. Ha, *Appl. Cat. B: Env.*, 34 (2001) 267.

[42]] J. Barrault, J-M. Tatibouet, N. Papayannakos, *C. R. Academie des Science Paris, Serie IIc Chimie*, 3 (2000) 777.

[43] J. Barrault, C. Bouchoule, K. Echachoui, N. Frini-Srasra, M. Trabelsi, F. Bergaya, *Appl. Cat. B: Env.*, 15 (1998) 269.

[44] C. Wang, Y. Feng, P. Gao, N. Ren, B.-L. Li, *Sci. Total Environ.*, 431 (2012) 366.

[45] A. Santos, P. Yustos, A. Quintanilla, F. Garcia-Ochoa, J. A. Casas, J. J. Rodriguez, *Envir. Sci. Tech.*, 38 (2004) 133.

[46] L.C. Cotet, A. Roig, I.C. Popescu, V. Cosoveanu, E. Molins, V. Danciu, *Rev. Roum. Chim.*, 52 (2007) 1077.

[47] X. Chang, D. Chen, X. Jiao, *Polymer*, 51 (2010) 3801.

[48] B.S. Girgis, A.A. Attia, N.A. Fathy, *Desalination* 265 (2011) 169.

[49] G. Zhou, H. Tian, H. Sun , S. Wang, C.E. Buckley, *Chem. Eng. J.*, 171 (2011) 1399.

[50] X. Wu, D. Wu, R. Fu, W. Zeng, *Dyes Pigmemts*, 95 (2012) 689.

[51] R.S. Ribeiro, N.A. Fathy, A.A. Attia, A.M.T. Silva, J.L. Faria, H.T. Gomes, *Chem. Eng. J.*, 195–196 (2012) 112.

[52] C.A. Paez, M.S. Contreras, A. Leonard, S. Blacher, C.G. Olivera-Fuentes, J.-P. Pirard, N. Job, *Adsorption* 18 (2012) 199.

[53] S.A.C. Carabineiro, T. Thavorn-Amornsri, M.F.R. Pereir, P. Serp, J.L. Figueiredo, *Catalysis Today*, 186 (2012) 29.

[54] F.J. Maldonado-Hodar, C. Moreno-Castilla, F. Carrasco-Marın, A.F. Perez-Cadena, *J. Hazard. Mater.*, 148 (2007) 548.

[55] S.M.; Chathoth, E.; Mamontov, Y.B.; Melnichenko, M. Zamponi, *Microp. Mesop. Mater.*, 132 (2010) 148.

[56] K.; Kadirvelu, J.;Goel, C. Rajagopal, *J. Hazard. Mater.*, 153 (2008) 502.

[57] A.K. Meena, G.K. Mishra, P.K. Rai, C. Rajagopal, P.N. Nagar, *J. Hazard. Mater.*, 122 (2005) 161.

[58] K. Watson, M.J. Farré, N. Knight, *J. Env. Manag.*, 110 (2012) 276.

[59] C. Moreno-Castilla, F.J. Maldonado-Hodar, *Carbon,* 43 (2005) 455.

[60] H. Sun , H. Tian, Y. Hardjono, C.E. Buckley, S. Wang, *Catalysis Today,* 186 (2012) 63.

[61] F.J. Maldonado-Hodar, H. Jirglova, A.F. Perez-Cadenas, S. Morales-Torres, *Carbon,* 51 (3013) 213.

[62] Y. Hardjono, H. Sun , H. Tian, C.E. Buckley, S. Wang, *Chem. Eng. J.*, 174 (2011) 376.

[63] S.A. Al-Muhtaseb, J.S. Ritter, *Adv. Mater.*, 15 (2003) 101.

[64] L.C. Cotet, C.I. Fort, V. Danciu, *Stud. U. Babes-Bol. Che.*, 3 (2012) 73.

[65] D. Gligor, L.C. Cotet, V Danciu, *J. New Mat. Electroch. Syst.,* 16 (2013) 97.

[66] C.I. Fort, L.C. Cotet, V. Danciu, G.L. Turdean, I.C. Popescu, Iron doped carbon aerogel-new electrode material for electrocatalytic reduction of H_2O_2, *Mater. Chem. Phy.*, 138(2013)893-898.

[67] L.C. Cotet, M. Gich, A. Roig, I.C. Popescu, V. Cosoveanu, E. Molins, V. Danciu, *J. Non-Crist. Solids,* 352 (2006) 2772.

[68] M. Baia, L.C. Cotet, L. Baia, L. Barbu-Tudoran, V. Cosoveanu, V. Danciu, J. Popp, *J. Optoel. Adv. Mater.-Symp.* 2 (2010) 9.

[69] M. Gich, C. Fernandez-Sanchez, L.C. Cotet, P. Niu, A. Roig, Facile, *J. Mater. Chem. A*, 1 (2013) 11410.

[70] A. Maicaneanu, L.C. Cotet, V. Danciu, M. Stanca, *Stud. U. Babes-Bol. Che.*, 54 3 (2009) 117.

[71] J.G.M. van Bercum, A.C. Vermeulen, R. Delhez, T.H. de Keijser, E.M. Mittemeijer, *J. Appl. Phys.*, 27 (1994) 345.

[72] N. Aldea, E. Indrea, Comput. *Phys. Commun.*, 60 (1990) 155.

[73] E. Indrea, A. Barbu, *Appl. Surf. Sci.*, 106 (1996) 498.

[74] L.C. Cotet, M. Baia, L. Baia, I.C. Popescu, V. Cosoveanu, E. Indrea, J. Papp, V. Danciu, *J. All. Comp.,* 434-435 (2007) 854.

[75] L.C. Cotet, A. Maicaneanu, C.I. Fort, V. Danciu, *Sep. Sci. Tech.*, 2013, accepted. DOI :10.1080/01496395.2013.805782.

In: Comprehensive Guide for Mesoporous Materials. Volume 1 ISBN: 978-1-63463-958-3
Editor: Mahmood Aliofkhazraei

Chapter 10

SYNTHESIS AND ELECTROCATALYTIC APPLICATIONS OF POLYANILINE

Rajendra Srivastava, Anu Prathap M. U. and Martin Francis Pulikottil
Department of Chemistry, Indian Institute of Technology Ropar, Punjab, India

ABSTRACT

Organic porous materials are a class of advanced materials, which have shown promising potential in many technological applications. Polyaniline (a conducting polymer) is a novel organic semiconducting material, which has attracted considerable attention because of its remarkable optical, electrochemical, and conducting properties. Nanostructured polyaniline combines the properties of low-dimensional organic conductors with high surface area materials. Polyaniline has been investigated in a number of catalytic and bio-technological applications because of the unique flexibility with respect to morphological controlled synthesis, tunable pore size, and ion-exchange properties. This chapter covers the state -of-the-art in the synthesis, chemical structure and properties, and electrocatalytic activity of polyaniline for sensing application. Single and simultaneous detection of clinically and pharmaceutically important molecules using polyaniline based electrodes is discussed in detail.

1. INTRODUCTION

Today modern world cannot imagine life without polymeric materials. Polymer is a large organic molecule, which is composed of many repeating subunits (known as monomers). A variety of natural (known as bio-polymers such as DNA and proteins) and synthetic polymers (such as polystyrene, polyethylene, polycarbonates etc.) are known [1], which play an essential and ubiquitous role in everyday life from plastic containers, home appliance to the body implant material. Due to high demand and wide applications, polymers are produced in large scale. Today synthetic polymers are produced in larger quantity than any other class of materials and compounds .

Discovery of polymerization catalyst for the production of polymer was awarded Nobel Prize [2], because with this development, whole world was benefited, and it was a boon for the modern society . Polythiazyl (SN)x crystals discovered in 1973 actually had conductivity as high as metals . Conducting polymers have stimulated a lot of interest from scientists all around the world. Conducting polymers find numerous applications in various interdisciplinary research areas and device development (Scheme 1). The chronology of the development of some important conducting polymers and their applications are given in Table 1 [3-45].

1.1. Conducting Polymers

A variety of polymers discovered in the early stages are good electrical insulators . These insulated polymers cannot be used for electronic devices due to negligible conductivity . For a polymer to exhibit electrical conducting, its electrons must be free or mobile and not bound to the atoms . This mobility occurs if the polymer backbone is having alternate single and double bonds i.e; conjugated double bonds . Polymeric materials that exhibit the electrical and optical properties of metals or semiconductors are known as conducting polymers [46]. Alan MacDiarmid, Hideki Shirakawa, and Alan J. Heegar with talented young research group members discovered first such conducting polymers and the ability to dope these polymers over the full range from insulator to metal, for which they have been awarded Nobel Prize in the year 2000 [47].

In conjugated polymers , the chemical bonding leads to one unpaired electron (the π electron) per carbon atom. In the conjugated system, carbon orbital is in sp^2 configuration and overlap with other carbon orbital along the polymer backbone leading to electronic delocalization. So the electronic structure in conducting polymers is determined by the chain symmetry , which explains their semi conducting or even metallic properties [48, 49]. The classic example is polyacetylene $(-CH=CH)_n$, the π band is divided into π and π^* bands.

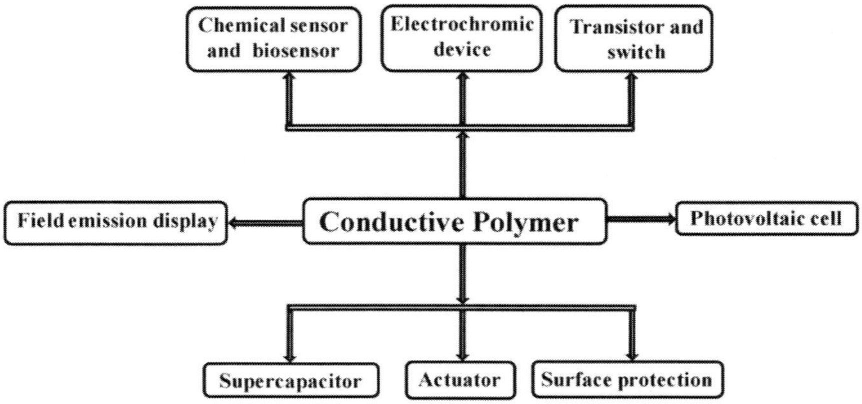

Scheme 1. Applications of electronic polymers .

Table 1. The chronological development of some important conducting polymers

Year	Title	Ref
1910	(SN)x synthesized	[3]
1973	Metallic conductivity of (SN)x	[4]
1971	Semiconductivity of polyaniline (PANI)	[5]
1977	Doping of PANI	[6]
1979	Electrochemical synthesis of polypyrrole (PPy)	[7]
1979	Polyacetylene battery	[8]
1979	Conducting polyparaphenylene	[9]
1980	Conducting polyparaphenylene sulphide	[10]
1982	Polythiophene (PTh)	[11]
1983	Conducting poly para-phenylene sulfide	[12]
1984	PPy films used in neurotransmitter timed release into the brain	[13]
1985	PPy for EMI shielding	[14]
1985	Polymeric heterojunctions based on polypyrrole deposited on silicon for solar cells	[15]
1986	PTh, PPy, PANI for electrochromic displays and thermal smart windows	[16]
1992	PANI for EMI shielding	[17]
1993	Conjugated polymer light-emitting diodes	[18]
1993	PANI based electrochromic windows	[19]
1994	Light-emitting diodes fabricated with conjugated polymers	[20]
1997	Polymer electroluminescent Devices	[21]
1998	Conjugated polymer for optical applications	[22]
1999	PTh, PPy, PANI electrode materials for rechargeable batteries	[23]
1999	Electro-optical polythiophene devices	[24]
2000	Blue light emitting polymers	[25]
2000	Fluorene-based polymers	[26]
2000	Poly(aryleneethynylene)s	[27]
2000	Semiconducting polymers based solid-state lasers	[28]
2001	Polyfluorenes: twenty years of progress	[29]
2002	Semiconducting polyfluorenes	[30]
2003	Conjugated polymers as molecular materials	[31]
2003	Carbazole containing polymers	[32]
2004	Conjugated oligomers and polymers: the organometallic way	[33]
2004	(poly(3,4-ethylene dioxythiophene))/polystyrene sulfonate ink developed for inkjet printing	[34]
2005	Conducting PPy-coated textile based on nylon lycra as wearable biomechanical sensors to monitor human motion	[35]
2005	Conjugated polymers based diodes, capacitors, resistors	[36]
2005	PPy based cooling fabrics	[37]
2006	An all-polymer battery based on PPy	[38]
2005-2007	PANI based conductive adhesive, conductive ink, conductive paints	[39]
2008	Undoped and doped conjugated polymers based gas sensing devices	[40]
2002-2010	Solving the solubility issues related to some conjugated polymers	[41]
2004-2011	Conducting polymers for solar cell application	[42,43]
2010-2011	Undoped and doped conjugated polymers for field effect transistors (FETs)	[44]
1999-2014	Biomedical applications of conducting polymers	[45]

Since each band can hold two electrons per atom (spin up and spin down), the π band is filled and the π* band is empty. The energy difference between the highest occupied state in the π band and the lowest unoccupied state in the π* band is the π - π* energy gap, E_g. The alternate double bond structured molecules are conjugated polymers (Scheme 2).

However, due to the presence of the extended conjugation in polymer backbone, there is a strong inter-chain interaction and the chains become rigid, which leads to form insoluble and infusible materials [49, 50]. Conductivity of conjugated polymers can be tailored by varying the chain length, because Eg depends upon the molecular structure of the repeating unit [46]. However, conductivity of material can be increased to semi-conductor range by increasing the chain length [46, 51].

Such semi-conducting polymeric materials have limited applications. Conductivity of these materials can be increased by doping process [51]. With the doping process, polymers having small conductivity (10^{-10} to 10^{-5} S/cm) are converted to polymer with metallic conductivity (1 to 10^4 S/cm) (Table 2) [52, 53]. The controlled addition of small amount of non-stoichiometric quantity of chemical species results in dramatic changes in the electronic, electrical, magnetic, and optical properties of the polymer.

Thanks to organic chemists to provide a wide range of organic monomers , which can be used to prepare a variety of conducting polymers (Scheme 2). As discussed above, conductivity can be increased by increasing the chain length or by doping process.

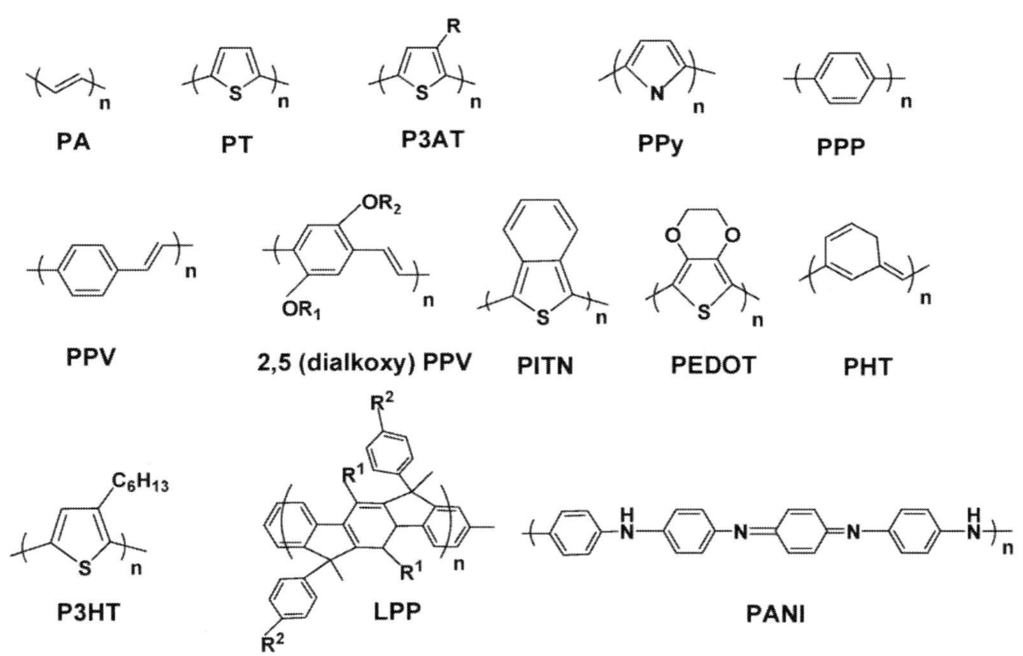

PA-Polyacetylene; PT-Polythiophene; P3AT-Poly(3-alkyl)thiophene (R = methyl, butyl, etc.); PPy-Polypyrrole; PPP-Polyparaphenylene; PPV-Polyparaphenylenevinylene; 2,5 (dialkoxy)PPV- Poly (2,5 dialkoxy)paraphenylenevinylene (e.g., MEH-PHV); PITN-Polyisothionaphthene; PEDOT-Polyethylenedioxythiophene; PHT-Polyheptadiyne; P3HT-Poly(3-hexyl) thiophene; LPP-Ladder-typepolyparaphenylene; PANI-Polyaniline .

Scheme 2. Examples of a few conjugated polymers .

Furthermore, if suitable organic functionality (especially heteroatom such as N and S) is introduced in the polymer chain , conducting and doping efficiency of the polymer can be increased significantly. By varying different monomers , a variety of conducting polymers can be prepared (Scheme 2) and by adopting a suitable doping process, a desired conductivity can be achieved (Table 2). Following section describes various doping methods used for a variety of conducting materials .

1.2. Doping Process

Doping involves a chemical reaction in which charge is injected into the macromolecular chain. Doping process influences the conductivity of the polymer . This reversible charge injection in a polymer chain can be accomplished in a number of ways, as shown in Scheme 3. The concept of doping is unique that distinguishes conducting polymers from the rest of polymers [48, 53]. During the doping process of an organic polymer, either an insulator or semiconductor having a small conductivity is converted into a polymer with metallic conductivity [48]. The controlled addition of dopants leads to dramatic change in the physico-chemical properties of the polymer. Doping process can be classified in four different categories and changes occur is shown in Scheme 3.

Table 2. Conductivity of conducting polymers and their similarity with metals and other known materials

Material	Conductivity (S/cm)	Conducting Polymer
Ag, Cu, Fe	10^{6}	
Mg	10^{4}	Doped Trans-(Poly acetylene)
In, Sn	10^{2}	Doped Polyaniline
	10^{0}	
Ge	10^{-2}	
	10^{-4}	
Si	10^{-6}	Trans-(Poly acetylene)
AgBr	10^{-8}	
Glass	10^{-10}	Polyaniline
Diamond	10^{-12}	
Nylon	10^{-14}	
Quartz	10^{-16}	

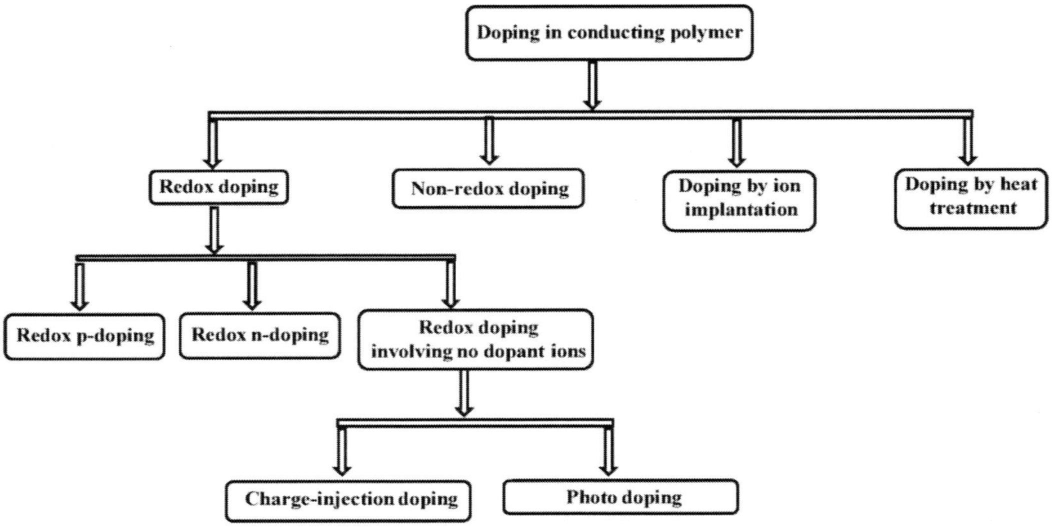

Scheme 3. Different methods for doping conducting polymers .

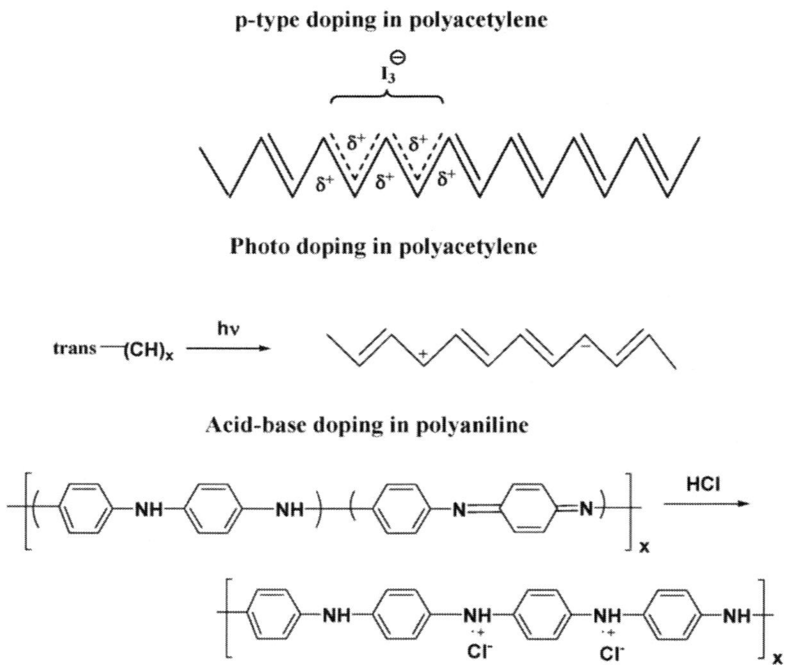

Scheme 4. Doping mechanism involved to increase the conductivity of polymers .

1.2.1. Redox Doping

Redox doping includes either p- and/or n-type doping by chemical and/or electrochemical processes by which the number of electrons in the polymeric matrix changes. This type of doping process was used to increase the conductivity in polyacetylene type of conducting materials [48]. By this process, conductivity of the polymers can be increased to 10^3 S/cm from 10^{-5} S/cm.

P-type of doping is the process in which the partial oxidation of organic polymer takes place by treating with a mild oxidizing agent such as I_2. During this process, approximately 85% of the positive charge is delocalized over 15 CH units of polyacetylene (Scheme 4, for simplicity only shown for five units) to give a positive soliton [The soliton is a domain boundary between the two possible degenerate ground-state configurations of trans-(-CH=CH-)$_n$]. p-Doping can also be accomplished by electrochemical anodic oxidation by $LiClO_4$ [54].

$$\text{trans-}(CH)_{x} + (xy)(ClO_4)^{-} \rightarrow [(CH^{+y}(ClO_4)^{y-}]_x + (xy)e^{-}$$

In n-doping , partial reduction of polymer takes place in the presence of reducing agent such as liquid sodium amalgam or preferably sodium naphthalene [48].

$$\text{trans-}(CH)_{x} + (xy)(Na^{+}\ Npth)^{-} \rightarrow [Na^{y+}(CH)^{-y}]_x + Npth$$

n-Doping can also be accomplished by electrochemical anodic reduction by LiClO [55].

1.2.2. Photo Doping

When sufficient energy is given to polymer directly or by photosensitization (greater than its band gap), electrons are promoted across the gap, and the polymer undergoes photo-doping . After irradiation to polyacetylene, the positive and negative solitons are actually delocalized over ca. 15CH units (Scheme 4, only shown for five units for simplicity) [56]. They disappear rapidly because of the recombination of electrons and holes when irradiation is discontinued. If electric field is applied during irradiation, then the electrons and holes separate and photoconductivity is observed.

1.2.3. Charge-Injection Doping

Electrons and holes can be injected from metallic contacts into the π^* or π orbital of conducting material [57]. Charge-injection doping is most conveniently carried out using a metal/insulator/semiconductor involving a metal, and a conducting polymer separated by a thin layer of a high dielectric strength insulator [58]. It was this approach, which resulted in the observance of superconductivity in a polythiophene derivative. In this method, holes are injected into a filled π orbital or electrons are injected into an empty π^* orbital of polymer. In the case of charge injection at a metal-semiconducting polymer interface , the polymer is oxidized or reduced (electrons are added to the π^* orbital or removed from the π orbital) [59].

1.2.4. Non-Redox (Acid-Base) Doping

In this doping process, acid or base is used to enhance the conductivity of polymers . This kind of doping process is widely used for polyaniline (Scheme 4). Number of electron associated with the polymer backbone does not change during the doping process, but the energy levels are rearranged during doping. Nine to ten folds enhancement in the conductivity was observed when emeraldine base (one form of polyaniline) was treated with aqueous protonic acids (Scheme 4) [60]. Protonic acid doping has subsequently been extended to systems such as poly(heteroaromatic vinylenes) [61].

During the doping process, original polymer can be reproduced with little or some degradation of the polymer backbone. Therefore, mechanical/thermal stability of the polymer is very important. For device fabrication , the mechanical/thermal stability plays a very important role. Polyaniline is the polymer of aniline , and it can be easily synthesized by the oxidative polymerization under acid medium [62].

Due to its high stability and tunable conductivity , it can be used in various device fabrications such as light -emitting diode, electrochemical sensors , solar cell etc. [63]. Therefore, significant attention has been paid for the synthesis of polyaniline.

This chapter is focused on the synthesis and electrocatalytic applications of polyaniline. Catalytic activity of any material is highly depended on its physico-chemical properties . With the doping process, conductivity of polyaniline can be tailored.

However, its textural properties such as morphology , size, and surface area are several other factors that play a very important role in the catalysis and electrocatalysis . Following section is focused on the synthesis of polyaniline. Reaction parameters and additive used in the polyaniline synthesis dramatically change the physico-chemical properties of the materials , which will be discussed in the next section.

2. Polyaniline : Structure, Properties, and Synthesis Mechanism

Polyaniline (also known as "aniline black") is a well known conducting polymer [62]. Polyaniline (hereafter represented as PANI) is the oldest and potentially one of the most useful conducting polymers because of its facile synthesis , environmental stability , and simple acid /base doping /de-doping chemistry [62, 63].

Nanostructured PANI has attracted a great deal of interest during the past few years, because a nano-size of this polymer could offer new properties or enhanced performance [64].

Before discussing about the synthesis of micro-/nano-PANI, it is important to know the chemical structure and synthesis mechanism of conventional PANI.

2.1. Chemical Structure and Properties

Polyaniline can be described as poly (1,4-phenylene-p-phenylenediamine-co-1,4-phenylene-p-benzo-quinodiimine) and its molecular formula is shown in Scheme 5. The average oxidation state can be varied continuously from $y=1$ to give the completely reduced polymer (known as leucoemeraldine), to $y=0.5$ to give the half-oxidized polymer (known as emeraldine), to $y=0$ to give the completely oxidized polymer (known as pernigraniline) and these forms are summarized in Scheme 5. Structural form of PANI and related properties depend on acidity of the medium, which is summarized in Table 3.

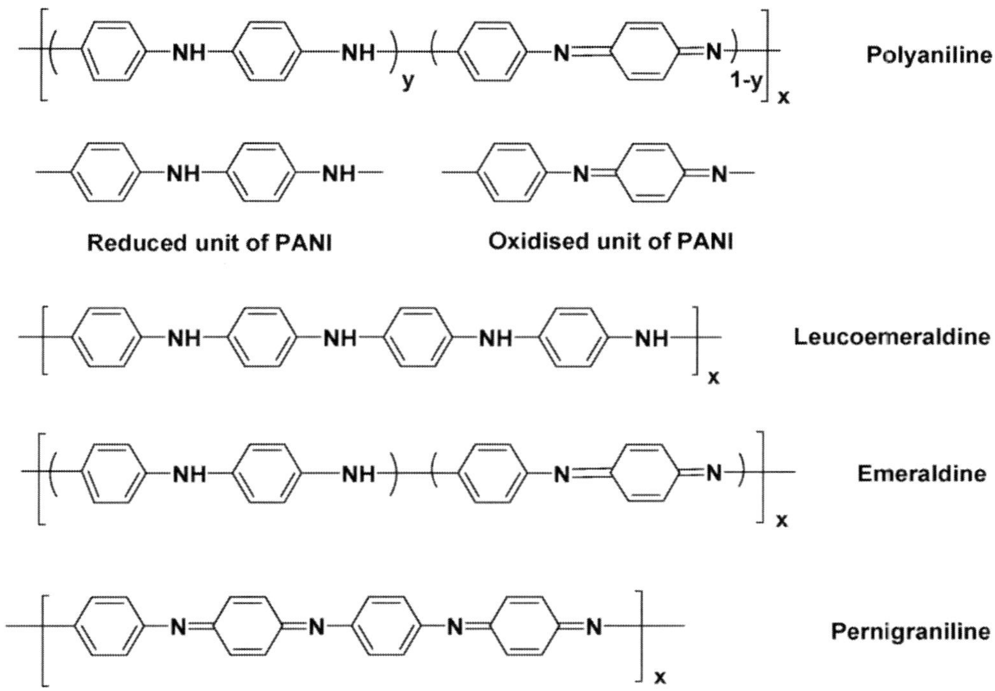

Scheme 5. Structure of polyaniline and its reduced and oxidized form.

2.2. Synthesis Mechanism

Synthesis of PANI depends on oxidant and reaction conditions. Conventional oxidants used for the PANI synthesis such as H_2O_2 and ammonium peroxydisulfate (APS) can acts as an oxygen donor as well as electron acceptor depending on reaction condition. Synthesis of PANI takes place in several steps. They are initiation , dimerization , oligomerization , and polymerization (Scheme 6). The initial process occurs not only due to the significant differences in the oxidizability of aniline and its protonated form, but also because of differences in redox/acid –base properties of its reactive species formed by single-electron (anilinium di-cation radical and aniline cation radical/neutral radical) and two-electron oxidation (aniline di-cation/nitrenium cation/nitrene) [65]. These results confirm that the generation of aniline nitrenium cation is the common feature of all two-electron oxidations of aniline in a very broad pH range [65-68], which was also verified by the computational method [65]. Second step in the PANI synthesis is dimerization. In this step, monomeric species formed in the initiation step react with other aniline molecules to form dimmers [69]. In this step also acidity plays a very important role in the formation of various dimmers (Scheme 7), which define the property of final polymer . Computational investigation of the reaction between aniline cation radical and aniline indicates the formation of 4-ADPA as the major aniline dimer [65]. Based on semi empirical quantum chemical calculations and experimental observation, it was proposed that 4-ADPA and *o*-aminophenyl sulfate are formed during the early stages of oxidative oligomerization/ polymerization of aniline with

APS via reactions of aniline nitrenium cation with aniline and in-situ generated sulfate anions, respectively (Scheme 7) [65].

Third and final step of the aniline polymerization is oligomerization and polymerization steps, respectively. In this step various dimmers formed in the second step react with itself/ aniline monomeric species /aniline to form trimmers/tetramers. Such sequence of reaction proceeds further to form PANI. It was revealed that, depending on pH of the reaction medium , linear and/or branched PANI can be formed [70, 71]. It was shown that the linear/ branched PANI ratio increases with increasing acidity [70]. The formation of branched aniline tetramer (which contains N-phenylphenazine moiety) was computationally predicted in the case oxidative polymerization of aniline in neutral /slightly acidic solutions [71]. During the oligomerization step at higher pH (pH>2), Michael addition reaction takes place to form other cyclic compounds . It was recently found that nonconducting PANI was produced that contain phenazine-like units, oxygen in semiquinonoid and/or quinoid constitutional units, and sulfur in sulfonic and/or sulfate groups, upon the oxidation of aniline with APS under alkaline conditions [72]. The aerial oxidation of aniline under protic acid medium led to form substituted phenazines as major products [73].

Table 3. Structural form of PANI produced at various pH and its property

Structural form of PANI	Acidity of the medium	Extent of oxidation	Colour	Electronic conductivity Scm^{-1}
Leuco-emeraldine salt	Highly acidic medium pH<3	Unoxidised, protonated polymer	Transparent Yellow	Insulator ca. 10^{-12}
Leuco-emeraldine base	Weakly acidic, neutral or basic medium pH>3	Unoxidised deprotonated polymer	Yellow	Insulator ca. 10^{-12}
Emeraldine base	Weakly acidic, neutral or basic medium pH>3	Partially oxidised deprotonated polymer	Blue	Insulator ca. 10^{-12}
Emeraldine salt	Highly acidic medium pH<3	Partially oxidised protonated polymer	Green	Electronic Conductor 10^{-4}-10^{2}
Pernigraniline base	Weakly aci.dic, neutral or basic medium pH>3	Fully oxidised deprotonated polymer	Violet	Insulator ca. 10^{-12}
Pernigraniline salt	Highly acidic medium pH<3	Fully oxidjsed protonated polymer.	Violet	Insulator ca. 10^{-12}

Aniline + oxidant

↓ **Initiation (Step 1)**

Aniline nitrenium cation

↓ **Dimerization (Step 2)**

Dimmers

↓ **Oligomerization (Step 3)**

Trimmers, tetramers etc.

↓ **Polymerization (Step 4)**

Polyaniline

Scheme 6. Steps involved in the synthesis of polyaniline.

3. SYNTHESIS OF NANOSTRUTRED POLYANILINE

The possible reactions of aniline during several polymerization stages are crucial to understand the mechanism of nanostructure formation since they form the basis for the self-assembly processes. PANI produced using APS under high acidic condition possesses a granular morphology of irregular shaped micron-sized particles.

Morphology of PANI can be tailored by varying the pH and reaction medium . Self-assembly and growth of PANI can be tailored by varying the pH of the medium, oxidant, and reaction medium. During the PANI synthesis , many reaction parameters can be varied. Changes in the synthetic conditions can produce various types of nanostructures (Figure 1).

The variety of nanostructures that can be produced offers a unique opportunity for developing an improved understanding of the evolution of nanostructures in precipitation polymerization process es . Selected examples of nanostructured polyanilines obtained by template free and template directed methods are summarized in Table 4.

Each of the established methods possesses advantages and disadvantages, but large scale production of well aligned arrays of PANI with controllable morphologies and sizes is still a challenge. Following section describe the detailed information about morphologically controlled synthesis of PANI. Particular emphasis is given on soft template methodology for the preparation of nanostructured PANI.

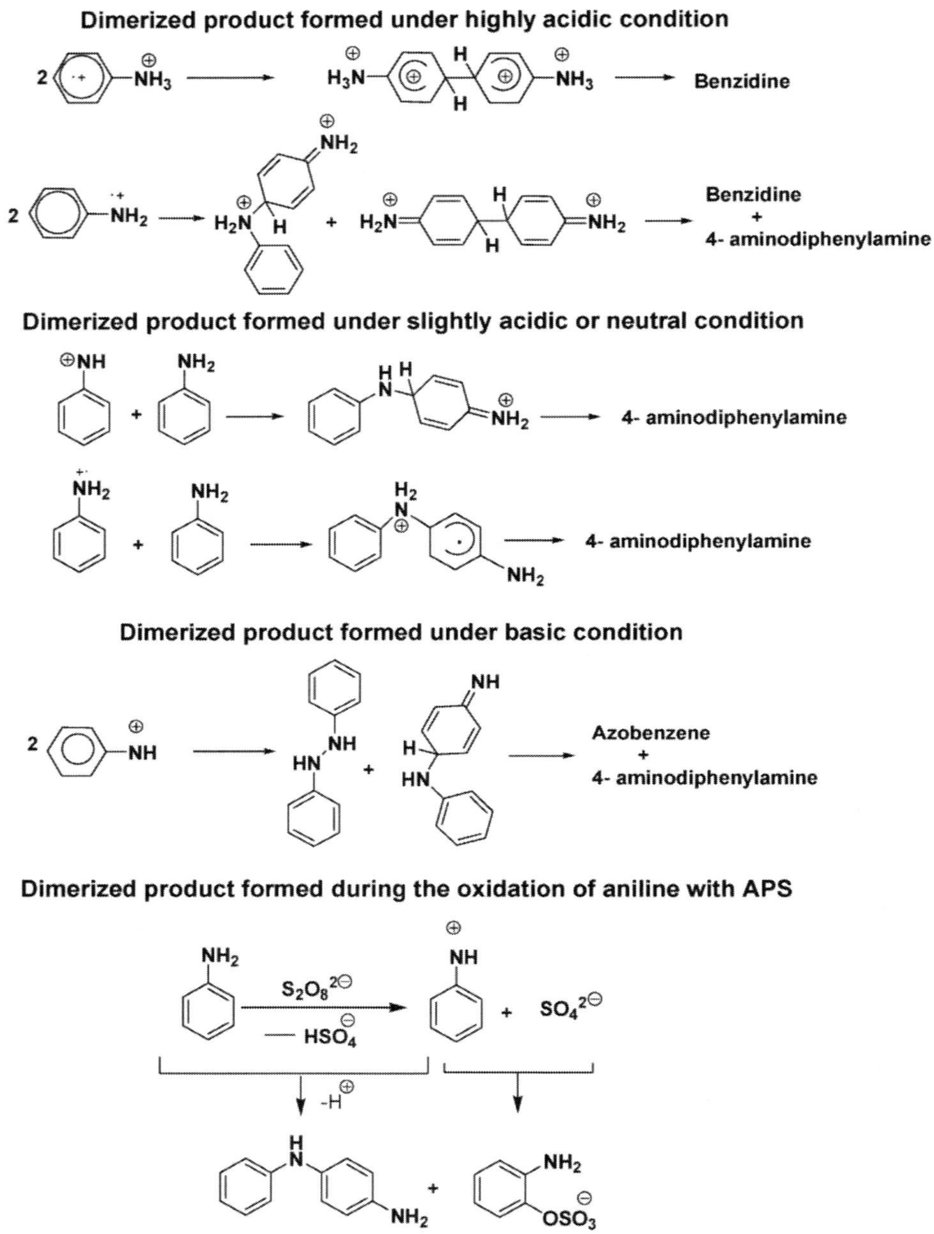

Scheme 7. Various dimerized products formed during the oxidation of aniline at different reaction condition.

3.1. Synthesis of Nanostructured Polyaniline without Using Structure Directing Agents

3.1.1. One Dimensional Polyaniline Nanostructures

One-dimensional PANI nanostructure includes nanofibers , nanowire, and nanotube morphology . Nanowires possess a non-entangled, 1-D nanostructured morphology, while nanofibers refer to entangled 1-D morphology. Interfacial polymerization is used for the synthesis of PANI nanofibers [74]. In this process, synthesis is performed in an immiscible organic/aqueous biphasic system. The oxidation of aniline occurs at the interface and PANI nanofibers are formed within several minutes, and the polymer then migrates into the aqueous phase. The yield of polymer formed under this condition is very low. Furthermore, a large volume of organic solvent is required in this synthesis approach.

Therefore, this method has some serious limitations. PANI nanofibers can also be produced with the assistance of an external energy source such as ultrasonic and sonochemical [76], and microwave irradiation [77]. The external energy based process accelerates the rate of polymerization of aniline , which leads to the formation of PANI nanofibers.

For the production of nanofibers , secondary growth of PANI particles should be suppressed. This can also be explained using classical nucleation theory.

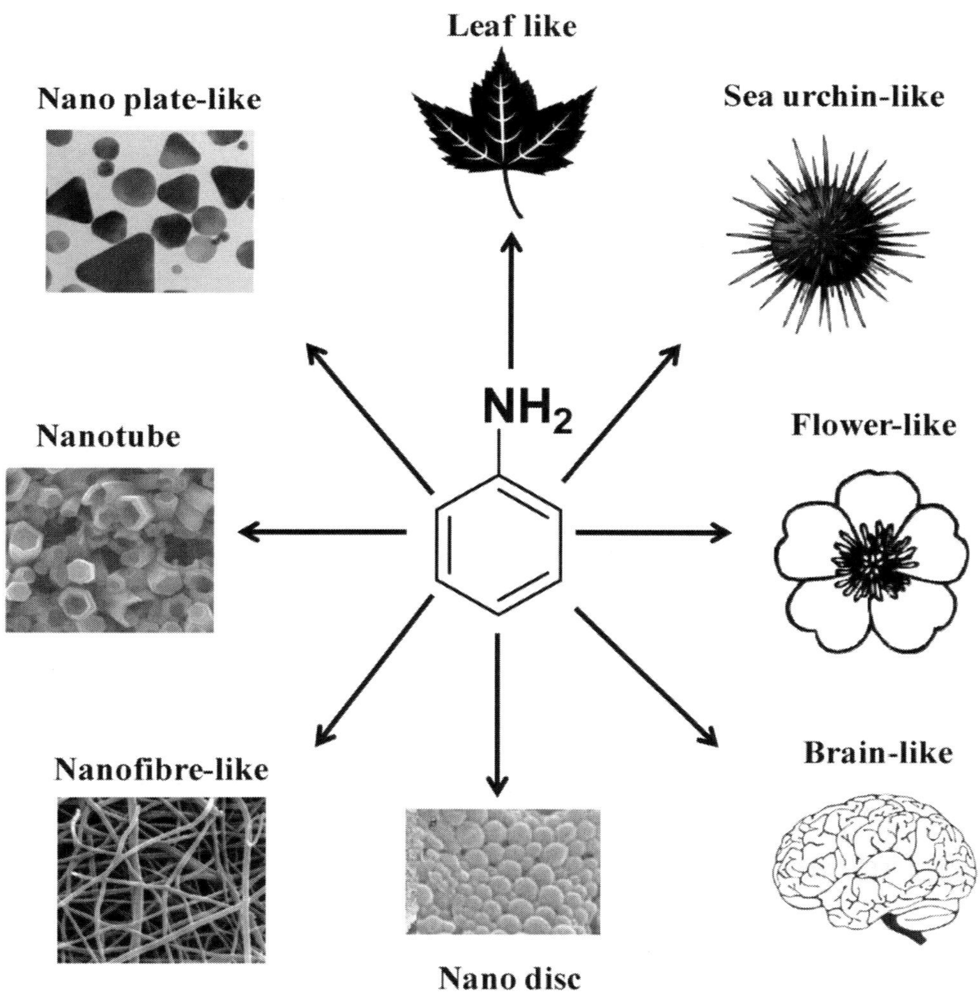

Figure 1. Morphology of various nanostructure polyaniline reported in literature.

Table 4. Selected examples of nanostructured polyaniline synthesized by template free and template directed methods

Category	Methods	Nanostructure	Mesoporosity	Reference
Template free synthesis	Interfacial polymerization	Nanowires	Nil	[74]
	ultrasonic and sonochemical	Nanowires	Nil	[76]
	microwave irradiation	Nanowires	Nil	[77]
	p-amino substituted monomeric or oligomeric additive induced method	Nanowire	Nil	[79]
	Grown on solid surface	Nanowire	Nil	[80]
	o-phenylenediamine induces method	Nanotube	Nil	[82]
	Acid induced emulsion type polymerizations	Hollow sphere	Nil	[84, 85]
Template-directed synthesis	Cetyltrimethylammonium chloride	Nanowire	Nil	[86]
	Bulky organic acids	Nanotube and nanowires	Nil	[87, 88]
	Tobacco mosaic virus	Nanotube	Nil	[89]
	Ethylene glycol/polyethylene glycol additive at pH 7	Curved flower	Nil	[91]
	Polyethylene glycol additive at pH 3	Nanorod	Nil	[91]
	Block co-polymer (P123) at pH 7	Leaf	Nil	[91]
	Block co-polymer (P123) at pH 3	Nanorod	Nil	[91]
	Sodium dodecylsulfate + Block co-polymer (P123)	Granular	Yes	[92]

Because PANI is a water -insoluble polymer , its growth is accompanied by a precipitation event; i.e., a nucleation event in which liquid turns into solid. Based on a number of experimental observations, it has been suggested that the shape of PANI can be correlated to its type of nucleation. Homogeneous nucleation of PANI results in nanofibers , while heterogeneous nucleation leads to granular particulates [78]. Rapid mixing of oxidant and monomer solution (homogenous mixing) or reaction performed at an elevated temperature can thus lead to the production of high-quality nanofibers.

Aniline appears to produce a polymer that readily forms the nanofiber morphology . However, when substituted anilines are used, nanofibers did not form instead only agglomerated or large spherical structures are observed [75].

Recently, it was reported that nanofibers of PANI can be produced by introducing *p*-amino substituted monomeric or oligomeric additives into the oxidation of substituted anilines [79]. For example, when ethyl aniline is dissolved in an aqueous solution of acid and rapidly mixed with an aqueous acidic solution of APS, only agglomerated, micron-sized particles were observed [79]. However, when the same reaction is performed in the presence of 1–2 mol% (with respect to the monomer concentration) of pre-dissolved *p*-aniline dimer or p-phenylenediamine, nanofibers of polyethylaniline are observed [79]. Only a very small amount of oligomer or monomeric additive is typically required in order to change the morphology of the entire product from agglomerates to nanofibers.

By carefully controlling the nucleation and growth of PANI on a solid surface, oriented arrays of nanostructured PANI can be produced [80]. PANI nucleates heterogeneously on solid surfaces before nucleating in the bulk solution . After a short induction period in which PANI grows on solid surfaces, PANI nanofibers begin forming in the bulk solution, thereby consuming reactive intermediates necessary for continued PANI growth on the solid substrate . It is believed that the suppression of PANI growth on the solid substrate allows PANI to grow vertically from the solid substrate into the bulk solution, which results in the production of oriented arrays of 1-D nanostructured PANI [80].

3.1.2. Polyaniline with Nanotube Morphology

It is possible to prepare nanotube morphology of PANI at optimized reaction condition [81]. To achieve the nanotube morphology, reaction condition is set in such a way that *o*-coupled aniline is initially formed, which subsequently oxidizes into phenazine-like moieties. The phenazine units are believed to aggregate into a template-like structure to promote the formation of nanotube [82]. Phenazinium units provide the necessary organization and order that predetermines the growth of nanotubes [82]. The presence of phenazine moieties in PANI nanotubes has been observed using various spectroscopic techniques [83]. Nanotube morphology can be selectively obtained upon the addition of very small quantity of *o*-phenylenediamine. Addition of small amounts of *o*-phenylenediamine induces the formation of phenazine-like moieties, which can aggregate into structures that dictate nanotube formation; this subsequently serves as the platform from which aniline can grow into nanotubes.

3.1.3. Polyaniline with Spherical Morphology

Nanotube and nanofiber morphology is suitable for device fabrication but spherical morphology especially hollow spheres are useful for their potential application in drug delivery and other biological and catalytic applications. PANI hollow spheres have often been observed when oxidizing aniline in emulsion type polymerizations under weak acidic condition using naphthalene sulfonic acid [84]. It is believed that the anilinium salts form two kinds of particles in the emulsion—micelles and monomer droplets. The aniline droplets in the emulsion are believed to be composed of aniline monomer and aniline salts that aggregate into micelles to form spherical morphology. Other acid such as salicylic acid has also been used to generate hollow spheres [85]. Lowering the dopant/monomer ratio from 1.0 to 0.1 changes the morphology of the resulting polymer from hollow spheres to nanotubes [85]. Type of aniline monomer and co-additive also lead to form spherical morphology. Oxidizing methoxy-aniline under acid free conditions can readily produce hollow spheres when o-phenylenediamine was added as an additive [84].

However, under this condition small amount of nanotube was also observed. Only hollow spheres were observed, when polymerizing anisidine [84]. From the differences in morphology observed between aniline and anisidine, one can conclude that the basicity and solubility of the monomers has an effect on which hollow structures are observed.

3.2. Synthesis of Nanostructured Polyaniline Using Ionic and Bulky Acidic Compounds As Structure Directing Agents

Structure directing agents are known to tailor the morphology and porosity of materials . Soft template and hard templates both can tailor the morphology of materials. The role of hard template is limited because of two reasons: (1) removal of template after synthesis , and (2) limited availability of hard template. Soft templates are more versatile because it generates a variety of mesophase under different synthesis condition at different concentrations. Soft-templates can self-assemble into mesophase structures that induce the formation of various nanostructured materials .

Surfactants, such as cetyltrimethylammonium chloride or hexadecyltrimethylammonium chloride can self-assemble into transient micelles when their concentration reaches a critical value . One-dimensional cylindrical micelles can be achieved under appropriate conditions [86]. These one-dimensional nanoscale structures can subsequently serve as a scaffold for aniline to grow during polymerization , which results in nanofibers [86]. In addition to these ionic structure directing agents, bulky organic acids [87] also act as structure directing agents for the synthesis of nanofibers [87] and nanotubes [88]. The bulky organic acids are believed to play a dual role: guiding the morphology of the polymer and serving as the anion dopant in order to produce PANI in its conductive state . Biological templates such as the tobacco mosaic virus (TMV) have also been used as a pre-existing scaffold in order to grow PANI nanotubes [89].

3.3. Synthesis of Nanostructured Polyaniline Using Non-Ionic Structure Directing Agents

Non-ionic structure directing agents can also be used in the synthesis of PANI nanostructure. Non-ionic surfactants possess the hydrophile–lipophile balance, which varies with the relative lengths of the hydrophilic and hydrophobic moieties in the surfactant [90]. Such hydrophile–lipophile moieties are responsible for the formation of micelle in aqueous medium [90]. One-step synthetic protocol for PANI micro-/nano-structured materials by using non-ionic soft-templates such as ethylene glycol , polyethylene glycol , and block copolymers (pluronic 123) have been reported [91]. It was found that morphology changed by varying the templates and pH of the medium. Different morphology of PANI was obtained at different pH condition using ethylene glycol (EG). A curved flower like, mixed rod like, and granular morphologies were obtained at pH 7, pH 3, and pH 1, respectively (Figure 2). At pH 7, during the early stage of polymerization , aniline exists mainly as a neutral molecule, which can be easily oxidized to form *ortho*-coupled and *para*-coupled aniline molecules by the reduction of APS. The oxidation of *ortho*-coupled aniline molecules yielded oligomers that are composed of phenazine-like segments.

Figure 2. Various morphology of PANI obtained by using (a) EG at pH 7, (b) EG at pH 3, (c) EG at pH 1, (d) PEG at pH7, (e) PEG at pH3, and (f) PEG at pH1.

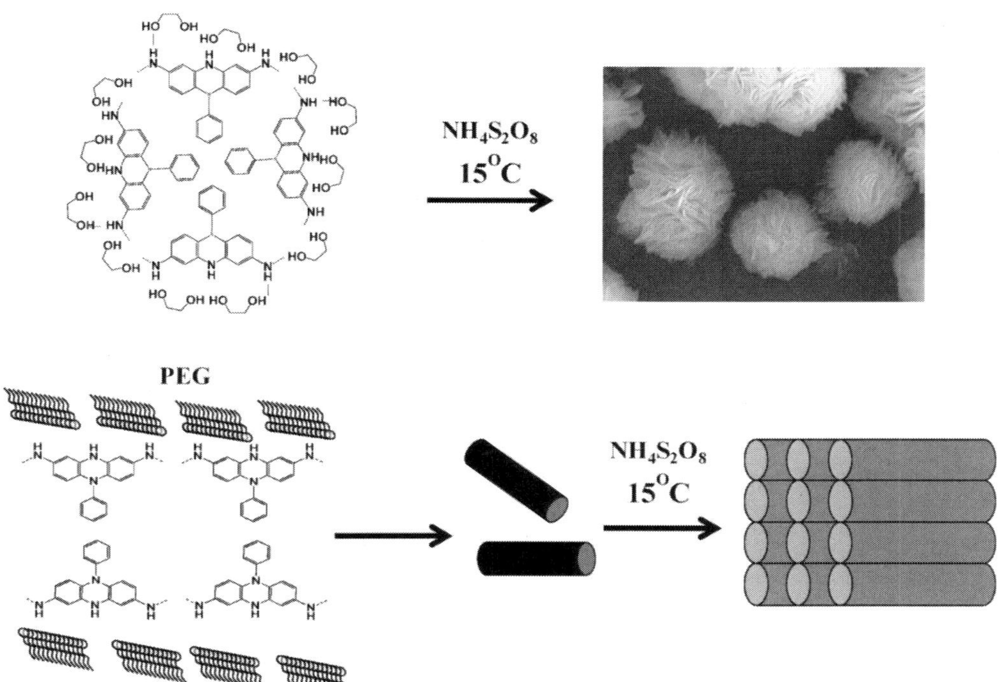

Figure 3. Oriented growth of curved flower like and nanorod like morphology using EG and PEG.

The newly formed phenazine like segments forms intra-/inter-molecular hydrogen bonding with EG. These hydrogen bonded phenazine-ethylene glycol moiety self-assemble and form curved three- dimensional flower-like network structure (Figure 3). However, it may be noted that polymerization take place only when pH <3 due to the formation of pernigraniline.

When the pH of the reaction mixture gradually reached a value where pernigraniline started to be protonate (this increase in pH is due to the formation of H_2SO_4 by the reduction of APS), the concentration of electrons with unpaired spins on pernigraniline chains dramatically increased. The unpaired electrons at the end of oligomer molecules became re-localized and, because of that, were re-activated so that they could again participate in polymerization reactions . Not well-aligned nanorod like structure was obtained when the synthesis was performed at pH 3 (Figure 2). Although, here also intra-/inter-molecular hydrogen bonding exists among pernigraniline like moiety and EG, but the pernigraniline keeps protonating from the beginning of the reaction (at the beginning of the reaction pH of the medium was 3). This protonation helps growing the chain in the elongated form, from the start of the reaction and produces a nanorod like structure. It may be noted that due to significant hydrogen bonding, the nanorods are not properly aligned but mixed with one another as shown in the SEM images .

When the synthesis was carried out in the presence of polyethylene glycol (PEG) at pH 7, flower like structure was obtained (Figure 2). However, it may be noted that the flower like structure obtained using PEG was less crosslinked than it was obtained using EG. This may be due to the fact that highly ordered hydrogen bonded oligomeric network was formed in EG to yield highly cross-linked curved 3D network structure, whereas in PEG, the polymeric backbone provides only limited hydrogen bonding and thus formed less ordered oligomeric structure to produce flowerlike structure with less cross-linked network 3D structure. Sample prepared at pH 3 produced highly ordered nanorod like structure, which is better in quality and size when compared to the sample synthesized from EG. The long PEG chains help forming the nanorod structure following the same mechanism as it was discussed for EG (Figure 3). Again, in this case also at pH 1, disordered granular like structure was obtained. When the synthesis was carried out in the presence of polyethylene glycol-poly propylene glycol-polyethylene glycol (Pluronic P123 block co-polymer) under neutral condition, a leaf-like structure was obtained (Figure 4). P123 (an amphiphilic molecule) forms micelles in aqueous solution in which EO chains are located in the outer parts of the micelles due to their hydrophilic features. When aniline is added, the hydrogen bonds among anilines and the EO groups of P123 drive the aniline molecules to be adsorbed on the EO chains. As the polymerization proceeds, the long EO chains of P123 may act as "structural directors " for the chemical polymerization.

Aniline has an amphiphilic structure of hydrophilic $-NH_2$ and hydrophobic $-C_6H_5$, therefore there would be droplets formed by aniline , and such droplets would be adsorbed on the EO chains of P123 owing to the hydrogen bonds among the aniline and EO groups (Figure 4). After adding APS to the reaction system, the droplets will act as a template for the formation of nanorods because EO chains in PANI have a tendency to elongate in the direction of the EO chains, which results in the formation of PANI nanorod. As the polymerization proceeds, the growth of building blocks of nanorods and the aggregation along with fusing process of micelles will lead to the formation of micrometer -sized leaf-like PANI. When the sample was prepared at pH 3, the resultant morphology of PANI is agglomerated nanorods (Figure 4). As commonly known, aniline may exist as aniline hydrochloride after the addition of HCl to the reaction system. Such a salt form of aniline is unfavourable for the formation of droplets that lead to the formation of nanorods (Figure 4). Using P123 at pH 1, disordered granular-like structure was obtained, which is similar to the case of EG/PEG (Figure 4).

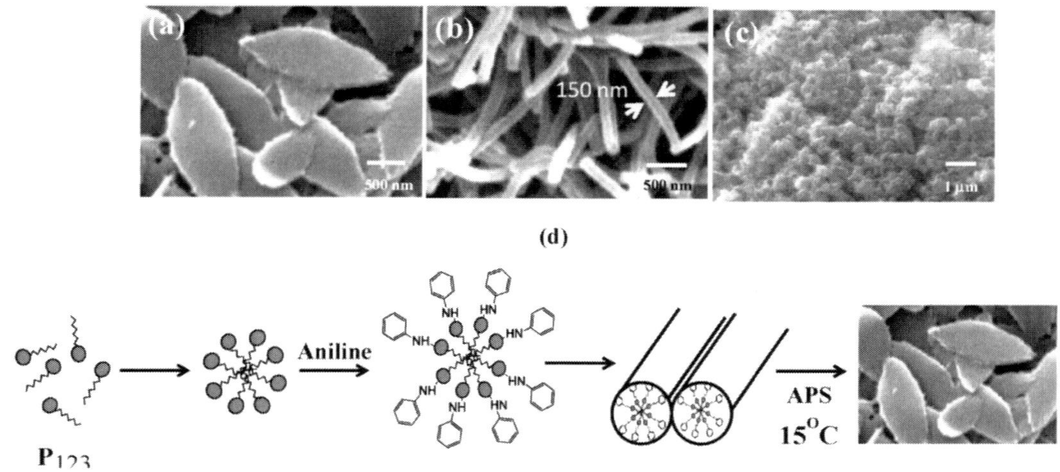

Figure 4. Various morphology of PANI obtained by using P123 at (a) pH 7, (b) pH 3, (c) pH 1. (d) Oriented growth of leaf like morphology obtained using P123.

3.4. Synthesis of Mesoporous Polyaniline Using Dual Surfactant

Previous section described the synthesis of a variety of nanostructured PANI. However, for catalytic and electrocatalytic applications in addition to high surface area , porosity is also very important. Previous study based on P123 show that various nanostructures can be obtained under different pH condition, however well defined porosity was not observed.

Mesostructured PANI was prepared by the self-assembly of a mixture of an anionic surfactant , sodium dodecylsulfate and a non-ionic polymeric surfactant (polyethylene glycol , and block-co-polymers such as Pluronic P123 and Brij-35) [92]. The basic structure of a P123 micelle consists of a well hydrated corona of PEO and a core of PPO blocks. When SDS is added, these micelles encapsulate SDS molecules in their cavity through hydrogen bonding . Upon introduction of hydrophobic aniline monomer, it will diffuse into the interior of micelle through hydrophobic–hydrophobic interaction.

Consequently, when APS is added to the reaction system, polymerization of aniline takes place resulting in uniform PANI layers (Figure 5). The idea behind taking mixed surfactant in our study is to organize these PANI frameworks to grow around its periphery through ionic interaction of the $C_{12}H_{25}SO_3^-$ and NH^+ moiety of growing PANI chain (Figure 5). After the synthesis , obtained materials were extracted in an aqueous solution containing ammonium acetate to remove the surfactant.

A very sharp low-angle peak at $2\theta = 2.45^\circ$ was observed for PANI synthesized using SDS and P123 (hereafter represented as PANI-SDS-P123), whereas a broad low-angle reflection cantered at $2\theta = 2.3^\circ$ was observed for PANI synthesized using SDS and Brij-35/PEG (Figure 6a). Low-angle reflection confirmed the presence of the ordered mesostructure. A sharp low-angle peak for PANI-SDS-P123 confirmed the existence of the highly ordered mesostructured PANI, which was synthesized by using SDS and P123. Wide angle XRD pattern of various PANI well matched with the conventional PANI (Figure 6b). PANI samples synthesized in this study exhibited type IV isotherm but with different hysteresis loop s (Figure 7) [98].

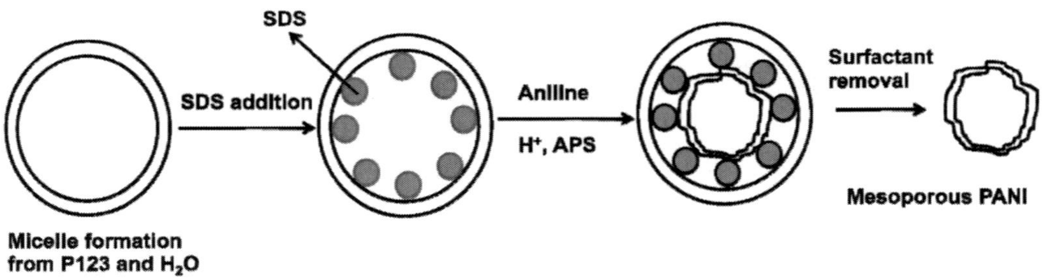

Figure 5. Schematic representation of the proposed mechanism for the formation of mesostructured PANI by the self-assembly of SDS and tri-block copolymer .

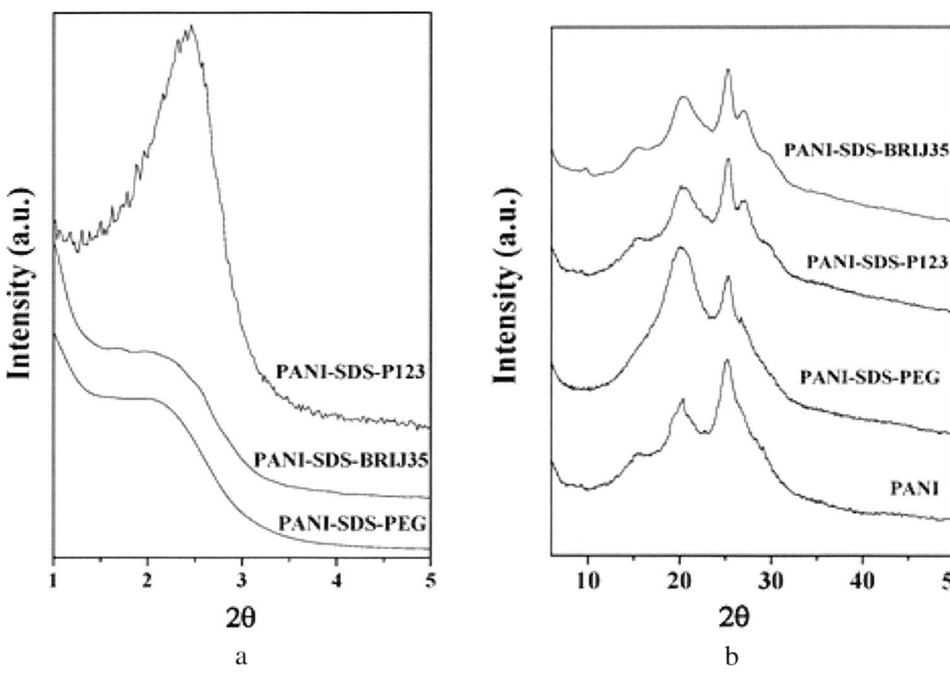

Figure 6. (a) Low angle and (b) wide angle XRD patterns of various PANI materials synthesized using various non-ionic surfactant and SDS.

Conventional PANI exhibited H4 hysteresis loop , which shows that the material contains both, micropores and mesopores (Figure 7a). PANI synthesized using dual templates exhibited H3 hysteresis loop, which shows that materials are having non-rigid aggregates like particles and signify the presence of slit shape pores (Figure 7a).

Materials exhibited broad pore size (4–10 nm) distribution (Figure 7b). Conventional PANI has the minimum surface area (35.4 m^2/g), whereas PANI-SDS-P123 has the highest surface area (83.6 m^2/g).

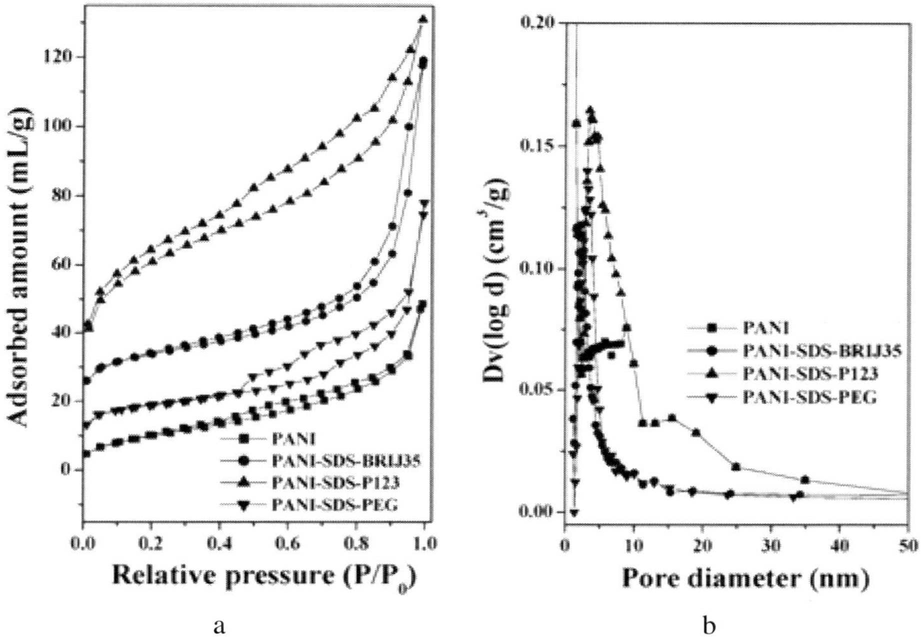

Figure 7. (a) N_2 adsorption isotherms and (b) pore size distributions of conventional PANI, PANI-SDS-P123, PANI-SDS-PEG and PANI-SDS-BRIJ35.

4. Electrocatalytic Applications of Polyanilines

PANI is interesting material for electrocatalytic applications because it can act as an effective mediator for electron transfer in redox reactions [93]. This is possibly due to the effective delocalization of redox charges over a series of conducting phases of PANI [94]. PANI is considered to be an attractive polymer matrix because it exhibits two redox couples in appropriate potential range to facilitate the charge transfer and thereby acts as self-contained electron transfer mediator. Electrocatalytic applications can be improved by tailoring the physico-chemical properties . High surface area , mesoporosity, and suitable morphology can effectively improve the electrocatalytic activity. PANI nanostructures have been exploited in many organic electronic applications, including non-volatile memory [95], electrochromics [96], and sensors [97]. Electrocatalytic applications of PANI can be further improved and tailored by incorporating suitable materials on the surface of PANI and making composite materials. Following section outline some of the important application of PANI based electrodes in bio-sensing and non enzymatic sensing applications.

4.1. Electrocatalytic Activity towards Enzymatic and Non-Enzymatic Sensing

4.1.1. Electrocatalytic Oxidation of Hydrogen Peroxide for Sensing

Hydrogen peroxide is widely used in food industries . It is generated in several enzyme catalyzed reactions , including those of glucose oxidase , uricase, cholesterol oxidase, alcohol oxidase , sarcosine oxidase, galactose oxidase, and L-amino acid oxidase, etc.

However, hydrogen peroxide is extremely toxic to cellular life. Due to their enzymatic action and wide application in food industries , they are present in waste water effluent . Therefore, it's accurate determination is very important for biochemistry , clinical diagnosis , and environmental chemistry point of view [98].

The amperometric sensors based on horseradish peroxidase and soya bean peroxidase are extensively used for the determination of H_2O_2 [99]. Peroxidase is an iron -containing enzyme , which can catalyze the oxidation of H_2O_2. The most common form of iron in peroxidase is heme , in which iron exists in two oxidation states : Fe^{2+} and Fe^{3+}. They can reversibly oxidize and reduce between the Fe^{2+} and Fe^{3+} oxidation states. The reversible redox plays an important role in the catalytic oxidation of H_2O_2.

PANI nanofibers doped with 4-toluenesulfonic acid were used for the loading of horseradish peroxidase [100]. The use of PANI nanofibers led to efficient enzyme loading and rapid electron transfer rate between the active centers of enzymes and electrodes . Horseradish peroxidase entrapped in the PANI nanofibers could keep its native bioactivity and effectively catalyze the reduction of hydrogen peroxide . PANI doped with ferrocenesulfonic acid was used for the detection of H_2O_2 [101]. Ferrocenesulfonic acid can effectively catalyze the oxidation of hydrogen peroxide , so ferrocenesulfonic acid is used as a sensor to determine the concentration of H_2O_2. Iron in the ferrocenesulfonic acid exists in two oxidation states : Fe^{2+} and Fe^{3+}. They can reversibly oxidize and reduce between the Fe^{2+} and Fe^{3+} oxidation states, which play an important role in the catalytic oxidation of H_2O_2. However, this sensor did not show specificity, because some compounds , such as phenol , catechol, formaldehyde , and ascorbic acid etc., interfere with the determination of H_2O_2. These compounds can also be oxidized at the electrode to a certain extent.

Mesoporous PANI (discussed in section 3.4) modified Pt electrode was used in the electrochemical oxidation of H_2O_2 [92]. The amperometric current obtained was proportional to the concentration of H_2O_2 (Figure 8a). Mesoporous PANI exhibits good sensitivity to H_2O_2 in a wide concentration range. The value of the current was plotted against the concentration of H_2O_2 to obtain a calibration plot (Figure 8a).

Mesoporous PANI exhibited higher sensitivity (2.27 μA/mM) when compared with PANI (0.77 μA/mM). Mesoporous PANI exhibited a linear dependence of the response current on the H_2O_2 concentrations in the range from 1 to 45 mM, whereas PANI showed a linear response up to 20 mM indicating a less restricted diffusion of H_2O_2 through the porous mesoporous PANI layer. The above CV and amperometry studies indicate that mesoporous PANI has high sensing capability than PANI electrode due to its better permeability and electrocatalytic activity.

4.1.2. Electrocatalytic Oxidation of Glucose for Sensing Application

The determination of glucose concentrations is important in several aspects for food , microorganism or medicine [102]. Several approaches are known to detect glucose and similar bio-molecules .

In clinical practice, diabetes mellitus is one of the leading causes of death and disability in the world. This metabolic disorder results from insulin deficiency and hyperglycaemia is reflected by blood -glucose concentrations higher than the normal range of about 3.9–6.2 mM (empty stomach) or 3.9–7.8 mM (2 h after food). The quantitative monitoring of blood glucose is of great clinical importance, which can greatly reduce the risk of diabetes mellitus induced heart disease , kidney failure , or blindness .

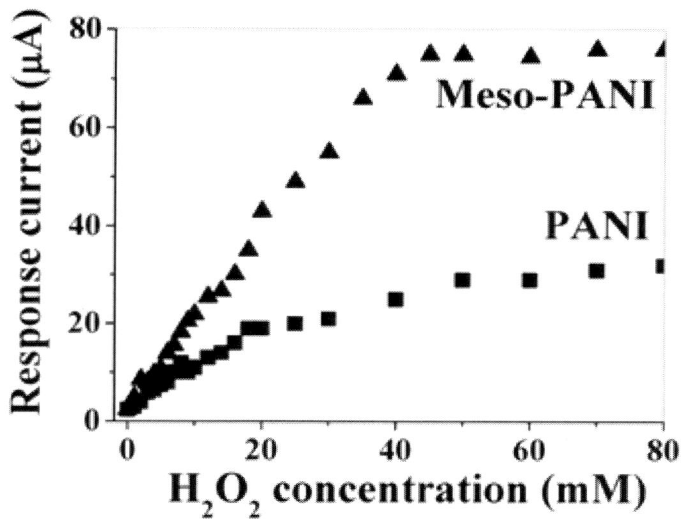

a

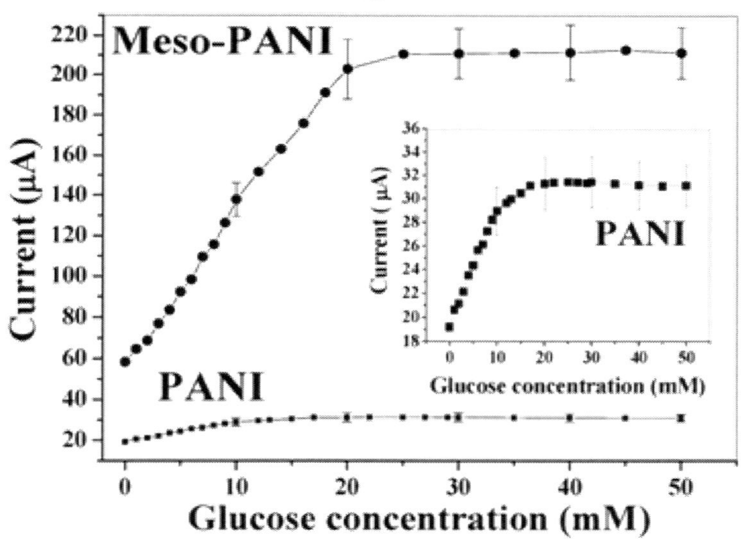

b

Figure 8. (a) Calibration curve for the biosensor response to H_2O_2 and (b) Calibration curve for the biosensor response to glucose for PANI and Meso-PANI (PANI-SDS-P123) modified electrodes . Inset: magnified plot for the response of PANI/GOx electrodes to glucose.

Glucose oxidase (GOx)-based amperometric enzyme electrodes is simple and easy-to-use for blood -glucose monitoring, since GOx is relatively low in price as well as having high bioactivity and stability [103]. Glucose dehydrogenase (GDH) is another kind of enzyme used in glucose sensing and is also utilized in fabricating commercial test strips for blood-glucose [104]. GDH-based amperometric biosensors are advantageous because it is able to be operated at lower detection potentials than the GOx-based sensors , and their performance is not influenced by the oxygen level in the analyte solution .

The advantages of PANI as potentiometric transducer for detection of glucose have been developed [105, 106].

The potentiometric detection of enzyme activity is based on the measurement of change in pH in the enzymatic layer on the sensor surface. The pH change is caused by the oxidation of H_2O_2 produced by the enzymatic reactions for potentiometric detection of glucose . These electrodes are found to perform well in terms of dynamic range of detection and short response time . GOx enzyme immobilized on the mesoporous PANI is known to catalyze the oxidation of glucose to gluconic acid and hydrogen peroxide [115].

Continuous amperometry method for the detection of glucose using PANI/GOx and mesoporous PANI/GOx confirmed that the immobilized GOx was stable in the polymer matrix during these investigations.

Calibration curve of mesoporous PANI/GOx exhibited a linear dependence of the response current on the glucose concentration in the range from 1 to 20 mM, whereas PANI/ GOx showed the linear response up to 10 mM (Figure 8b), which show that mesoporous PANI/GOx exhibited high sensing capability than PANI/GOx electrode [98].

To develop non-enzymatic sensor for glucose , Cu nanoparticles supported on mesoporous PANI (Cu/Meso-PANI) was synthesized [107]. Besides reflection corresponding to Meso-PANI [92], several more reflections were also observed in the XRD pattern of Cu/ Meso-PANI. The diffraction peaks located at 2θ values of 43.3°, 50.4°, 74.0°, and 90.0° corresponds to (111), (200), (220), and (311) respectively, for metallic Cu in the fcc lattice (JCPDS, File No. 85–1326) (Figure 9a). Uniform dispersion of Cu nanoparticles in the Meso-PANI matrix was confirmed using TEM (Figure 10), however, in the TEM image , it was noticed that the particles were not clearly monodispersed. The selected area electron diffraction patterns show a diffuse ring pattern for the Cu nanoparticles present in Cu/Meso-PANI (Figure 10). CV of Cu/Meso-PANI exhibited six peaks/waves, which matches well with the reported literature on bare Cu electrode (Figure 9b) [108].

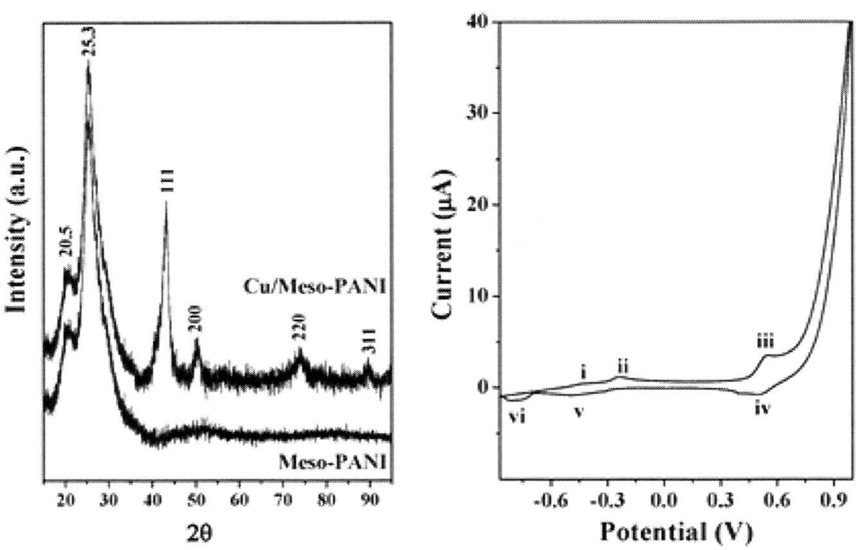

Figure 9. (a) XRD patterns of Meso-PANI and Cu/Meso-PANI. (b) CV of Cu/Meso-PANI modified electrode obtained in 0.1 M NaOH at a scan rate of 50 mV/s.

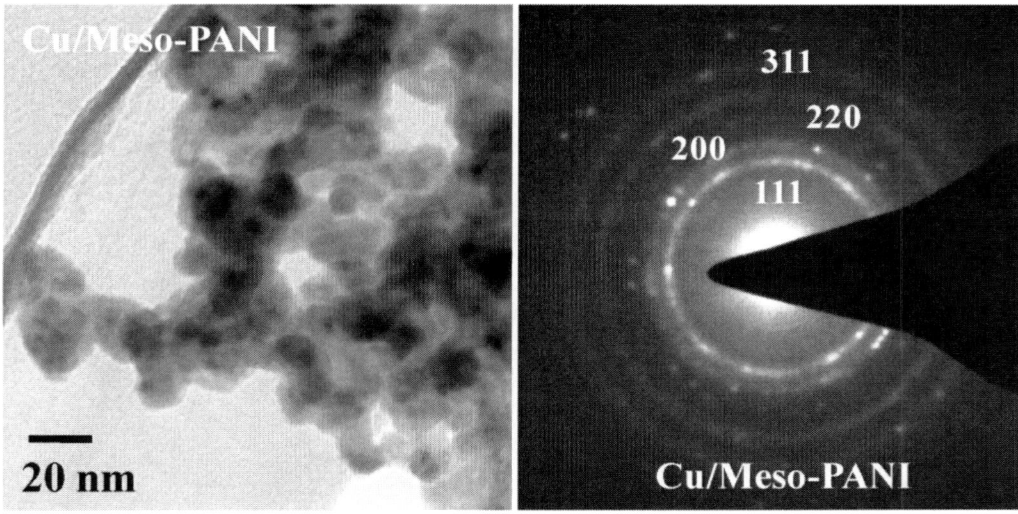

a b

Figure 10. TEM micrograph of Cu/Meso-PANI and corresponding diffraction pattern of Cu nanoparticles .

In the presence of glucose , a single forward oxidative wave was observed at approximately 0.45 V in CV , which corresponds to the irreversible glucose oxidation reaction for Cu/Meso-PANI modified electrode (Figure 11a). The currents continue to increase up to a potential of 0.60 V. This agrees with the proposed involvement of copper (II) and copper (III) surface species in the oxidation of glucose [109]. This indicated that glucose could be irreversibly oxidized at the Cu/Meso-PANI modified electrode.

Similar increase in the oxidation currents on the successive addition of glucose was also observed for Cu/PANI modified electrode (Figure 11b). However, the oxidation currents were smaller when compared with Cu/Meso-PANI modified electrode. This improved analytical performance of Cu/Meso-PANI compared to Cu/PANI was attributable to highly dispersed Cu nanoparticles on Meso-PANI, which facilitate the electron transfer to the electrode.

4.1.3. Electrocatalytic Oxidation of Ascorbic Acid for Sensing Application

Ascorbic acid is a vital component in the diet of human , and it is known to take part in several biological reactions and is present in the brain [110]. Recent clinical studies have demonstrated that the content of ascorbic acid in biological fluids can be used to assess the amount of oxidation stress in human metabolism , and excessive oxidative stress has been linked to cancer , diabetes mellitus and hepatic disease [111]. Ascorbic acid undergoes a two-electron oxidation to dehydro-L-ascorbic acid. The process is quasi-reversible and at neutral pH , it has an E^0 of +0.058V vs. reference hydrogen electrode. It has two pK_a values of 4.30 and 11.57 in aqueous solution and therefore it exists in the mono-deprotonated form at neutral pH. The direct electrochemical oxidation of ascorbic acid is possible but requires high potentials. The equilibrium potential of the couple being 0.185V vs. SCE but oxidation at bare glassy carbon or platinum electrodes requires potentials of +0.4 and +0.6V vs. SCE, respectively. However, real analyte solutions often contain many species that can be oxidized anodically at such high electrode potentials.

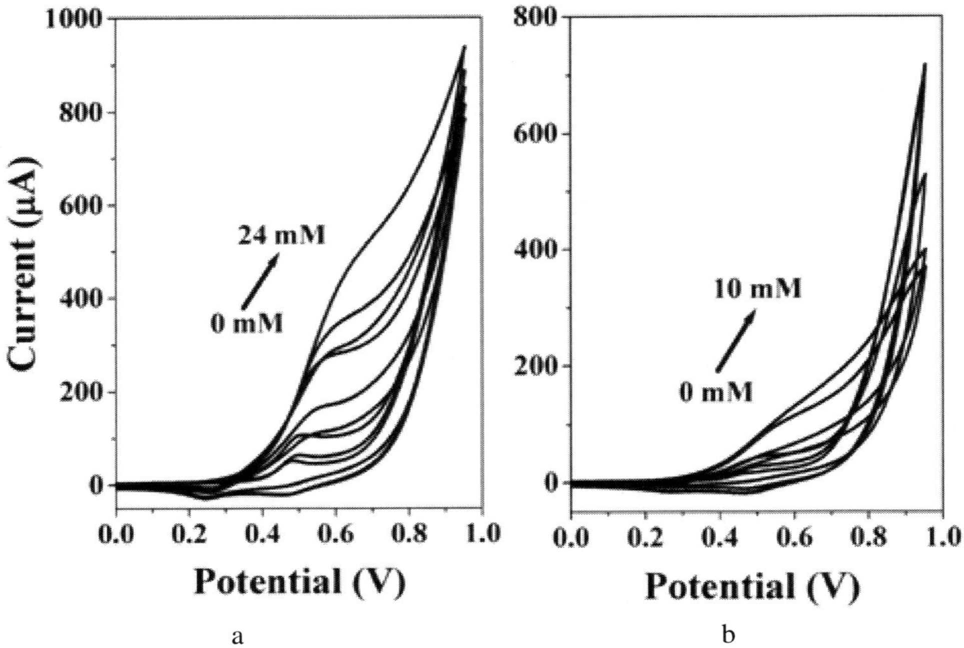

Figure 11. (a) CV of Cu/Meso-PANI modified electrode in the presence of glucose in 0.1 M NaOH solution at a scan rate of 50 mV/s. (b) CV performances of Cu/PANI modified electrode in the presence of glucose in 0.1 M NaOH solution at a scan rate of 50 mV/s.

Thus, the anodic current, derived from the electrooxidation of these substances can result in equivalent, or in some cases, even greater anodic current responses than the ascorbic acid itself. Because of this high over-potential, poor reproducibility, low selectivity and poor sensitivity on the conventional electrodes , it is nearly impossible to determine ascorbic acid by direct oxidation of it on these electrodes. Thus, much interest has been focused on the use of mediators and modified electrodes to catalyse the oxidation of ascorbic acid.

PANI–organosulfonic acid composite film on the platinum electrode was used in the electrochemical oxidation of ascorbic acid [112]. It was found that the doping of PANI with organosulfonic acid extends the electroactivity of PANI in neutral and even in alkaline media [112]. The PANI–organosulfonic acid composite film coated platinum electrodes exhibited the best result in phosphate buffer solution (PBS) at pH 7.0. The anodic peak potential of ascorbic acid shifts from 0.63 V at the bare platinum electrode to 0.34 V at the PANI–organosulfonic acid composite modified platinum electrode with a very high current response . PANI doped with ferrocenesulfonic acid was also used for the oxidation of ascorbic acid [113]. For the electrolysis of ascorbic acid at pH 5.64, the anodic peak potential of ascorbic acid shifts from 0.32 V (vs. SCE) on the bare platinum electrode to 0.05 V on both PANI and PANI doped with ferrocenesulfonic acid (PAnFc) modified electrodes.

4.1.4. Electrocatalytic Oxidation of DNA Bases for Sensing Application

Deoxyribonucleic acid (DNA) carries heritage information and instructs the biological synthesis of proteins and enzymes through the process of replication and transcription of genetic information and hence it plays major role in the life process. DNA is a natural polymer , which consist of a phosphate acid group, a basic group (guanine (G), adenine (A), thymine (T), and cytosine (C)), and a sugar unit. Change in the concentration of DNA bases may affect the activity of catabolic, anabolic, and inter-conversion of enzymes and causes various diseases , which alters the normal purine and pyrimidine metabolic pathways [114]. The abnormal change in the DNA bases lead to the deficiency and mutation of the immunity system, which are responsible for various diseases. Concentration levels of DNA bases are considered to be an important parameter for the diagnosis of cancer , AIDS , myocardial cellular energy status, disease progress, and therapy responses [115]. Hence, determination of individual concentrations of G, A, T, C, and their ratio in DNA is important. The electro-active groups in the DNA are purine (G, A) and pyrimidine (T, C) bases. These bases exhibit low sensitivity due to slow electron transfer at the bare electrode. Therefore, a variety of chemically modified electrodes were fabricated to investigate the direct electrochemistry of nucleic bases and its related nucleosides [116]. The electrocatalytic oxidation of purine bases has been extensively investigated in the literature [117].

However, there are only a few reports for the electrochemical detection of pyrimidine bases [118]. It is difficult to obtain accurate oxidation signals for pyrimidine bases because of their high positive oxidation potentials and slow electron transfer [119]. In order to overcome these limitations, modified electrodes with a wide potential window, high electrocatalytic activity, and excellent antifouling property are highly required. Only a few electrochemical methods are reported for the simultaneous determination of all four DNA bases [120]. For DNA sensing , in general DNA probes are anchored on the surface of host materials and then these composite material based electrodes are fabricated and used in the electrochemical detection. This method is known as DNA hybridization technique. A novel DNA biosensor based on oxidized graphene and PANI nanowires modified glassy carbon electrode was developed [121]. The resulting graphene/PANI layers exhibited good DPV current response for the complementary DNA sequences. The good electron transfer activity was attributed to the effect of graphene and PANI. The immobilization of the probe DNA on the surface of electrode was largely improved due to the unique synergetic effect of graphene and PANI. An efficient DNA immobilization matrix based on PANI nanotube and gold nanoparticle films modified at the glassy carbon electrode was developed for the fabrication of a DNA electrochemical biosensor [122]. This matrix incorporated the advantages of PANI nanotube and gold nanoparticle, which enhanced greatly the immobilization of the DNA probe on the films and the detection sensitivity of the target DNA.

A novel DNA electrochemical biosensor was fabricated for the detection of specific gene sequences (Figure 12) [123]. In this study following DNA were used: ssDNA: 5'-TCT TTG GGA CCA CTG TCG-3'; target DNA: 5'-CGA CAG TGG TCC CAA AGA-3; double-base mismatched DNA: 5'-CGA **A**AG TGG TCC **G**AA AGA-3'; ncDNA: 5'-GCA TCG AGC GAG CAC GTA-3'. Electrochemically reduced graphene oxide (ERGNO) was prepared on PANI nanofibers modified glassy carbon electrode. ssDNA/graphene/PANI nanocomposites based electrochemical DNA sensing platform was fabricated to recognize the specific DNA hybridization (Figure 12). The conductive PANI nanofibers played a significant role in the formation of the nanocomposites.

It not only interacted with the basal plane of graphene oxide sheets through π-π stacking, but also promoted the electrochemical reduction of graphene oxide. Due to the non-covalent assembly of ERGNO with DNA bases, it was anticipated that ERGNO could bind ssDNA probe . Therefore, ERGNO/PANI nanocomposites were applied to bind ssDNA probe via the non-covalent assembly.

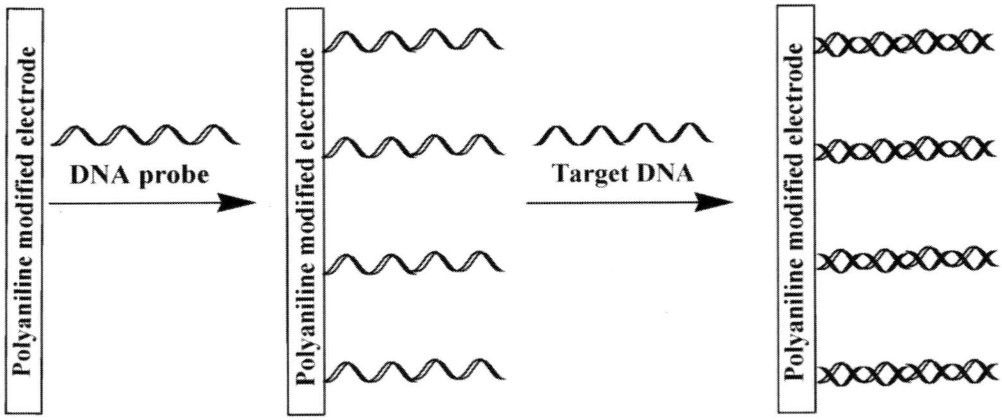

Figure 12. Conventional method for the immobilization and hybridization of DNA probe on PANI modified electrode.

The surface density of ssDNA was calculated by voltammetric studies of redox cations ($[Ru(NH_3)_6]^{3+}$), which were bound to the surface via electrostatic interaction with negative charged phosphate backbone of the DNA . A similar strategy was used for the detection of DNA by fabricating DNA biosensor based on electrochemically fabricated PANI nanowire but it used methylene blue for DNA hybridization detection [124]. DNA hybridization biosensor based on PANI electrochemically deposited onto Pt disc electrode was also fabricated in which biotin –avidin was used as an indirect coupling agent to immobilize single-stranded 5'-biotin end-labeled polydeoxycytidine (BdC) probes and 5'-biotin end-labeled 35 base -long oligonucleotide probe (BdE) to detect complementary target [125].

A simple and non-DNA probe based, PANI and metal oxide based nanocomposite material was developed for the DNA sensing , which is discussed in the next section.

4.2. Electrocatalytic Activity of Polyaniline Based Nanocomposite Materials

Although pure PANI is a versatile material for chemical sensing but several challenges still remain, including low sensitivity to certain analytes and poor selectivity . Modification of PANI with selected additives may provide an effective strategy to address these issues . If a target analyte has weak interaction s with PANI, the sensitivity can be improved by incorporating suitable functional dopants or catalysts to effectively convert target molecules into easily detected species . In this section, application of transition metal ions /metal oxides incorporated PANI based nanocomposite is discussed.

4.2.1. Transition Metal Ion Incorporated Mesoporous Polyaniline for the Simultaneous Determination of Bio-Molecules

Ion-exchange property and mesoporosity of PANI were exploited in the simultaneous detection of ascorbic acid (AA), dopamine (DA), uric acid (UA), and tryptophan (Trp), which is very important in the clinical diagnosis . For this purpose, Fe^{2+} exchanged mesoporous PANI (hereafter represented as Fe-Meso-PANI) were prepared along with other transition metal ion exchanged PANI materials (M-PANI, where M= Cu, Ni, Co, and Mn) [126].

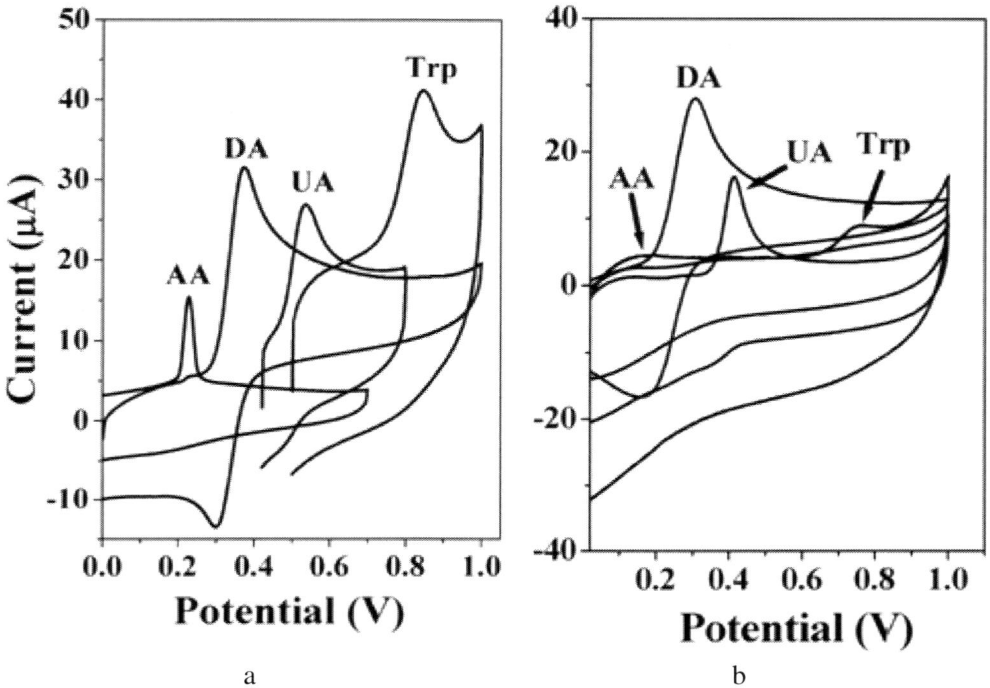

Figure 13. (a) CVs of AA, DA, UA, and Try 60 μM each, at the Fe-Meso-PANI modified electrode in buffer solution (pH 3.5). Scan rate: 50 mV/s. (b) CVs at glassy carbon electrode in 10 mL buffer solution (pH 3.5) containing 100 μM each of AA, DA, UA, and Trp at a scan rate of 50 mV/s.

The Fe-Meso-PANI modified electrode was employed to study the electrocatalytic oxidation of AA, DA, UA, and Trp, individually (Figure 13). The results show that all four compounds were oxidized with well-defined and distinguishable sharp oxidation peaks with peak potentials at 225, 375, 535, and 847 mV for AA, DA, UA, and Trp, respectively on Fe-Meso-PANI modified electrode (Figure 13a). On the other hand, bare glassy carbon electrode (GCE) failed to distinguish the oxidation peaks of AA and merged voltammetric anodic peaks appear at 340 mV, 450 mV, and 800 mV for DA, UA, and Trp, respectively (Figure 13b). At traditional solid electrodes , AA and UA are oxidized at potentials close to that of DA, resulting in an overlapping voltammetric response [127]. Therefore, the merged voltammetric peaks at bare GCE may be attributable to the sluggish electron kinetics , restricted diffusion of the analytes, and electrode fouling [127]. Compared with the bare GCE, Fe-Meso-PANI modified electrode show significant increase in the oxidation peak current, which indicates that the modified electrode plays a catalytic effect on the oxidation of AA, DA, UA, and Trp. The enlarged separation of the anodic peak potential, coupled with the increased sensitivity , makes it possible to effectively determine the concentration of AA, DA, UA, and Trp at Fe-Meso-PANI modified electrode. Furthermore, this study also deals with the simultaneous determination of AA, DA, UA, and Trp, from their mixture solution (Figure 14 a). One of such real system is the blood serum , in which all these four species coexist. Linear sweep voltammetry (LSV) studies show that the anodic peak responses for quaternary mixture containing AA, DA, UA, and Trp, were well separated from each other with a potential difference of 150 mV, 160 mV, and 312 mV for the oxidation peak potentials of AA–DA, DA–UA and UA–Trp, respectively, which was large enough to simultaneously determine the concentration of AA, DA, UA, and Trp in their mixture solution. These results demonstrate that individual or simultaneous determination of these bio-molecules at Fe-Meso-PANI modified electrode can be achieved with high sensitivity and selectivity (Figure 14 a). A comparison of the activity of different M-Meso-PANI modified electrode towards electrocatalytic oxidation of AA, DA, UA, and Trp is summarized (Figure 14b). Based on the experimental evidence , one can conclude that Fe-Meso-PANI exhibited superior sensing ability and current sensitivity compared to other M-Meso-PANI modified electrodes and Fe-PANI modified electrode investigated in this study [126]. This improved analytical performance is attributable to inherent oxidation capability of Fe^{2+}, highly dispersed Fe^{2+} active centers on large surface area Meso-PANI, which facilitates the electron transfer to the electrode.

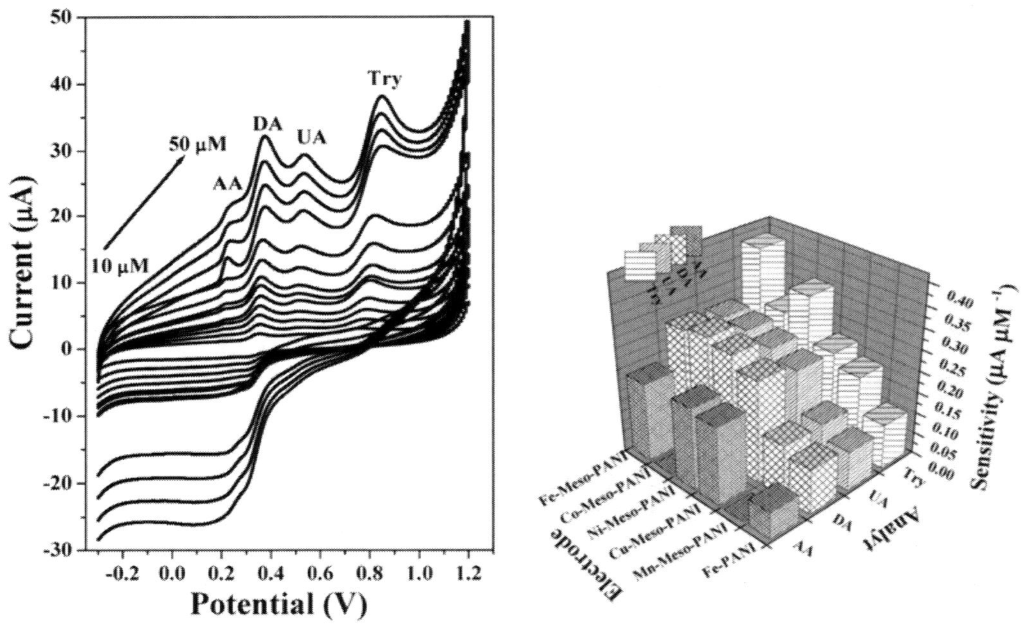

(a) (b)

Figure 14. (a) CVs of the quaternary mixtures (having equal concentrations of AA, DA, UA, and Try) obtained at Fe-Meso-PANI modified electrode in buffer solution (pH 3.5) in the concentration range of 10-50 μM. (b) Comparison of the sensitivity for various analyte at different M-Meso-PANI modified electrodes .

4.2.2. Polyaniline /MnO_2 Nanofibers and Its Electrochemical Application in the (A) Simultaneous Determination of Catechol, Hydroquinone, and Resorcinol, and (B) Simultaneous Determination of DNA Bases

PANI nanofibers were treated with aqueous solution of $KMnO_4$ to obtain PANI/MnO_2 nanofibers [128]. XRD investigation confirmed the formation of nanocomposite. SEM image of the PANI/MnO_2 confirmed that the nanofiber morphology of the PANI was preserved after the MnO_2 coating [128] (Figure 15). TEM image of PANI/MnO_2 further

confirmed that PANI nanofiber was coated by a thick layer of MnO_2 and form a core–shell nanostructure
(Figure 15).

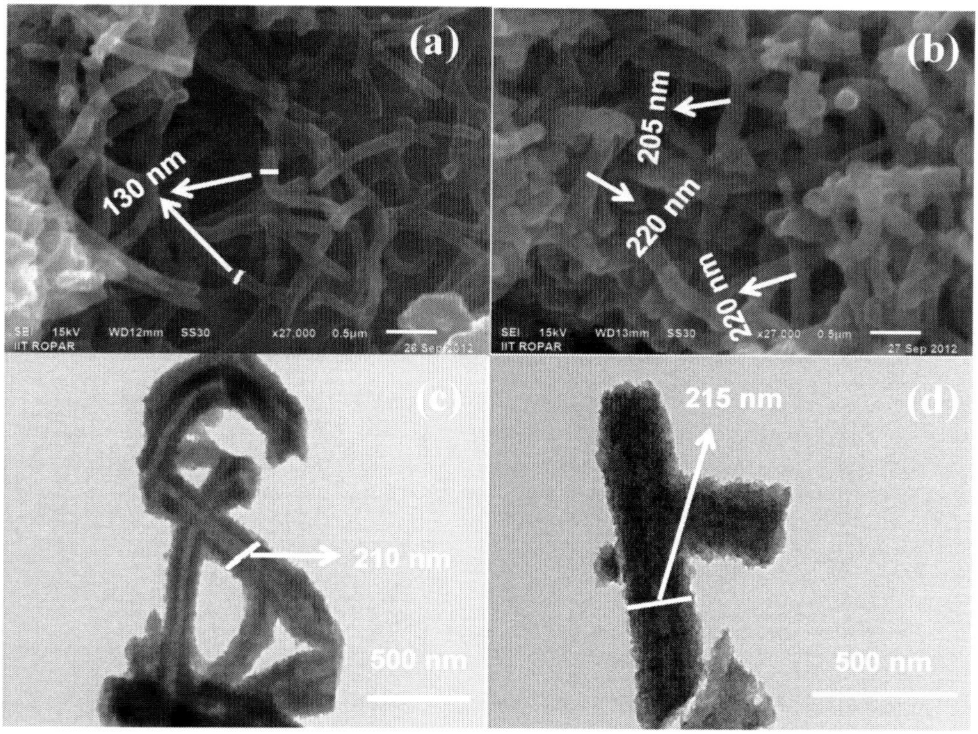

Figure 15. (a) SEM images of PANI, (b) SEM images of PANI/MnO_2, and (c,d) TEM images of PANI/ MnO_2.

The spatial distributions of the atomic contents across the PANI/MnO_2 nanofibers were obtained using energy dispersive X-ray spectroscopy (EDX), which confirmed the presence of O, Mn, and N atoms (Figure 16) [128]. HAADF analysis provided the Z-contrast image , where the intensity of scattered electrons is proportional to the square of the atomic number Z. Figure 16 depicts the HAADF-STEM image of a core–shell nanofiber. Comparison of concentrations across the structure was made by EDX. The representative EDX line (marked by 2 in Figure 16) scans taken across the positions showing atomic distribution of Mn and O as shown in Figure 16.

The PANI/MnO_2 modified electrode was employed to study the individual electrocatalytic oxidation of catechol (CC), hydroquinone (HQ), and resorcinol (RC) [143]. The results show that all three compounds are oxidized with well-defined and distinguishable sharp oxidation peaks with peak potentials at 50, 155, and 565 mV for HQ, CC, and RC, respectively on PANI/MnO_2 modified electrode (Figure 17a). On the other hand, bare GCE exhibited merged voltammetric anodic peaks, which appeared at 200 mV and 203 mV for HQ and CC and a broad (less indistinguishable) oxidation peak of RC was obtained at 760 mV (Figure 17b). The merged voltammetric response at bare GCE may be attributable to the sluggish electron kinetics , restricted diffusion of the analytes, and electrode fouling [129].

The distinguished separation of the anodic peak potential, coupled with the increased sensitivity is very important for the precise determination of the concentration of HQ, CC, and RC at PANI/MnO_2 modified electrode.

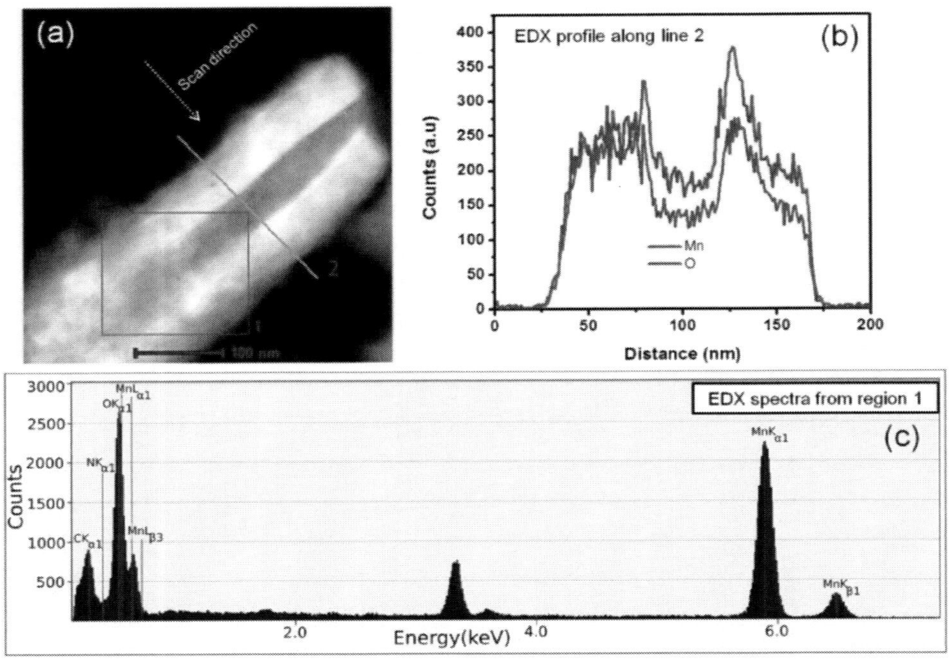

Figure 16. (a) HAADF-STEM image of a nanofiber; (b) EDX profiles for Mn and O across the line 2 as shown in (a); (c) EDX spectra from a region marked by a box 1 in (a).

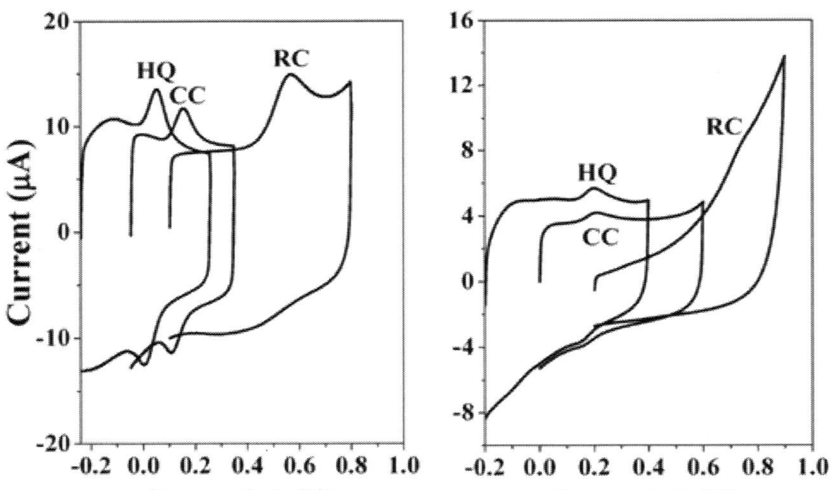

(a) (b)

Figure 17. (a) CVs of HQ (500 nM), CC (500 nM), and RC (500 nM), at the PANI/MnO_2 modified electrode in buffer solution (pH 7) at a scan rate of 50 mV/s. (b) CVs at glassy carbon electrode in buffer solution (pH 7) containing 100 μM each of HQ, CC, and RC at a scan rate of 50 mV/s.

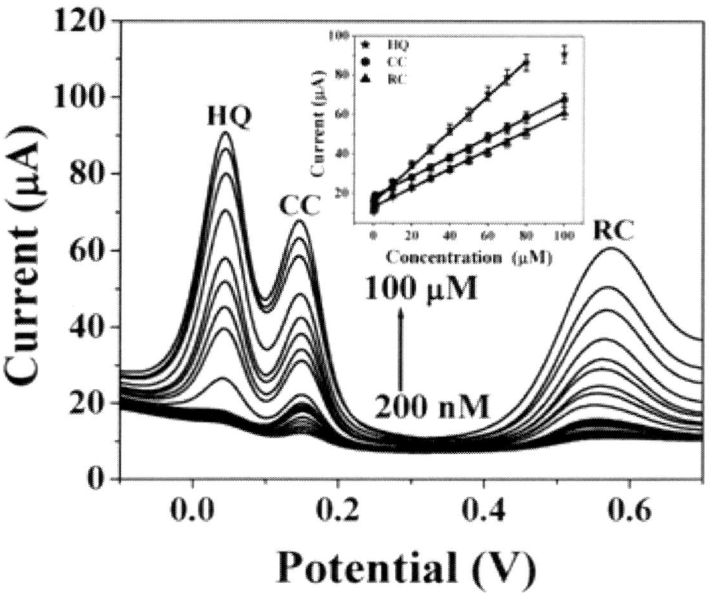

a

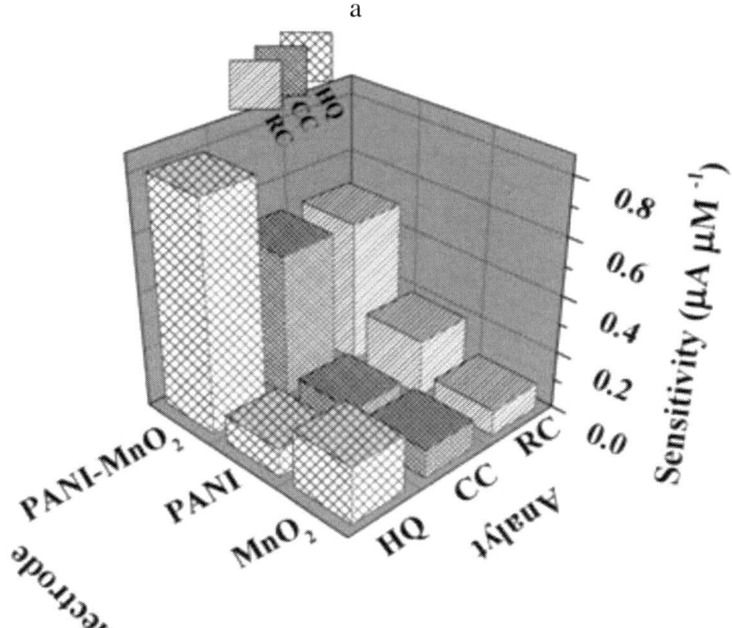

b

Figure 18. (a) DPVs of the ternary mixtures (having equal concentrations) of HQ, CC, and RC at the PANI/MnO_2 modified electrode. DPV parameters were selected as: pulse amplitude : 50 mV, pulse width: 50 ms, scan rate: 20 mV/s. Inset shows the calibration plot. (b) Comparison of the sensitivity for HQ, CC, and RC at different modified electrodes .

DPV studies show that the anodic peak responses for ternary mixture containing HQ, CC, and RC were well separated from each other with a potential difference of 105 mV and 410 mV for the oxidation peak potentials of HQ–CC and CC–RC, respectively.

With an increase in the concentration of the ternary mixture, the response current corresponding to the analyte oxidation increases. The current obtained was proportional to the concentration of analytes in the reaction vessel. A linear response was obtained between the peak currents and analytes concentrations (Figure 18a). The anodic peak current obtained was found to be linearly dependent on the concentration of analytes (in the range of 200 nM to 100 μM for HQ, CC, and RS) with the sensitivity of 0.8, 0.5, and 0.5 μA/μM and the lower detection limit of 0.13, 0.16, and 0.09 μM for HQ, CC, and RC, respectively. The results demonstrate that individual or simultaneous determination of these analytes at PANI/MnO_2 modified electrode can be achieved with high sensitivity and selectivity . Although, the potential response at PANI and MnO_2 modified electrodes for different species were similar to PANI/MnO_2 modified electrode, but the oxidation peak currents corresponding to the oxidation of HQ, CC, and RC bases were found to be significantly low compared to PANI/MnO_2 modified electrode (Figure 18b). This improved analytical performance is attributable to the highly dispersed MnO_2 active centers on PANI matrix owing to its unique mesoporous structure and large surface area .

Furthermore, PANI/MnO_2 modified electrode show very good electrocatalytic activity in the simultaneous determination of all four DNA bases (Figure 19) [130]. PANI/MnO_2 modified electrode exhibited promising activity in the electrochemical oxidation of guanine (G), adenine (A), thymine (T), and cytosine (C) with high sensitivity , selectivity , and antifouling properties. The voltagrammes for the quaternary mixture containing G, A, T, and C were well separated from each other with a potential difference of $\Delta E_{G\text{–}A}$ = 270 mV, $\Delta E_{A\text{–}T}$ = 210 mV, and $\Delta E_{T\text{–}C}$ = 170 mV, which are large enough to simultaneously determine the concentration of G, A, T, and C. PANI/MnO_2 modified electrode exhibited the sensing abilities for G, A, T, and C in the linear range of 10 μM to 100 μM; with the sensitivity of 0.2, 0.2, 0.06 and 0.1 μA/μM and the lower detection limit of 4.8, 2.9, 1.3 and 1.3 μM for G, A, T, and C, respectively (Figure 19a). PANI/MnO_2 modified electrode exhibited a well resolved voltametric response with high-current sensitivity compared to PANI and MnO_2 modified electrodes , which may be due to the synergistic contribution provided by both species (PANI and MnO_2) in composite material (PANI/MnO_2) (Figure 19b). Moreover, the proposed method was used to find the actual concentration of G, A, T, and C in a calf thymus DNA sample with high accuracy.

CONCLUSION

The oxidative polymerization of aniline and substituted anilines can produce a variety of micro- and nano-structures of PANI with different sizes and shapes including nanofibers , nanowires , nanotubes, hollow spheres, flowers , etc. A variety of nanostructured PANI was obtained by varying synthetic conditions such as reagent concentration, pH , homogeneous or heterogeneous polymerization, addition rate of reagents . To understand the effect of these synthesis conditions on final product, mechanism of the oxidative polymerization of PANI is discussed. Mechanism shows that a variety of monomers , dimmers, trimmers, tetramers can be formed depending on synthesis condition and types of precursor. A significant progress was made to understand the mechanism of aniline polymerization and the formation of PANI supramolecular structures, especially regarding the structural characterization of oligoanilines (tetraanilines, etc.) formed during the early stages of the chemical oxidative polymerization of aniline, which govern further PANI chain-growth and formation of PANI micro/ nanostructures . Having understood the synthesis mechanism, synthesis of a variety of nanostructured PANI by tailoring the synthesis condition, especially with respect to pH of the medium and aniline precursor is described.

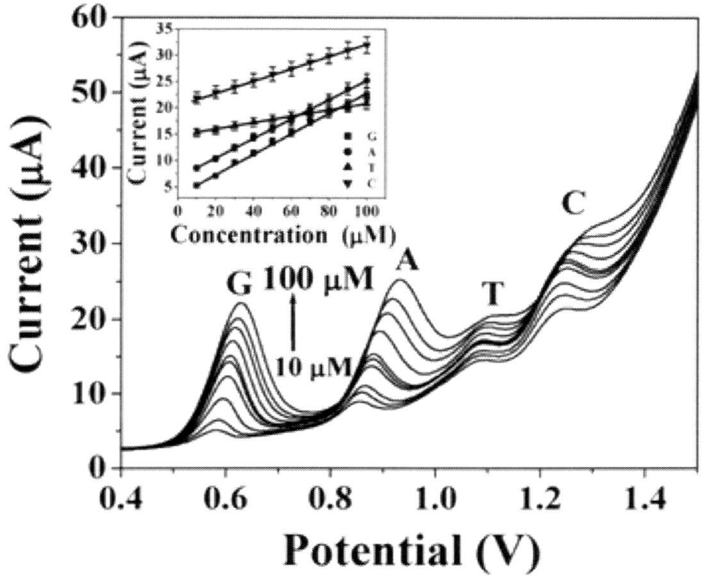

a

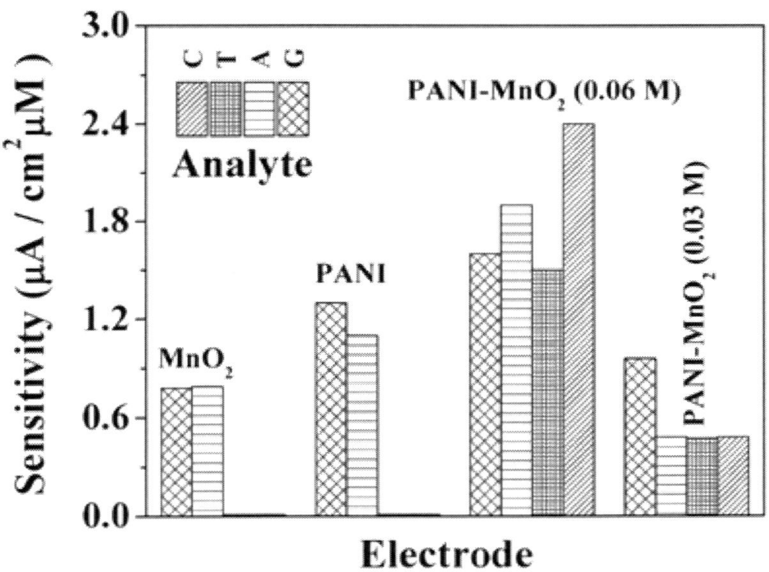

b

Figure 19. DPVs of the quaternary mixtures (having equal concentrations) of G, A, T, and C at the PANI/MnO_2 modified electrode. DPV parameters were selected as: pulse amplitude : 50 mV, pulse width: 50 ms, scan rate: 20 mV/s. (b) Comparison of the sensitivity for all four DNA bases at different modified elec-trodes.

Morphology of the PANI can also be tailored by incorporating suitable structure directing agents in the synthesis medium. It is reported that even by varying the structure of oxidizing agents, different nanostructured PANI can be formed. Ionic and neutral templates have also been investigated in the synthesis of PANI. A variety of morphology was obtained by varying the structure directing agent and pH of the medium. Morphology, size, and porosity play very important role in catalytic and electocatalytic application. Nanostructured PANI exhibited much higher activity than conventional PANI in the electrochemical oxidation of H_2O_2, glucose , ascorbic acid , and nucleobases. Variety of enzymatic and non-enzymatic electrochemical sensing based on PANI is highlighted. Property of PANI can be tailored by incorporating different inorganic or organic functionality to achieve desired activity. All these characteristics suggest that PANI and other conjugated polymers will likely lead to even more exciting future discoveries and intriguing applications. It is expected that continuing research efforts will be made in this rapidly developing field. Some progress has been made in the past is partly reviewed in this chapter; however, numerous scope of development still exists. Given the wide variety of nanostructured PANI that are currently available, it is hardly surprising that most current research devoted to tailor the physicochemical properties of PANI is concerned with development of novel routes for their synthesis with a variety of structure directing agents. Exploration of entirely new approaches for the preparation of porous PANI promises to open up new research opportunities for device fabrication .

ACKNOWLEDGMENTS

Authors are pleased to acknowledge Council of Scientific and Industrial Research (CSIR), New Delhi, for financial support under CSIR (01(2423)/10/EMR-II). We acknowledge Director, IIT Ropar for constant encouragements. Authors thank Dr. Biswarup Satpati, Saha Institute of Nucelar Physics, Kolkata, India for TEM analysis.

REFERENCES

[1] Nicholson, J. W. *The Chemistry of Polymers*; 3rd ed.; The Royal Society of Chemistry: Thomas Graham House , Cambridge, UK , 2006; pp. 1-22.
[2] Natta, G.; Mazzanti, G.; *Tetrahedron*, 1960, 8, 86.
[3] Burt, F. P. *J. Chem. Soc.* 1910, 1171.
[4] Greene, R. L.; Street, G. B.; Suter, L. J. *Phys. Rev. Lett.* 1975, 34, 577.
[5] Diaz, A. F.; Kanazawa, K. K.; Gardini, G. P. *J. Chem. Soc. Chem. Commun.* 1979, 635.
[6] Chang, C. K.; Fincher, C. R.; Park, Y. W.; Heeger, A. J.; Shirakawa, H.; Louis, E. J.; Gan, S. C.; Mac-Diarmid, A. G. *Phys. Rev. Lett.* 1977, 39, 1098.
[7] Gardini, G. P. *Adv. Heterocyclic Chem.* 1973, 15, 67.
[8] Nigrey, P. J.; MacDiarmid, A. G.; Heeger, A. J. *J. Chem. Soc. Chem. Commun.* 1979, 594.
[9] Rabolt, J. F.; Clarke, T. C.; Kanazawa, K. K.; Reynolds, J. R.; Street, G. B. *J. Chem. Soc. Chem. Commun.* 1980, 347.

[10] Tourillou, G. T.; Garnier, F. *J. Electronal Chem.* 1982, 135, 173.
[11] Former, J. E.; Elsenbaumer, R. C.; Chance, R. R. *Org. Coat. Appl. Polym. Sci. Proc.* 1983, 48, 552.
[12] A. F. Diaz and J. Bargon, In: T. A. Skotheim (ed.), *Handbook of Conducting Polymers*, Vol. 1, Marcel Dekker, New York, 1986, pp. 81.
[13] Zinger, B.; Miller, L. L. *J. Am. Chem. Soc.* 1984, 106, 6861.
[14] Yoshino, K.; Tabata, M.; Kaneto, K.; Ohsawa, T. *Jpn. J. Appl. Phys.* Part 2: Lett. 1985, 24, 693.
[15] Audebert, P.; Bidan, G. J. *Electroanal. Chem.* 1985, 190, 129.
[16] Gazard, M., Skotheim, T. A. (Ed.), *Handbook of Conducting Polymers*, Marcel Dekker, US 1986, Vol. 1, pp. 673.
[17] Colaneri, N. F.; Shacklette, L. W. *IEEE Trans. Instrum. Meas.* 1992, 41, 291.
[18] Brown, A. R.; Greenham, N. C.; Gymer, R. W.; Pichler, K.; Bradley, D. D. C.; Friend, R. H.; Burn, P. L.; Kraft, A.; Holmes A. B. *Intrinsically Conducting Polymers: An Emerging Technology*, NATO ASI Series, Series E: Applied Sciences 1993, 246, 87.
[19] Jelle, B. P.; Hagen, G.; Sungle, S.; Obegard, R. *Synth. Met.* 1993, 54, 315.
[20] Baigent, D. R.; Greenham, N. C.; Gruener, J.; Marks, R. N.; Friend, R. H.; Moratti, S. C.; Holmes, A. B. *Synth. Met.* 1994, 67, 3.
[21] Y. Yang, *MRS Bulletin*, 1997, 22, 31.
[22] Harrison, M. G.; Friend, R. H. *Electronic Materials: The Oligomer Approach* (Müllen, K., Wegner G., Eds.; Wiley-VCH, Weinheim 1998, 515.
[23] Saraswathi, R.; Gerard, M.; Malhotra, B. D. *J. Appl. Polym. Sci.* 1999, 74, 145.
[24] Granström, M.; Harrison, M. G.; Friend R. H. *Handbook of Oligo- and Polythiophenes,* D. Fichou, ed.), Wiley-VCH, Weinheim 1999, pp. 405.
[25] Kim, D. Y.; Cho, H. N.; Kim, C. Y. *Prog. Polym. Sci.* 2000, 25, 1089.
[26] Bernius, M.; Inbasekaran, M.; Woo, E.; Wu, W.; Wujkowski, L. *J. Mater. Sci.: Mater. Electronics* 2000, 11, 111.
[27] Bunz, U. H. F. *Chem. Rev.* 2000, 100, 1605.
[28] McGehee, M. D.; Heeger, A. J. *Adv. Mater.* 2000, 12, 1655.
[29] Leclerc, M. *J. Polym. Sci. Part A: Polym. Chem.* 2001, 39, 2867.
[30] Scherf, U.; List, E. J. W. *Adv. Mater.* 2002, 14, 477.
[31] Schwartz, B. J. *Annual Rev. Phys. Chem.* 2003, 54, 141.
[32] Grazulevicius, J. V.; Strohriegl, P.; Pielichowski, J.; Pielichowski, K. *Progr. Polym. Sci.* 2003, 29, 1297.
[33] Babudri, F.; Farinola, G. M.; Naso, F. *J. Mater. Chem.* 2004, 14, 11.
[34] Wang, J. Z.; Zheng, Z. H.; Li, H. W.; Huck, W. T. S.; Sirringhaus, H. *Nature* 2004, 3, 171.
[35] Hu, E.; Kaynak, A.; Li, Y. *Synth. Met.* 2005, 150, 139.
[36] Min, G. *Synth. Met.* 2005, 153, 49.
[37] Hu, E.; Kaynak, A.; Li, Y. *Synth. Met.* 2005, 150, 139.
[38] Song, H. K.; Palmore, G. T. R. *Adv. Mater.* 2006, 18, 1764.
[39] (a) Barros, R. A.; Martins, C. R.; Azevedo, W. M. *Synth. Met.,* 2005, 155, 35. (b) Hosoda, M.; Hino, T.; Kuramoto, N. *Polym. Int.* 2007, 56, 1448.
[40] Waghuley, S. A.; Yenorkar, S. M.; Yawale, S. S.; Yawale, S. P. *Sens. Actuat. B* 2008, 128, 366.

[41] (a) MacDiarmid, A. G. *Synth. Met.*, 2002, 125, 11. (b) Posudievsky, O. Y.; Goncharuk, O. A.; Pokhodenko, V. D. *Synth. Met.* 2010, 160, 47.
[42] Krinichnyi, V. I.; Yudanova, E. I. *Sol. Energy Mater. Sol. Cells* 2011, 95, 2302.
[43] (a) Bundgaard, E.; Krebs, F. C. *Sol. Energy Mater. Sol. Cells* 2007, 91, 954. (b) Jang, S. -Y.; Lim, B.; Yu, B.-K.; Kim, J.; Baeg, K.-J.; Khim, D. Y.; Kim, D.-Y. *J. Mater. Chem.* 2011, 21, 11822. (c) Haick, H.; Cahen, D. *Prog. Surf. Sci.* 2008, 83, 217.
[44] Zhao, X.; Zhan, X. *Chem. Soc. Rev.* 2011, 40, 3728.
[45] (a) Peng, H.; Soeller, C.; Vigar, N. A.; Caprio, V.; Travas-Sejdic, J. *Biosens. Bioelectron.* 2007, 22, 1868. (b) Cosnier, S. *Biosens. Bioelectron.* 1999, 14, 443. (c) Deshpande, M. V.; Amalnerkar, D. P. *Prog. Polym. Sci.* 1993, 18, 623.
[46] Salaneck, W. R.; Lundstro¨m, I.; Rånby, B. *Conjugated Polymers and Related Materials*; Oxford University Press: Oxford, 1993.
[47] Shirakawa, H.; Louis, E. J.; MacDiarmid, A. G.; Chiang, C. K.; Heeger, A. J. *Chem. Commn.* 1977, 578.
[48] Chiang, C. K.; Fincher, C. R. Jr.; Park,Y. W.; Heeger, A. J.; Shirakawa, H.; Louis, E. J. *Phys. Rev. Lett.* 1977, 39, 1098.
[49] Heeger, A. J. *Angew. Chem., Int. Ed.* 2001, 40, 2591.
[50] (a) Rao, P. S.; Subrahmanya, S.; Sathyanarayana, D. N.; *Synth. Met.* 2003, 139, 397. (b) Park, S. Y.; Cho, M. S.; Choi, H. J. *Curr. Appl. Phys.* 2004, 4, 581.
[51] (a) Skotheim, T. A.; Elsenbaumer, R. L.; Reynolds, J. R. *Handbook of Conducting Polymers*, 2nd ed.; Marcel Dekker: New York, 1997 Vol. 1 and 2. (b) Kanatzidis, M. G.; *Chem. Eng. News* 1990, 68, 36.
[52] MacDiarmid, A. G. *Angew. Chem., Int. Ed.* 2001, 40, 2581.
[53] Chiang, C. K.; Druy, M. A.; Gau, S. C.; Heeger, A. J.; Louis, E. J.; MacDiarmid, A. G. *J. Am. Chem. Soc.* 1978, 100, 1013.
[54] Nigrey, P. J.; MacDiarmid, A. G.; Heeger, A. J. *J. Chem. Soc. Chem. Commn.* 1979, 594.
[55] MacInnes, Jr. D.; Druy, M. A.; Nigrey, P. J.; Nairns, D. P.; MacDiarmid, A. G.; Heeger, A. J., *J. Chem. Soc., Chem. Commn.*, 1981, 317.
[56] Heeger, A. J.; Kivelson, S.; Shrieffer, J. R.; Su, W. P. *Rev. Mod. Physics* 1988, 60, 781.
[57] Pei, Q.; Yu, G.; Zhang, C.; Yang, Y.; Heeger, A. J., *Science* 1995, 269, 1086.
[58] Parker, J. D. *J. Appl. Phys.* 1994, 75, 1656.
[59] Burroughes, J. H.; Bradley, D. D. C.; Brown, A. R.; Marks, R. N.; Friend, R. H.; Burns, P. L.; Holmes, A. B. *Nature* 1990, 347, 539.
[60] (a) Chiang, J. C.; MacDiarmid, A. G. *Synth. Met.* 1986, 13, 193. (b) MacDiarmid, A. G.; Chiang, J. C.; Richter, A. F.; Epstein, A. J. *Synth. Met.* 1987, 18, 285. (c) MacDiarmid, A. G.; Epstein, A. J. *Faraday. Discuss. Chem. Soc.* 1989, 88, 317. (d) MacDiarmid, A. G.; Epstein, A. J. In: *Science and Applications of Conducting Polymers*; Salaneck, W. R.; Clark, D. T., Samuelsen, E. J., Eds.; Adam Hilger, Bristol, 1990; 117.
[61] Han, C. C.; Elsenbaumer, R. L. *Synth. Met.* 1989, 30, 123.
[62] (a) Genies, E. M.; Boyle, A.; Lapkowski, M.; Tsintavis, C. *Synth. Met.* 1990, 36, 139. (b) MacDiarmid, A. G.; Chiang, J. C.; Richter, A. F.; Epstein, A. J. *Synth. Met.* 1987, 18, 285. (c) Li, D.; Huang, J.; Kaner, R. B. *Acc. Chem. Res.* 2009, 42, 135.
[63] Marjanović, G. Ć. *Synth. Met.* 2013, 177, 1.

[64] (a) Tran, H. D.; Li, D.; Kaner, R. B. *Adv. Mater.* 2009, 21, 1487. (b) Wan, M. X. *Adv. Mater.* 2008, 20, 2926. (c) Wan, M. X. *Macromol. Rapid. Commn.* 2009, 30, 963. (d) Huang. J. X.; Kaner, R. B. *Chem. Commn.* 2006, 367. (e) Jang. J. *Adv. Polym. Sci.* 2006, 199, 189.

[65] Marjanović, G. Ć.; Trchova. M.; Stejskal. J. *Collect. Czech. Chem. Commun.* 2006, 71, 1407. (b) Marjanović, B.; Juraniĉ, I.; Marjanović, G. Ć. *J. Phy. Chem. A* 2011, 115, 3536.

[66] Qin, L.; Tripathi, G. N. R.; Schuler, R. H. Z.; Zeitschrift fur, N. A. *J. Phy. Sci.* 1985, 40, 1026.

[67] McClelland, R. A.; Kahley, M. J.; Davidse, P. A.; Hadzialic, G. *J. Am. Chem. Soc.* 1996, 118, 4794.

[68] Junker, K.; Zandomeneghi, G.; Guo, Z.; Kissner, R.; Ishikawa, T.; Kohlbrecher, J.; Walde, P. *RSC Advances* 2012, 2, 6478.

[69] (a) Mohilner, D. M.; Adams, R. N.; Argersinger, W. J. Jr. *J. Am. Chem. Soc.* 1962, 84, 3618. (b) Bacon, J.; Adams, R. N. *J. Am. Chem. Soc.* 1968, 90, 6596. (c) Ohsaka, T.; Ohnuki, Y.; Oyama, N.; Katagiri, G.; Kamisako, K. *J. Electroanalytical Chemistry.*, 1984, 161, 399.

[70] Marjanović, G. Ć.; Konyushenko, E. N.; Trchova, M.; Stejskal, J. *Synth. Metal.* 2008, 158, 200.

[71] Marjanović, G. Ć.; Trchova, M.; Stejskal, J. *Int. J. Quantum Chem.* 2008, 108, 318.

[72] (a) Krîz, J.; Trchova, E. N.; Stejskal, J., *Polymer Inter.* 2011, 60, 1296; (b) Stejskal, J.; Trchova, M. *Polymer Inter.* 2012, 60, 240.

[73] Roy, S. K.; Samanta, S.; Sinan, M.; Ghosh, P.; Goswami, S. *J. Org. Chem.* 2012, 77, 10249.

[74] Huang, J. X.; Kaner, R. B., *J. Am. Chem. Soc.* 2004, 126, 851. (b) Huang, J. X.; Virji, S.; Weiller, B. H.; Kaner, R. B. *J. Am. Chem. Soc.* 2003, 125, 314. (c) Zhang, X.; Chan-Yu-King, R.; Jose, A.; Manohar, S. K. *Synth. Met.* 2004, 145, 23. (d) Xing, S.; Zheng, H.; Zhao, G. *Synth. Met.* 2008, 158, 59. (e) Ameen, S.; Akhtar, M. S.; Kim, Y. S.; Yang, O. B.; Shin, H. S. *J. Phys Chem.* C 2010, 114, 4760.

[75] Huang, J. X.; Kaner, R. B. *Chem. Commun.* 2006, 367.

[76] (a) Jing, X. L.; Wang, Y. Y.; Wu, D.; Qiang, J. P. *Ultrason. Sonochem.* 2007, 14, 75. (b) Li, Y.; Wang, Y.; Wu, D.; Jing, X. *J. Appl. Polymer. Sci.* 2009, 113, 868. (c) Wang, Y.; Jing, X.; Kong, J. *Synth. Met.* 2007, 157, 269.

[77] Nikolaidis, G. M. R.; Stanisavljev, D. R.; Easteal, A. J.; Zujovic, Z. D. *Macromol. Rapid Commun.* 2010, 31, 657.

[78] (a) Li, D.; Kaner, R. B. *J. Am. Chem. Soc.* 2006, 128, 968. (b) Li, D.; Kaner, R. B. *J. Mater. Chem.* 2007, 17, 2279.

[79] (a) Tran, H. D.; Kaner, R. B. *Chem. Commun.* 2006, 3915. (b) Tran, H. D.; Norris, I.; D'Arcy, J. M.; Tsang, H.; Wang, Y.; Mattes, Kaner, R. B. *Macromolecules* 2008, 41, 7405.

[80] (a) Liu, J.; Lin, Y.; Liang, L.; Voigt, J. A.; Huber, D. L.; Tian, Z. R.; Coker, E.; Mckenzie, B.; Mcdermott, M. J. *Chem. Eur J.* 2003, 9, 605. (b) Chiou, N. R.; Lui, C. M.; Guan, J. J.; Lee, L. J.; Epstein, A. J. *Nat. Nanotechnol.* 2007, 2, 354.

[81] (a) Trchova, M.; Sedenkova, I.; Konyushenko, E. N.; Stejskal, J.; Holler, P.; Marjanović, G. Ć. *J. Phys. Chem. B* 2006, 110, 9461. (b) Zujovic, Z. D.; Laslau, C.; Bowmaker, G. A.; Kilmartin, P. A.; Webber, A. L.; Brown, S. P.; Sejdic, T. J.

Macromolecules 2010, 43, 662. (c) Laslau, C.; Zujovic, Z. D.; Zhang, L. J.; Bowmaker, G. A.; Sejdic, T. J. *Chem. Mater.* 2009, 21, 954.

[82] (a) Stejskal, J.; Sapurina, I.; Trchova, M.; Konyushenko, E. N. *Macromolecules* 2008, 41, 3530. (b) Konyushenko, E. N.; Stejskal, J.; Sedenkova, I.; Trchova, M.; Sapurina, I.; Cieslar, M.; Prokes. *J. Polym. Int.* 2006, 55, 31.

[83] Stejskal, J.; Sapurina, I.; Trchova, M.; Konyushenko, E. N.; Holler, P. *Polymer* 2006, 47, 8253.

[84] (a) Venancio, E. C.; Wang, P. C.; MacDiarmid, A. G. *Synth. Met.* 2006, 156, 357. (b) Tan, Y.; Bai, F.; Wang, D.; Peng, Q.; Wang, X.; Li, Y. *Chem. Mater.* 2007, 19, 5773.

[85] (a) Zhang, L.; Wan, M. *Adv. Funct. Mater.* 2003, 13, 815. (b) Zhang, L.; Long, Y.; Chen, Z.; Wan, M. X. *Adv. Funct. Mater.* 2004, 14, 693.

[86] Li, J.; Jia, Q.; Zhu, J.; Zheng, M. *Polym. Int.* 2008, 57, 337. (b) Li, G. C.; Zhang, Z. K. *Macromolecules* 2004, 37, 2683. (c) Lindman, B.; Wennerstrom, H. Micelles: Amphiphile aggregation in aqueous solution ; Springer-Verlag: Berlin; New York, 1980.

[87] (a) Zhang, L.; Wan, M. X.; Wei, Y. *Macromol. Rapid Commun.* 2006, 27, 366. (b) Zhang, L.; Wan, M. X.; Wei, Y. *Synth. Met.* 2006, 156, 454. (c) Zhang, Z.; Wei, L.; Zhang, L.; Wan, M. X. *Acta. Mater.* 2005, 53, 1373. (d) Wei, Z.; Zhang, Z.; Wan, M. X. *Langmuir* 2002, 18, 917. (e) Huang, K.; Meng, X. H.; Wan, M. X. *J. Appl. Polym. Sci.* 2006, 100, 3050. (f) Huang, K.; Wan, M. X. *Chem. Mater.* 2002, 14, 3486.

[88] (a) Zhang, Z.; Wan, M.; Wei, Y. *Adv. Funct. Mater.* 2006, 16, 1100. (b) Wei, Z.; Zhang, L.; Yu, M.; Yang, Y.; Wan, M. X. *Adv. Mater.* 2003, 15, 1382. (c) Qiu, H.; Wan, M. X. *Macromolecules* 2001, 34, 675.

[89] (a) Niu, Z.; Liu, J.; Lee, L. A.; Bruckman, M. A.; Zhao, D.; Koley, G.; Wang, Q. *Nano Lett.* 2007, 7, 3729. (b) Niu, Z. W.; Bruckman, M. A.; Li, S. Q.; Lee, L. A.; Lee, B.; Pingali, S. V.; Thiyagarajan, P.; Wang, Q. *Langmuir* 2007, 23, 6719. (c) Niu, Z. W.; Bruckman, M.; Kotakadi, V. S.; He, J. B.; Emrick, T.; Russell, T. P.; Yang, L.; Wang, Q. *Chem. Commun.* 2006, 3019.

[90] Zhao, D.; Feng, J.; Huo, Q.; Melosh, N.; Fredrickson, G. H.; Chmelka, B. F.; Stucky, G. D. *Science* 1998, 279, 548.

[91] Prathap, M. U. A.; Srivastava, R. *J. polymer Res.* 2011, 18, 2455.

[92] Prathap, M. U. A.; Bhawana T.; Shilpa, N. S.; Srivastava, R. *Colloids and Surfaces: B Biointerfaces* 2012, 89, 108.

[93] Shi, L.; Xiao, Y.; Willner, I. *Electrochem. Commun.* 2004, 6, 1057.

[94] Wolter, A.; Rannou, P.; Travers, J. P.; Gilles, B.; Djurado, D. *Phys. Rev. B* 1998, 58, 7637.

[95] Kim, Y. H.; Kim, M. K.; Oh, S. W.; Jung, H. S.; Kim, Y. J.; Yoon, T. S.; Kim, Y. S.; Lee, H. H. *Appl. Physics Lett.* 2012, 100, 163301.

[96] Wei, J.; Xiong, S. X.; Bai, Y.; Jia, P. T.; Ma, J.; Lu, X. H. *Sol. Energy Mater. Sol. Cells* 2012, 99, 141.

[97] Spain , E.; Kojima, R.; Kaner, R. B.; Wallace, G. G.; O'Grady, J.; Lacey, K.; Barry, T.; Keyes, T. E.; Forster, R. J. *Biosens. Bioelectron.* 2011, 26, 2613.

[98] (a) Trojanowicz, M.; Hitchman, M. L. *Trends Anal. Chem.* 1996, 15, 38 (b) Kriz, K.; Anderlund, M.; Kriz, D. *Biosens. Bioelectron.* 2001, 16, 363.

[99] Wang, B.; Li, B.; Wang, Z.; Xu, G.; Wang, Q.; Dong, S., *Anal. Chem.* 1999, 71, 1935.

[100] Du, Z.; Li, C.; Li, L.; Zhang, M.; Xu, S.; Wang, T. *Mater. Sci. Engg. C* 2009, 29, 1794.

[101] Yang,Y.; Mu, S. *Biosens. Bioelectron.* 2005, 21, 74.
[102] (a) Heller, A.; Feldman, B. *Chem. Rev.* 2008, 108, 2482. (b) Lee, S. R.; Lee, Y. T.; Sawada, K.; Takao, H.; Ishida, M. *Biosens. Bioelectron.* 2008, 24, 410. (c) Newman, J. D.; Turner, A. P. F. *Biosens. Bioelectron.* 2005, 20, 2435. (d) Wu, L.; Zhang, X.; Ju, H. *Biosens. Bioelectron.* 2007, 19, 1461.
[103] Wilson, R.; Turner, A. P. F. *Biosens. Bioelectron.* 1992, 7, 165.
[104] (a) Saleh, F. S.; Mao, L.; Ohsaka, T. *Analyst* 2012, 137, 2233. (b) Zafar, M. N.; Wang, X. J.; Sygmund, C.; Ludwig, R.; Leech, D.; Gorton, L. *Anal. Chem.* 2012, 84, 334. (c) Yamazaki, T.; Kojima, K.; Sode, K. *Anal. Chem.* 2000, 72, 4689.
[105] Karyakin, A. A.; Vuki, M.; Lukachova, L. V.; Karyakina, E. E.; Orlov, A. V.; Karpachova, G. P.; Wang, J. *Anal. Chem.* 1999, 71, 2534.
[106] Gaikwad, P. D.; Shirale, D. J.; Gade, V. K.; Savale, P. A.; Kharat, H. J.; Kakde, K. P.; Shirsat, M. D. *Int. J. Electrochem. Sci.* 2006, 1, 425.
[107] Prathap, M. U. A.; Pandiyan, T.; Srivastava, R. *J. Polymer Res.* 2013, 20, 83.
[108] (a) Marioli, J. M.; Kuwana, T. *Electrochim. Acta.* 1992, 37, 1187. (b) Kaur, B.; Prathap, M. U. A.; Srivastava, R. *Chem. Plus Chem.* 2012, 77, 1119. (c) Brisard, G. M.; Rudnicki, J. D.; McLarnon, F.; Cairns, E. J. *Electrochim. Acta.* 1995, 40, 859.
[109] (a) Prathap, M. U. A.; Kaur, B.; Srivastava, R. *J. Colloid. Interface. Sci.* 2012, 370, 144. (b) Farrell, S. T.; Breslin, C. B. *Electrochim. Acta.* 2004, 49, 4497. (c) Wei, H.; Sun , J. J.; Guo, L.; Li, X.; Chen, G. N. *Chem. Commun.* 2009, 2842.
[110] Martin, D. W. Jr. In: *Harper's Review of Biochemistry*; Martin, D. W. Jr., Mayes, P. A., Rodwell, V. W., Eds.; Lange, Los Altos, CA, 1983; 19th ed.
[111] Koshiishi, I.; Imanari, T. *Anal. Chem.* 1997, 69, 216.
[112] (a) Zhang, L.; Dong, S. *J. Electroanal. Chem.* 2004, 568, 189. (b) Zhang, L. *Electrochim. Acta* 2007, 52, 6969. (c) Ambrosi, A.; Morrin, A.; Smyth, M. S.; Anthony, J. K. *Anal. Chim. Acta.* 2008, 609, 37.
[113] Mu, S.; Kan, J. *Synth. Met.* 2002, 132, 29.
[114] Sheng, R. S.; Ni, F.; Cotton, T. M. *Anal. Chem.* 1991, 63, 437.
[115] Yang, F. Q.; Guan, J.; Li, S. P. *Talanta* 2007, 73, 269.
[116] (a) Teh, H. F.; Yang, X. H.; Gong, H. Q.; Tan, S. N. *Electroanalysis* 2004, 16, 769. (b) Huang, W.; Zhang, S.; Wu, Y. Russ. *J. Electrochem.* 2006, 42, 153. (c) Goyal, R. N.; Gupta,V. K.; Oyama, M.; Bachheti, N. *Talanta* 2007, 71, 1110. (d) Wu, K. B.; Fei, J. J.; Bai, W.; Hu, S. S. *Anal. Bioanal. Chem.* 2003, 376, 205.
[117] (a) Ye, Y. K.; Ju, H. X. *Biosens. Bioelectron.* 2005, 21, 735. (b) Shen, Q.; Wang, X. *J. Electroanal. Chem.* 2009, 632, 149. (c) Zhu, Z. H.; Li, X.; Zeng, Y.; Sun , W. *Biosens. Bioelectron.* 2010, 25, 2313. (d) Ferancová, A.; Rengaraj, S.; Kim, Y.; Labuda, J.; Sillanpää, M. *Biosens. Bioelectron.* 2010, 26, 314.
[118] (a) Brett, A. M. O.; Piedade, J. A. P.; Silva, L. A.; Diculescu, V. C. *Anal. Biochem.* 2004, 332, 321. (b) Wang, P.; Wu, H.; Dai, Z.; Zou, X. *Biosens. Bioelectron.* 2011, 26, 3339.
[119] Zhou, M.; Zhai, Y.; Dong, S. *Anal. Chem.* 2009, 81, 5603.
[120] (a) Zeng, H. J.; Yang, R.; Wang, Q. W.; Li, J. J.; Qu, L. B. *Food Chem.* 2011, 127, 842. (b) Qiu, H.; Xi, Y.; Lu, F.; Fan, L.; Luo , C. *Spectrochim. Acta, Part A* 2012, 86, 456.
[121] Bo, Y.; Yang, H.; Hu, Y.; Yao, T.; Huang, S. *Electrochim. Acta* 2011, 56, 2676.
[122] Feng, Y.; Yang, T.; Zhang, W.; Jiang, C.; Jiao, K. *Anal. Chimica. Acta* 2008, 616, 144.
[123] Du, M.; Yang, T.; Li, X.; Jiao, K. *Talanta* 2012, 88, 439.

[124] Zhua, N.; Changb, Z.; Heb, P.; Fang, Y. B. *Electrochimica. Acta* 2006, 51, 3758.

[125] Arora, K.; Prabhakar N.; Chand, S.; Malhotra, B. D. *Biosens. Bioelectron.* 2007, 23, 613.

[126] Prathap, M. U. A.; Srivastava, R. *Sens. Actuators. B* 2013, 177, 239.

[127] (a) Huang, J.; Liu, Y.; Hou, H.; You, T. *Biosens. Bioelectron.* 2008, 24, 632. (b) Safavi, A.; Maleki, N.; Moradlou, O.; Tajabadi, F. *Anal. Biochem.* 2006, 359, 224. (c) Dayton, M. A.; Ewing, A. G.; Wightman, R. M. *Anal. Chem.* 1980, 52, 2392.

[128] Prathap, M. U. A.; Satpati, B.; Srivastava, R. *Sens. Actuators, B* 2013, 186, 66.

[129] Carvalho, M. R.; Mello, C.; Kubota, L. T. *Anal. Chim. Acta.* 2000, 420, 109.

[130] Prathap, M. U. A.; Satpati, B.; Srivastava, R. *Electrochimica Acta* 2013, 114, 285.

In: Comprehensive Guide for Mesoporous Materials. Volume 1 ISBN: 978-1-63463-958-3
Editor: Mahmood Aliofkhazraei

Chapter 11

SELF-ASSEMBLY AND TRANSPORT THROUGH MESOPOROUS COLLOIDAL SHELLS

Nily Dan*
Department of Chemical and Biological Engineering,
Drexel University, Philadelphia , PA, US

ABSTRACT

Self-assembly of colloidal particles at liquid/liquid interfaces leads to the formation of mesoporous shells: Pickering emulsions result from assembly at interface between immiscible hydrophilic/hydrophobic phases, while shells separating an aqueous core from an aqueous continuous phase give rise to 'colloidosomes'. In both cases, the unique properties of the mesoporous shells provide selective and/or controlled release with unique transport properties. This chapter reviews the characteristics and properties of colloidal assembly at the interfaces of Pickering emulsions and colloidosomes, and the factors controlling transport through the colloidal shells.

1. INTRODUCTION

The ability to control the transport rate of diffusants into or out of micro or nano scale droplets (100nm to 100μm) is of interest in applications ranging from food technology to cosmetics and pharmaceutics [1-4]. In these systems, a component encapsulated in the droplet core is released into the environment in a controlled manner dictated by the needs of the particular application. Therefore, effective utilization hinges on the ability to form coatings , or shells, with controllable permeability , that can provide a constant flux over a period of time for some applications, or triggered release where the encapsulant remains in the droplet core until a specific change in environmental conditions occurs in others.

* Corresponding author: Nily Dan. Department of Chemical and Biological Engineering, Drexel University, Philadelphia , PA 19104, US. Phone: 1 (215) 895 6624; Fax: 1 (215) 895 5837; E-mail: dan@coe.drexel.edu.

Coating micro and nano-scale drops requires shells that self-assemble at the drop interface . Typically, amphiphilic molecules are used for shell formation , since the dual hydrophilic/hydrophobic nature of these molecules drives the formation of interfacial films : Small molecule surfactants are used for the stabilization of emulsions , namely drops of oil in water or water in oil [5], while aqueous drops in a continuous aqueous phase are coated by lipid(liposomes) [6] or block copolymer (polymersomes) [7] bilayers. The advantage of such molecules is that they easily self-assemble at interfaces, and—due to their amphiphilic nature—suppress drop coalescence.

Unlike amphiphilic molecules whose clearly defined hydrophobic and hydrophilic regions lead to self-assembly , colloidal particles are not expected to exhibit amphiphilic behavior. However, they have been found to aggregate at interfaces between hydrophobic and hydrophilic phases, lowering the interfacial tension and thereby stabilizing emulsions . These 'Pickering emulsions' [8-12] are composed of oil in water or water in oil drops, stabilized by a shell of colloidal particles. A similar process can also be used to produce aqueous drops in an aqueous medium, termed 'colloidosomes' [13, 14].

Self-assembled colloidal shells are composed of impenetrable regions (the particles) and penetrable ones (the voids, or pores between them), thereby forming a mesoporous structure . The rate of transport through the colloidal shells is controlled by three parameters: The shell thickness, the colloidal packing density (or, alternately, the void fraction) and the colloidal particle size [8-14]. These parameters set the characteristic pore size , which determines the maximal size of diffusants that can cross through the shell. They also set the shell permeability to smaller molecules , a complex function of the volume available for transport and the pore tortuosity.

As a self-assembled system, shell parameters cannot be controlled directly; Both shell thickness and packing density can be set only indirectly using the colloidal surface characteristics, shell processing procedure, and system parameters such as the colloidal particle concentration.

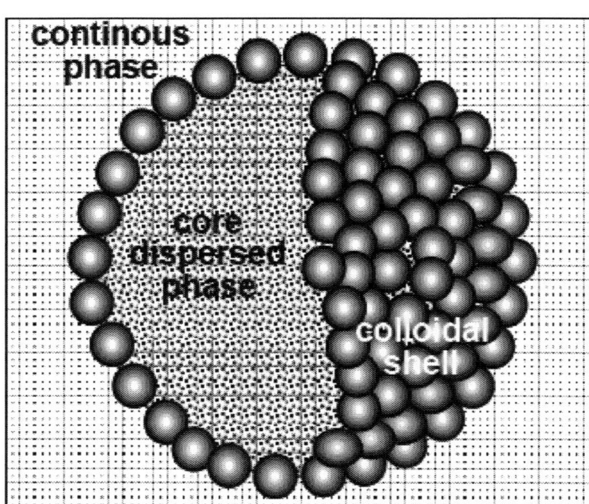

Figure 1. Colloidal particles assembled at the interface of micro-scale drops into a colloidal shell. Shells assembled at microemulsion interfaces (oil /water) form Pickering emulsions , while those with an aqueous core in a continuous aqueous medium form colloidosomes.

This chapter discusses the properties of Pickering emulsions and colloidosomes: How they are synthesized, methods for controlling the shell properties, and measurements of the rate of transport through the resulting composite structure . Also reviewed are theoretical models for colloidal shell self-assembly and transport through these mesoporous shells.

2. Pickering Emulsions

The tendency of amphiphilic molecules to assemble at interfaces between hydrophobic and hydrophilic liquids has been known for centuries. Driven by the 'schizophrenic' nature of molecules that have a hydrophobic and a hydrophilic region, they can stabilize emulsions [5, 15]. Similar assembly was found in amphiphilic 'Janus' colloidal particles that are partly hydrophobic and partly hydrophilic [16].

Colloidal particles with *homogeneous* surfaces are not expected to display amphiphile-like behavior. Yet, in the early 1900s Ramsden [17] and Pickering [18] found that colloidal particles with homogenous surfaces accumulate at liquid interfaces , thereby stabilizing emulsions . About 15 years later, the process was linked to partial wetting of the colloidal particles by the two liquid phase s [19]. Particles that are strongly hydrophilic or hydrophobic remain solubilized in their preferred phase, while ones with moderate solubility and/or finite contact angles in both phases accumulate at interfaces [20]. The type of emulsion (oil in water or water in oil) depends on the particle wettability : Generally, hydrophobic particles form oil in water emulsions, while hydrophilic particles form water in oil ones [8, 9] (see also Section 5 for details).

The concentration of colloidal particles in solution affects the emulsion properties [21-26]. In the limit of low concentrations where the particles cannot form a complete interfacial monolayer , Pickering emulsion drops are expected to be unstable since the 'bare' patches can act as loci for drop coagulation [22]. Above this concentration, full surface coverage is expected to impart stability .

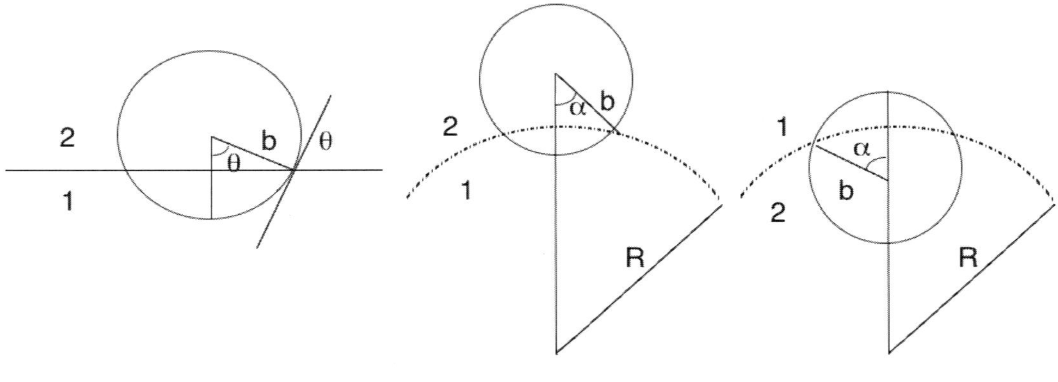

Figure 2. Spherical colloidal particles at liquid/liquid interfaces . The particles partition to the interface to reduce the interfacial energy , as set by Equation (1). The preferred contact angle between the particles and phase 2 (taken to be the aqueous phase) is given by θ (left). However, interface curvature leads to a deviation from this optimal value, so that the contact angle is $\alpha \neq \theta$. The definition of α depends on the type of emulsion, namely, whether phase 2 is the continuous (center) or dispersed (right) phase.

Table 1. Emulsion type as a function of the particle oil /water contact angle (see Figure 2), in systems with 1:1 oil/water volume ratios

Particle	Oil	Oil/water p article contact angle (^{0})	Emulsion type
Hydrophilic silica	Dodecane	38	O/W
Hydrophilic silica	PDMS 50cS	81	O/W
Hydrophobic silica	Dodecane	135	W/O
Hydrophobic silica	PDMS 50cS	172	W/O
Partially hydrophobic silica	Dodecane	83	O/W

Data from reference [8].

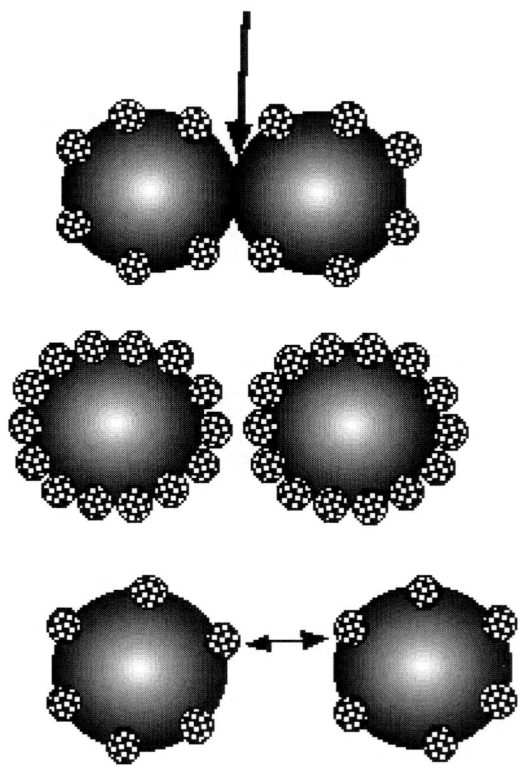

Figure 3. Effect of particle concentration on emulsion stability . In partially covered emulsions (top), bare patches allow coalescence of the drops and the formation of larger assemblies. Surface saturation (middle) imparts steric stability. However, if the particles are electrostatically charged (bottom), long-range repulsion can inhibit coalescence even between sparsely covered interfaces.

However, in several cases Pickering emulsions were found to be stable even when interface coverage was incomplete [23-26]. As demonstrated by Horozov and Binks [24], this is due to particle-particle repulsion : Charged particles that display long-range repulsion stabilized the emulsion drops even when the surface coverage was relatively low, while hydrophobic particles that do not exhibit such interactions stabilized emulsion drops only when interface layer was highly packed [24]. It should be noted, though, that highly-charged particles may not localize at the oil /water interface [20], so that a balance must be achieved.

Particle concentration also affects the size of the emulsion drops, which was found to decrease almost linearly with particle concentration [21, 23]. The fraction of particles adsorbed at the interface decreased sharply with particle concentration [22] while the ratio of particles at the interface to those in solution remained constant [21].

Above a certain concentration, however, further addition of colloidal particles did not affect the size of the emulsion drops, which remained nearly constant [21]. As a result, in this limit the excess of colloidal particles increased. This dependence of emulsion drop size on concentration is reminiscent of that observed for molecular amphiphiles [5].

Emulsion drop size is also sensitive to the colloidal particle size. Binks and Lumsdon [27] found that the size distribution of emulsion drops stabilized by hydrophobic latex particles, which formed water in oil emulsions , was relatively large. The median drop size increased with increasing particle size until reaching a limiting, maximal value. In contrast, emulsions stabilized by hydrophilic particles displayed quite a different behavior. Increasing particle size led to a decrease in the water in oil emulsion drop size, until reaching a critical value at which inversion to oil in water occurred. Above this value, drop size increased with increasing particle size [27]. Although many studies of Pickering emulsions utilized silica or latex particles, recent work focused on the emulsifying properties of other types of colloidal particles. In particular, the use of these emulsions in food and biotechnology applications has led to interest in emulsions stabilized by bio-colloids such as cellulose [28], starch [29] or chitin nanocrystals [30]. The properties of these emulsions largely resemble that of emulsions stabilized by 'classical' colloids. Specifically, increasing the concentration of particles was shown to decrease the drop size, until reaching a constant value [28, 30].

Pickering emulsions stabilized by non-spherical nanoparticles such as rod or sheet-like colloids have also been examined [31-33]. Shape anisotropy was found to greatly affect the stability and properties of the emulsion drops. Kalashnikova, et al. [32] found that increasing the aspect ratio of rod-like particles by an order of magnitude did not affect the drop size, suggesting that the rod properties do not directly affect the preferred interfacial curvature. However, the aspect ratio did affect the packing density of particles at the interface : Rods with a smaller aspect ratio achieved high surface coverage, forming drops.

However, longer rods did not fully cover the interface , thereby allowing coalescence that led to a network of interconnected droplets [32]. In a study using nano-sheets for foam stabilization , Guevara, et al. [33] found that the effect of the sheet aspect ratio on emulsion stability depended on the degree of particle hydrophobicity : Low aspect ratio was effective for highly hydrophobic particles, while high aspect ratio required, for stability, an intermediate degree of hydrophobicity [33].

Table 2. Effect of particle size on the emulsion drop size in water -toluene (1:1 by volume) mixtures stabilized by 1 wt % hydrophilic latex particles in pure water

D particle (μm)	Emulsion type	Drop size – arithmetic mean (μm)
0.03	W/O	35
0.15	W/O	10
0.5	O/W	30
1	O/W	40
4	O/W	60

From [27].

Synthesis procedures for Pickering emulsions tend to follow those of 'traditional' emulsification [34-38]. Interfacial tension is overcome through the application of shear via sonication or high pressure, which break large droplets into smaller ones. Colloidal particles are introduced in one phase and are driven to the interface where they form a barrier to drop coalescence.

Another methodology for emulsion formation is by extrusion of a dispersed phase through a porous medium such as membranes or capillaries [39-41] into a continuous phase that contains the emulsifying agent. This method has been adapted to the formation of Pickering emulsions in microfluidic devices [42-45], yielding highly monodisperse drops.

3. Colloidosomes

The spontaneous self-assembly of amphiphilic molecules such as lipids into bilayer structures yields micro-capsules containing an aqueous core, suspended in an aqueous solution : Liposomes [6] or polymersomes [7]. Colloidosomes are the colloidal equivalent [46, 47], where a colloidal shell separates an inner aqueous core from a continuous aqueous solution. (In principle, the core and continuous media could also be hydrophobic). However, although colloidal particles assemble at hydrophobic/hydrophilic interfaces, they do not spontaneously form interfaces in a single solvent. Therefore, unlike amphiphilic layers or Pickering emulsions , colloidosome synthesis requires different methodologies.

Many studies of colloidosomes used Pickering emulsions as a template for colloidosome shells [46-53]. In this method, an emulsion—stabilized by colloidal particles—of a hydrophobic core in an aqueous continuous medium, is formed. The particles are locked in place using sintering , fusion , or through binding molecules such as polymers [13, 14]. Once the shell is fixed in place, the inner core is removed [46, 48, 49, 50, 51]. The inverse process has also been used, where the inner core is an aqueous solution and the outer phase is a hydrophobic one that is removed after shell formation [47, 52, 53].

The removal of the hydrophobic core phase is a non-trivial step, which can lead to shell destabilization. Furthermore, in oil in water emulsion-based colloidosomes, introduction of the preferred aqueous solution into the core after the shell is locked in place requires transport through the shell. However, since the purpose of the shell is to reduce the transport rate of these components, or suppress it altogether, this is an ineffective process. Using water in oil emulsions eliminates this issue, since the core's aqueous phase can incorporate the desired components during the emulsification stage. Replacement of the continuous oil phase by another aqueous phase does not affect, then, the encapsulants. However, during the removal of the oil phase the shells may coagulate [52].

Another way to form colloidosomes using Pickering emulsions is based on double emulsions. In this method, water -in oil -in water emulsions are formed, where the colloids are suspended in the middle, oil phase [47, 54]. Removal of the oil locks the particles in place through either van der Walls interactions or by using a binding agent. The relative dimensions of the core and middle phase, and the concentration of the colloids in the oil phase, set the thickness and density of the shell. The advantages of this method are that the content of the colloidosome is introduced initially in the inner phase, as in the case of water in oil single emulsions, and shell destabilization or coagulation is minimized.

Shell fusion is an essential step in all emulsion-based colloidosome synthesis . van der Waals interactions may become dominant when the particle-particle spacing decreases, thereby fixing the particles in place [47]. However, since these forces dominate only at extremely close range, any type of particle stabilization —whether steric or electrostatic— may interfere with this process. Indeed, Velev, et al. [46] found that electrostatic stabilization of the colloidal particles had to be partly neutralized for the shells to assemble. Neutralization of particle stabilization can lead , however, to aggregation in solution —which could interfere with shell formation .

Sintering has been found to effectively fix polymeric shell particles in place [47, 51]. Furthermore, sintering conditions could be used to control the packing density of particles in the shells [51]. However, this method is not practical when using inorganic particles such as silica , and the sintering temperature (in the range of 50-100°C, depending on the particle properties) may adversely affect encapsulated biomolecules .

A common methodology for locking colloidosome shells in place is through the use of some bridging agent, dispersed in either the core or the continuous phase. These binding agents, which are usually polymeric, coat the colloidal particles and form bridges that hold them in place [46, 47, 48, 54, 74]. Investigating the effect of binder length on shell stability , Croll and Stover [48] found that low MW polymeric chains were ineffective in shell stabilization . Increasing the chain MW yielded stable shells, but there was no discernible difference between moderately long polymers (M_n~6500) and chains an order of magnitude longer (M_n~60,000). They also found that the rate of binder addition affects the shell properties [48]: Rapid addition led to thick, porous shells, while slow addition yielded thin, dense shells. This was attributed to particle rearrangement: Rapid addition of binder 'quenches' the particles in place. Slow mixing enables particle movement and densification.

Colloidosomes have also been synthesized using a non-emulsion route, where a microgel sphere is utilized as a 'scaffold' [55-62]. The colloidal particles assemble and bind to the surface of the microgel through specific interactions that are, typically, electrostatic. As in the single emulsion route, the shell is initially monolayer thick, but the binding of the particles to the hydrogel surface eliminates the need for an additional 'locking' step. Furthermore, thicker shells can be produced through partial collapse of the microgel core: Kim, et al. constructed colloidal shells on temperature -sensitive microgel cores, where a relatively small change in temperature induced core collapse and the formation of thick, multi-layered and densely packed shells [59]. This methodology could be extended to other environmentally-responsive microgel cores, such as ones composed of pH sensitive hydrogels .

Since the assembly of colloidal shells on microgel cores is driven by colloidal adsorption onto the gel surface, controlling the shell properties can be done only indirectly through the adsorption process. Serizawa, et al. [63] found that, as may be expected, particle adsorption on an oppositely charged gel formed monolayer -thick layers. Surface coverage increased with the particle concentration in solution .

The adsorption process generally followed the Langmuir isotherm, where at low concentrations increasing the particle concentration led to a significant increase in coverage, while at higher concentrations the effect is reduced. The saturation density of particles on the surface was found to increase significantly with particle size [63].

The maximal packing density of monodisperse particles is limited by the random, close-packing value for hard spheres. The use of polydisperse particles can increase the packing density [65-67].

The size distribution of particles on the surface was found to differ, significantly from the distribution in solution [64], and the packing density increased with increasing size asymmetry in both two dimensions [65] and three dimensions [66, 67].

4. TRANSPORT THROUGH COLLOIDAL SHELLS

The transport of materials out of, or into, Pickering emulsions and colloidosomes is typically measured by monitoring the release rate of components encapsulated in the core. This process is driven by diffusion of the encapsulant from a high-concentration region (the core, or dispersed phase) to a low concentration one (the continuous phase), through a barrier—the mesoporous colloidal shell. Thus, the rate of transport, or flux (mass or mole per unit area per time) depends on the diffusion of the component in the core medium, as well as through the shell and in the continuous phase. Characterization of colloidal shell permeability must distinguish then between these different regions.

The driving force to transport in these systems is the effective concentration gradient between the core and continuous phase. When the concentration in the continuous phase reaches the solubility limit, transport is arrested. However, if the continuous phase is kept at 'infinite sink' (high dilution) conditions, solubility should not affect transport. Indeed, Simovic and Prestidge [69] found significant differences between systems kept at 'infinite sink' conditions and those that were not. Binks, et al. [68] examined the transport of a hydrophobic components solubilized in the dispersed core of oil in water emulsions , finding that the transport rate of components with higher water solubility was faster than that of components with lower water solubility as expected. In contrast, the rate of release of hydrophilic components from the aqueous core of colloidosome shells did not correlate with the component solubility in water, most likely because the continuous solution was kept at 'infinite sink' conditions so that the solubility limit was not approached [62].

Another parameter that can affect the rate of transport is diffusion in the core medium. Even if the continuous phase is maintained at sink conditions, the component must transverse the core first before reaching the shell. In a study of transport through colloidosome shells, differences in the rate of transport between several components was found to arise solely from the diffusion coefficient of those components in the aqueous core phase [62].

Table 3. Effect of mixture composition on maximal two-dimensional surface coverage at air/water interfaces, in bi-disperse silica particle systems of 0.75μm and 2μm particles, i.e., size ratio of 0.375

Mixture composition (by fractional surface area of the larger particles)	Mixture composition (by number fraction of larger particles)	Maximal surface coverage
0.5	0.07	0.57-0.58
1	0.14	0.57-0.58
2	0.28	0.57-0.58
3.5	0.49	0.62-0.64
7	0.98	0.62-0.64

From [65].

Understanding the effect of shell properties on transport requires elimination of the effects of the core and continuous phases. This can be achieved by comparing release from emulsions stabilized by colloidal shells—Pickering, to release from emulsions stabilized by 'traditional' surfactants . The results are contradictory. Frelichowska, et al. [70] found that the release of caffeine from Pickering emulsions stabilized by silica particles was about twice as fast as the release from emulsions stabilized by a mix of commercial amphiphiles. Tikekar, et al. [71] found that the release of curcumin from silica-stabilized emulsions was somewhat faster than the release of that same compound from emulsions stabilized by polymeric amphiphiles. In contrast, Binks, et al. [68] found that the release rate from Pickering emulsions stabilized by silica particles was slower by an order of magnitude than the release from emulsions stabilized by a small molecule surfactant , sodium dodecyl sulfate (SDS) . It should be noted that the particles used by Binks, et al. [68] were similar in size to those used by Frelichowska, et al. [70], so that the difference could not be attributed to particle dimensions. The release rate of components from emulsions stabilized by classical amphiphiles is highly sensitive to the amphiphile properties: The permeability of small molecule surfactant films depends on their molecular structure and whether the surfactant is non-ionic or carries a charge [15]. In polymeric amphiphile films, the molecular weight and structure significantly affect permeability, while in Pickering emulsions, transport will depend on the packing density [21-26]. Thus, comparisons between amphiphile-stabilized and Pickering emulsions may be of practical interest for particular systems, but does not yield insights regarding the transport mechanism and how it can be controlled.

As noted earlier, colloidal shells are mesoporous composites of an impenetrable phase the particles, and a penetrable one the voids, or pores. As a result, colloidal shells offer the potential for size selection, where components smaller than the void size can cross the shell, but transport of components larger than the voids, or pores, is suppressed. Thompson, et al. [72] find that a small-molecule dye diffused through colloidal shells composed of relatively large particles (~1μm) at a rate that was practically identical to transport from uncoated cores. Dinsmore, et al. [47] found that colloidosome shells were impermeable to particles of the same order of magnitude as the particles composing the shell, but allowed free transport to diffusants that were an order of magnitude smaller. However, in some cases the retardation effects of colloidal shells can significantly affect transport of molecules that are much smaller than the expected pore size : for example, Zhao, et al. [73] found that the silica shells of Pickering emulsions significantly reduce the rate of peroxyl radicals , molecules that are much smaller than the pore size. Transport through colloidal shells was found to vary with the particle density and shell thickness. Simovic and Prestidge [69] found (under sink conditions) that the rate of release from Pickering emulsions (as measured by the amount released per unit time, initially) was faster for partially coated drops than for those coated by thick, multi-layered shells, while uncoated drops 'dumped' their load immediately. This is as may be expected. After a long period of time, when no more release was observed, the amounts released from the partially-coated drops were identical to those from the uncoated ones. This suggests that in either case, the entire amount of encapsulant was released. However, quite surprisingly, the amounts released from thick, multi-layered shells after a long period of time were much lower, indicating that not all the encapsulant was released. In contrast, Simovic and Prestidge [69] found that release under non-sink conditions was slowest for uncoated drops, and the amount released from those after a long period of time was lower than in the case of the coated drops.

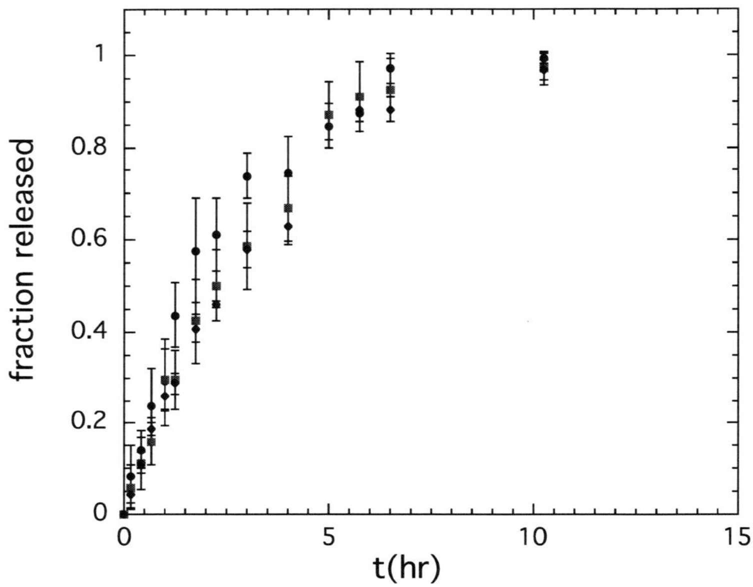

Based on data from [60].

Figure 4. Release of caffeine from colloidosomes, as a function of colloidal particle size: Circles-20nm, squares-1μm, diamonds -2.2μm.

Annealing colloidosome shells—which was thought to increase packing density—reduced the rate of release when compared to non-annealed shells [72]. Similarly, Yow and Routh [51] found that longer sintering time increased the packing of particles in colloidosome shell, significantly reducing transport .

Few studies examined, quantitatively, the effect of colloidal particle size transport through colloidal shells. The results were conflicting: In multi-layered colloidosome shells of equal thickness, the rate of transport of small molecules was found to decrease with decreasing colloidal particle size [59]. In contrast, the diffusion of small molecules through colloidal monolayers was found to be independent of the colloid size over a broad range of particle diameters [60, 61].

5. Colloidal Shell Formation

The driving force for colloidal aggregation at fluid interfaces was elucidated by Finkle et al. [19], who noted that hydrophobic and hydrophilic liquids wet colloidal particles differently, as measured by differences in the contact angle. The result is an asymmetry that can cause the particles to partition to interfaces, as sketched in Figure 2. The difference in free energy per unit area, F_p/A, between a colloidal particle immersed in phase *1* and that particle at the interface between phase *1* and phase *2*is given by [15]

$$\frac{F_p}{A} = \gamma_{12} + \gamma_{c1} - \gamma_{c2} - \sin^2\theta(\gamma_{c1} - \gamma_{c2}) = \gamma_{12} + \cos^2\theta(\gamma_{c1} - \gamma_{c2}) \qquad (1)$$

where θ is the contact angle between phase *1* and the particle, and γ_{12} the interfacial energy between phase *1* and phase *2*.γ_{c1} and γ_{c2} are the interfacial energies between the colloidal particle and phase *1* and *2*, respectively. F_p/A can be thought of as the 'adsorption energy' of a particle at the liquid/liquid interface : If it is positive, particles would favor the interface, while anegative value indicates that particles favor immersion in one phase. Applying Young's equationleads to the following relationship [15, 75]

$$\frac{F_p}{A} = 4\gamma_{2c} - \gamma_{12}\left(1-\cos\theta\right)^2 \tag{2}$$

Although Pickering emulsions may be stable for very long periods of time, they are not a true equilibrium system like microemulsions stabilized by molecular amphiphiles. Kralchevsky, et al. [75] developed a detailed model that included the effect of emulsion drop size on the assembly of particles at the liquid/liquid interface . Taking the particles to be initially solubilized in phase *2*, the energy to form the emulsion drops was calculatedto be

$$F \approx 4\pi R^2 \gamma_{12} N_d \left\{1 - N_p \frac{\alpha^2}{4R^2}\left(1-\cos\theta\right)^2\right\} \tag{3}$$

where α is the effective contact angle between the particles and the interface (perturbed from the preferred value θ due to interfacial curvature), a is the radius of the drop. N_d is the number of drops, and N_p the number of particles per drop, so that N_dN_p is the total number of particles adsorbed at the interface. F is calculated relative to the case where there is no liquid/liquid interface and all particles are immersed in phase *2*. Thus, if no drops are formed ($N_d=0$), $F=0$. If no particles adsorb at the interface, F is equal to the mechanical energy required to form N_d drops of size R.

Kralchevsky, et al. [75] considered two cases that differ both in their geometry and in their interfacial thermodynamics ; One where the phase the particles are immersed in, phase *2*, is the continuous phase ('Bancroft') and one where this phase is the dispersed phase ('anti-Bancroft'). The difference between the two cases are expressed via the definition of α, as shown in Figure 2. In both cases the drop size was found to vary linearly with the work of emulsification, namely, the energy of mixing , except for at one critical angle ($\theta \approx 93^0$) [75]. Catastrophic inversion was predicted to occur when the volumes of the two phases were equal. This prediction contradicts findings such as Binks and Lumsdon's [27] that in some systems changing particle size could cause phase inversion .

In a subsequent paper, Golemanov, et al. [76] incorporated particle concentration into the calculations for the drop size of anti-Bancroft emulsions , namely ones where the particles are immersed in the dispersed phase. Assuming that all particles adsorb at the interface , the drop size was predicted to decrease with particle concentration until reaching a limiting, smallest drop size [76]. This limit value was taken to be set by the mixing energy , a function of the mechanical energy input (per unit time and volume), γ_{12}, and the viscosity and density of the continuous phase [77].

It should be noted, however, that measurements show that a fraction of the particles remains in solution , even when the overall concentration of particles is low so that only partial coverage is expected [8, 22], thus invalidating the basic model assumption.

The maximal packing density of particles at interfaces, ϕ_{max} is a critical parameter that sets, to a large degree, the properties of colloidal shells. In randomly packed, monodisperse and uncharged hard spheres, ϕ_{max} ranges between *0.5* and *0.65* [78] and is independent of the particle dimensions. Thus, the shells of Pickering emulsions formed by uncharged particles in the limit of excess particle concentration are expected to fall in this range. Similar densities are expected in colloidosome shells formed via double emulsions [54], where evaporation of the middle phase enforces close-packing.

In charged particle systems, the long range of electrostatic forces is expected to reduce the maximal packing density in colloidal shells, either Pickering emulsions or colloidosomes. ϕ_{max} can be estimated by using an effective particle radius that accounts for the range of electrostatic interactions [79-81]:

$$\phi_{\max} \approx \phi^* \frac{R^2}{R_{eff}^2} \tag{4}$$

where ϕ^* is the packing limit of hard, uncharged spheres (of order *0.6)* and R is the particle radius . The effective radius, R_{eff} accounts for the effective particle radius, can be approximated by $R_{eff}=R+1/\kappa$, where $1/\kappa$ is the Debye screening length, although the exact form is more complex, dependingnon-linearly on $1/\kappa$ and the Bjerrum length [82]. In the limit where electrostatic interactions are negligible due to screening by high salt concentration s, $1/\kappa$ is small and does not affect the packing density. However, in systems where salinity is low $1/\kappa$ can greatly increase the effective particle radius, thereby reducing the maximal packing density significantly. This is in agreement with the findings of Binks and Lumsdon [27], who found that increasing the salt concentration, which reduces R_{eff}, had a similar effect on Pickering emulsions formed by charged latex particles as decreasing the particle size.

In either uncharged or charged spherical, monodisperseparticles, the maximal packing density in the shell is capped by the random close-packing limit [78]. The density at interfaces may be increased by varying the particle anisotropy : Both the maximal random close packing and the surface density of ellipsoidal particles were found to reach a maximum at an aspect ratio of order *2* [87, 88]. The values at this aspect ratio were approximately 10% higher than that of spheres [87, 88, 89]. However, it should be noted that increasing the aspect ratio beyond 2 led to a significant decrease in the packing density and surface coverage, below that of spheres. The packing density can also be increased, quite significantly, by mixing particles of different sizes, or using polydisperse colloidal suspensions [81, 90].

6. Models for Transport through Colloidal Shells

Colloidal shells, whether those of Pickering emulsions or of colloidosomes, separate an inner core volume from a continuous phase. Therefore, transport in such systems occurs in three corresponding regions: The core, the shell and the outer phase. Under sink conditions, transport in the outer, continuous phase can be neglected. Analysis must therefore consider the diffusion of the encapsulant in the core and the finite amount of diffusant in the core, as well asthe permeability of the shell (a function of the colloidal particle size, packing density, shell thickness).

Several studies simplified the system by assuming that the concentration in the core varies only slowly with time (pseudo-steady state). The flux j is then given by [51]

$$j = -4\pi D_S c \frac{a+h}{h} \tag{5}$$

where c here is the average diffusant concentration in the core, D_S the diffusion coefficient through the shell, a the core radius , and h the shell thickness. In the limit where the shell thickness is much smaller than the core radius, the concentration in the core is predicted to decreaseexponentially with time. The fraction of component in the core is then

$$f(t) = \frac{M(t)}{M_0} = 1 - \exp\left(\frac{-3tD_S\{a+h\}}{hR^2}\right) \tag{6}$$

where $M(t)$ denote the amount of diffusant (in whichever units appropriate) at time t, and M_0 is the initial amount of diffusant. (Equation 6 may be inverted to yield the fraction of overall diffusant *entering* a spherical core.)

The release of components from colloidosome shells was indeed found to follow the exponential function predicted by Equation (6) [51, 83]. Fitting the release profile also enables estimating D_S, and therefore shell permeability . However, its application poses some issues . First, by assuming that the core is well mixed, it neglects the development of concentration gradients in the core that arise from diffusion in the core medium. While this assumption may be correct for small, nano-scale drops, it can fail in larger micro-scale ones. For example, differences in the rate of release of caffeine , aspirin, theobromine and theophilene from colloidosomes was found to be due solely to the differences in their diffusion coefficient in the aqueous core, rather than to transport through the shell (or solubility) [62]. A potentially more significant limitation is the fact that this approach is descriptive rather than predictive: It cannot directly provide correlation between parameters such as colloidal particle size and the shell permeability.

The effect of shell parameters on transport was taken into account in a model that accounts for the mesoporous structure of colloidal shells [61, 84]. The model includes three dominant retardation effects:

1 The finite size of the voids, or pores, between the colloidal particles. This induces size selectivity , where diffusants with a dimension R_D that is larger than the characteristic size of the pore R_p cannot cross the shell. However, diffusants whose size is smaller than R_p can cross freely. The pore radius , R_p depends on the particle radius b and the packing volume fraction, approximated by $b\{(\pi/6\phi)^{1/3}-1\}$ [85].

2 A reduction in the *surface area* available for transport due to the impenetrable phase. This reduction depends not only on the overall fraction of the impenetrable surface, but also on the size and distribution of the penetrable phase [86].
3 The composite-like structure of the shell, where thetortousity of the diffusive path reduces the rate of transport [82].

Combining these contributions yields an effective permeability . For systems where the diffusant is much smaller than the pore dimensions [61, 84]

$$L=\frac{RD_S}{hD}=\frac{\pi b(1-\phi)}{2\alpha R^2 h\phi} \tag{7}$$

where D is the diffusion coefficient through the voids phase, b the colloidal particle radius , and ϕ the volume fraction of the particles in the shell. α accounts for the pore structure , and is of order $1/(1+\phi)$ for highly packed layers [82].

For uncoated cores where $\phi=0$ and there is no barrier to transport , L is infinite. In the limit where there are no pores, or voids so that $\phi\approx 1$ the permeability $L=0$ and transport is inhibited.

It is interesting to note that in shells that are monolayer thick so that $h=2b$, the permeability is predicted to be independent of particle size, in agreement with experimental observations [60-62]. In the case of shells of a given thickness, the permeability decreases linearly with particle size (except for, indirectly, through the packing density), in agreement with the experiments of Kim, et al. [59].

The amount of component released, as a function of time, is given then by

$$f(t)=\frac{M(t)}{M_\infty}=1-\frac{\sum_{n=1}^{\infty}\frac{e^{-D\beta_n^2 t/a^2}}{\beta_n^2\left(\beta_n^2+L^2-L\right)}}{\sum_{n=1}^{\infty}\frac{1}{\beta_n^2\left(\beta_n^2+L^2-L\right)}} \tag{8}$$

where the eigenvalues β_n are given by the roots of the expression [82]

$$\beta_n \cot\beta_n + L - 1 = 0 \tag{9}$$

The exponential form is similar to that of Equation (6); However, rather than lump all the effects of the shell into one, undetermined parameter D_S, this model provides a correlation between the mesoporous shell parameters and transport .

In Figure 5 the predictions of Equations (7) for the release of a compound through a colloidal shell is plotted. As may be expected, the initial rate of release is lower for thick shells than for thin ones, and for dense shells when compared to partially coated drops. This is in qualitative agreement with the measurements of Simovic and Prestidge [69].

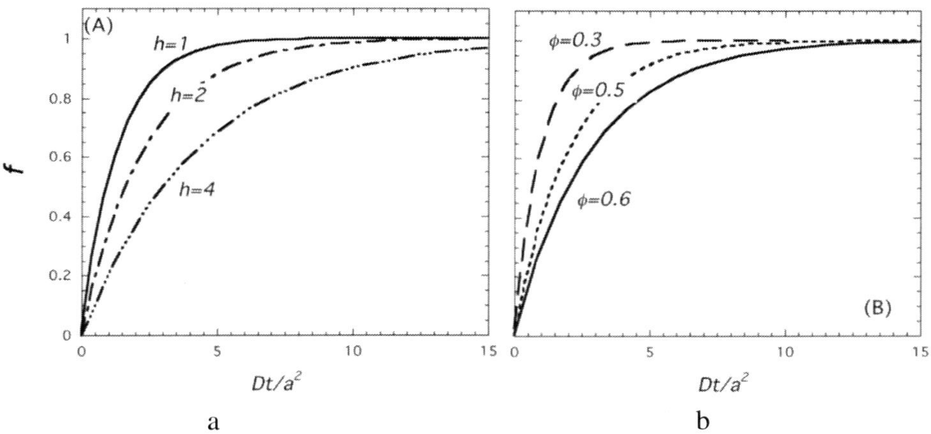

Figure 5. Effect of system parameters on the release of components through colloidal shells, based on Equation (7). *f* is the fractional amount released from the core. (A) The release as a function of shell thickness*h*, in systems where the colloidal particle size and packing density are fixed. We see that increasing the thickness reduces the rate of release. (B) Effect of the packing density ϕ on release, for shells with the same thickness and particle size. Increasing the packing density reduces the rate of release. However, even in the limit of the random close-packing density (*~0.6)* the rate of release is non-zero. Dt/a^2 is the dimensionless time, where *D* is the diffusion coefficient in the core, *t* time and *a* the core radius .

CONCLUSION

The ability of colloidal particles to aggregate at liquid/liquid interfaces is somewhat unexpected. Yet, despite the homogeneity of the colloidal particle surfaces, they can partition to hydrophobic/hydrophilic interfaces, driven by differences in the wetting contact angle between the phases [19]. As a result, colloidal particles can form interfacial films , shells, in a manner similar to that of amphiphilic molecules : in mixtures of hydrophilic and hydrophobic phases, they can stabilize Pickering emulsions [8, 9], while in aqueous phases they can form colloidosomes [13, 14]. The characteristics of Pickering emulsions are set by several parameters: The particle surface hydrophobicity generally determines the type of emulsion (oil in water or water in oil) [8, 9], although, quite surprisingly, in some systems changing the particle size can cause an inversion [27]. The drop size is sensitive to the concentration of particles: In low particle concentrations, the drops are only partially covered, and coalescence is likely to occur unless some additional steric stabilization (e.g., via electrostatics) is present [22]. As the concentration of particles increases, so does surface coverage, until reaching close-packing. The size of the emulsion drops decreases linearly with particle concentration [21, 23, 28, 31], until reaching a lower limit below whichit remains constant. Due to their homogeneous surface, colloidal particles cannot self-assemble into shell structures in a single phase. However, methods utilizing Pickering emulsions can yield colloidosomes, where an inner aqueous core is separated by a colloidal shell from a continuous aqueous solution [46, 47]. Typically, particles are locked in place by sintering , fusion , or binding molecules such as polymers [13, 14].

Unfortunately, using colloidosomes formed in this way for encapsulation applications is hindered by the need to remove the hydrophobic phase and the particle 'locking' step, processes that typically involve harsh conditions that can degrade encapsulants.

These issues have led to the development of another methodology for colloidosome synthesis , where the colloidal shells are assembled on the surface of a microgel core [55-62]. The colloids are held in place by specific interactions (usually electrostatic) between the particles and the gel . The resulting shells are monolayer thick, but partial collapse of the core can lead to thicker, multi-layered structures [59].

In either Pickering emulsions or colloidosome shells, the packing density is limited by the random close packing limit for hard spheres. In systems of charged particles, this is further reduced by repulsive electrostatic interactions between the particles [27]. The packing density of anisotropic particles can exceed this limit [87, 88], although it should be noted that the anisotropy should be moderate, close to the optimal ratio *(1.6-2)*: The packing density of highly anisotropic particles is much lower than that of spherical particles. Mixing particles of different sizes can also increase the packing density [81], depending on the size ratio and mix composition .

Interest in Pickering emulsions and colloidosomes is largely driven by the selectivity of the colloidal shells. The unique structure of the shell—a mesoporous composite of an impenetrable phase (the particles) and penetrable voids, allows a degree of control that is not available in other types of surface films . Components larger than the void, or pore, size are hindered from crossing the shell, while those smaller than the pore size can diffuse through it [47]. However, the transport rate of the small molecules is sensitive to the properties of the shell, and can be greatly reduced when compared to transport through uncoated interfaces.

The most significant parameter controlling the rate of transport through colloidal shells—whether those of Pickering emulsions or colloidosomes—was found to be the packing density of the particles [51, 69, 72]. Other parameters, such as the shell thickness, play a lesser role [69]. In multi-layered shellsthe rate of transport was found to decrease with decreasing colloidal particle size [59], while in monolayers particle size did not seem to affect the rate [60, 61]. These findings are consistent with the predictions of a transport model, based on two-film theory [61, 84] that accounts for the composite nature of the mesoporous colloidal shells.

In summary, colloidal shells assembled at liquid/liquid interfaces provide a unique tool for the control of transport into or out of micro or nano scale droplets, with applications to different technologies. Unlike amphiphilic films , the mesoporous nature of colloidal shells allows size-selectivity , as well as the ability to finely control transport rate through accessible system parameters.

REFERENCES

[1] Stuart, M. A. C., Huck, W. T. S., Genzer, J., Muller, M., Ober, C., Stamm, M., Sukhorukov, G. B., Szleifer, I., Tsukruk, V. V., Urban, M., Winnik, F., Zauscher, S., Luzinov, I., Minko, S., Emerging applications of stimuli-responsive polymer materials , *Nature Materials* 2010, 9, 101-113.

[2] Ma, G., Su, Z. G. (Eds.) *Microspheres and Microcapsules in Biotechnology: Design, Preparation and Applications*, Pan Stanford Publishing, 2013.

[3] De Geest, B. G., Sanders, N. N., Sukhorukov, G. B., Demeester, J., De Smedt, S. C., Release mechanisms for polyelectrolyte capsules, *Chem. Soc. Rev.* 2007, 36, 636-649.

[4] Guzey, D., McClements, D. J., Formation, stability and properties of multilayer emulsions for application in the food industry , *Adv. Coll. Int. Sci.* 2006, 128, 227-248.

[5] Emulsions and Emulsion Stability: Surfactant Science Series/61, *Sjoblom J.* (Ed.), 2005, CRC Press.

[6] Torchilin, V. P., Recent advances with liposomes as pharmaceutical carriers, *Nature Rev. Drug Discovery* 2005, 4, 145-160.

[7] Discher, D. E., Eisenberg, A., Polymer vesicles, *Science* 2002, 297, 967-973.

[8] Aveyard, R., Binks, B. P., Clint, J. H., Emulsions stabilised solely by colloidal particles, *Adv. Coll. Int. Sci.* 2003, 100, 503–546.

[9] Dickinson, E., Food emulsions and foams : Stabilization by particles, *Curr. Op. Coll. Int. Sci.* 2010, 15, 40-49.

[10] Zeng, C., Bissig, H., Dinsmore, A. D., Particles on droplets: From fundamental physics to novel materials , *Solid St. Comm.* 2006, 139, 547-556.

[11] Binks, B. P., Particles as surfactants - similarities and differences, *Curr. Op. Coll. Int. Sci.,* 2002, 7, 21-41.

[12] Rousseau, D., Trends in structuring edible emulsions with Pickering fat crystals , *Curr. Op. Cool Int. Sci.* 2013, 18, 283-291.

[13] Dan, N., Transport through self-assembled colloidal shells (colloidosomes), *Curr. Op. Coll. Int. Sci.* 2012, 17, 141-146.

[14] Shilpi, S., Jain, A., Gupta, Y., Jain, S. K., Colloidosomes: An emerging vesicular system in drug delivery , *Crit. Rev. Therap. Drug Carr. Sys.* 2007, 4, 361-391.

[15] Israelachvili, J., *Intermolecular and Surface Science*, 1992, Academic Press.

[16] Kumar, A., Park, B. J., Tu, F. Q., Lee, D., Amphiphilic Janus particles at fluid interfaces, *Soft Matter.* 2013, 9, 6604-6617.

[17] Ramsden, W., Separation of Solids in the Surface-Layers of Solutions and 'suspensions ', *Proc. Roy. Soc.* 1903, 72, 156-164.

[18] Pickering, S. U., Emulsions, *J. Chem. Soc.* 1907, 91, 2001-2012.

[19] Finkle, P., Draper, H. D., Hildebrand, J. H., The theory of emulsification, *J. Am. Chem. Soc.* 1923, 45, 2780-2788.

[20] Schulman, J. H., Leja, J., Control of contact angles at the oil -water -solid interfaces. Emulsions stabilized by solid particles (BaSO4), *Trans. Faraday Soc.* 1954, 50, 598-605.

[21] Binks, B. P., Whitby, C. P., Silica particle-stabilized emulsions of silicone oil and water : Aspects of emulsification, *Langmuir* 2004, 20, 1130-1137.

[22] Levine, S., Bowen, B. D., Partridge, S. J. Stabilization of emulsions by fine particles I. Partitioning of particles between continuous phase and oil /water interface . *Colloids Surf.* 1989, 38, 325–343.

[23] Arditty, S., Schmitt, V., Giermanska-Kahn, J., Leal-Calderon, F. Materials based onsolid-stabilized emulsions . *J. Colloid Interface Sci.* 2004, 275, 659–664.

[24] Horozov, T. S., Binks, B. P. Particle-stabilized emulsions : a bilayer or a bridging monolayer ? *Angew. Chem. Int. Ed.* 2006, 45, 773–776.

[25] Vignati, E., Piazza, R., Lockhart, T. P. Pickering emulsions : interfacial tension , colloidal layer morphology , and trapped-particle motion. *Langmuir* 2003, 19, 6650–6656.

[26] Gupta, R., Rousseau, D. Surface-active solid lipid nanoparticles as Pickering stabilizers for oil -in-water emulsions . *Food Funct*. 2012, 3, 302–311.

[27] Binks, B. P., Lumsdon, S. O., Pickering Emulsions Stabilized by Monodisperse Latex Particles: Effects of Particle Size, *Langmuir* 2001, 17, 4540-4547.

[28] Kalashnikova, I., Bizot, H., Cathala, B., Capron, I., New Pickering emulsions stabilized by bacterial cellulose nanocrystals . *Langmuir* 2011, 27, 7471–7479.

[29] Li, C., Sun , P. D., Yang, C., Emulsionsstabilizedbystarch nanocrystals . *Starch/Starke* 64, 2012, 497–502.

[30] Tzoumaki, M. V., Moschakis, T., Kiosseoglou, V., Biliaderis, C. G., Oil-in- water emulsions stabilized by chitin nanocrystal particles. *Food Hydrocolloids* 25, 2011, 1521–1529.

[31] Kim, J., Cote, L. J., Kim, F., Yuan, W., Shull, K. R., Huang, J. X., Graphene Oxide Sheets at Interfaces, *JACS* 2010, 132, 8180-8186.

[32] Kalashnikova, I., Bizot, H., Bertoncini, P., Cathala, B., Capron, I., Cellulosic nanorods of various aspect ratios for oil in water Pickering emulsions , *Soft Matt*. 2013, 9, 952-959.

[33] Guevara, J. S., Mejia, A. F., Shuai, M., Chang, Y. W., Mannan, M. S., Cheng, Z. D., Stabilization of Pickering foams by high-aspect-ratio nano-sheets, *Soft Matt*. 2013, 9, 1327-1336.

[34] Priest, C., Reid, M. D., Whitby, C. P., Formation and stability of nanoparticle-stabilised oil -in-water emulsions in a microfluidic chip, *J. Coll. Int. Sci*. 2011, 363, 301–306.

[35] Madivala, B., Vandebril, S., Fransaer, J., Vermant, J., Exploiting particle shape in solid stabilized emulsions , *Soft Matter* 2009, 5, 1717-1727.

[36] Kaiser, A., Liu, T. T., Richtering, W., Schmidt, A. M., Magnetic Capsules and Pickering Emulsions Stabilized by Core-Shell Particles, *Langmuir* 2009, 25, 7335-7341.

[37] Arditty, S., Schmitt, V., Lequeux, F., Leal-Calderon, F., Interfacial properties in solid-stabilized emulsions , *Euro. Phys. J. B* 2005, 44, 381-393,

[38] Kohler, K., Santana, A. S., Braisch, B., Preis, R., Schuchmann, H. P., High pressure emulsification with nano-particles as stabilizing agents, *Chem. Eng. Sci*. 2010, 65, 2957-2964.

[39] Omi, S., Preparation of monodisperse microspheres using the Shirasu porous glass emulsification technique, *Colloids Surf. A* 1996, 109, 97-107.

[40] Umbanhowar, P. B., Prasad, V., Weitz, D. A., Monodisperse emulsion generation via drop break off in a coflowing stream, *Langmuir* 2000, 16, 347-351.

[41] Liu, X. Q., Nakajima, M., Nebetani, H., Xu, Q. Y., Ichikawa, S., Sano, Y., Stability characteristics of dispersed oil droplets prepared by the microchannel emulsification method, *J. Coll. Int. Sci*. 2001, 233, 23-30.

[42] Yuan, Q. C., Cayre, O. J., Manga, M., Williams, R. A., Biggs, S., Preparation of particle-stabilized emulsions using membrane emulsification, *Soft Matter* 2010, 6, 1580-1588.

[43] Nie, Z., Park, J. I., Li, W., Bon, S. A. F., Kumacheva, E., An "Inside-Out" Microfluidic Approach to Monodisperse Emulsions Stabilized by Solid Particles, *J. Am. Chem. Soc.* 2008, 130, 16508–16509.

[44] Subramaniam, A. B., Abkarian, M., Stone, H. A., Controlled assembly of jammed colloidal shells on fluid droplets, *Nature Mater.* 2005, 4, 553-556.

[45] Xu, Q., Nakajima, M., Binks, B. P., Preparation of particle-stabilized oil -in-water emulsions with the microchannel emulsification method, *Colloids Surf. A* 2005, 262, 94-100.

[46] Velev, O. D., Furusawa, K., Nagayama, K. Assembly of Latex Particles by using emulsion droplets as templates. 1. Microstructured hollow spheres, *Langmuir* 1996, 12, 2385-2391.

[47] Dinsmore, A. D., Hsu, M. F., Nikolaides, M. G., Marquez, M., Bausch, A. R., Weitz, D. A. Colloidosomes: selectively permeable capsules composed of colloidal particles, *Science* 2002, 298, 1006-1009.

[48] Croll, L. M., Stöver, H. D. H. Formation of tectocapsules by assembly and cross-linking of poly(divinylbenzene-alt-maleic anhydride) spheres at the oil –waterinterface, *Langmuir* 2003, 19, 5918–5922.

[49] He, X. D., Ge, X. W. W., Wang, M. Z., Zhang, Z. C. The preparation of composite micro-sphere with hollow core/porous shell structure by self-assembling of latex particles at emulsion interface , *J. Coll. And. Int. Sci.* 2006, 299, 791–796.

[50] Croll, L. M., Stöver, H. D. H. Mechanism of self-assembly and rupture of cross- linked microspheres and microgels at the oil –water interface , *Langmuir* 2003, 19, 10077–10080.

[51] Yow, H. N., Routh, A. F. Release Profiles of Encapsulated Actives from Colloidosomes Sintered for Various Durations, *Langmuir* 2009, 25, 159-166.

[52] Velev, O. D., Nagayama, K. Assembly of latex particles by using emulsion droplets .3. Reverse (water in oil) system, *Langmuir* 1997, 13, 1856-1859.

[53] Hsu, M. F., Nikolaides, M. G., Dinsmore, A. D., Bausch, A. R., Gordon, V. D., Chen, X., Hutchinson, J. W., Weitz, D. A., Marquez, M. Self-assembled shells composed of colloidal particles, Fabrication and characterization, *Langmuir* 2005, 21, 2963-2970.

[54] Lee, D., Weitz, D. A. Double emulsion-templatednanoparticle colloidosomes with selective permeability , *Adv. Mat.* 2008, 20, 3498–3503.

[55] Ngai, T., Behrens, S., Auweter, H. Novel emulsions stabilized by pH and temperature sensitive microgels , *Chem. Comm.* 2005, 3, 331–333.

[56] Cejkova, J., Hanus, J., Stepanek, F. Investigation of internal microstructure and thermo-responsive properties of composite PNIPAM/silica microcapsules, *J. Coll. Int. Sci.* 2010, 346, 352-360.

[57] Liu, H. X., Wang, C. Y., Gao, Q. X., Chen, J. X., Ren, B. Y., Liu, X. X., Tong, Z. Facile fabrication of well-defined hydrogel beads with magnetic nanocomposite shells, *Int. J. Pharma.* 2009, 376, 92-98.

[58] Texter, J. Templatinghydrogels, *Coll. and Pol. Sci.* 2009, 287, 313-321.

[59] Kim, J. W., Fernandez-Nieves, A., Dan, N., Utada, A. S., Marquez, M., Weitz, D. A. Colloidal assembly route for responsive colloidosomes with tunable permeability , *Nano Lett.* 2007, 7, 2876-2880.

[60] Rosenberg, R. T., Dan, N. R. Controlling surface porosity and release from hydrogels using a colloidal particle coating, *J. Coll. Int. Sci.* 2010, 349,498-504.

[61] Rosenberg, R. T., Dan, N. R. Diffusion through colloidosome shells, *J. Coll. Int. Sci.* 2011, 354,478-482.

[62] Rosenberg, R. T., Dan, N. Self-Assembly of Colloidosome Shells on Drug-Containing Hydrogels. *J. Biomat. Nano Biotech.* 2011, 2, 1-7.

[63] Serizawa, T., Takeshita, H., Akashi, M. Electrostatic adsorption of polystyrene nanospheres onto the surface of an ultrathin polymer film prepared by using an alternate adsorption technique, *Langmuir* 1998,14,4088-4094.

[64] Hanarp, P., Sutherland, D. S., Gold, J., Kasemo, B. Influence of polydispersity on adsorption of nanoparticles , *J. Coll. Int. Sci.* 2001, 241, 26–31.

[65] Hur, J., Mahynski, N. A., Won, Y. Y. Crystallization of Bidisperse Repulsive Colloids in Two-Dimensional Space, A Study of Model Systems Constructed at the Air-Water Interface, *Langmuir* 2010,14, 11737-11749.

[66] Clarke, A. S., Wiley, J. D. Numerical Simulation of the dense random packing of a mixture of hard spheres-amorphous metals , *Phys. Rev. B* 1987, 35,7350-7356.

[67] Farr, R. S., Groot, R. D. Close packing density of polydisperse hard spheres, *J. Chem. Phys.* 2009, 131,10.1063/1.3276799.

[68] Binks, B. P., Fletcher, P. D. I., Holt, B. L., P. Beaussoubre, and K. Wong, Selective Retardation of Perfume Oil Evaporation from Oil-in-Water Emulsions Stabilized by Either Surfactant or Nanoparticles, *Langmuir* 2010, 26, 18024–18030.

[69] Simovic, S., Prestidge, C. A., Nanoparticle layers controlling drug release from emulsions , *Europ. J. Pharma and Biopharma* 2007, 67, 39–47.

[70] Frelichowska, J., Bolzinger, M. A., Valour, J. P., Mouaziz, H., Pelletier, J., Chevalier, Y., Pickering w/o emulsions : Drug release and topical delivery, *Int. J. Pharma.* 2009, 368, 7–15.

[71] Tikekar, R. V., Pan, Y., Nitin, N., Fate of curcumin encapsulated in silica nanoparticle stabilized Pickering emulsion during storage and simulated digestion , *Food Res. Int.* 2013, 51, 370–377.

[72] Thompson, K. L., Armes, S. P., Howse, J. R., Ebbens, S., Ahmad, I., Zaidi, J. H., York, D. W., Burdis, J. A. Covalently Cross-Linked Colloidosomes. *Macromolecules* 2010, 43, 10466–10474.

[73] Zhao, Y., Dan, N., Pan, Y. J., Nitin, N., Tikekar, R. V., Enhancing the barrier properties of colloidosomes using silica nanoparticle aggregates, *J. Food Eng.* 2013, 118, 421-425.

[74] Wang, H., Zhu, X., Tsarkova, L., Pich, A., Moller, M. All-Silica Colloidosomes with a Particle-Bilayer Shell, *ACS Nano* 2011, 5, 3937–3942.

[75] Kralchevsky, P. A., Ivanov, I. B., Ananthapadmanabhan, K. P., Lips, A. On the thermodynamics os particle stabilized emulsions : Curvature effects and catastrophic phase inversion . *Langmuir* 2005, 21, 50-63.

[76] Golemanov, K., Tcholakova, S., Kralchevsky, P. A., Ananthapadmanabhan KP, Lips A. Latex particle stabilized emulsions of anti-Bancroft type. *Langmuir* 2006, 22, 4968-4977.

[77] Walstra, P., Principles of emulsion formation . *Chem. Eng. Sci.* 1993, 48, 333-349.

[78] Berryman, J. G. Random close packing of hard spheres. *Phys. Rev. A* 1983, 27, 1053-1061.

[79] Adamczyk, Z., Szyk-Warszyn, L., Siwek, B., Weronski, P. Fluctuations in the number of irreversibly adsorbed particles. *J. Chem. Phys.* 2000,113, 11336-11342.

[80] Adamczyk, Z. Particle adsorption and deposition , role of electrostatic interactions. *Adv. Coll. Int. Sci.* 2003, 100, 267–347.
[81] Rosenberg, R. T., Dan, N. Effect of Colloidal Particle Size on Adsorbed Monodisperse and BidisperseMonolayers, *Langmuir* 2011, 14, 8729-8734.
[82] Crank, J., *Mathematics of Diffusion*. Second ed. New York, Oxford Science Publications, 1975.
[83] San Miguel, A., Behrens, S. H. Permeability control in stimulus -responsive colloidosomes, *Soft Matter* 2011, 7, 1948-1956.
[84] Rosenberg, R. T., Dan, N. Designing Nano-Particle Colloidal Shells for Selective Transport. *Soft Materials* 2011, 10.1080/1539445X.2011.591867.
[85] Rintoul, M. D., Torquato, S. Hard-sphere statistics along the metastable amorphous branch. *Phys. Rev. E.* 1998, 58,532-537.
[86] Berg, H. C. *Random walks in biology*. Princeton, Princeton University Press, 1983.
[87] Sacanna, S., Rossi, L., Wouterse, A., Philipse, A. P., Observation of a shape -dependent density maximum in random packings and glasses of colloidal silica ellipsoids, *J. Phys. Cond. Matt.* 2007, 19, Article Number, 376108 DOI , 10.1088/0953-8984/19/37/376108.
[88] Basavaraj, M. G., Fuller, G. G., Fransaer, J., Vermant, J., Packing, flipping, and buckling transitions in compressed monolayers of ellipsoidal latex particles, *Langmuir*, 2006, 22, 6605–6612.
[89] Dugyala, V. R., Daware, S. V., Basavaraj, M. G., Shape anisotropic colloids, synthesis , packing behavior, evaporation driven assembly, and their application in emulsion stabilization , *Soft Matt.* 2013, 9, 6711-6725.
[90] Monteux, C., Jung, E., Fuller, G. G., Mechanical properties and structure of particle coated interfaces, Influence of particle size and bidisperse 2D suspensions , *Langmuir* 2007, 23, 3975-3980.

In: Comprehensive Guide for Mesoporous Materials. Volume 1 ISBN: 978-1-63463-958-3
Editor: Mahmood Aliofkhazraei

Chapter 12

TEXTURAL, CHEMICAL AND ENTHALPIC CHARACTERIZATION OF ACTIVATED CARBON MONOLITHS FOR CO_2 ADSORPTION

Diana. P. Vargas[1], L. Giraldo[1] and J. C. Moreno-Piraján[2,*]
[1]Universidad Nacional de Colombia , Facultad de Ciencias,
Departamento de Química, Bogotá, Colombia
[2]Universidad de los Andes, Facultad de Ciencias, Departamento de Química,
Grupo de Investigación en Sólidos Porosos y Calorimetría,
Bogotá, Colombia

ABSTRACT

The development of materials with potential application in CO_2 capture is a topic of scientific interest. In this respect it was established that by adding nitrogen groups is achieved increasing adsorption of this gas. Chemical modification was performed in gas phase with ammonia and ammonium hydroxide solution at 30%, of various activated carbon monoliths obtained by activation of African palm stone with H_3PO_4, $ZnCl_2$ and $CaCl_2$ solutions. The original and modified monoliths were characterized by N_2 adsorption at 77K, infrared spectroscopy , Boehm titration and immersion calorimetry in benzene and water . Was evaluated the ability of CO_2 adsorption in a volumetric equipment at a temperature of 273 K and until 1 bar, was obtained values between 120-334 $mgCO_2g^{-1}$. The experimental results indicated that trough both methods of chemical modification used in this work, is achieved the increasing the content of nitrogen groups in the monolith and thus increases the CO_2 adsorption capacity.

Keywords: Activated Carbon Monoliths, functionalization , ammonium , CO_2 adsorption , immersion enthalpy

---

* Corresponding autor: jumoreno@uniandes.edu.co.

INTRODUCTION

The formation of carbon dioxide , CO_2 , and its accumulation in the atmosphere , are a widespread concern in most industrialized countries , mainly due to the contribution of this gas to global climate change . A variety of gases contribute to the greenhouse effect. But among them, the CO_2 has been identified as the gas individually scaled contributes to this effect. The explanation of the increase in global temperature caused by the accumulation of CO_2 in the atmosphere is simple from a physical standpoint.

Currently, the source of energy in modern societies is based on the use of fossil fuels such as: coal , oil and natural gas . Complete combustion of any of these substances generates CO_2 , but until a few decades ago, it was thought that this gas was not harmful. Indeed, CO_2 plays an important role in the carbon cycle on Earth, and is an essential component for the life cycle of animals and plants . However, the continuous and uncontrolled increase of atmospheric CO_2 concentrations, at this time disturbs the balance of the planet and contributes strongly to global climate change . Because of this was increased the concern by the development of technologies to reduce the impact of greenhouse gas es in the atmosphere [1,2].

The solid adsorbents have been seen as a possible solution alternative that allows for a suitable management of the flue gas es and reduce their emissions directly to the atmosphere . Within these solids, the activated carbon due to its high surface area , porous structure and its ability to exhibit different surface groups is one of the materials more studied to design their texture and chemical features according to the needs required for increasingly specific applications [3,4]. Table 1 presents results of CO_2 adsorption capacity for different solids at diverse temperature and pressure.

Table 1. Adsorption capacity of adsorbents for CO_2 at 100 and 1000 kPa and ambient temperature

Adsorbent	Type	CO2 capacity (mol kg-1)	T (K)	P (kPa)	References
Mg-MOF-74	MOF	8.50	298	100	[5] Bao et al.
5A (Sinopec)	Zeolite	4.55	298	100	[6] Saha et al.
13X (Sinopec)	Zeolite	3.30	298	100	[5] Bao et al.
Natural chabazite	Zeolite	3.30	303	100	[7] Jensen et al.
Na-Beta	Zeolite	2.70	303	100	[8] Xu et al.
TRI-PE-MCM-41	Amine-silica	2.50	298	100	[9] Belmabkhout et al.
PCB, Calgon Corp.	AC	2.41	296	100	[10] Ritter and Yang
Norit RB1 extra	AC	2.20	298	100	[11] Dreisbach et al.
MSC-3K-161	CMS	2.15	298	100	[12] Watson et al.
PET- DC-0	AC/CMS	2.08	298	100	[13] Cansado et al.
Sutcliffe Speakman carbon	AC	1.83	298	100	[14] Esteves et al.
CORK-DC-0	AC/CMS	1.81	298	100	[13] Cansado et al.
H-Beta	Zeolite	1.70	303	100	[8] Xu et al.

Adsorbent	Type	CO2 capacity (mol kg-1)	T (K)	P (kPa)	References
Beta	Zeolite	1.67	308	100	[15] Huang et al.
MOF-177	MOF	1.59	298	100	[6] Saha et al.
MSC-3K-162	CMS	1.30	303	100	[16] Bae and Lee
ZIF-100	ZIF	1.05	298	100	[17] Wang et al.
MOF-5	MOF	0.91	298	100	[6] Saha et al.
Cu-MOF	MOF	0.65	298	100	[18] Bao et al.
Maxsorb	AC	13	298		[19] Himeno et al.
Sutcliffe Speakman carbon	AC	8.01	298	1000	[20] Esteves et al.
Norit RB1 extra	AC	7.60	298	1000	[11] Dreisbach et al.
PCB, Calgon Corp.	AC	7.20	296	1000	[21] Ritter and Yang
13X	Zeolite	6.40	298	1000	[22] Cavenati et al.
1C' –Li	MOF	5.30	298	1000	[23] Bae et al.
2L' -Li	MOF	4.60	298	1000	[23] Bae et al.
MSC-3K-161	CMS	4.03	298	1000	[24] Watson et al.
Natural chabazite	Zeolite	3.95	303	1000	[25] Watson et al.
MSC-3K-162	CMS	3.20	303	1000	[16] Bae and Lee
H-Mordenite	Zeolite	3.10	293	1000	[26] Delgado et al.
Beta	Zeolite	3.05	308	1000	[15] Huang et al.
Na-Mordenite	Zeolite	3.00	293	1000	[26] Delgado et al.
1M'-Li	MOF	2.30	298	1000	[23] Bae et al.

Traditionally it has been considered that the porosity and surface area were the parameters that define the quality of the activated carbon , but today it is known that surface chemistry plays an important role in its adsorption capacity. Actually the porosity is the necessary conditions for an activated carbon can perform its mission of adsorb, but in many cases the adsorption is not possible if the surface is not modified properly to interact with the compound that desired adsorb. In the case of the CO_2 adsorption, several studies have that established by incorporating nitrogen groups is achieved to increase the selectivity of the surface of the carbonaceous materials for this molecule [27-29].

Considering the above, in this work were prepared, characterized and modified activated carbon monoliths for the carbon dioxide adsorption . African palm stone (*Elaeis guineensis*) was used as lignocellulosic precursor which was activated with three types of agents: H_3PO_4, $ZnCl_2$ and $CaCl_2$ at different concentrations. All materials were characterized by physical adsorption of nitrogen at 77 K, infrared spectroscopy , Boehm titration s and immersion calorimetry in benzene and water . Were selected in each series activated carbons with the best textural character istics and these were subjected to a modification process in gas phase with ammonia and in liquid phase with an ammonium hydroxide solution . Finally was studied in the original and modified monoliths the CO_2 adsorption capacity at 273 K and until 1 bar.

Preparation and Characterization of Activated Carbon Monoliths

Figure 1 shows the procedure diagram used in the preparation of activated carbon monoliths.

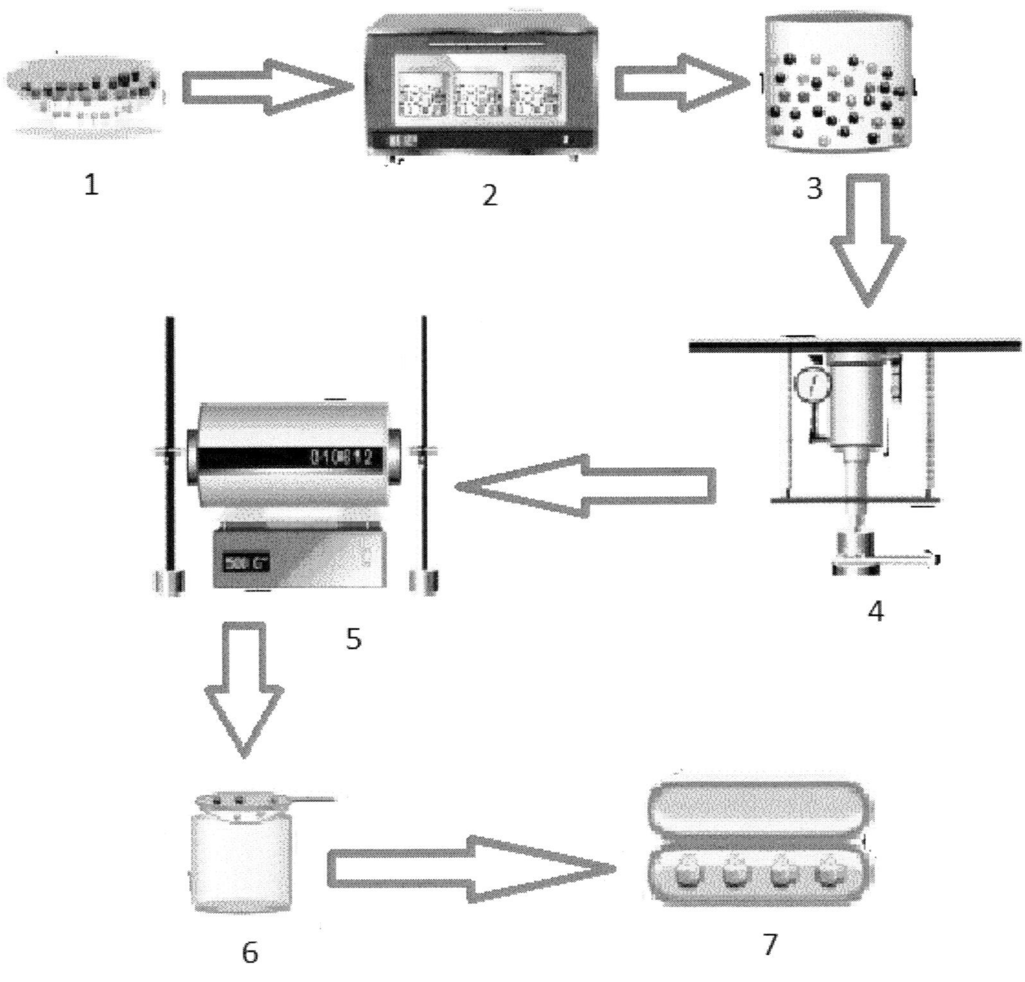

Figure 1. Diagram of activated carbon monoliths preparation .

The steps are:

1. The palm stone are milled and sieved
2. The precursor are mixed with solutions the activating agents is kept at temperature constant for a given time
3. The impregnated material are dried
4. The impregnated material is compacted under pressure
5. The heat treatment is performed under N_2 atmosphere
6. Activated carbon monoliths are washed

7. Activated carbon monoliths

Preparation of Materials

Activated carbon monoliths were prepared using as precursor lignocellulosic African Palm stone, which was carried at 4 mm in size. The impregnation was conducted in relation 1g of precursor material per 2 mL of H_3PO_4 (MP Series), $ZnCl_2$ (Series MZn) and $CaCl_2$ (Series MCa) solutions at different concentrations, these abbreviations are used throughout the written to name the samples, followed by concentration of the activating agent used. The preparation of honeycomb monoliths was performed using a compacting process in which the impregnated precursor was put in a uniaxial press and subjected to a pressure of 4500 psi and a temperature of 423 K. The monoliths obtained have an outer diameter of about 1.6 cm, a height of 8 mm, and are traversed longitudinally by six parallel channels of 3 mm each, in Figure 2 a scheme with the dimensions of the monoliths are shown.

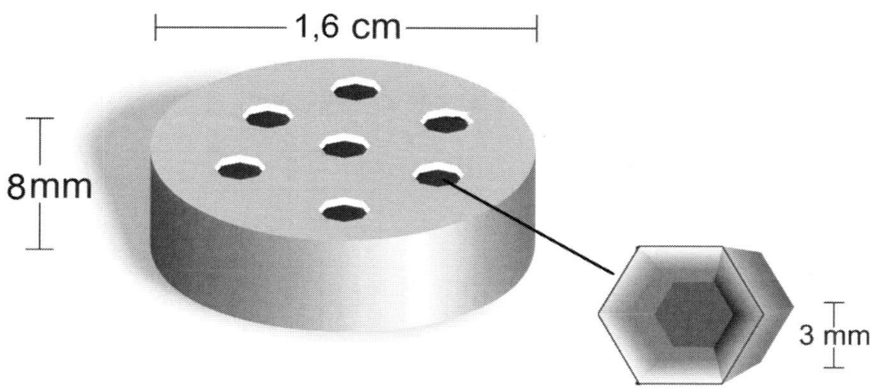

Figure 2. Dimensions of activated carbon monoliths prepared.

Tables 2-4 show the preparation conditions used for each series of activated carbon monoliths, MP, MZn and MCa respectively.

Table 2. Preparation conditions of activated carbon monoliths activated with H_3PO_4 solutions

Parameter	Activated carbón monoliths MP
Activating agent	H_3PO_4
Concentrations	32-36-40-48% p/v
Impregnation	358 K, 2 hours
Drying	393 K, 6 hours
Carbonization	723 K, 2 hours N_2 (80 mLmin^{-1}) 1 K min^{-1}
Washing	Hot distilled water until neutral pH
Storage	Plastic containers N_2 atmosphere

Table 3. Preparation conditions of activated carbon monoliths activated with $ZnCl_2$ solutions

Parameter	Activated carbón monoliths MZn
Activating agent	$ZnCl_2$
Concentrations	32-36-40-48% p/v
Impregnation	358 K, 7 hours
Drying	393 K, 6 hours
Carbonization	773 K, 2 hours N_2 (80 $mLmin^{-1}$) 1 K min^{-1}
Washing	Hot HCl 0.01M after Hot distilled water until neutral pH
Storage	Plastic containers N_2 atmosphere

Table 4. Preparation conditions of activated carbon monoliths activated with $CaCl_2$ solutions

Parameter	Activated carbón monoliths MZn
Activating agent	$CaCl_2$
Concentrations	2-3-5-7 % p/v
Impregnation	358 K, 7 hours
Drying	393 K, 6 hours
Carbonization	1073 K, 2 hours, 6 hours CO_2 (100 $mLmin^{-1}$) 3 K min^{-1} N_2, 873 K, 2 hours
Washing	Hot HCl 0.01M after Hot distilled water until neutral pH
Storage	Plastic containers N_2 atmosphere

Functionalization

Based on the results of the characterization that was performed on all activated carbon monoliths obtained and that are shown in a previous work [30], were chosen the materials higher CO_2 adsorption capacity in each series and are subjected to the processes described in Table 5.

Functionalized samples are identified along written with the same conventions used in each series, followed by the acronym: *FAL* (functionalization with ammonium hydroxide) and *FAG* (functionalization with gas ammonia) as appropriate.

Table 5. Preparation conditions of the modified samples

Conditions	Impregnation method	Gasification method
Reagent	NH_4OH	NH_3
Concentration	28-30%	99%
Temperature	353 K	1073 K
Duration	24 hours	5 hours
Washing	Distilled water u ntil constant pH	Do not make that
Drying	383 K	Do not make that
Storage	Plastic containers N_2 atmosphere	Plastic containers N_2 atmosphere

Characterization

Gas Adsorption

The textural properties of the prepared monoliths were analysed using N_2 adsorption measurements at 77 K in a volumetric system (Quantachrome, Autosorb 3-B). Before the adsorption experiments, samples were submitted to an outgassing treatment at 523 K for 24h. The volume of micropores, V_0 (N_2) and the volume of narrow micropores ($V_n(CO_2)$; pores <0.7 nm) was obtained by application of the Dubinin-Radushkevich equation to the nitrogen adsorption data (N_2 liquid density = 0.808 gcm^{-1}) and dioxide carbon adsorption data ($C0_2$ (273 K) liquid density = 1.023 gcm^{-1}), respectively. The total pore volume, V_t, was obtained from the amount adsorbed at a relative pressure P/P^0 of 0.99 while the mesopore volume, V_{meso}, was obtained from the difference between the total pore volume and the micropore volume [31].

Infrared Spectroscopy

We also performed an analysis of infrared spectroscopy for this purpose 0.1 g of the activated carbon s were ground and mixed with potassium bromide (to remove dispersers effects of large crystals). This powder mixture is compressed into a mechanical press to form a translucent disc through which can pass the light ray spectrometer. Then we read on a Thermo-Nicolet 6700 DRIFTS [32].

Boehm Titrations

We determined the content of surface groups of the activated carbon s by Boehm titration s, for this analysis about of 100 mg adsorbent are contacted with 25 mL of NaOH, Na_2CO_3, $NaHCO_3$ and HCl 0.1 M solutions, using plastic containers of 50 mL, the solutions are left at 298 K, are manually agitated twice daily for 48 hours and occasionally on the solutions are bubbled N_2 in order to remove atmospheric CO_2 and finally were titrated aliquots of 10 mL with previously standardized acid or base solutions [33].

Zero Charge Point

Determination of the zero charge point on the monoliths was performed by means of mass titration method of Noh and Schwarz [34]. Was a weighed amount of the materials in a range of 0.050 to 0.300 g, each of which was placed in a glass container of 50 mL and was added 10 mL of NaCl 0.1 M solution . The containers were capped and left under stirring in constant temperature at 298 K during 48 hours, and then the pH was measured for each of the solutions. The zero charge point was determined as the pH to which tends the pH curve in function of the mass of the sample in suspension.

Immersion Calorimetry

Additionally were made the immersion calorimetry in benzene and water , for determining the immersion enthalpy, were weighed about 100 mg of sample and placed in a glass ampoule coupled to the calorimetric cell designed for this purpose. Were then added 8 mL of the solvents mentioned. The cell was placed in the main reservoir of heat, and then was left to stabilize the temperature of the calorimetric set for about 1 hour, or the time needed for stabilization of the calorimeter, after was performed of immersing the sample in solution and were recorded thermal changes until that the baseline stabilized again, finally we recorded an after-period during 20 minutes. And the measure ended with the electrical calibration experience [35-37].

CO_2 Adsorption Capacity

CO_2 adsorption capacity in the different activated carbon monoliths at 273K and until 1 bar was measured using a volumetric system (Quantachrome, Autosorb 3-B). Before the adsorption experiments, the different samples were outgassed at 423 K for 4 h.

Results and Discussion

Textural Characteristics

The textural character istics of the activated carbon s were determined by fisiadsorción N_2 at 77 K. Each series presented different behaviors associated with the type of activating agent employed, the surface area s are between 587-1320 m^2g^{-1} and the pore volume between 0.24 to 0.47 cm^3g^{-1}. Within the series was found that the MP48, MZn48 and MCa2 samples have the best adsorption capacity and for this reason were used to perform the functionalization process.

Figure 3 shows the adsorption isotherms for the original and modified carbonaceous materials . In Figure 3a is observed that the isotherms for the samples obtained are type I according to the IUPAC classification [38], the MP48, MP48FAG and MP48FAL materials have the open elbow a relative low pressure (P/P0 <0.1), indicating a wide pore size distribution . Meanwhile in Figure 3b the isotherms obtained for the original and modified samples: MZn48, MZn48FAG and MZn48FAL are type I, with closed elbow, characteristic of materials with narrow porosity . The isotherms of MCA2, and MCa2FAL MCa2FAG materials, Figure 3c, are type I with a closed elbow own of narrow pores distribution.

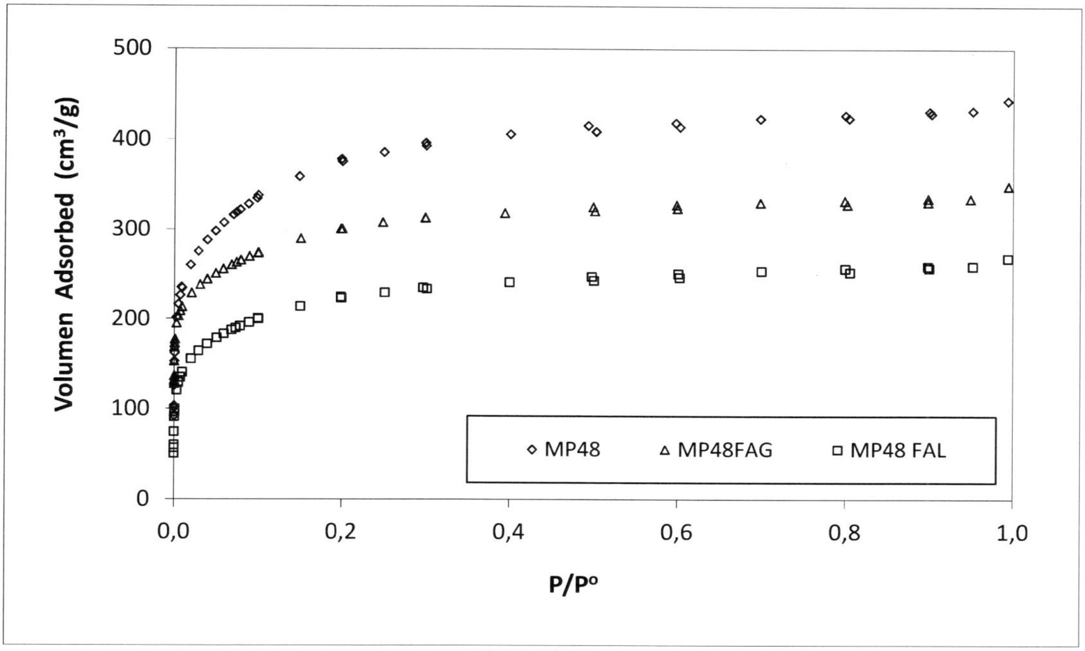

(a) MP48

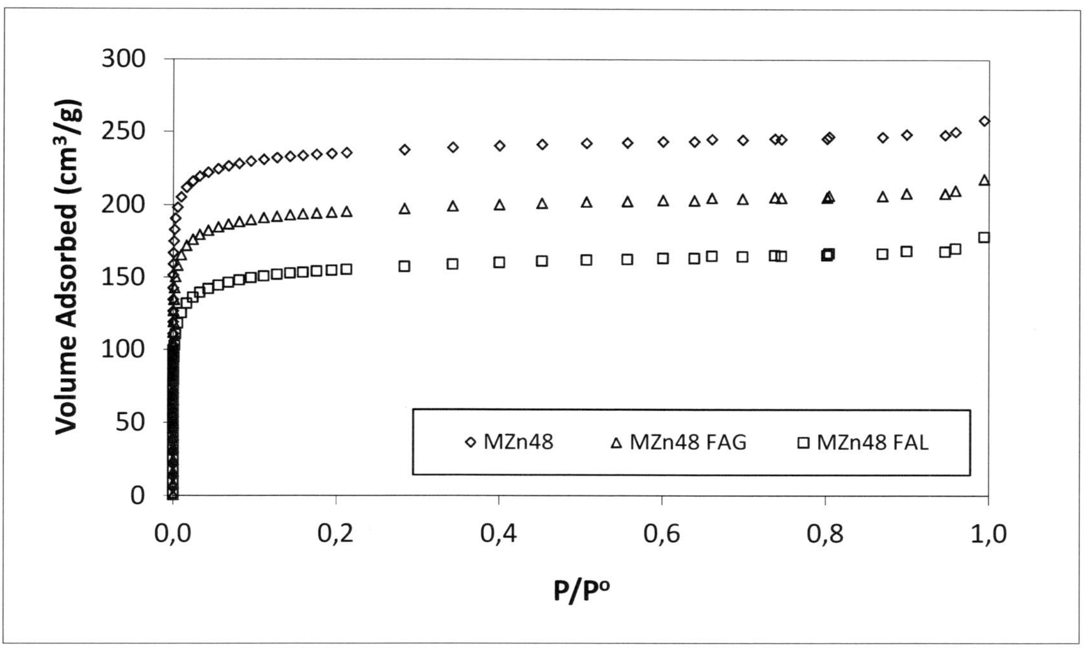

(b) MZn48

Figure 3. (Continued)

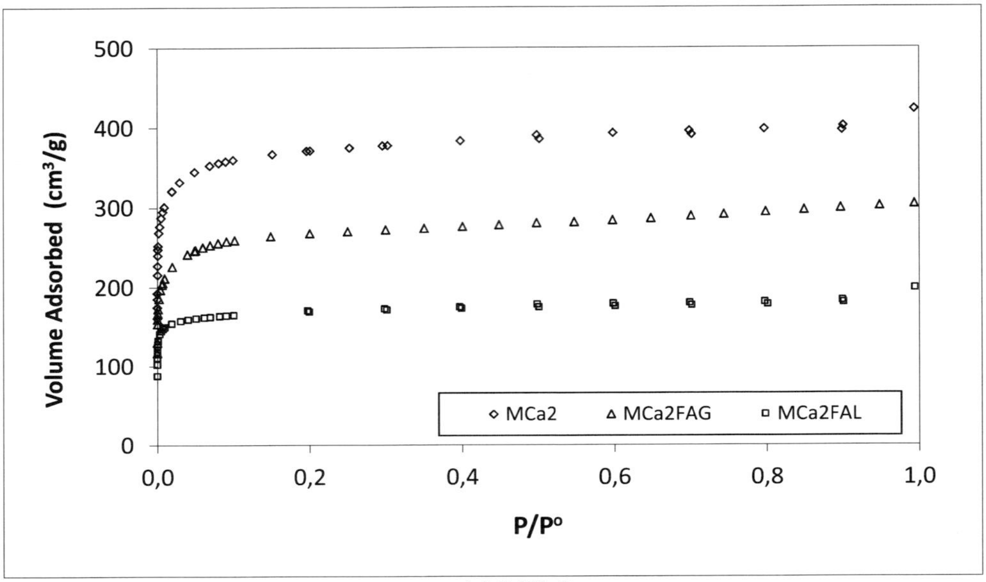

(c) MCa2

Figure 3. Nitrogen adsorption isotherms for original and modified activated carbon s (a) MP48 (b) MZn48 (c) MCa2.

For all sets of activated carbon s the functionalization processes generate a decrease in BET area and pore volume. From the nitrogen adsorption isotherms can be seen that functionalization with ammonia gas modifies in minor proportion the textural character istics of the original samples, between 20-30%, while the functionalization with ammonium hydroxide decreases in 36 - 40% the surface area and pore volume of the monoliths. This decrease can be attributed to a blockade in the carbon matrix produced by the nitrogen surface groups that are generated during the reaction with ammonia and ammonium hydroxide.

Table 6. Textural Parameters for carbonaceous materials obtained from the N_2 adsorption isotherm at 77 K and CO_2 adsorption isotherm at 273 K

N_2 adsorption data at 77 K						**CO_2 adsorption data at 273 K**
Sample	**S_{BET} (m^2g^{-1})**	**V_O (cm^3g^{-1})**	**V_{meso} (cm^3g^{-1})**	**$V_{0.99}$ (cm^3g^{-1})**		**V_n (cm^3g^{-1})**
MP48	1320	0.47	0.20	0.67		0.24
MP48FAG	916	0.32	0.15	0.47		0.28
MP48FAL	785	0.28	0.13	0.41		0.31
MZn48	924	0.36	0.01	0.38		0.31
MZn48FAG	737	0.30	0.02	0.32		0.33
MZn48FAL	587	0.24	0.02	0.26		0.34
MCa2	1425	0.53	0.12	0.65		0.50
MCa2FAG	1030	0.38	0.08	0.46		0.53
MCa2FAL	658	0.25	0.05	0.30		0.64

Table 6 shows the textural character istics of the carbonaceous materials , as stated above there is a decrease in the surface area , volume of micropores, mesopores and the total pore volume. By contrast with the functionalization of materials, narrow micropore volume (Vn) obtained from the CO_2 adsorption isotherm at 273 K, corresponding to pore volume of below 0.7 nm ~ [31], increase. For monoliths MP48FAL, MZn48FAL, MZn48FAG, MCa2FAG and MCa2FAL this value is greater than the total micropore volume (V_0) determined by N_2 adsorption that measured all pores 2 nm below, this behavior reflects the existence of kinetic restrictions for N_2 access in the narrow micropores.

Chemical Characteristics

Infrared Spectroscopy

Figure 4 shows the DRIFT spectra of the original and modified samples. The spectra of Figure 4a belonging to the monoliths MP48, MP48FAG and MP48FAL, is possible to distinguish several bands of interest, for the three samples there is a band located between 3200 to 3600 cm^{-1}, associated to O-H stretching vibration of alcohols and carboxylic acid s , the band intensities of the samples are similar in MP48 and MP48FAG but in MP48FAL sample is shows a decrease, also in the three monoliths shows a band at ~ 1600 cm^{-1} characteristic of C = O and C = N stretching of aromatic ring whose intensity and definition becomes more noticeable in the MP48FAL sample, between 800 cm^{-1} and 1400 cm^{-1} the samples show significant differences in their spectra in this region is difficult to assign the bands as there is overlap bands of some groups, however this region is associated with C-O stretching. Additionally the MP48FAL sample has a doublet at 1275 cm^{-1} and 1300 cm^{-1} corresponding to C-N and C-O stretching [39-43]. Figure 4b shows the spectra of the samples MZn48, MZn48FAG and MZn48FAL in all three samples there are a band located between 3200 to 3600 cm^{-1}, associated to O-H stretching vibration of alcohols and carboxylic acids, the intensity of this band decreases in MZn48FAG and MZn48FAL monoliths, this is can be associated with a decrease of carboxyl groups due to the treatments that were subjected the carbonaceous materials , in the three spectra shows a band at ~ 1600 cm^{-1} characteristic of C = O and C = N stretching of aromatic ring, moreover in the MZn48FAG and MZn48FAL samples appears a low intensity band at 1400 cm^{-1} of amino group [44].

Finally Figure 4c shows the MCA2, MCa2FAG and MCa2FAL monoliths spectra having bands between 900 to 1450 cm^{-1} between 3100-3700 cm^{-1} and a peak at 1580 cm^{-1} characteristic of carbonaceous materials . Similar to other functionalized samples prepared in this study, it is clear that the aforementioned peaks have different intensity and this decreased with the modifications made. Additionally the MCa2FAL presents a small signal at 1190 cm^{-1} characteristic of the CN group.

According to the spectra is observed that the modification with ammonium hydroxide generates greater changes in the surface chemistry than the modified materials with ammonia , such changes can be associated with the reactions between the surface groups in the original carbon and ammonium hydroxide used in the modification. The results obtained by FTIR are consistent with groups determined by Boehm titration s that are presented below.

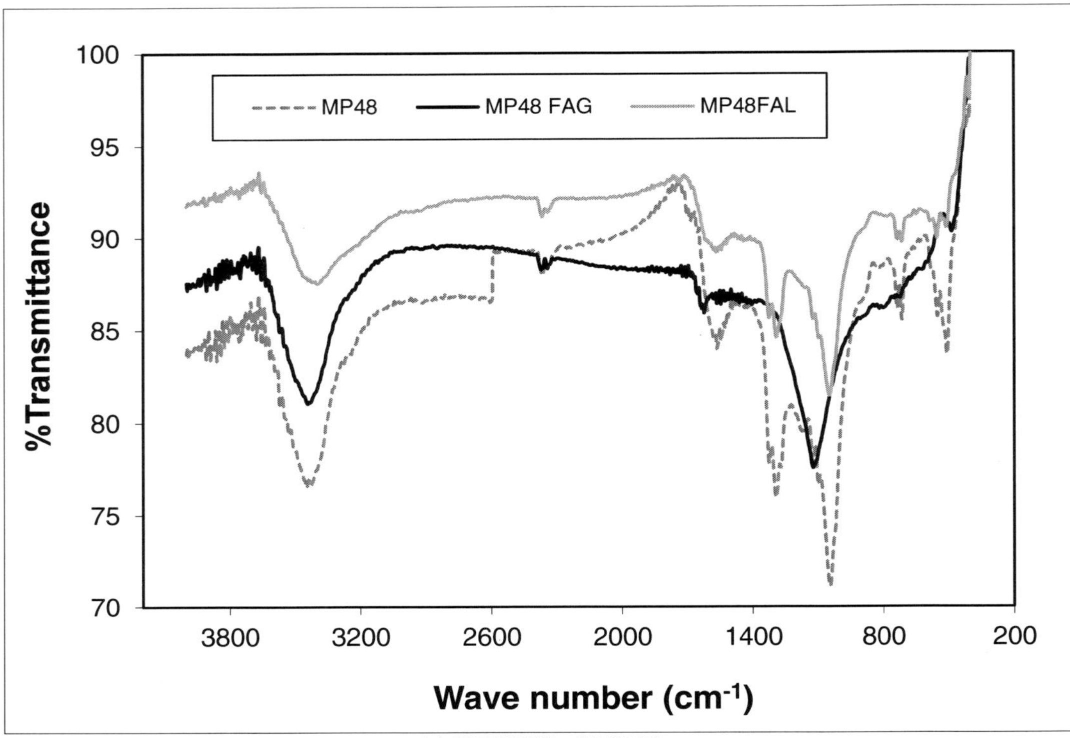

(a) DRIFTS of MP48

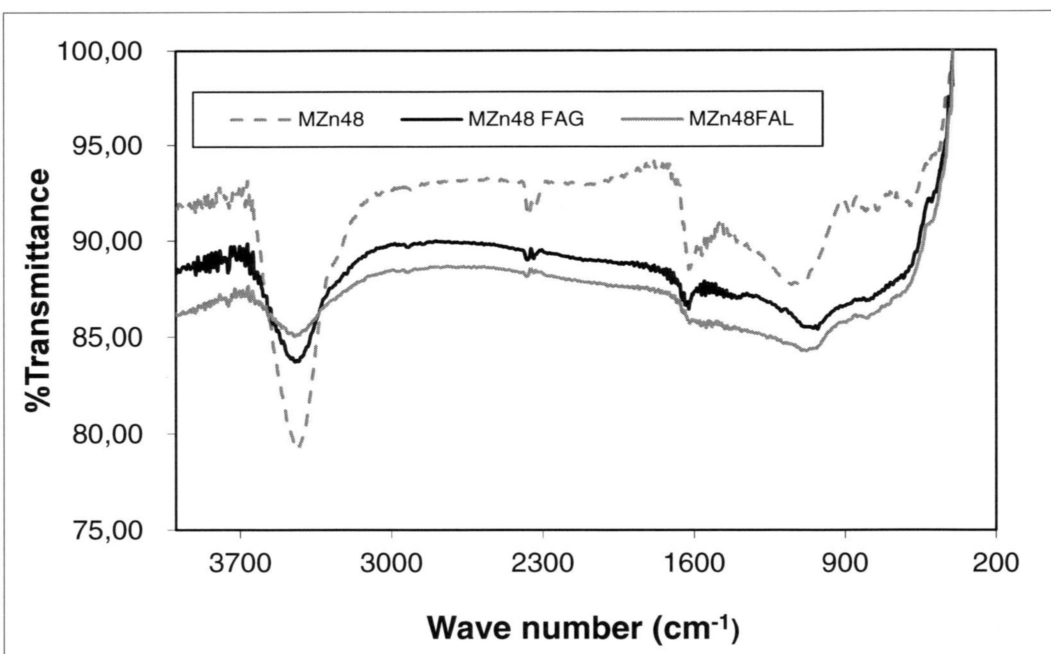

(b) DRIFTS of MZn48

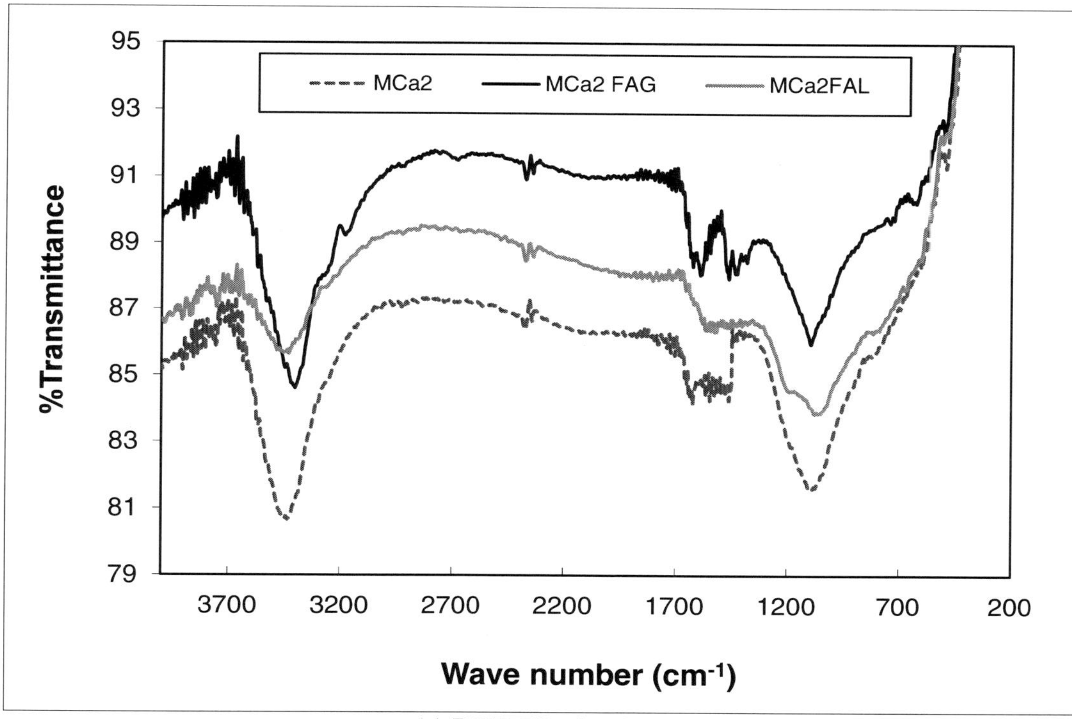

(c) DRIFTS of MCa2

Figure 4. DRIFT spectra for the original and modified samples (a) MP48 (b) MZn48 (c) MCa2.

Boehm Titrations

Table 7 shows the content of the surface groups in the monoliths, determined by Boehm titration s. From the results are evaluated changes in surface chemistry of each sample according to the modification treatment which was applied.

The original samples MP48, MZn48 and MCa2 present a variety of surface groups, was observed a greater amount of carboxylic groups and higher total acidity with respect to the total basicity , so that the character of the surface is acid such as is evidenced a $pH_{PZC} < 7$, subsequently with modifying, these groups decreased in all samples, this behavior can be attributed to that carboxylic groups to be more acids (pKa = 2-5) that lactonic (pKa = 8.2) and phenolic (pKa = 10), react in higher proportion with ammonia as shown in the following reaction [44]:

$$-COO^-NH_4^+ \xrightarrow{-H_{2O}} -CO-NH_2 \xrightarrow{-H_{2O}} C\equiv N \quad (1)$$

With the reduction of the carboxyl groups in the modified samples, the acidity decreases between 20-26% and simultaneously increases the basicity in more than 75%, despite the increase of acidity values, these still exceed the basicity and therefore the acidic nature of the surface changes only slightly as evidenced by the small increase suffered by pH_{PZC}.

Likewise it is evident that incorporation of basic groups is favored with the acidity of the original material, due to as with the modification with ammonium hydroxide in the sample

MZn48 having an acidity of 187 μmolg^{-1}, is possible increased in more than twofold its basicity (MZn48FAL sample), whereas that in MP48 sample which has an acidity of 145μmol g^{-1}, the basic character is increased by less than double the original value (MP48FAL sample). Additionally it is noted that for both samples the treatment with ammonium hydroxide is more effective than the reaction with gaseous ammonia .

For monoliths MCA2 series is observed that the modifications increase the amount of basic groups on the surface, with a value greater than 174 g mol^{-1} for the liquid phase modification.

Table 7. Content of surface groups determined by titration Boehm and pH in the point of zero charge

Sample	Carboxylic acids (μmol g^{-1})	Lactone (μmol g^{-1})	Phenol (μmol g^{-1})	Total Acidity (μmol g^{-1})	Total Bacisity (μmol g^{-1})	pH_{PZC}
MP48	145	27	126	298	96	5.7
MP48FAG	80	0	158	238	137	6.1
MP48FAL	54	0	164	218	168	6.4
MZn48	187	21	169	377	82	5.2
MZn48FAG	125	0	205	330	98	5.4
MZn48FAL	106	0	194	300	180	5.6
MCa2	363	161	28	553	53	2.5
MCa2FAG	105	0	78	184	129	6.7
MCa2FAL	81	0	52	133	174	6.9

Immersion Calorimetry

Immersion calorimetry is a technique that allows to obtain the thermodynamic information regarding interactions, that occur between a solid and differents immersion liquids [36-38]. Immersion enthalpy is associated with the formation of an adsorbed layer of fluid molecules on the solid surface, together with the wet of adsorbed layer. The intensity of the effect leads to comparisons and relations between different characteristics of porous solids. Thus, the determination of the immersion enthalpy, under specific conditions, is an additional parameter to characterize the solids, which is proportional to surface area , pore volume, the active site s content, among others.

By calorimetric technique is possible to evaluate the porous structure and some parameters associated with the chemical surface of solid. In the evaluation of the porous structure are used nonpolar liquids (benzene , carbon tetrachloride , α pinene etc) with which the solid does not have specific interactions such that the immersion enthalpy is proportional to the surface area accessible to the liquid molecule. Meanwhile when it is desired to obtain information about the surface chemistry of materials is used solutions with different solutes that may be acids, bases, metal ion s and water as immersion liquid, in this case the thermal effect resulting involves the interactions of liquid with surface functional groups. The water immersion calorimetry allows to evaluate the polarity of the surface of the activated carbon , under the assumption of that the water molecules interacting with the surface groups mainly ox ygenated located in polar sites at the edges of the graphene layers [45], as a result can be established relations between the immersion enthalpy and surface groups determined by Boehm titration s.

Figure 5 shows the thermograms of immersion of the MCA2 series monoliths in water , it is observed that the solid MCa2FAL presents the greatest interaction with the solvent because the enthalpy that is generated in the contact is proportional at area under the curve of the peak produced.

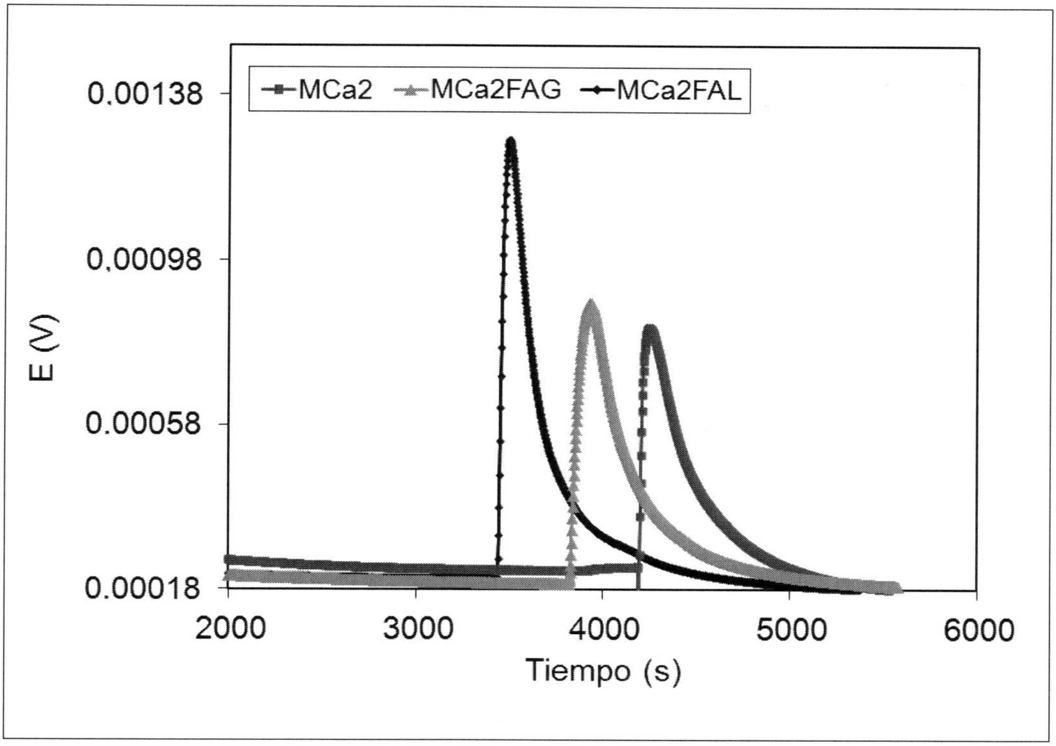

Figure 5. Thermogram of immesrsion of monoliths MCA2 series in water .

Table 8 shows the immersion enthalpy in benzene and water together with the hydrophobic factor, calculated as the ratio between the immersion enthalpy in benzene and immersion enthalpy in water; it is observed that all enthalpies are exothermic. The enthalpies in benzene are among -51.85 and -173.2 Jg^{-1} and in water are between -15.98 and -67.01 Jg^{-1}. It is appreciated that with the functionalization process decreases the enthalpy value of the monoliths in benzene, which is indicative of a decreased accessible area in the solid, such as show the nitrogen adsorption isotherms , this behavior has been shown in other studies [37] these two techniques despite having different considerations, employing molecules of similar size 0.36 nm for nitrogen and 0.37 nm for benzene, which allows identify the solid accessibility and some surface characteristics. Meanwhile the water immersion enthalpy decreases with modification processes, indicating a change in the surface chemistry of solids studied.

Table 8. Immersion enthalpies of the original and modified samples in benzene and water

Sample	$-\Delta H_{imm^*}$ C_6H_6 (Jg^{-1})	$-\Delta H_{imm^*}$ H_2O (Jg^{-1})	Hydrofobic factor $\frac{-\Delta H_{imC_6H_6}}{-\Delta H_{imH_2O}}$
MP48	134.5	67.01	2.01
MP48FAG	102.1	30.64	3.33
MP48FAL	79.57	27.82	2.86
MZn48	90.51	40.06	2.26
MZn48FAG	81.05	34.21	2.37
MZn48FAL	51.85	33.87	1.53
MCa2	173.2	61.35	2.82
MCa2FAG	112.5	19.65	5.73
MCa2FAL	69.54	15.98	4.35

* The standard deviation s of immersion enthalpies are between 0.1 – 1.3 J/g.

Figure 6 shows the correlation between the immersion enthalpies of activated carbon s monoliths in water and the content of total acidic and basic surface groups, in these graphs it is evident that the immersion enthalpy in water is determined by interactions of ox ygenated surfaces groups and by Lewis basic sites, to correspond mainly to delocalized π electrons of the grahene layers [45].

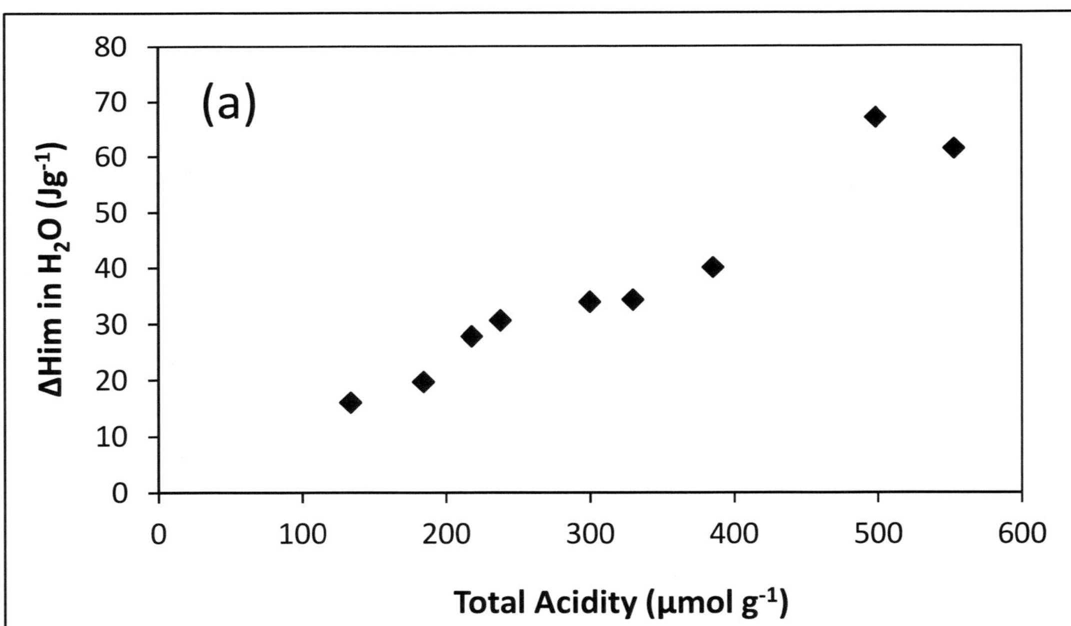

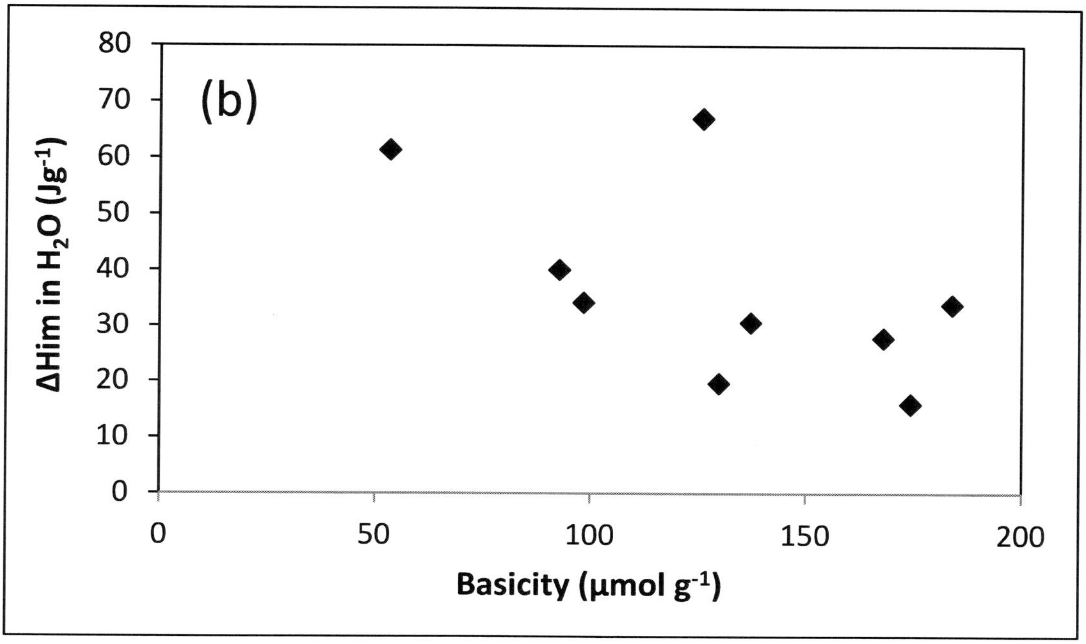

Figure 6. Relationship between the immersion enthalpy in water and the total (a) acidity and (b) basicity .

Was found that with increasing the surface oxygen groups concentration the immersion enthalpy in water increase due to established specific interactions between these groups in the polar sites located at the edges of the graphene layers and water molecules . Likewise there is a decrease of the enthalpy in water with the increasing basicity of the monoliths, due to with the modification process performed, much of the surface groups with character acid as the carboxylic groups , are consumed by reacting with ammonia and therefore the solid surface resulting has a lower ability to establish specific interactions with water molecules.

Figure 7 shows the relationship between the pH in the point of charge zero, pH_{PZC}, and hydrophobic factor, is observed there is a direct relationship between these two parameters. The increased hydrophobic factor is closely related to the decrease of the acid groups and increased in the basicity , so is understandable that the pH_{PZC} (pH at which the number of positive charges equals the negative charges such that the net charge on the carbon surface is zero) in this study increases slightly with the modifications , due to that was achieved increase the basic nature of the surface [46,47].

CO_2 Adsorption at 273 K

After the modification process of activated carbon monoliths, was evaluated CO_2 adsorption capacity of the original and modified carbonaceous materials , for this was used the volumetric equipment at 273 K and until 1 bar.

Figure 8 shows the CO_2 adsorption isotherms obtained for the materials . It is appreciated that in the modified samples there is an increase in the gas adsorption capacity. In agreement with previous work the shape of the CO_2 adsorption isotherms at 273 K is related to the adsorption mechanism. Accordingly, isotherms the original and modified samples, having one trend near rectilinear can be associated with a CO_2 adsorption mechanism in which the surface coverage is associated with pore size s greater than two times the CO_2 molecular dimension [48].

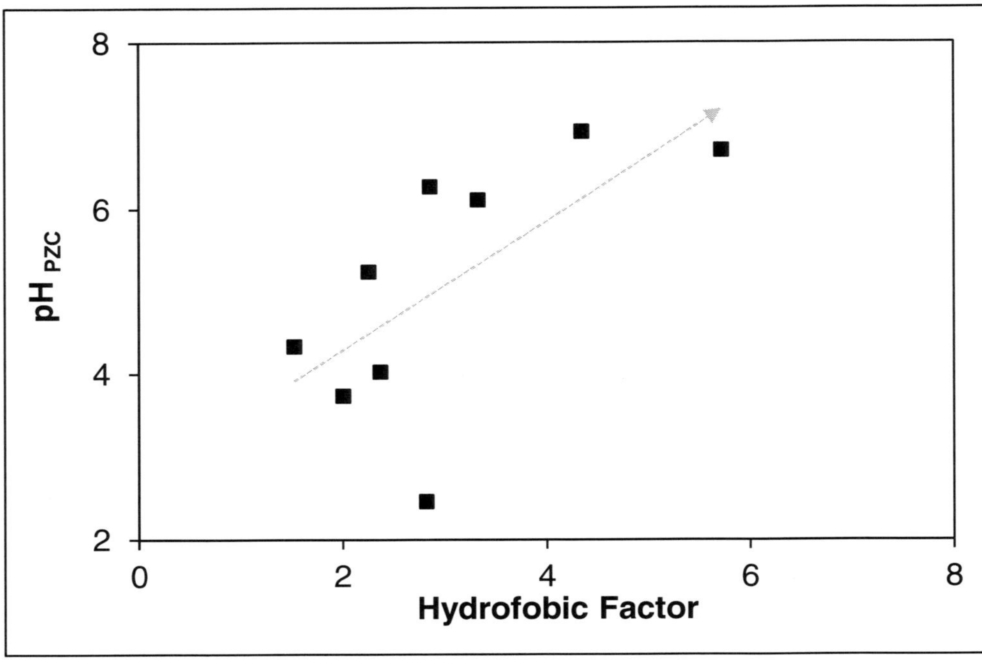

Figure 7. Relationship between the pHPZC and hydrophobic factor.

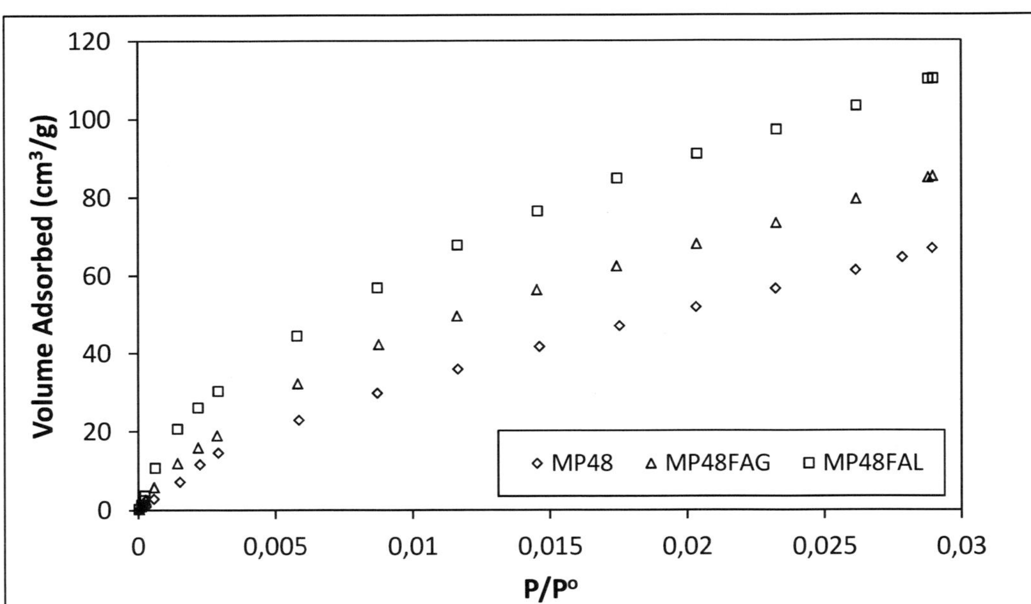

(a) MP48

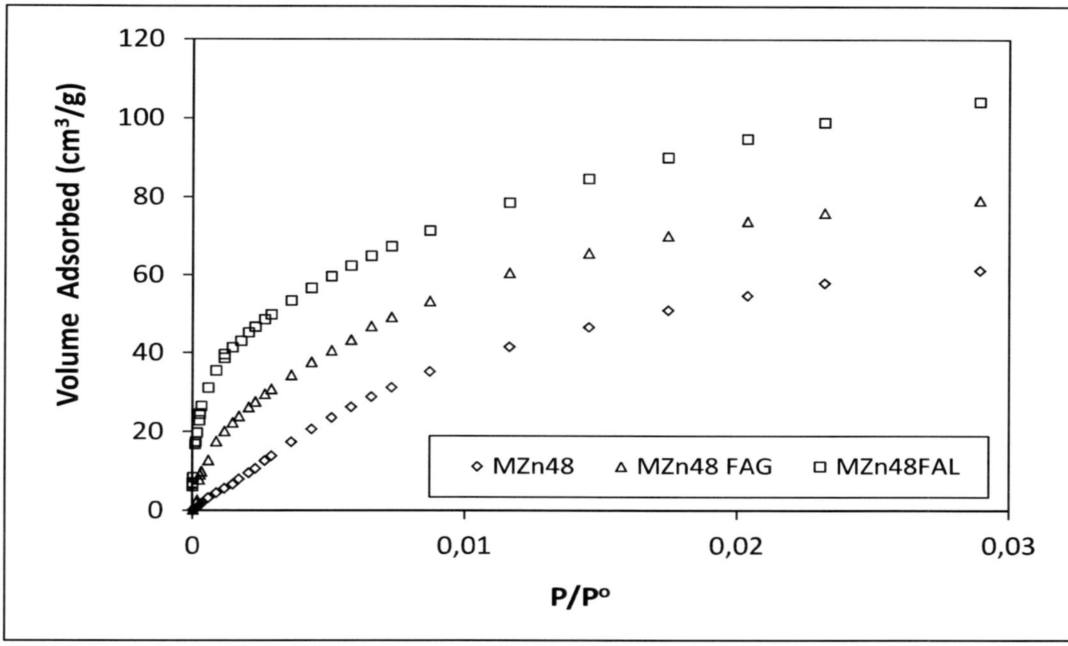

(b) MZn48

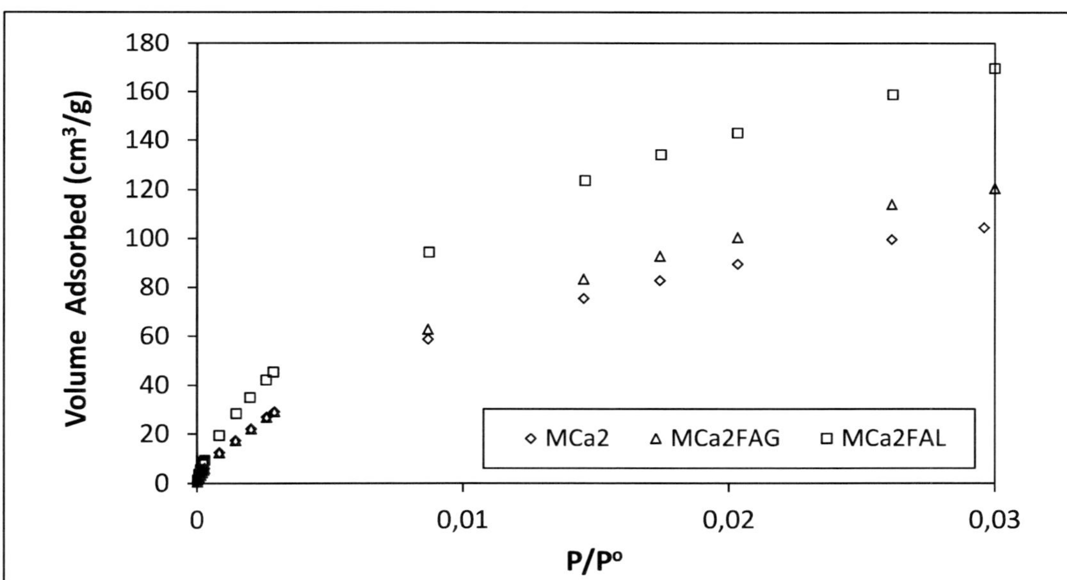

(c) MCa2

Figure 8. CO_2 adsorption isotherms at 273K and until 1 bar.

Table 9 shows the amounts of CO_2 adsorbed. An adsorption capacity between 120-334 mg CO_2 g^{-1} is observed. The highest retention of gas is obtained for MCa2FAG and MCa2FAL samples. The functionalization with ammonia generates an increase in CO_2 adsorption capacity between 10-30 %, while with ammonium hydroxide the increase is greatest, was achieved a 20-70 % more CO_2 adsorbed.

Table 9. CO_2 adsorbed amount (mgCO$_2$g-1) at 273 K and 1 bar of samples

Sample	***CO_2 adsorbed*** **mg CO_2 g^{-1}**
MP48	138
MP48FAG	170
MP48FAL	220
MZn48	120
MZn48FAG	153
MZn48FAL	200
MCa2	203
MCa2FAG	230
MCa2FAL	334

Activated carbon monoliths MP48FAL, MZn48FAL and MCa2FAL adsorb CO_2 quantities of: 220 mg CO_2 g^{-1}, 200 mg CO_2 g^{-1} and 334 mg CO_2 g^{-1} respectively. These values are satisfactory when are compared with the adsorption capacity of CO_2 at 273 K and 1 bar than for other solid are obtained and are shown in Table 10, which shows that have been achieved adsorbed quantities up to 400 mg CO_2 g^{-1} in a Metal-Organic Framework (MOF)-Co-MOF-74 (M) enriched with Co with a surface area of 1314 m^2g^{-1} and Vn 0.51 cm^3g^{-1} and 380 mg CO_2 g^{-1}, for a carbon molecular sieve VR-5-M with a surface area of 3100 m^2g^{-1} and Vn 1.40 cm^3g^{-1} [49].

Table 10. Comparison of the CO_2 adsorption capacity of different porous materials at 273K and 1 bar with activated carbon monoliths

Solid	**CO_2 adsorption capacity (mg/g)**	**S_{BET} (m^2g^{-1})**	**V_O (cm^3/g)**	**Reference**
CPM-5 MOF	132	2187	--	Sabouni et. al 2013 [50]
Microporous polymelamine network	174	827	--	Liu et. al 2013 [51]
Carbon Fiber P-R	189	998	0.26	An et. al 2013 [52]
PAC- grass cuttings	176	841	0.28	Hao et. al 2013 [53]
ATS-2-700	255	1330	0.54	Sevilla y Fuertes 2012 [54]
Zeolita MCM-22	123	481	0.16	Yang et. al 2012 [55]
CAGC M-30	167	2350	0.79	Sevilla y Fuertes 2012 [54]
Template Carbon CMK-3	140	970	0.12	Sevilla y Fuertes 2012 [54]
Carbon nitride–carbón MCN/C	134	338	0.057	Deng et. al 2012 [56]
TMC C-1012	232	2000	0.73	Wabhy et. al 2012 [57]
Co-MOF-74(M)	400	1314	0.51	Cho et. al 2012 [58]
Zr-MOF	155	1434	0.65	Abid et. al 2012 [59]
Molecular sieve carbón ACS-1	235	1030	0.40	Carruthers et. al 2012 [60]
TMC VR-5-M	380	2450	1.42	Wabhy et. al 2010 [49]
Norit RNO300	176	950	0.37	Plaza et. al 2010 [61]
MP48FAL	220	785	0.31	En este trabajo
MCa2FAL	334	658	0.64	En este trabajo

Figure 9 shows the relationship between immersion enthalpy of activated carbon monoliths in water and the CO_2 adsorption capacity, as the modification of solid by the nitrogen groups addition presents higher values in the gas adsorption and such modification affects the immersion enthalpy in water of the monoliths, since water interacting with the solid because is a polar solvent.

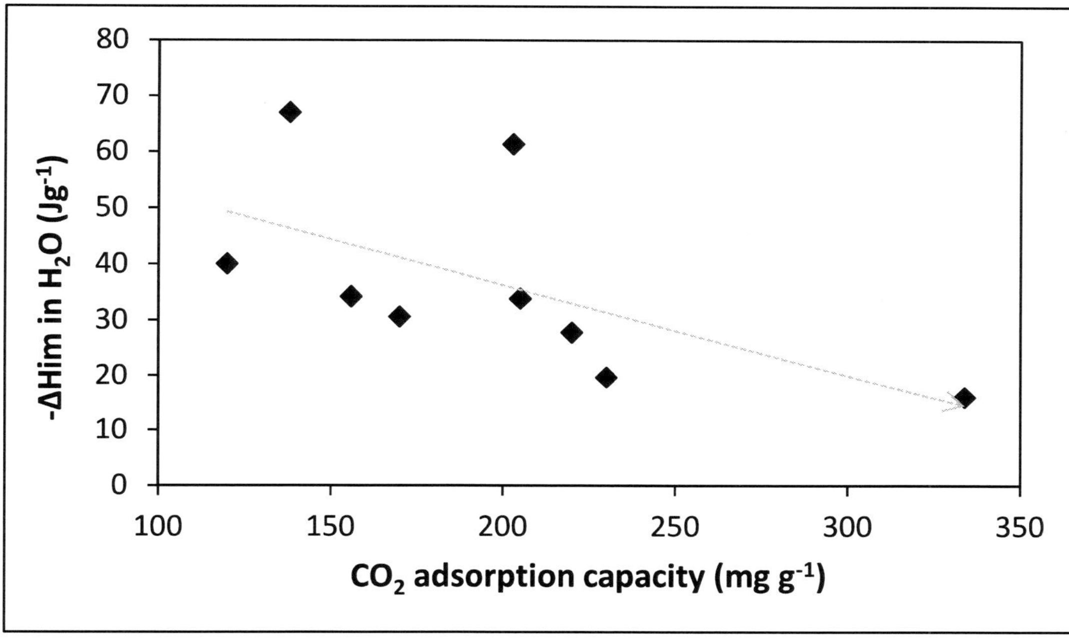

Figure 9. Relation between immersion enthalpy activated carbon monoliths and CO_2 adsorption capacity.

A decreasing in the immersion enthalpy of the monoliths in water that occurs with a greater addition of nitrogen in the surface chemistry of monolith is observed and too a increases in CO_2 adsorption capacity, so that immersion enthalpy in water that characterizes the interaction solid-liquid allows establish the solid that will have a greater interaction with the CO_2.

Conclusion

The chemical modification of the activated carbon monoliths by gasification and impregnation technique allowed the increase of the basicity of materials with including nitrogen groups. These groups increased between 10-70 % the affinity of surface carbonaceous by the CO_2 molecule, as was demonstrated in the adsorption isotherms of the gas. The monoliths obtained by chemical activation with calcium chloride have CO_2 adsorption capacity greater than the monoliths prepared using phosphoric acid and zinc chloride, due to that have better surface area , a large pore size distribution that facilitates the diffusion of gas into the micropores and a surface chemistry that conducive the adsorption process of the interest molecule.

The activated carbon monoliths have surface area between 587-1425 m^2g^{-1} and poro volumen between 0.24-0.53 cm^3g^{-1}. Moreover the immersion enthalpies in benzene are among -51.85 and -173.2 Jg^{-1} and in water are between -15.98 and -67.01 Jg^{-1}.

The modification process caused changes in texture and chemical characteristics of the monoliths. The functionalization produces in both cases a decrease of the N_2 adsorption capacity, surface area and pore volumes attributed to a blockade of the porosity by the groups generated by the modification, meanwhile the content of acid group was decreased and consequently the basicity of the surface is increased. Furthermore it was established that the most effective method of functionalization, in this study, is carried out in the liquid phase with ammonium hydroxide , reaching a maximum adsorption capacity of 334 mg CO_2g^{-1} in the MCa2FAL sample.

ACKNOWLEDGMENTS

The authors thank the Framework Agreement between Universidad de los Andes and Universidad Nacional de Colombia , as well as the Agreement Statement (Acta de Acuerdo) between the Departments of Chemistry of both Universities. Special thanks go to the Faculty of Science and the Vice-Rectory of Research at the University of the Andes (Bogotá, Colombia) for funding this research.

REFERENCES

[1] Plaza M.G, García S, Rubiera F, Pis J.J, Pevida C. Post-combustion CO_2 capture with a commercial activated carbon : Comparison of different regeneration strategies. *Chem. Eng. J.* 2010; 163: 41-47.

[2] Drage T.C, Smith K.M, Pevida C, Arenillas A, Snape C.E. Development of adsorbent technologies for post-combustion CO_2 capture Original. *Energ. Proced.* 2009; 1: 881-884.

[3] García S., Gil M.V, Martín C.F, Pis J.J, Rubiera F, Pevida C. Breakthrough adsorption study of a commercial activated carbon for pre-combustion CO_2 capture. *Chem. Eng. J.* 2011; 171:549-556.

[4] Plaza M.G, Pevida C, Pis J.J, Rubiera F. Evaluation of the cyclic capacity of low-cost carbon adsorbents for post-combustion CO_2 capture. *Energ. Proced.* 2011; 4: 1228-1234.

[5] Bao Z.B, Yu L.A, Ren Q.L, Lu X.Y, Deng S.G. Adsorption of CO_2 and CH4 on a magnesium -based metal organic framework. *J. Colloid Interface Sci.* 2011; 353: 549-556.

[6] Saha D, Bao Z, Jia F, Deng S. Adsorption of CO_2 , CH4, N2O, and N2 on MOF-5, MOF-177, and Zeolite 5A. *Environ. Sci. Technol.* 2010; 44: 1820-1826.

[7] Jensen N.K, May E.F, Watson G, Rufford T.E, Zhang D, Chan K.I. Development of a method to efficiently screen adsorbents for gas separation applications involving methane , nitrogen , and carbon dioxide . In: 9th Inter- national Symposium on Characterisation of Porous Solids (IX COPS). 2011. Dreseden, Germany .

[8] Xu X, Zhao X, Sun L, Liu X. Adsorption separation of carbon dioxide , methane , and nitrogen on Hb and Na-exchange b-zeolite . *J. Natural Gas Chem.* 2008;17: 391-396.
[9] Belmabkhout Y, De Weireld G, Sayari A. Amine-bearing mesoporous silica for CO_2 and H2S removal from natural gas and biogas . *Langmuir.* 2009; 25: 13275-13278.
[10] Ritter J.A, Yang R.T, Equilibrium adsorption of multicomponent gas mixtures at elevated pressures. *Ind. Eng. Chem. Res.* 1987; 26: 1679-1686.
[11] Dreisbach F, Staudt R, Keller J.U. High pressure adsorption data for methane , nitrogen , carbon dioxide and their binary and ternary mixtures on activated carbon . *Adsorption.* 1999; 5: 215-227.
[12] Watson G.C.Y, Jensen N.K, Rufford T.E, Chan K.I, May E.F. Volumetric adsorption measurements of N2, CO_2 , CH4, and a CO_2/CH4 mixture on a natural chabazite from 5 to 3000 kPa. *J. Chem. Eng. Data.* 2012; 57: 93-101.
[13] Cansado I.P.P, Maura, P.A.M, Ribeiro Carrott M.L, Carrott P.J.M. Activated carbons prepared from natural and synthetic raw materials with potential applications in gas separations. *Adv. Mater. Res.* 2010; 107: 1-7.
[14] Esteves I.A.A.C, Lopes M.S.S, Nunes P.M.C, Mota J.P.B. Adsorption of natural gas and biogas components on activated carbon . *Sep. Purif. Technol.* 2008; 62: 281-296.
[15] Huang Z, Xu L, Li J.-H, Guo G.-M, Wang Y. Adsorption equilibrium of carbon dioxide and methane on b-zeolite at pressures of up to 2000 kPa using a static volumetric method. *J. Chem. Eng. Data.* 2009; 55: 2123-2127.
[16] Bae Y.-S, Lee C.-H. Sorption kinetics of eight gases on a carbon molecular sieve at elevated pressure. *Carbon.* 2005; 43: 95-107.
[17] Wang B, Cote A.P, Furukawa H, O'Keefe M, Yaghi O.M. Colossal cages in zeolitic imidazolate frameworks as selective carbon dioxide reservoirs. *Nature.* 2008; 453: 207-211.
[18] Bao Z, Alnemrat S, Yu L, Vasiliev I, Ren Q, Lu X, Deng S. Kinetic separation of carbon dioxide and methane on a copper metal-organic frame-work. *J. Colloid Interface Sci.* 2011; 357: 504-509.
[19] Himeno S, Komatsu T, Fujita S. High-Pressure adsorption equilibrua of methane and carbon dioxide on several activated carbon s. *JCED.* 2005; 50: 369-376.
[20] Esteves I.A.A.C, Lopes M.S.S, Nunes P.M.C, Mota J.P.B. Adsorption of natural gas and biogas components on activated carbon . *Sep. Purif. Technol.* 2008; 62: 281-296.
[21] Ritter J.A, Ebner A.D. State-of-the-art adsorption and membrane separation processes for hydrogen production in the chemical and petrochemical industries . *Sep. Sci. Technol.* 2007; 42: 1123-1193.
[22] Cavenati S, Grande C.A, Rodrigues A.E. Adsorption equilibrium of methane , carbon dioxide , and nitrogen on zeolite 13X at high pressures. *J. Chem. Eng. Data.* 2004; 49: 1095-1101.
[23] Bae Y.-S, Hauser B.G, Farha O.K, Hupp J.T, Snurr R.Q. Enhancement of CO_2 /CH4 selectivity in metal-organic frameworks containing lithium cations. *Micropor. Mesopor. Mater.* 2011; 141: 231-235.
[24] Watson G, May E.F, Graham B.F, Trebble M.A, Trengove R.D, Chan K.I. Equilibrium adsorption measurement of pure nitrogen , carbon dioxide , and methane on a carbon molecular sieve at cryogenic temperature sand high pressures. *J. Chem. Eng. Data.* 2009; 54: 2701-2707.

[25] Watson G.C.Y, Jensen N.K, Rufford T.E, Chan K.I, May E.F. Volumetric adsorption measurements of N2, CO_2 , CH4, and a CO2/CH4 mixture on a natural chabazite from 5 to 3000 kPa. *J. Chem. Eng. Data.* 2012; 57: 93-101.

[26] Delgado J, Uguina M.A, Gómez J.M, Ortega L. Adsorption equilibrium of carbon dioxide , methane and nitrogen onto Na-and H-mordenite at high pressures. *Sep. Purif. Technol.* 2006; 48: 223-228.

[27] Zhang Z, Xu M, Wang H, Li Z. Enhancement of CO_2 adsorption on high surface area activated carbon modified by N_2, H_2 and ammonia . Chem Eng J. 2010; 160: 571-577.

[28] Plaza M.G, Pevida C, Pis J.J, Rubiera. Ammoxidation of carbon materials for CO_2 capture. *Appl. Surf. Sci.* 2010; 256: 6843-6849.

[29] Plaza M.G, García S, Rubiera F, Pis J.J, Pevida C. Evaluation of ammonia modified and conventionally activated biomass based carbons as CO_2 adsorbents in postcombustion conditions. *Sep. Purif. Technol.* 2011; 80: 96-104.

[30] Vargas D.P, Giraldo L, Moreno-Piraján J.C. CO_2 adsorption on granular and monolith carbonaceous materials . *J. Anal. App. Pyrol.* 2012; 96: 146–152.

[31] Garrido J, Linares-Solano A, Martín-Martínez J. M, Molina-Sabio M, Rodriguez-Reinoso F, Torregrosa, R. Use of N_2 vs. $C0_2$ in the Characterization of Activated Carbons. *Langmuir*. 1987; 3: 76-81.

[32] Kohl S, Drochner A, Vogel H. Quantification of oxygen surface groups on carbon materials via diffuse reflectance FT-IR spectroscopy and temperature programmed desorption . *Catal Today*. 2010; 150: 67-70.

[33] Boehm H.P. Surface oxides on carbon and their analysis: a critical assessment . *Carbon*. 2002; 40: 145–149.

[34] Noh J.S, Schwarz J.A. Effect of HNO_3 treatment on the surface acidity of activated carbon s. *Carbon*. 1990; 28: 675-682.

[35] Giraldo L, Moreno J.C. Determination of the Immersion Enthalpy of activated carbon by Microcalorimetry of the Heat Conduction. Instrum Sci Technol. 2000; 28: 171-178.

[36] Stoeckli F, Centeno T.A. On the characterization of microporous carbons by inmersion calorimetry alone. *Carbon*. 1997; 35: 1097-1100.

[37] Vargas D.P, Girado L, Moreno J.C. Relación entre la entalpía de inmersión de monolitos de carbón activado y parámetros texturales. *Química Nova*. 2011; 34: 196-199.

[38] Sing K. S. W.; Everett D. H..; Haul R.A.W.; Moscou L.; Pierotti R. A.; Rouquerol J.; Siemieniewska T. Reporting physisorption data for gas/solid systems with Special Reference to the Determination of Surface Area and Porosity. *Pure Appl. Chem*. 1985; 57: 603-619.

[39] Mangun C.L, Benak K.R, Economy J, Foster K.L. Surface chemistry, pore size s and adsorption properties of activated carbon fibers and precursors treated with ammonia . *Carbon*. 2001; 39: 1809–1820.

[40] Shafeeyan M.S, Ashri W.M, Daud W, Houshmand A, Arami-Niya A. Ammonia modification of activated carbon to enhance carbon dioxide adsorption : Effect of pre-oxidation . *Appl. Surf. Sci*. 2011; 257: 3936–3942.

[41] Przepiòrski J, Skrodzewicz M, Morawski A.W. High temperature ammonia treatment of activated carbon for enhancement of CO_2 adsorption . *Appl. Surf. Sci*. 2004; 225:235–242.

[42] Figueiredo J.L, Pereira M.F.R, Freitas M.M.A, Orfao, J.J.M. Modification of the surface chemistry of activated carbon s. *Carbon.* 1999; 37: 1379–1389.

[43] Moreno-Castilla C, López-Ramón M.V, Carrasco-Marín F. Changes in surface chemistry of activated carbon s by wet oxidation . *Carbon.* 2000; 38: 1995–2001.

[44] Plaza, M.G.; Pevida, C.; Arias, B.; Fermoso, J.; Rubiera, F.; Pis, J.J. (2007). CO_2 capture by adsorption with nitrogen enriched carbons. *Fuel.* 2007; 86: 2204–2212.

[45] M. Victoria López-Ramón, Fritz Stoeckli, Carlos Moreno-Castilla, Francisco Carrasco-Martín. Specific and non-specific interactions of water molecules with carbon surfaces from immersion calorimetry . *Carbon.* 2000; 38: 825–829.

[46] Faur- Brasquet C, Kadirvelu K, Le Cloirec P. Removal of metal ion s from aqueous solution by adsorption onto activated carbon cloths: adsorption competition with organic matter . *Carbon.* 2002; 40: 2387-2392.

[47] Grzegorz S, Stanisław B, Gerhard R. Carbon surface polarity from immersion calorimetry . *Fuel Process Technol.* 2002; 79: 217-223.

[48] Martin-Martinez J.M, Torregosa-Macia R, Mittelmeijer-Hazeleger, M.C. Mechanisms of adsorption of CO_2 in the micropores of activated anthracite. *Fuel.* 1995; 74: 111–114.

[49] Wahby A, Ramos-Fernández J.M, Martínez-Escandell M, Sepúlveda-Escribano A, Silvestre-Albero J, Rodríguez-Reinoso F. High-Surface-Area Carbon Molecular Sieves for Selective CO_2 Adsorption. *Chem. Sus. Chem.* 2010; 3: 974–981.

[50] Sabouni R, Kazemian H, Rohani S. Carbon Dioxide Adsorption in Microwave Synthesized Metal Organic Framework CPM-5: Equilibrium and Kinetics study. *Micropor. Mesopor. Mat.* 2013; 175: 85-91

[51] Liua L, Lia P.Z, Zhua L, Zoub R, Zhaoa Y. Microporous polymelamine network for highly selective CO_2 adsorption . *Polymer.* 2013; 54: 596-600

[52] Ana H, Fenga B, Sub S. Effect of monolithic structure on CO_2 adsorption performance of activated carbon fiber –phenolic resin composite: A simulation study. *Fuel.* 2013; 103: 80-86.

[53] Hao W, Björkman E, Lilliestråle M, Hedin N. Activated carbons prepared from hydrothermally carbonized waste biomass used as adsorbents for CO_2 . *Appl. Energ.* 2013; 112: 526-532

[54] Sevilla M, Fuertes A.B. CO_2 adsorption by activated templated carbons. *J. Colloid Interface Sci.* 2012; 366: 147-154.

[55] Yang S.T, Kim J.Y, Kim J, Ahn W.S. CO_2 capture over amine -functionalized MCM-22, MCM-36 and ITQ-2. *Fuel.* 2012; 97: 435-442.

[56] Deng Q.F, Liua L, Lina X.Z, Dub G, Liua Y, Yuana Z.Y. Synthesis and CO_2 capture properties of mesoporous carbon nitride materials . *Chem. Eng. J.* 2012; 203: 63-70.

[57] Wahby A, Silvestre-Albero J, Sepúlveda-Escribano A, Rodríguez-Reinoso F. CO_2 adsorption on carbon molecular sieves. *Micropor. Mesopor. Mat.* 2012; 164: 280-287.

[58] Cho H.Y, Yang D.A, Kim J, Jeong S.Y, Ahn W.S. CO_2 adsorption and catalytic application of Co-MOF-74 synthesized by microwave heating . Catal. *Today.* 2012; 185: 35-40.

[59] Abid H.R, Tian H, Ang H-M, Tade M.O, Buckley C.E, Wang S. Nanosize Zr-metal organic framework (UiO-66) for hydrogen and carbon dioxide storage . *Chem. Eng. J.* 2012; 187: 415-420.

[60] Carruthers J-D, Petruska M.A, Sturm E.A, Wilson S.M. Molecular sieve carbons for CO_2 capture. *Micropor. Mesopor. Mat.* 2012; 154: 62-67.

[61] Plaza M.G, Rubiera F, Pis J.J, Pevida C. Ammoxidation of carbon materials for CO_2 capture. *Appl. Surf. Sci.* 2010; 256: 6843-6849.

In: Comprehensive Guide for Mesoporous Materials. Volume 1 ISBN: 978-1-63463-958-3
Editor: Mahmood Aliofkhazraei

Chapter 13

CHARACTERIZATION OF THE STRUCTURE OF MESOPOROUS MATERIALS BY X-RAY SCATTERING

Pierre-Antoine Albouy
and Marianne Impéror-Clerc
Laboratoire de Physique des Solides, Université Paris-Sud,
Orsay, France

ABSTRACT

X-ray scattering is a very important tool for the investigation of mesoporous materials . In this chapter, we give an overview of their use for the study of periodic porous networks. First, we introduce the three types of textures: monoliths, powders and thin films . Then we review all the known possible structures encountered for mesoporous materials . The basic notions of X-ray scattering are explained: diffraction , reciprocal space and space-group determination. They are detailed for the different types of structures: 2D or 3D networks, monolith, powders and thin films . More advanced analysis methods based on the measurement of the Bragg peaks intensities in relationship with the form factor contribution are introduced, allowing to a refined structural description of the porosity . Lastly, some examples of in-situ studies under controlled atmosphere or during material formation are given. These examples illustrate the wide range of experimental possibilities offered by X-ray scattering analysis.

INTRODUCTION

Mesoporous/nanoporous materials are essentially characterized by the periodicity of their porous network. Among the different characterization techniques , X-ray scattering and electron microscopy are certainly the most adequate to study the structure of these materials. As electron microscopy presents the great advantage to provide a direct imaging of a material, X-ray scattering is also very powerful for all structural aspects. Scattering experiments give access to the spatial average structure of a sample. Their main advantage is to be very versatile, as they can be used in almost any experimental condition s (varying temperature , for thin film or bulk materials , in solution , under an external field …) and can be coupled in-situ with many other types of measurements like controlled atmosphere conditions.

In this chapter, we give an overview of the use of X-ray scattering methods for the study of periodic porous networks. In PART 1, we introduce the three types of textures: monoliths, powders and thin films (Figure 1). Then we review all the known possible structures encountered for mesoporous materials (Figure 2). PART 2 is devoted to basic notions about X-ray scattering methods: diffraction , reciprocal space and space-group determination. These notions are detailed for the different types of structures: monoliths (3D and 2D networks) (PART 2), powders and thin films (PART 3). More advanced analysis methods based on the measurement of the Bragg peaks intensities in relationship with the form factor contribution are introduced in PART 4, allowing to a refined structural description of the porosity . Lastly, in PART 5, some examples of in-situ studies under controlled atmosphere or during material formation are given. These applications illustrate the wide range of experimental possibilities offered by X-ray scattering analysis.

PART 1:
STRUCTURES AND TEXTURES

Mesoporous/nanoporous materials are all solid replica of soft-matter mesophases, which are obtained using various templating strategies. Consequently, the palette of all possible structures for mesoporous materials is the same as in many other soft-matter systems like Liquid Crystals of various chemical natures (block-copolymers , dendrimers, surfactants , lipids …) or Colloidal Crystals. All these systems share in common the same geometrical packing constraints and the same general rules of self-assembly govern the formation and stability of the different structures.

Among all the different structures, the most common and widespread structure is the 2D-hexagonal phase. This material consists in a 2D-hexagonal lattice of parallel cylindrical pores. These pores may be very long, up to micron sizes. Their main feature is that their diameter is very well-controlled in size and can be adjusted in a tunable range (1-30 nm), depending on the size of the templating agent. A related structure, often referred as a 'worm-like' structure in the literature, consists in a disordered version of the 2D-hexagonal phase. The pores are still locally cylindrical and monodisperse in size, but are curved with only a local order between them without forming any periodic arrangement.

For the structural analysis, three basis situations are encountered, as shown in Figure 1. Materials can be obtained as macroscopic monoliths [1,2] typically a few millimeters in size, and single-crystal analysis is possible (PART 2). More usually, materials are powder samples made of micron-size grains (PART 3). The structural analysis then includes an average over all possible orientations in space of the grains, assuming that all orientations have the same probability . The third situation is the deposition in thin film (PART 3). In this situation, some preferred orientation with respect to the substrate is often observed.

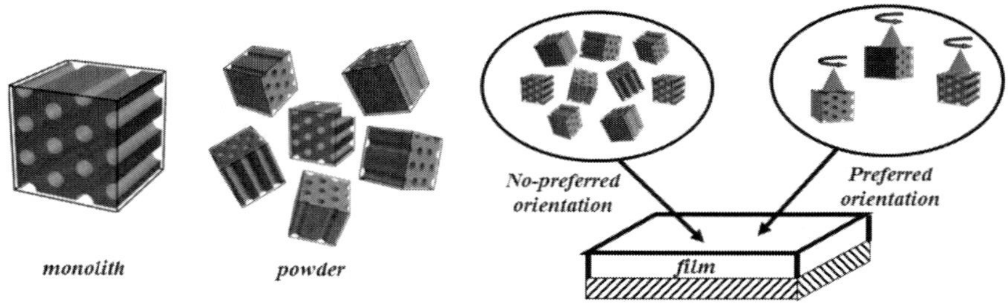

Figure 1. Scheme of the different textures for mesoporous materials : Monolith, powder and thin film.

All the possible types of structures for mesoporous materials are presented in Figure 2. They can be sorted in different families according to the packing parameter of the templating agent, ranging from spherical, cylindrical to lamellar pores. [3] The 2D-hexagonal phase (cylindrical pores) corresponds to intermediate values of the packing parameter.

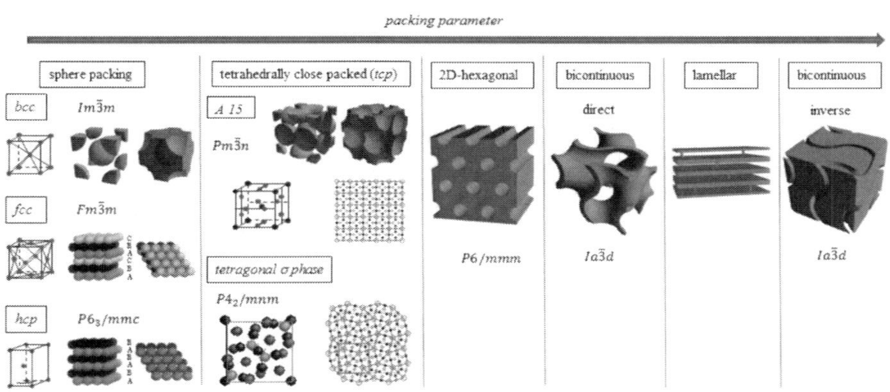

Figure 2. The different families of structures for mesoporous/nanoporous materials depending on the surfactant packing parameter. Sphere packing: bcc (cubic space-group $Im\bar{3}m$), fcc (cubic space-group $Fm\bar{3}m$), and hcp (hexagonal space-group $P6_3/mmc$). Tetrahedrally close packed (tcp) structures with the A15 structure (cubic space-group $Pm\bar{3}n$) and the tetragonal σ phase (tetragonal space-group $P4_2/mnm$)). Dodecahedral cages are depicted in blue, 14 facets polyhedrons in red and 15 facets polyhedrons in green. Drawings of the projections of these two structures are adapted from reference5. 2D-hexagonal phase (space-group $P6/mmm$), consisting in a 2D-hexagonal lattice of cylindrical pores. Bicontinous structures, illustrated with the Gyroid phase (cubic space-group $Ia\bar{3}d$), either direct or inverse. Lamellar structures, consisting in a 1D packing of parallel sheets. They need to be stabilized by defects like junctions between two consecutive sheets.

For higher values of the packing parameter, bicontinous structures can be obtained. Two versions of these materials can be synthesized. In the direct case, the material has the shape of a periodic curved film of almost constant thickness, which separates two porous periodic networks. The inverse case corresponds to the complementary structure : The porosity is shaped along the periodic curved film, which separates two continuous solid sub-phases. In both versions, direct and inverse, the structures exhibit highly-open tridimensional porous networks. A related structure is the lamellar structure, obtained from a lamellar or smectic mesophase. However, this structure often does not give a very stable 3D architecture, as a sufficient amount of junctions/defects between consecutive sheets is needed to avoid its collapse.

Smaller values of the packing parameter correspond to three-dimensional lattices of spherical pores/cages. These lattices can be sorted in two families : sphere packing and tetrahedrally close packed (tcp) structures. Their ideal representation is a three-dimensional lattice of holes (like the holes in a piece of cheese , but periodically distributed ...). But very often in real materials , secondary pores give rise to connections between adjacent holes, and the porosity is in fact a connected tridimensional network. There are many possibilities for the symmetry of such three-dimensional lattices of holes. Simple lattices like the body-centered-cubic lattice (bcc) are observed. Close packing of spheres is also encountered: the face centered cubic (fcc) and the hexagonal close packing (hcp) [4]. These two structures are closely related and both correspond to the densest packing of identical spheres. The sequence of 2D dense planes is the so-called ABC sequence in the fcc structure and the ABAB sequence in hcp. Quite often, a disordered stacking of the A, B and C planes may be obtained instead of completely periodic packing (statistical crystal).

The last family contains the tetrahedrally close packed (tcp) structures, also known as "Frank and Kasper" phases. [5] One of their main features is that different types of holes/cages coexist in the same structure , unlike the previously mentioned sphere packing (bcc, fcc and hcp), where all cavities are identical and have essentially a spherical symmetry . As a counterpart, when fcc/hcp packings contain two different types of interstices, tetrahedral and octahedral ones, a tcp structure exhibits only tetrahedral interstices. Note that the tetrahedrons are always slightly distorted in a tcp structure because it is not possible to fill the space only with regular tetrahedrons. The most commonly observed phase in this family is the so-called A15 structure with the cubic space-group $Pm\bar{3}n$ (see PART 2.3). Two kinds of cages are present in the structure: Dodecahedral cages and polyhedrons with 14 facets. They are expected to have slightly different sizes and shapes. The dodecahedral cages have essentially a spherical symmetry, whereas the 14 facets polyhedrons are slightly ellipsoidal. Moreover, in the real porous structure, connections between adjacent cages may be present, and again an open tridimensional porosity may be present. It has also been shown that the porous network of the A15 structure can be alternatively described by a periodic surface. [6]

In the near future, more and more other tcp structures are expected to be observed in soft matter in general. [7-10] Concerning mesoporous/nanoporous materials , a first example of the tetragonal σ phase has been already discovered, [3,11,12] which exhibits three different types of cages in coexistence. Last but not least, the first example of a quasi-crystal with a dodecagonal symmetry was found recently, [13] illustrating the relation between quasi-crystalline structures and tcp ones. [5] No doubt that more and more new complex structures will be elaborated, like chiral materials [14] or hierarchical materials. [15,16]

PART 2: X-RAY SCATTERING: GENERALITIES

2.1. Scattering of X-Rays by Matter

X-rays are electromagnetic waves with wavelengths ranging typically from 0.05 nm to 0.25 nm for crystallographic applications, which corresponds to energies between 5 keV and

25 keV. They are especially interesting for structural investigations because they are highly penetrating and their wavelengths are of the order of inter-atomic distances found in condensed matter . They may interact with the electrons in matter by three different mechanisms: (i) the photo-electric effect which is essentially responsible for absorption and fluorescence phenomena, (ii) Compton scattering , where the scattering of the incident X-ray photon is accompanied by some energy transfer to the electron it is interacting with (inelastic scattering) and (iii) Thomson scattering where the scattered X-ray photons have the same energy as in the incident beam (elastic scattering). Compton scattering is incoherent, which means that Compton photons present no phase relation. On the contrary, Thomson effect is coherent and the scattered photons present phase relation: they will give thus rise to interference effects. From a classical point of view, it may be described as the oscillation of the electrons under the influence of the electric field associated to the incident X-ray beam. Each electron reemits a wavelet of similar wavelength whose amplitude is proportional to the exciting field. Central to the theory of X-ray diffraction is the calculation of the phase relation between these scattered waves. It is depicted in Figure 3.

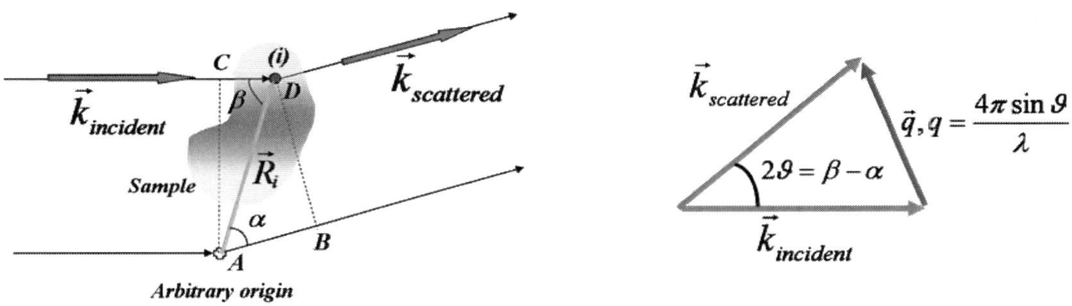

Figure 3. Thomson scattering for X-rays.

The directions of the incoming and scattered X-rays with wavelength λ are characterized by the wave vector s $\vec{k}_{incident}$ and $\vec{k}_{scattered}$ with equal modulus $2\pi/\lambda$ because Thomson scattering is elastic. The phase shift of the wavelet scattered by the electron located at C with respect to an arbitrary origin O is given by $\varphi = (2\pi/\lambda)\cdot(BC - OB) = (2\pi/\lambda)\cdot R\cdot(\cos\beta - \cos\alpha) = \vec{q}\cdot\vec{R}$ where

$\vec{q} = \vec{k}_{scattered} - \vec{k}_{incident}$ is the scattering vector ($q = 4\pi\sin\vartheta/\lambda$). The amplitude scattered by the whole sample $A(\vec{q})$ is simply the sum of the amplitudes scattered by each electron . The contribution of the electrons of the ith atom may be grouped and this summation can be written $A(\vec{q}) = \sum_i f_i(q)\exp\left(-i\vec{q}\cdot\vec{R}_i\right)$. Coefficients $f_i(q)$ are called atomic scattering factors and only depend on the modulus q due to the spherical symmetry of atoms . Analytical expressions can be found in the International Tables for Crystallography. [17] This scalar expression does not include the polarization of the electric field reemitted by the atoms. Indeed, polarization effects play no significant role in the scattering range of interest for mesoporous materials (see later in PART 4). Care must be taken that X-ray detectors only measure an intensity, that is an amount of electro-magnetic energy received during a measurement : $I(\vec{q}) \propto \left\langle |A(\vec{q})|^2 \right\rangle = \left\langle \sum_i \sum_j f_i(q) f_j(q) \exp\left(-i\vec{q} \cdot \vec{R}_{ij}\right) \right\rangle$. It is readily seen that if some characteristic distance R gets repeated in the sample, scattering will occur for directions such that $\vec{q} \cdot \vec{R} \approx 2\pi$. This means that the larger the distances investigated, the smaller the modulus of the associated $\vec{q}$-vectors.

One usually distinguish three different domains for the scattering vector : Wide-angle X-ray scattering (WAXS), small-angle (SAXS) and ultra-small angle X-ray scattering (USAXS). The typical limits of these three regions are illustrated in Figure 4 along with the corresponding distance values in real space. The distances spanned by SAXS typically correspond to inter-pores distances found in mesoporous materials : this is the most useful scattering domain. Inter-atomic distances are associated to the WAXS domain and can be of interest to analyse the structure of the pore walls or of nanoparticles included into the pores [18] (see Figure 4). Lastly, USAXS provides information at larger length-scales, like the shape of the grains for powder materials .

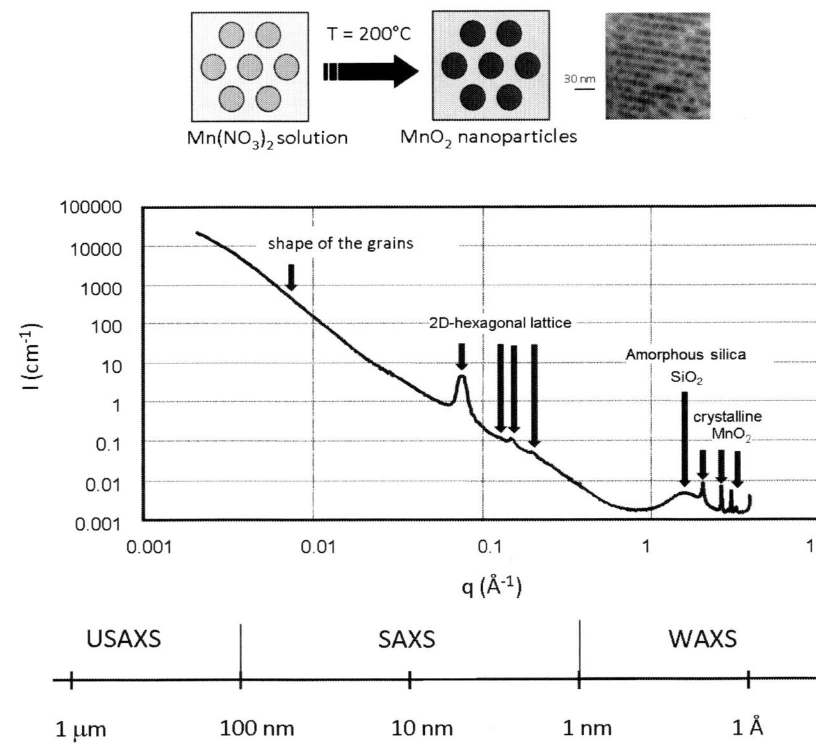

Figure 4. Illustration of the different X-ray scattering ranges by a powder sample of a 2D-hexagonal SBA-15 material loaded with MnO_2 nanoparticles . Adapted from [18].

$A(\vec{q})$ may also be expressed in terms of electron density $\rho_{el}(\vec{r})$ as an integral $A(\vec{q}) = \int_{V_{tot}} \rho_{el}(\vec{r}) e^{-i\vec{q}\cdot\vec{r}} d^3r$. As explained above, SAXS is not sensitive to atomic scale. In that scattering range, the electron density $\rho_{el}(\vec{r})$ describes matter at the nano-scale as a continuous medium and is simply equal to the number of electrons per unit volume (taking for the atomic scattering factors their values at $q = 0$). Only relative variations of the electron density $\rho_{el}(\vec{r})$ with respect to a constant electron density level contribute to the SAXS intensity. This can be explicated by writing $\rho_{el}(\vec{r}) = \rho_0 + \Delta\rho(\vec{r})$ where ρ_0 is a constant, such as the solvent electron density.

Then

$$A(\vec{q}) = \int_{V_{tot}} \rho_{el}(\vec{r})e^{-i\vec{q}\cdot\vec{r}}d\vec{r} = \int_{V_{tot}} \Delta\rho(\vec{r})e^{-i\vec{q}\cdot\vec{r}}d\vec{r} + \rho_0\int_{V_{tot}} e^{-i\vec{q}\cdot\vec{r}}d\vec{r} \cong \int_{V_{tot}} \Delta\rho(\vec{r})e^{-i\vec{q}\cdot\vec{r}}d\vec{r}\ ,$$

as the total sample volume V_{tot} is always assumed to be macroscopic and the second integral can be neglected.

A very convenient way to understand and visualize diffraction effects is to introduce the notion of *reciprocal space* and *Ewald sphere*. The expression $A(\vec{q}) = \int_{V_{tot}} \rho_{el}(\vec{r})e^{-i\vec{q}\cdot\vec{r}}d^3r$ makes it clear that the space of amplitudes $(\vec{q}, A(\vec{q}))$ is the Fourier transform of the direct space associated to the sample and defined by the electron density $\rho_{el}(\vec{r})$ at any point $\vec{r}$. The space of amplitudes is called the reciprocal space. As depicted in Figure 5, accessible scattering vectors at a given wavelength λ have a modulus smaller than $4\pi/\lambda$. Their extremities are located on a sphere of radius $2\pi/\lambda$, the so-called Ewald sphere, which is a very convenient geometrical representation of elastic scattering conditions. In case of structural determinations of conventional molecules where precision is required on inter-atomic distances , it is better to work with low wavelengths (hard X-rays) that give access to $\vec{q}$-vectors of high modulus. On the other hand, for a given $q = 4\pi\sin\vartheta/\lambda$ value, the lower the wavelength, the lower the diffraction angle 2ϑ. In the case of SAXS , this means working closer to the incident beam direction and soft X-rays are generally preferred.

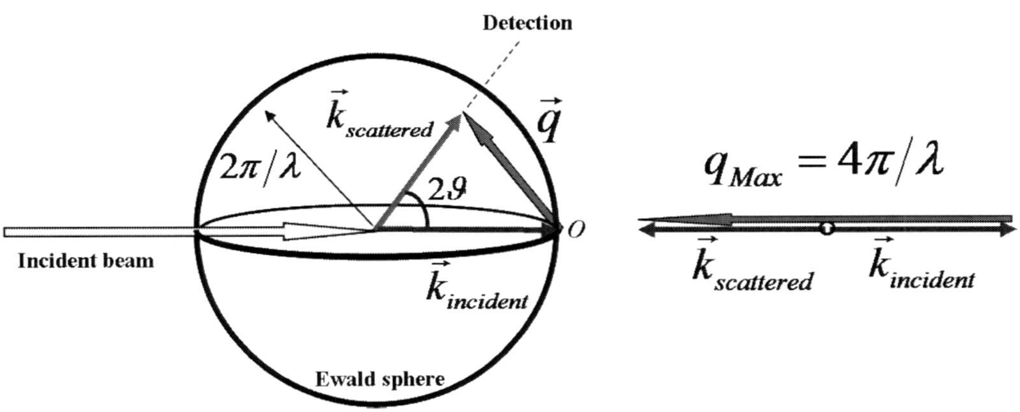

Figure 5. The Ewald sphere of reflection.

2.2. X-Ray Scattering By Ordered Mesoporous Materials

In this section, we first show how to construct the reciprocal space associated to ordered single-domains (3D and after 2D) that may be illustrated at a macroscopic scale by monoliths. It is then discussed how to use the Ewald sphere to determine diffraction conditions.

3D single domains: as shown in Figure 2, there is a great variety of 3D ordered structures that may be observed as monoliths. All these are obtained by the periodic repetition of a motif with translations $\vec{R}_{u,v,w} = u\vec{a} + v\vec{b} + w\vec{c}$ where *u*, *v* and *w* are integers and $\vec{a}$, $\vec{b}$ and $\vec{c}$ the unit cell vectors. The extremities of the $\vec{R}_{uvw}$ vectors define the lattice: placing the origin of the lattice on any point of the sample helps to visualize all other points that are equivalent by translation . The confusion between the mathematical lattice and the real structure must be avoided. The motif can be constituted of one or two identical pores (*bcc* or *fcc* and *hcp* structures), the association of pores of different sizes (*tcp* structures) or by a curved layer (bicontinuous phase). For the sake of simplicity, we assume the single-domain to be a parallelepiped with dimensions $L_a = N_a a$, $L_b = N_b b$ and $L_c = N_c c$. The electron density of the sample is given by the convolution of the motif electron density by the sum $\sum_{u=1}^{N_a}\sum_{v=1}^{N_b}\sum_{w=1}^{N_c}\delta(\vec{r}, \vec{R}_{u,v,w})$. The scattered amplitude is then the product of the Fourier transform of both quantities: $A(\vec{q}) = F(\vec{q})\sum_{u=1}^{N_a}\sum_{v=1}^{N_b}\sum_{w=1}^{N_c} e^{-i\vec{q}\cdot\vec{R}_{uvw}}$, where the factor $F(\vec{q})$ is the Fourier transform $\int_{V_c}\rho_{el}(\vec{r})e^{-i\vec{q}\cdot\vec{r}}d^3r$ of the electron density on the motif volume V_c. Each sum constitutes a geometric series so that: $\sum_{u=0}^{N_a-1} e^{-iu\vec{q}\cdot\vec{a}} = \frac{1-e^{-iN_a\vec{q}\cdot\vec{a}}}{1-e^{-i\vec{q}\cdot\vec{a}}} = e^{-i\frac{N_a-1}{2}\vec{q}\cdot\vec{a}}\frac{\sin\left(\frac{\vec{q}\cdot\vec{a}N_a}{2}\right)}{\sin\left(\frac{\vec{q}\cdot\vec{a}}{2}\right)}$.

The phase term plays no role as only intensities are finally considered. One has thus finally:

$$A(\vec{q}) = F(\vec{q})\frac{\sin\left(\frac{\vec{q}\cdot\vec{a}N_a}{2}\right)}{\sin\left(\frac{\vec{q}\cdot\vec{a}}{2}\right)}\frac{\sin\left(\frac{\vec{q}\cdot\vec{b}N_b}{2}\right)}{\sin\left(\frac{\vec{q}\cdot\vec{b}}{2}\right)}\frac{\sin\left(\frac{\vec{q}\cdot\vec{c}N_c}{2}\right)}{\sin\left(\frac{\vec{q}\cdot\vec{c}}{2}\right)}.$$

Sharp maxima in reciprocal space are observed if the scattering vector $\vec{q}$ fulfils the three Laue conditions $\vec{q}\cdot\vec{a} = 2\pi h$, $\vec{q}\cdot\vec{b} = 2\pi k$ and $\vec{q}\cdot\vec{c} = 2\pi l$ where *h*, *k* and *l* are integers. These integers are the Miller indices of the so-called *hkl* Bragg reflection. These conditions are equivalently written $\vec{q} = h\vec{a}^* + k\vec{b}^* + l\vec{c}^*$ if the vectors $\vec{a}^*$, $\vec{b}^*$ and $\vec{c}^*$ are chosen so that $\vec{a}\cdot\vec{a}^* = 2\pi$, $\vec{a}^*\cdot\vec{b} = 0$, $\vec{a}^*\cdot\vec{c} = 0$ and so on. Vectors $\vec{a}^*$, $\vec{b}^*$ and $\vec{c}^*$ define the reciprocal lattice. Out of the reflection position, one may write $\vec{q} = (h+dh)\vec{a}^* + (k+dk)\vec{b}^* + (l+dl)\vec{c}^*$ and

$A(\vec{q})$ goes first to zero for $dh = 1/N_a$, $dk = 1/N_b$ and $dl = 1/Nc$. In reciprocal space, the widths of the reflections are inversely proportional to the sizes of the single–domain (L_a , L_b and L_c) (see Figure 6-left). Therefore, large single-domains give rise to very confined intensities around the nodes of the reciprocal space. For any orientation of the sample, only those parts of the reciprocal space that intersect the Ewald sphere give rise to scattering (Figure 6-right).

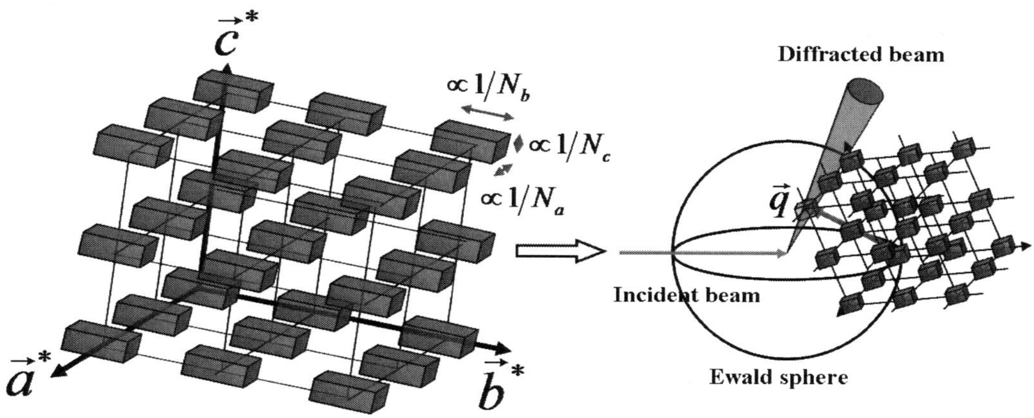

Figure 6. Left. Schematization of the reciprocal space of a 3D single-domain. Right. Illustration of the reflection conditions.

The intensity that may be scattered at any accessible $\vec{q}$ vector is given by $I(\vec{q}) \propto |F(\vec{q})|^2 L\left(\frac{\vec{q}\cdot\vec{a}}{2}, N_a\right) L\left(\frac{\vec{q}\cdot\vec{b}}{2}, N_b\right) L\left(\frac{\vec{q}\cdot\vec{c}}{2}, N_c\right)$, where we define the Laue function $L(x,N) = \dfrac{\sin^2 Nx}{\sin^2 x}$. It displays an approximate Gaussian shape around each main maximum $x = n\pi$ with a half-width at half-maximum of $\dfrac{0.9\pi}{2N}$. The scattered intensity is modulated by the term $|F(\vec{q})|^2$ that depends upon the motif structure (see PART 4).

Lattice planes: The lattice may be partitioned into families of equidistant and parallel planes called lattice planes. The association of the structural motif to such a family is a way to subdivide the sample into physical parallel planes (not to be confused with lattice planes that are mathematical objects). A family of planes is characterized by the Miller indices *(h,k,l)* and the interplanar distance d_{hkl} may be calculated knowing the unit cell parameters. The reflection of X-rays on these planes was first used to explain diffraction phenomena in crystals . Although not necessary to treat X-ray diffraction in single domains, the introduction of such planes is useful in some situations. A given *(hkl)* family of planes is perpendicular to vectors $\vec{q}_{nhnknl}$ where *n* is an integer and $q_{nhnknl} = 2\pi n/d_{hkl}$ (see Figure 7).

The Bragg law $2d_{hkl}\sin\vartheta = n\lambda$ is indeed equivalent to the Laue conditions on scattering vectors.

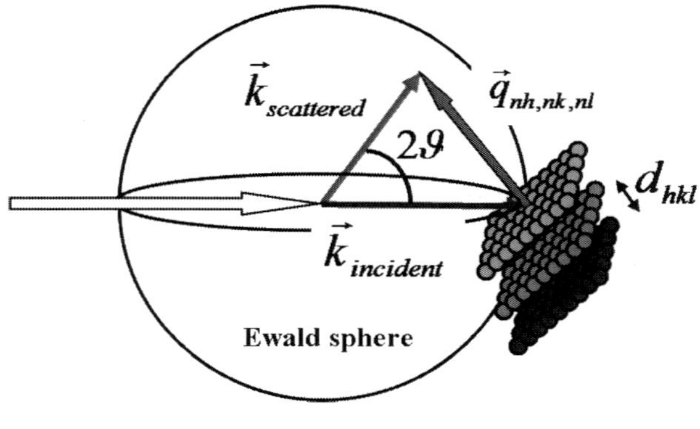

Figure 7. Reflections of X-rays on a family of planes: the Bragg law.

2D single domains: 2D structures in case of monoliths are essentially dense 2D- packing of cylindrical pores as depicted in Figures 1 and 8. It is obtained by the 2D-periodic repetition of a motif constituted by a pore and the surrounding walls by translations $\vec{R}_{u,v} = u\vec{a} + v\vec{b}$, where u and v are integers and $|\vec{a}| = |\vec{b}| = a$; the angle $(\vec{a},\vec{b})$ is chosen by convention equal to 120° in direct space. The discussion closely follows the same line as for 3D-ordered single-domains. The reciprocal lattice is now defined by two vectors $\vec{a}^*$ and $\vec{b}^*$ (with the angle $(\vec{a}^*,\vec{b}^*)$ = 60°) and it is easily shown that $|\vec{a}^*| = |\vec{b}^*| = 4\pi/a\sqrt{3}$. We suppose that the sample dimensions are $L_a = N_a \cdot a$ and $L_b = N_b \cdot b$ along $\vec{a}$ and $\vec{b}$ directions, and L_c in the pore axis direction (z axis).

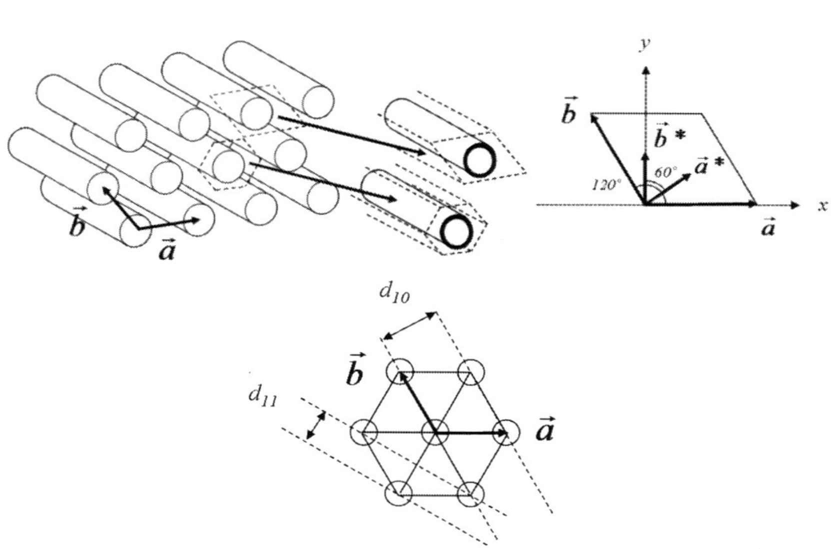

Figure 8. Left. Scheme of the 2D hexagonal packing. Notice that different unit cells may be used.

The scattered amplitude now reduces to

$$A(\vec{q}) = F(\vec{q}) \frac{\sin\left(\frac{\vec{q}\cdot\vec{a}N_a}{2}\right)}{\sin\left(\frac{\vec{q}\cdot\vec{a}}{2}\right)} \frac{\sin\left(\frac{\vec{q}\cdot\vec{b}N_b}{2}\right)}{\sin\left(\frac{\vec{q}\cdot\vec{b}}{2}\right)}.$$

$F(\vec{q})$ can be written as

$$F(\vec{q}) = \int_{-L_c/2}^{L_c/2} e^{-iq_z z} dz \int_{\Sigma_C} \rho_{el}(\vec{r}_{x,y}) e^{-i\vec{q}_{x,y}\cdot\vec{r}_{x,y}} d^2 r = \frac{\sin(q_z L_c/2)}{(q_z/2)} F_{x,y}(\vec{q}_{x,y}),$$ where Σ_C is

the unit-cell area. The left-hand term depends on q_z only. It presents a sharp maximum at $q_z = 0$ and goes to zero for $\delta q_z = 2\pi/L_c << 2\pi/\lambda$. As L_c is generally much larger than a, $A(\vec{q})$ is thus essentially concentrated around the nodes of the 2D reciprocal lattice defined by the translations $h\vec{a}^* + k\vec{b}^*$ (see Figure 9, left side). The intersection of the reciprocal space with the Ewald sphere is depicted in Figure 9-right for the {10} reflections that are usually the most intense. The lattice planes associated to these reflections are depicted in Figure 8-Right.

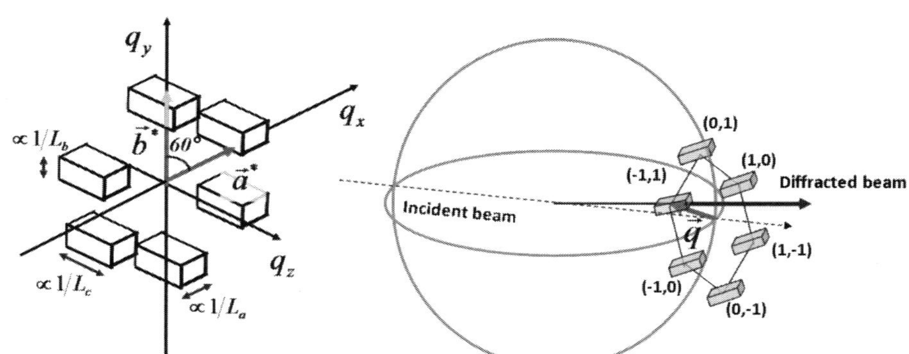

Figure 9. Left. Schematization of the reciprocal space of a 2D hexagonal single-domain. Right. Illustration of the reflection conditions.

The scattered intensity is

$$I(\vec{q}) = \left|F_{x,y}(\vec{q}_{x,y})\right|^2 \left(\frac{\sin(q_z L_c/2)}{(q_z/2)}\right)^2 \left(\frac{\sin\left(\frac{\vec{q}\cdot\vec{a}N_a}{2}\right)}{\sin\left(\frac{\vec{q}\cdot\vec{a}}{2}\right)}\right)^2 \left(\frac{\sin\left(\frac{\vec{q}\cdot\vec{b}N_b}{2}\right)}{\sin\left(\frac{\vec{q}\cdot\vec{b}}{2}\right)}\right)^2.$$

In Figure 10, the diffraction patterns corresponding to two perpendicular orientations of the same 2D hexagonal monolith are shown. [1] Figure 10-left corresponds to the incident beam parallel to the pore axis and hence to the z-axis (see Figures 8-9). It must be realized that the unit cell dimensions of such mesoporous materials are usually much bigger that the X-ray wavelength, that is $a >> \lambda$. In these conditions, the reciprocal cell dimension $a^* = 4\pi / a\sqrt{3}$ is much smaller than $2\pi/\lambda$ and the Ewald sphere looks almost flat close to the origin of the reciprocal space. Nevertheless, for a fixed position of a single-domain, the intersection with the Ewald sphere gives rise to asymmetric patterns on a 2D-detector. This is because, even if the scattering angles are small, Bragg peaks at $\vec{q}_{hkl}$ and $-\vec{q}_{hkl}$ are not simultaneously in reflection position. One usually needs to give the sample a slight oscillation in order to collect symmetric diffraction patterns. In Figure 10-left, all six first reflections are simultaneously recorded in reflection position. It is due to the fact that the monolith is not completely perfect and is made of domains slightly turned from each other along the hexagonal axis. It is classical in crystallography to describe the degree of perfection of the lattice translations in terms of so-called mosaic blocks, each block being a perfect single-domain. The mosaicity is ca.1° in the present case. [1] In Figure 10-right the sample has been turned by 90° so that the incident beam is perpendicular to the pore axis and only reflections of indices *(0,1)* and *(0,-1)* are visible.

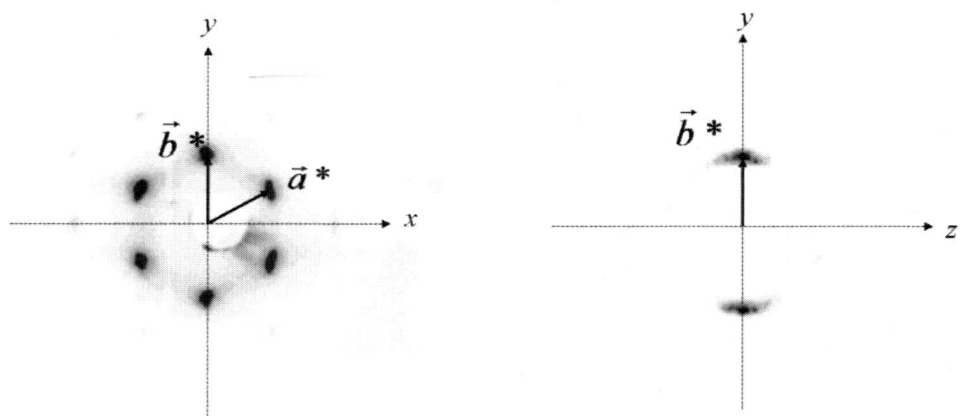

Figure 10. X-ray patterns obtained on a 2D-hexagonal monolith. Left. Incoming beam is parallel to the pore axis. Right. Incoming beam is perpendicular to the pore axis. Adapted from [1].

2.3. Space-Group Determination

The information about the pores organization is derived from the SAXS region (length scales between about 1 and 100 nm). The main feature is the presence of several Bragg peaks , because of the long-range order organization between the pores. Usually, only a very small number (typically 5 of even less) of Bragg reflections is recorded, like for example in Figures 4, 10 and after. From their positions in reciprocal space, one can deduce the unit cell dimensions that give the translation symmetry of the material. Other symmetry elements are reflection, rotation, roto-inversion , screw axis and glide plane. All these define the space group of the material. For each possible phase (see Figure 2), the list of most intense Bragg reflections is given in Table 1. The space-group identification is usually quite straightforward because of the small number of reflections. The first step is the indexation of the Bragg reflections. It consists to attribute Miller indices to each Bragg reflections. The scattering vector modulus q is a simple combination of Miller indices. Depending on the symmetry of the phase, notations are as follows, where m is the multiplicity (see PART 3): 3D-cubic (lattice parameter a), $q = 2\pi s/a$, hkl ($s^2 = h^2 + k^2 + l^2$, m); 3D-hexagonal (lattice parameters a and c) : $q = 2\pi s/a$, hkl ($s^2 = (4/3)(h^2 + k^2 + hk) + (1/R^2)l^2$, m) with $R = c/a \approx (8/3)^{1/2}$; 2D-hexagonal (lattice parameter a) : $q = 4\pi s/a\sqrt{3}$, hk ($s^2 = h^2 + k^2 + hk$, m); 3D-tetragonal (lattice parameters a and c) : $q = 2\pi s/a$, hkl ($s^2 = h^2 + k^2 + (1/R^2)l^2$, m) with $R = c/a$.

As only a small number of Bragg peaks is observed, the space-group attribution is often based on some additional knowledge of the system, like the phase diagram of the surfactant used as templating agent. When different space-groups are possible, the space-group with the highest symmetry is chosen. For example, for cubic phases, one always assumes that the point group is $m\bar{3}m$ and this corresponds to the presence of an inversion center and mirrors perpendicular to the 4-fold and 2-fold axis. However, the knowledge of the space-group may not be sufficient for the structure identification . For example in Table 1, two different phases have the same space-group $Im\bar{3}m$: the *bcc* phase and the P-surface bi-continuous cubic phase. To identify the structure, one has to take into account not only the positions but also the intensities of the Bragg peaks. Indeed, because the dimensions of the basic structural units (size of the pores) are comparable to the size of the unit cell, their form factor gives rise to strong modulations of the Bragg peaks intensities, as detailed in PART 4.

Table 1. Indexation of the Bragg reflections. The space-group and the sequence of the Bragg reflections are provided for all different possible structures. For each Bragg reflection, the Miller indices are given with the value of s^2 and the multiplicity *m* inside brackets. See PART 2.3 for notations. Miller indices are written in bold script when they correspond to a usually strong Bragg reflection

Type	Space-group	Bragg reflections	References
2D-hexagonal	P6/*mmm*	***10*** (1, 6), *11* (3, 6), *20* (4, 6), *21* (7, 12), *30* (9, 6), *22* (12, 6), *31* (13, 12) …	[19]
sphere packing	I$m\bar{3}m$ (*bcc*)	***110*** (2, 12), *200* (4, 6), *211* (6, 24), *220* (8, 12), *310* (10, 24), *222* (12, 8), *321* (14, 48), *400* (16, 6) …	
	F$m\bar{3}m$ (*fcc*)	***111*** (3, 8), *200* (4, 6), *220* (8, 12), *311* (11, 24), *222* (12, 8), *400* (16, 6), *331* (19, 24), *420* (20, 24), *422* (24, 24), *511* (27, 24)), *333* (27, 8), *440* (32, 12), …	
	P63/mmc (hcp)	***100 (4/3, 6), 002 (4/R2, 2), 101 (4/3+1/R2, 12), 102 (4/3+4/R2, 12), 110 (4, 6), 103 (4/3+9/R2, 12), 200 (16/3, 6), 112 (4+4/R2, 12), 201 (16/3+1/R2, 12) …***	[20-22]

Table 1. (Continued)

Type	Space-group	Bragg reflections	References
tetrahedral-close-packing (*tcp*)	$Pm\bar{3}n$	*110* (2, 12), ***200*** (4, 6), ***210*** (5, 24), ***211*** (6, 24), *220* (8, 12), *310* (10, 24), *222* (12, 8), *320* (13, 24), *321* (14, 48), *400* (16, 6) …	
	$Fd\bar{3}m$	*111* (3, 8), ***220*** (8, 12), ***311*** (11, 24), ***222*** (12, 8), ***400*** (16, 6), *331* (19, 24), *422* (24, 24), *511* (27, 24), *333* (27, 8), *440* (32, 12) …	[23]
	$P4_2/mnm$	***311*** (10+1/R^2, 16), ***002*** (4/R^2, 2), *112* (2+4/R^2, 8), ***410*** (17, 8), ***330*** (18, 4), ***202*** (4+4/R^2, 8), ***212*** (5+4/R^2, 16), ***411*** (17+1/R^2, 16), ***331*** (18+1/R^2, 8), ***222*** (8+4/R^2, 8), ***312*** (10+4/R^2, 16) …	[9,10]
bicontinuous cubic	$Ia\bar{3}d$ (G-surface)	***211*** (6, 24), ***220*** (8, 12), *321* (14, 48), *400* (16, 6), *420* (20, 24), *332* (22, 24), *422* (24, 24), *431* (26, 48), *521* (30, 48), *440* (32, 12) …	[19]
	$Pn\bar{3}m$ (D-surface)	***110*** (2, 12), ***111*** (3, 8), *200* (4, 6), *211* (6, 24), *220* (8, 12), *221* (9, 24), *310* (10, 24), *311* (11, 24)…	
	$Im\bar{3}m$ (P-surface)	***110*** (2, 12), ***200*** (4, 6), ***211*** (6, 24), *220* (8, 12), *310* (10, 24), *222* (12, 8), *321* (14, 48), *400* (16, 6), …	

PART 3. POWDER AND THIN FILMS

Powder: Mesoporous materials are quite often produced under the form of powders that are collections of randomly oriented micrometric monoliths (see Figure 1). The reciprocal space for a powder is thus obtained by averaging the reciprocal space of a single domain over all possible orientations. Each zone of the reciprocal space of a single domain located around a scattering vector $\vec{q}_{hkl} = h\vec{a}^* + k\vec{b}^* + l\vec{c}^*$ gives rise to a sphere of average radius $|\vec{q}_{hkl}|$ with some thickness. Different indices combinations may correspond to a similar scattering vector modulus and the number of such combinations is called "reflection multiplicity" m_{hkl} . The intersection of these concentric spheres with the Ewald sphere consists in circles as depicted in Figure 11. Each circle is associated to a cone of diffraction . The thickness of a given *(hkl)* sphere is related to the extension δq_{hkl} along the $\vec{q}_{hkl}$ direction of the zone surrounding each $\vec{q}_{hkl}$ node in the reciprocal space of a single microdomain (see Figure 8). More precisely the average single-domain dimension L_{hkl} along the $\vec{q}_{hkl}$ direction may be related to the diffraction cone width by the Scherrer formula $L_{hkl} = 0.9\pi/\delta q_{hkl} = (0.9\lambda)/(\cos\vartheta \cdot \Delta 2\vartheta)$ where $q_{hkl} = 4\pi \sin\vartheta/\lambda$.

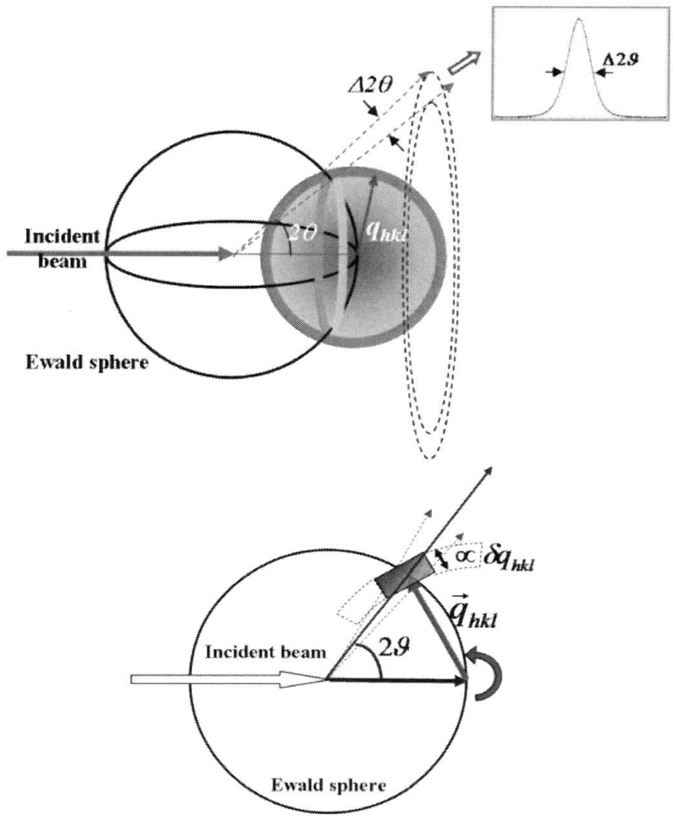

Figure 11. Left. Intersection of the reciprocal space of a powder sample with the Ewald Sphere. Right. Broadening of the reflection linked to finite domain size effects.

Films: Mesoporous materials may be prepared as thin films on solid substrate either by spin coating or by dip coating. A film has essentially a powder structure in that sense that it consists in independent micrometric single-domains. However two effects are specific to mesoporous materials . First, as detailed after, preferential orientation of the domains with respect to the substrate is frequent. Second, a further impact of the substrate is that strains are created during solvent evaporation and usually cannot relax due to the anchoring of domains at the substrate surface.

From a practical point of view, the substrate is in general opaque to X-rays which limits the accessible part of reciprocal space. As mentioned above, the maximal value for the *q*-modulus is $4\pi/\lambda$ which means that accessible $\vec{q}$ -vectors are all located within a sphere with this radius (Figure 12-left). The presence of the substrate directly eliminates the lower part of the sphere. Furthermore, scattering vectors located within the two half white spheres depicted in Figure 12-right correspond to forbidden transmission geometries (notice the orientation of the scattering direction given by $\vec{k}_{dif}$). In order to reduce shadowing by the substrate, one usually works with grazing incidence of the incident beam on the substrate (see Figure 12). This may lead to unwanted phenomena like reflected beam contamination or "phantom" reflections. [24] It must be signalled that using partially X-ray transparent substrates like ultra-thin silicon eliminates these restrictions .

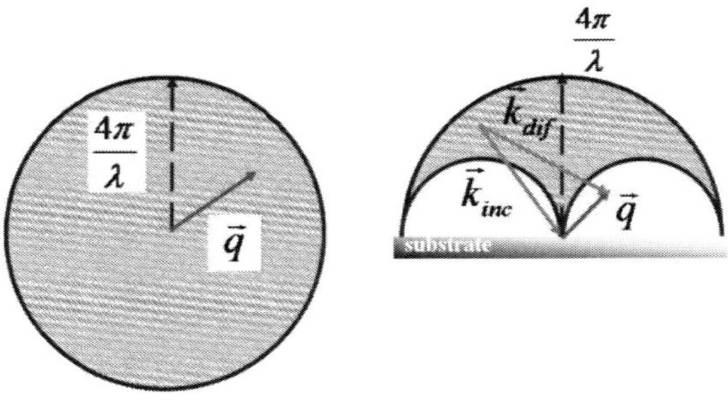

Figure 12. Left. Normally accessible range for $\vec{q}$ -vectors at a given wavelength. Right. Restrictions imposed by an opaque substrate .

Preferential orientation:. We introduced in PART 3 the notion of lattice planes with Miller indices *(hkl)*. They correspond to the subdivision of the single–domain into physical planes with spacing d_{hkl}. Smaller Miller indices are associated to smaller vectors $\vec{q}_{hkl} = h\vec{a}^* + k\vec{b}^* + l\vec{c}^*$ and hence larger interplanar distances $d_{hkl} = 2\pi/q_{hkl}$. As the bulk density is fixed, the 2D-density inside a lattice plane scales like $1/d_{hkl}$. Then, planes with large d_{hkl}. values are the densest planes, and have consequently the lowest surface energy . They are thus preferentially in contact with the substrate . A good example is given by films of 2D hexagonal mesoporous materials . The denser *{10}* planes usually lie flat on the substrate as depicted in Figure 13. The reciprocal space is thus obtained by rotating around the perpendicular to the substrate the reciprocal space associated to a single-domain. As a result, some Bragg peaks in reciprocal space are reinforced (for instance the *(0,1)* node located along the rotation axis) while others are transformed into circles.

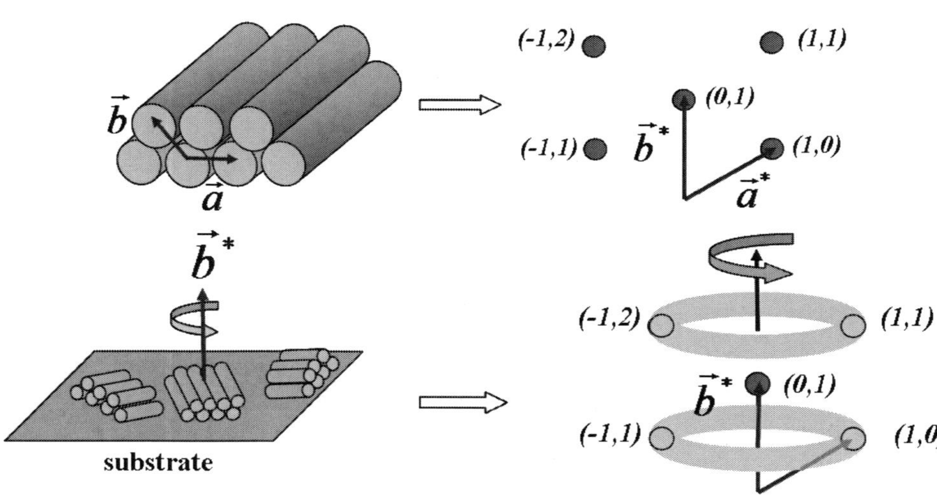

Figure 13. Schematization of the reciprocal space associated to a uniaxially oriented thin film.

It is shown in Figure 14 how the reciprocal space intercepts the Ewald sphere in case of grazing incidence geometry . The *(0,1)* "spot" of the reciprocal space is in reflection position only for a specific substrate orientation. On the other hand, circles intercept the Ewald sphere on a much wider angular range. Furthermore, the location of their intersection with the Ewald sphere depends upon the angular orientation of the substrate.

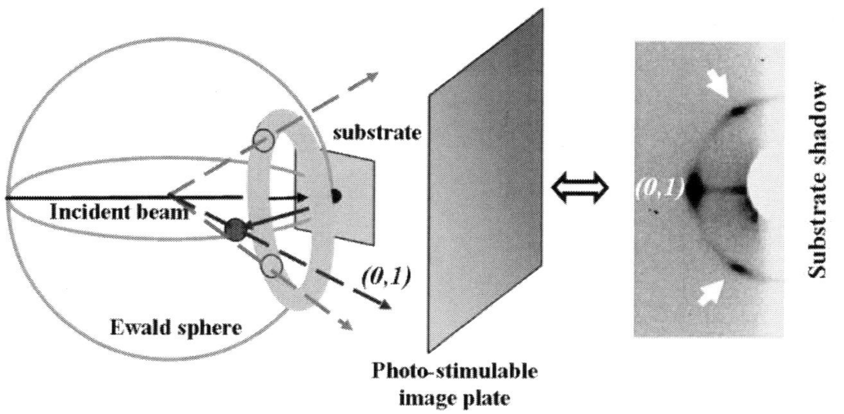

Figure 14. X-ray pattern obtained on a thin film of oriented 2D-hexagonal mesoporous material. Grazing incidence geometry .

Evaporation induced contraction. Evaporation of residual solvent during the last moments of the film formation generally leads to a unidirectional shrinkage as depicted in Figure 15 for material with 2D hexagonal structure . In-plane contraction is not possible, at least close to the substrate surface. It means that the in-plane component of any scattering vector $\vec{q}_{hkl}$ associated to planes of indices *(hkl)* is unaffected by the process.

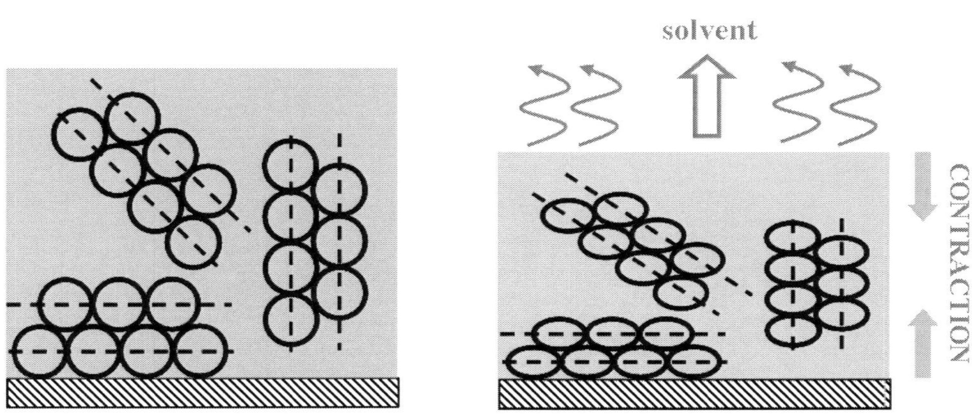

Figure 15. Effect of solvent evaporation on a thin film of 2D hexagonal mesoporous material.

This phenomenon is well-known in thin film analysis. [25] Let us suppose that residual stresses reduce to their in-plane components σ_{11} and σ_{22} (biaxial stress). The interplanar distance associated to a given family plane varies according to the orientation of the planes with respect to the substrate as $d_{\phi\varphi}^{hkl} = d_0^{hkl}\left(1+\left(\frac{1+\nu}{E}\right)\sigma_{\phi\varphi}\sin^2\psi - \frac{\nu}{E}(\sigma_{11}+\sigma_{22})\right)$

where ν and E are the material Young's modulus and Poisson ratio and $\sigma_{\phi\psi} = \sigma_{11}\cos^2\psi + \sigma_{22}\sin^2\psi$. In these conditions, concentric spheres that constitute the reciprocal space of powders are replaced by ellipsoids (see Figure 16).

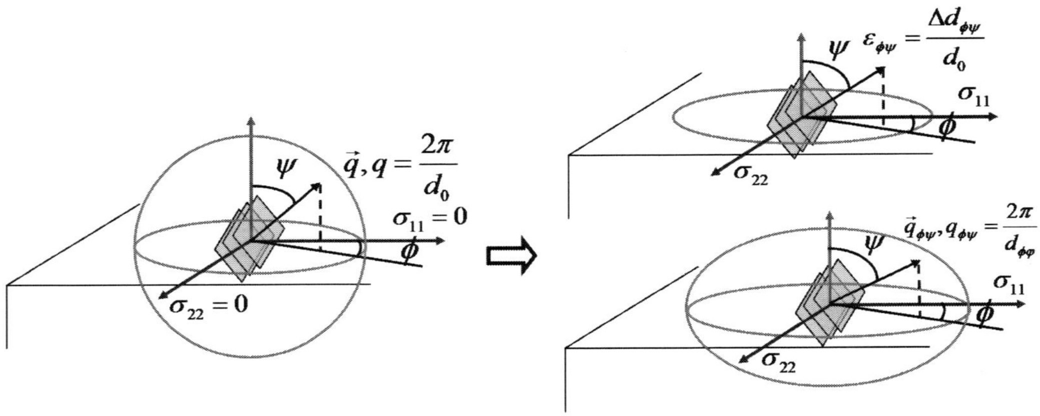

Figure 16. Effect of residual strain on the d-spacing and associated reciprocal sphere.

Mesoporous films are rather soft materials , especially in the early stages of their formation when wall condensation is uncompleted. Residual stresses lead thus to unusually strong effects. In general $\sigma_{11} = \sigma_{22}$ and the in-plane cross-section of ellipsoids is circular. In Figure 17, the scattering by a 2D hexagonal film deposited on ultra-thin (10µm) silicon is shown. [26] This type of substrate transmits more than 90% of X-rays photons at 8 keV which allows transmission geometry . The film was initially treated so that no preferential orientation is present. One clearly sees the transformation of the first sphere (associated to the distance d_{10} given in Figure 8) into an ellipsoid.

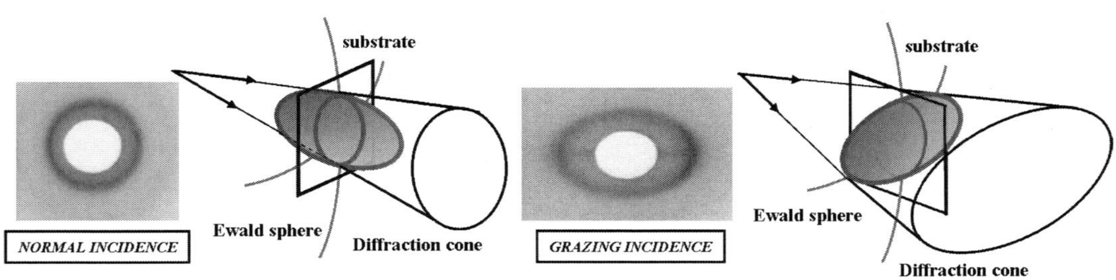

Figure 17. Effect of unidirectional contraction due to solvent evaporation on a thin film of randomly oriented 2D-hexagonal domains.

Remark: The technique of X-ray reflectivity is often used for the study of thin mesoporous films . The reader is referred to further references for a detailed description of the formalism. [27] It may be shortly described as SAXS in symmetric reflection geometry . It means that the $\vec{q}$ -vector remains perpendicular to the substrate surface during a scan. In this condition structural data contain no in–plane information. One may access the film thickness and interface rugosities, and the electron density of the film perpendicular to the substrate.

PART 4: QUANTITATIVE ANALYSIS OF SCATTERING DATA

As already mentioned in PART 2, a lot of structural information can be deduced from intensity measurements of the Bragg peaks . The essential fact is that, because the dimensions of the pores are close to the lattice size s, the form factor gives rise to strong modulations in the intensity. This contribution of the form factor gives precise information of the size and shape of the pores.

4.1. Intensities for SAXS

As seen previously, the Bragg peaks are present only in the SAXS region, where the intensity expression can be slightly simplified. Scattering angles are small (less than typically 5°) and polarization effects can be neglected. This is valid inasmuch as the direct beam delivered by modern multilayer optics is only very weakly polarized. As shown above, if any point $\left(\vec{q}, \vec{A}(\vec{q})\right)$ of the reciprocal space contacts the Ewald sphere, the diffracted intensity is proportional to $\left\langle |A(\vec{q})|^2 \right\rangle$. In SAXS, the intensity is usually expressed in absolute scale (cm^{-1} or mm^{-1} units): $I(\vec{q}) = \frac{r_e^2}{V_{tot}} < |A(\vec{q})|^2 >$, where $A(\vec{q}) = \int_{V_{tot}} \rho_{el}(\vec{r}) e^{-i\vec{q}\cdot\vec{r}} d\vec{r}$, r_e = 2.81794×10^{-15} m is the classical radius of the electron and V_{tot} is the sample volume irradiated by the X-ray beam. It corresponds to the differential scattering cross section, i.e., the number of photons scattered by a sample of unit volume per unit time, per unit solid angle, and divided by the incident flux. [28] After correction for the sample transmission , calibration in absolute units is usually obtained by measuring a standard sample of known scattering cross section, like liquid water ($I(0) = 1.6\,10^{-2} cm^{-1}$). [29]

For a perfectly 3D-ordered single-domain with a parallelepipedic shape , it reads $I(\vec{q}) = \frac{r_e^2}{V_{tot}} N |F(\vec{q})|^2 \frac{1}{N} L\left(\frac{\vec{q}\cdot\vec{a}}{2}, N_a\right) L\left(\frac{\vec{q}\cdot\vec{b}}{2}, N_b\right) L\left(\frac{\vec{q}\cdot\vec{c}}{2}, N_c\right)$ (see Part 2; $N = N_a N_b N_c$ is the number of unit cells). Generally, $I(\vec{q})$ is concentrated around each node $\vec{q}_{hkl} = 2\pi(h\vec{a}^* + k\vec{b}^* + l\vec{c}^*)$ of the reciprocal lattice into a small volume V^*, which indirectly reflects the single-domain shape. In the present case, V^* is a parallelepiped with typical dimensions ($1/(N_a a), 1/(N_b b), 1/(N_c c)$. The analysis of the reflection profile is thus informative on the single-domain shape. [17] This could be achieved normally by positioning the sample so that the Laue conditions are exactly fulfilled: in that case product of Laue functions is equal to N^2. This is often impossible in practice and one prefers to integrate the Laue functions by rotating the sample so that the whole volume V^* surrounding the $\vec{q}_{hkl}$ node crosses the Ewald sphere. The resulting measured intensity is proportional to the integrated intensity in reciprocal space, $I^{\text{int}}_{\sin gle}(\vec{q}_{hkl}) = \frac{r_e^2}{V_{tot}} \int_{V^*} |A(\vec{q}_{hkl} + \vec{q})|^2 d^3q \cong \frac{N}{V_{tot}} r_e^2 |F_{hkl}|^2 \frac{(2\pi)^3}{V_c}$, where V_c is the unit cell volume. A similar expression $I^{\text{int}}_{single}(\vec{q}_{hk}) = \frac{N}{V_{tot}} r_e^2 |F_{hk}|^2 (2\pi)^3 \frac{L_c}{\Sigma_c}$ is obtained for a 2D-hexagonal structure where Σ_c is the unit cell cross-section and L_c the pore length. [19]

For a single-domain, the proportionality factor between the measured intensity and this integrated intensity depends upon the velocity at which the domain is crossing the Ewald sphere (Lorentz correction). As an illustration, let us suppose that the sample is rotated around an axis perpendicular to a given $\vec{q}_{hkl}$ vector and to the incoming beam (see Figure 18). The node is approaching the sphere with a radial velocity $v_r = v_t \cos\vartheta = q_{hkl}\omega \cos\vartheta \propto \sin 2\vartheta$. Higher radial crossing velocities result in lower measured intensities, and the experimental data must be corrected by the factor $1/\sin 2\vartheta$.

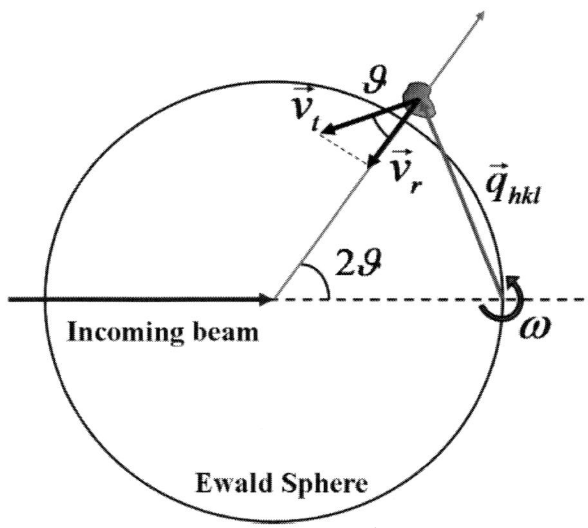

Figure 18. Illustration of the Lorentz correction during rotation of a single domain.

In case of powder samples, rotation of the sample is not necessary as all single-domain orientations are simultaneously present. Indeed reliable measurements of the integrated intensity are easier to obtain in that case than with a single-domain sample. As shown above (see Figure 19), for a powder sample, the intersection of the reciprocal space with the Ewald sphere consists in concentric circles. On can show that the powder intensity for SAXS is given by the formula $I^{\text{int}}_{powder}(q_{hkl}) = \frac{N}{V_{tot}} r_e^2 \frac{2\pi^2}{q_{hkl}^2 V_c} m_{hkl} |F_{hkl}|^2$ where m_{hkl} is the multiplicity of the *hkl* Bragg reflection (see Part 2). [19, 30] This value can be obtained either on a 2D detector by averaging the scattered intensity on a circle (dividing the total intensity by the circle's length) or using a conventional powder detector as depicted in Figure 19. In the later case, the punctual detector collects a given solid angle $d\Omega$ of the intensity $I^{\text{int}}_{powder}(q_{hkl})$ distributed over the perimeter $2\pi q_{hkl}\cos\vartheta$ of the intersection circle, but in SAXS the cosine term can be neglected. For a powder of 2D-hexagonal domains, one obtains a similar expression: $I^{\text{int}}_{powder}(q_{hk}) = \frac{N}{V_{tot}} r_e^2 \frac{2\pi^2}{q_{hk}^2} m_{hk} |F_{hk}|^2 \frac{L}{S_c}$. [19]

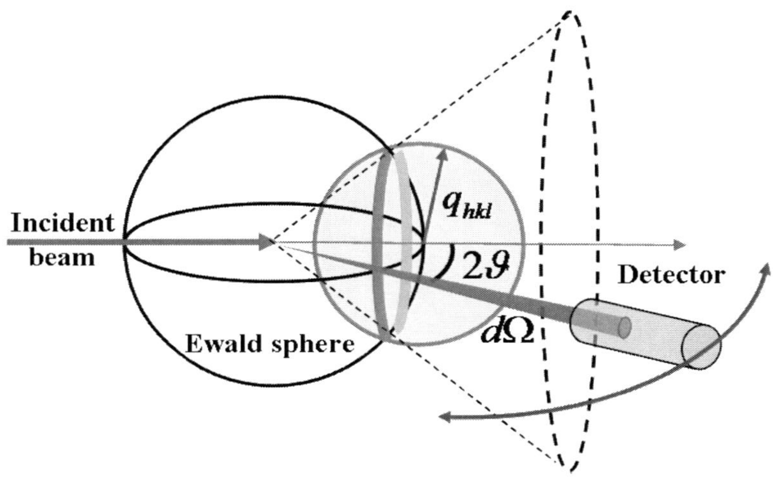

Figure 19. Scattering geometry for a conventional powder diffractometer.

4.2. Modeling of the Form Factor

We have shown how to deduce the absolute values of the Fourier coefficients F_{hkl} /F_{hk} of the electron density $\rho_{el}(\vec{r})$ from experimental measurements of the Bragg peaks intensities. Once the F_{hkl} /F_{hk} values are known, there are two ways to obtain a structural model in real-space, the so-called direct and inverse methods. With the direct method approach, the structural model is built directly in real space, the Fourier coefficients F_{hkl} are calculated from the model and compared to the experimental values. Essentially, core/shell models have been used for mesoporous materials , as illustrated in Figure 20.

Different parameters are introduced in the model: size, contrast, and possibly disorder terms. The values of these parameters are adjusted by a least-squared minimization procedure to fit the experimental intensity. [31] The main advantages of this method are: (i) to provide quantitative values of the parameters to describe the structural model, (ii) to be able to model the whole intensity curve (not only the Bragg peaks integrated intensities), and (iii) to introduce disorder terms in the description of the structure . The main drawback of a direct method is that a structural model has to be built already as a starting point, knowing that it may happen that several different models can fit as well the same experimental data.

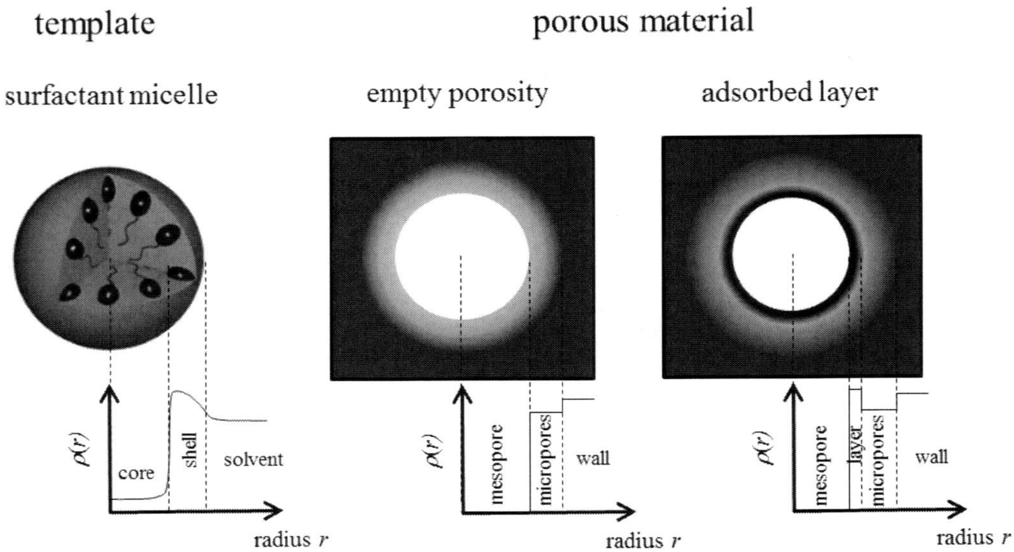

Figure 20. Different models used for structural characterization of pore and walls.

Inverse methods can be alternatively used. They consist in the reconstruction of the electron density function using an inverse Fourier transform from reciprocal space to real space. [32] Its main advantage is to provide the global features of the structure without assuming initially any models . Dealing with periodic structures, the key point is to solve the 'phases problem', that may be complicated for a large number of Bragg peaks . As the reconstruction is based only on the integrated intensities of the Bragg peaks, it provides a periodic average structure but includes no disorder terms. The electron density maps are obtained with a finite spatial resolution , fixed by the accessible q-range, and it is then more difficult to derive quantitatively the sizes of the structural units as with direct methods.

PART 5: APPLICATIONS

The main advantages of scattering experiments are the possibility to achieve time-resolved experiments and to combine the structural analysis in situ with a wide range of other types of measurements . In this last section, we will give a few examples to illustrate the variety of experiments that can be done.

5.1. Coupling with Adsorption/Desorption Experiments

Adsorption/desorption methods (BET) are very important techniques for studying porous materials . They are known to be crucial to investigate the porosity of mesoporous materials (after surfactant removal). The idea of coupling BET techniques and in situ SAXS is attractive as the incorporation of a liquid within the pores induces drastic changes in electron density contrast and thus in diffracted intensity. One of the first experiment of this type consisted in combining nitrogen adsorption at low temperature and SAXS measurement using a 2D-hexagonal silica material (CTAB template). [33] The powder sample fills an open glass capillary to as to allow nitrogen to flow in and out. The evolution of the most intense *(10)* diffraction line as a function of nitrogen vapor pressure is given in Figure 21. Capillary condensation at $p/p_S \approx 0.4$ is clearly manifested by a brutal drop in intensity. This comes from the fact that the electron density contrast between silica walls and liquid nitrogen is lower than with vacuum .

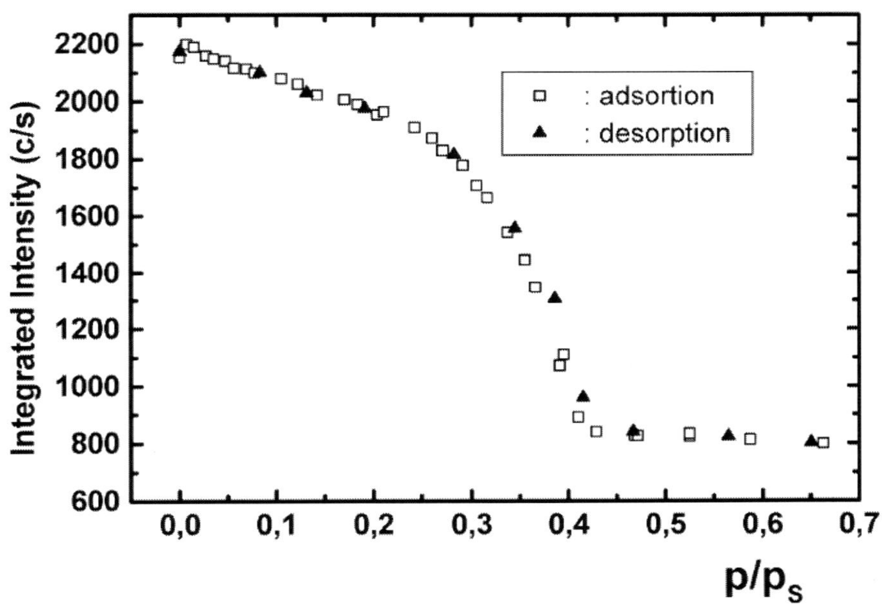

Figure 21. Evolution of the *(01)* reflection of a 2D-hexagonal mesoporous powder during nitrogen adsorption . Reprinted with permission from [33]. Copyright 2002, A.C.S., Chemistry of Materials.

The amplitude of the intensity drop allows access to the wall density knowing the density of adsorbed liquid nitrogen . The intermediate situation between zero vapor pressure and full pore filling was modeled by the addition of a layer of liquid nitrogen with variable thickness adsorbed at the pore surface (see Figure 20). The respective contribution of the mesopores and micropores could be separated this way. More recently, this type of approach has been refined and applied to other mesoporous silica materials like the 2D-hexagonal SBA-15 material [34] or to even more complex architectures with hierarchical porosity [35], and combining in situ adsorption measurements with both SAXS , [34, 35] and SANS [35]. The adsorption behavior of different fluids (nitrogen, [35] dibromomethane, [35] pentane/ perfluoropentane [34]) was fully quantitatively determined thanks to the modeling of the scattered intensity, based either on the chord-length distribution analysis [35] or on the direct modeling of the Bragg peaks intensities and the disordered micropores contribution. [34] It must be mentioned that X-ray reflectivity (see Part 2) has been similarly coupled to the adsorption of solvents for the study of thin mesoporous films . [36]

5.2. Evaporation Induced Self-Assembly (EISA)

The high flux delivered by synchrotron facilities allows real-time experiments that would be hardly feasible on laboratory X-ray sources. The formation and structuration of the mesoporous film deposited by dip-coating is usually a rapid process (a few minutes). Pioneering experiments in this domain were performed by D. Grosso and collaborators. [37] The principle of the experiment is summarized in Figure 22. The dip-coater is modified so that the beaker with the deposition solution is moved down while the substrate remains in position. The humidity level is carefully controlled. The beam illuminates the sample in grazing incidence geometry so as to minimize shadowing of the diagram by the substrate. The experiment depicted in Figure 22 corresponds to the deposition of a mesoporous silica film templated by CTAB, for which a succession upon time of intermediate structures was evidenced.

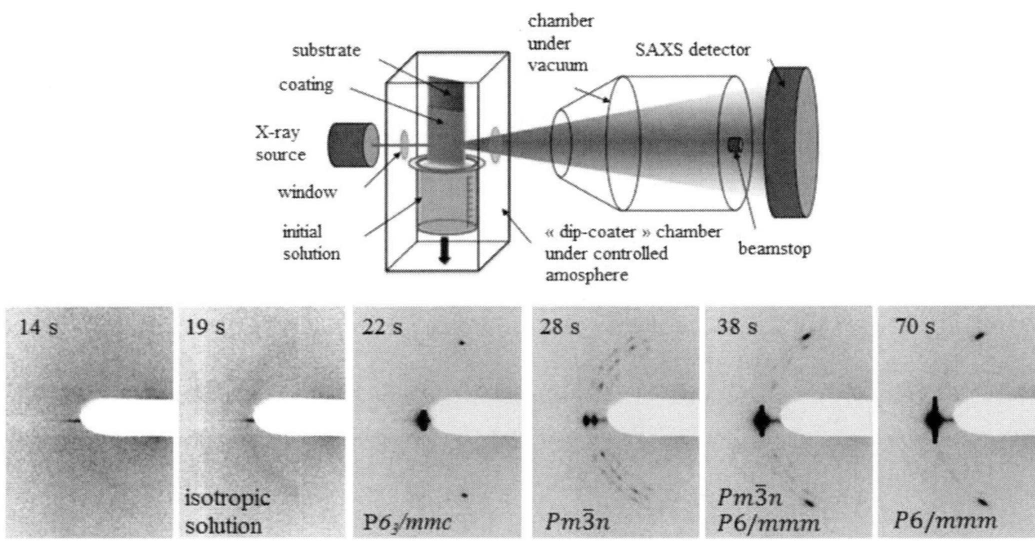

Figure 22. Structural evolution of CTAB-templated mesoporous film during initial steps of formation .

The observed phase sequence is better understood referring to Figure 2. The diffuse ring observed after 19s indicates that some short-range order starts to appear between the micelles . A 3D hexagonal compact phase forms after 22s followed by the formation an A15 structure (28s). This last is progressively replaced by a 2D hexagonal phase. The whole process corresponds to a global evolution from spherical micelles to cylindrical ones during silica condensation . [38] Experiments performed with a similar apparatus on titanium oxide films have shown the reversibility of the phase transformation s under the influence of changes in humidity . [39] The calcination of the template and the crystallization of the walls have been similarly followed at high temperatures. [40] This was done by combining simultaneously SAXS (to check the integrity of the pore organization) and WAXS (crystallization of the walls) experiments.

5.3. Self-Assembly in Solution of Powder Materials

Many materials are synthesized directly in water solution by adding an inorganic precursor to a dilute micellar solution of surfactant . In that case, a precipitation phenomenon occurs upon time, driven by the condensation reaction of the inorganic precursors. Using both in situ SAXS [41-46] and SANS [47, 48] experiments, the formation of the SBA-15 2D-hexagonal silica material has been extensively studied using TEOS or TMOS as a silica precursor. The surfactant which plays the role of the structure directing agent is a commercial non-ionic triblock copolymer (P123 [49] or P104 [50] - Pluronic@ BASF trademark) consisting of a central PPO hydrophobic block with two PEO hydrophilic blocks linked at its two extremities. Because the kinetic times for the formation of the material are long enough (typically 10 to 60 minutes), it was possible to follow in real time the evolution of the surfactant micelles upon time using synchrotron radiation [42, 43, 45, 47] or high-flux neutron sources. [47]

As shown in Figure 23, a simple experimental setup with a peristaltic pump was used to circulate the synthesis solution in a glass capillary /cell placed in front of the X-ray/neutron beam. [51]

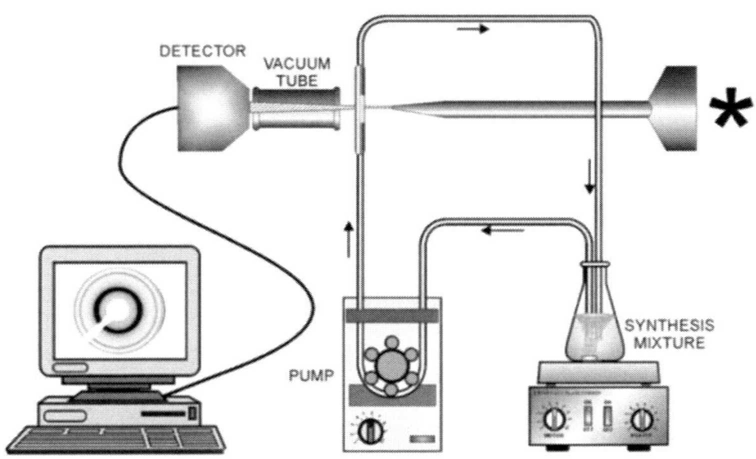

Figure 23. Schematic drawing of an in situ SAXS /SANS setup used for time-resolved analysis. The synthesis mixture is continuously flowing inside a quartz flow-cell. The asterisk (*) stands for the incoming X-rays/neutrons . Reprinted with permission from [51]. Copyright 2008, Elsevier, Advances in Colloid and Interface Science.

Typical acquisition times for one acquisition were 0.5 s with SAXS and 30 s with SANS , allowing recording of the scattering signals with a suitable temporal resolution as shown in Figure 24.

All the scattering curves were interpreted using a core/shell model to describe the micelles , as already introduced in PART 4 (Figure 20). [43, 45] As can be seen in Figure 24, at the beginning of the experiment, the scattering signals correspond essentially to the form factor of the micelles present in solution . The 2D-hexagonal phase is detected by the presence of its characteristic Bragg peaks , whose intensities are modulated by the form factor of the cylindrical micelles (see PART 4).

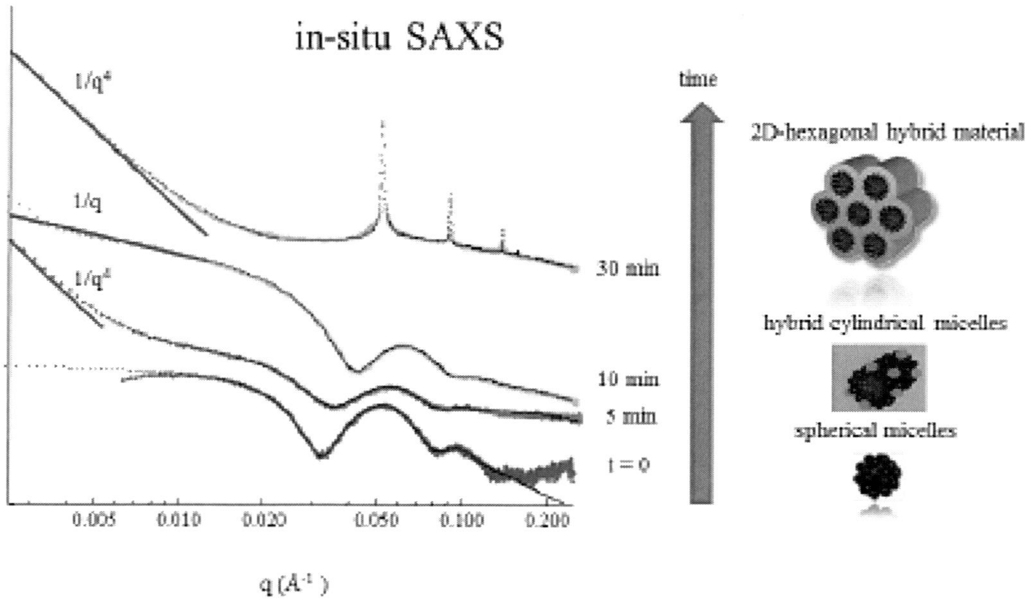

Figure 24. In-situ experiments. Evolution upon time of the 2D-hexagonal structure . Adapted from [45].

The main finding is that the micelles in solution undergo a sphere-to-rod shape transition upon time prior to the formation of the 2D-hexagonal hybrid material. This transition happens because an increasing amount of silica oligomers are present within the shell of the micelles, as derived from the modeling by a core/shell structure . [43] When a sufficient amount of silica oligomers are present within the shell of the micelles, the precipitation of the 2D-hexagonal hybrid material is observed. It can be explained by the fact that the attractive van der Waals interaction between the hydrophobic cores of the micelles becomes the dominant interaction, inducing the aggregation of the hybrid micelles into a 2D hexagonal lattice . Inside the hexagonal phase, the silica condensation continues and the progressive formation of the solid walls in-between the micelles can be followed by the variations of the Bragg peaks intensities, which are modulated by the form factor of the cylindrical micelles. [19] Using either SAXS [19, 41, 42] or SANS [48], different information is obtained. The silica wall formation in-between two adjacent micelles has been derived from SAXS results using either direct modeling[42] or electron density reconstruction . [41] This step is associated with a global shrinkage of the lattice parameter upon time because of the densification of the silica network during condensation. In addition, the concomitant decrease of the hydration of the wall upon time has been revealed by in situ SANS experiments, where the contrast is very sensitive to the amount of water within the walls. [48]

This type of in-situ experiments has been very useful to understand the influence of the many parameters of the SBA-15 synthesis and to consequently optimize them. For example, it has been shown that the nature of the acidic source [45] as well as the addition of salt [46] affect the whole kinetics of the material formation . It has been found that salting-out/kosmotropic anions (like $H_3PO_4^-$ or SO_4^{2-}) lead to better ordered materials than salting-in/chaotropic anions (like Br^-). [45] The influence of the nature of the silica sources (TMOS, TEOS , TPOS, [48] silicate [52]) has also been examined. In future works, the control over the morphology of the particles [53] obtained by precipitation will be investigated using USAXS experiments (see Figure 4). Indeed, it has been demonstrated recently that the stirring rate of the solution during the precursor hydrolysis step is a crucial synthesis parameter that controls directly the morphology of the grains, ranging from platelets to elongated rods or even to closed donuts-like shapes. [53]

Lastly, similar in-situ experiments have been recently performed with other surfactants as templating agents, CTAB, CTEAB and CTPAB, [54] and non-ionic per-fluorinated surfactants.[32] These studies revealed a rich variety of pathways towards the final hybrid materials , where different types of micelles (spherical of wormlike) may be involved, leading to 2D-hexagonal phases as well as $Pm\bar{3}n$ cubic phases.

CONCLUSION

In this chapter, the benefits of X-ray scattering methods in the investigation of periodic mesoporous materials have been underlined. Whatever the texture (monolith, powder or thin film), X-ray scattering is always the most straightforward way to establish the type of periodicity . Starting from basic notions in scattering theory, this chapter provides the rules to discriminate between all different possible types of periodic mesoporous materials and to have access to the texture of a material. Moreover, advanced methods for quantitative data analysis are explained, which give access to the shape and size of the pores. Many illustrative examples of applications are given, showing the high versatility for experimental condition s allowed with X-rays, especially in-situ conditions.

ACKNOWLEDGMENTS

We are grateful to Julien Schmitt (LPS, Orsay) for careful reading of the manuscript.

REFERENCES

[1] Melosh, N. A.; Davidson, P.; Chmelka, B. F., *J Am Chem Soc* 2000, *122*, 823-829.

[2] El-Safty, S. A.; Hanaoka, T.; Mizukami, F., *Chem Mater* 2005, *17*, 3137-3145.

[3] Liu, Z.; Fujita, N.; Miyasaka, K.; Han, L.; Stevens, S. M.; Suga, M.; Asahina, S.; Slater, B.; Xiao, C. H.; Sakamoto, Y.; Anderson, M. W.; Ryoo, R.; Terasaki, O., *J Electron Microsc.* 2013, *62*, 109-146.

[4] Ma, Y. H.; Han, L.; Miyasaka, K.; Oleynikov, P.; Che, S. N.; Terasaki, O., *Chem. Mater.* 2013, *25*, 2184-2191.

[5] Ungar, G.; Zeng, X. B., *Soft Matter* 2005, *1*, 95-106.

[6] Anderson, M. W.; Egger, C. C.; Tiddy, G. J. T.; Casci, J. L.; Brakke, K. A., *Angew Chem. Int. Edit.* 2005, *44*, 3243-3248.

[7] Iacovella, C. R.; Keys, A. S.; Glotzer, S. C., *P. Natl. Acad. Sci. USA* 2011, *108*, 20935-20940.

[8] Dotera, T., *Isr. J. Chem.* 2011, *51*, 1197-1205.

[9] Ungar, G.; Liu, Y. S.; Zeng, X. B.; Percec, V.; Cho, W. D., *Science* 2003, *299*, 1208-1211.

[10] Lee, S.; Bluemle, M. J.; Bates, F. S., *Science* 2010, *330*, 349-353.

[11] Garcia-Bennett, A. E.; Kupferschmidt, N.; Sakamoto, Y.; Che, S.; Terasaki, O., *Angew Chem. Int. Edit.* 2005, *44*, 5317-5322.

[12] Sakamoto, Y.; Terasaki, O., *Solid. State Sci.* 2011, *13*, 762-767.

[13] Xiao, C. H.; Fujita, N.; Miyasaka, K.; Sakamoto, Y.; Terasaki, O., *Nature* 2012, *487*, 349-353.

[14] Jin, H. Y.; Liu, Z.; Ohsuna, T.; Terasaki, O.; Inoue, Y.; Sakamoto, K.; Nakanishi, T.; Ariga, K.; Che, S. N., *Adv. Mater.* 2006, *18*, 593-+.

[15] Sanchez, C.; Arribart, H.; Guille, M. M. G., *Nat. Mater.* 2005, *4*, 277-288.

[16] Baldinozzi, G.; Muller, G.; Laberty-Robert, C.; Gosset, D.; Simeone, D.; Sanchez, C., *J. Phys. Chem. C* 2012, *116*, 7658-7663.

[17] *International Tables for Crystallography*; Kluwer Academic Publisher: Dordrecht, NL, 1996; Vol. C.

[18] Imperor-Clerc, M.; Bazin, D.; Appay, M. D.; Beaunier, P.; Davidson, A., *Chem. Mater.* 2004, *16*, 1813-1821.

[19] Imperor-Clerc, M., *Interface Focus* 2012, *2*, 589-601.

[20] Clerc, M., *Journal de Physique Ii* 1996, *6*, 961-968.

[21] Zeng, X. B.; Liu, Y. S.; Imperor-Clerc, M., *J. Phys. Chem. B* 2007, *111*, 5174-5179.

[22] Shearman, G. C.; Tyler, A. I. I.; Brooks, N. J.; Templer, R. H.; Ces, O.; Law, R. V.; Seddon, J. M., *J. Am. Chem. Soc.* 2009, *131*, 1678-+.

[23] Luzzati, V.; Spegt, P. A., *Nature* 1967, *215*, 701-704.

[24] Fall, S.; Kulij, M.; Gibaud, A., *J. Phys.-Condens. Mat* 2010, *22*.

[25] Birkholz, M. *Thin Film Analysis by X-Ray Scattering*; Wiley-VCH: Weinheim, Germany , 2006.

[26] Klotz, M.; Albouy, P. A.; Ayral, A.; Menager, C.; Grosso, D.; Van der Lee, A.; Cabuil, V.; Babonneau, F.; Guizard, C., *Chem. Mater.* 2000, *12*, 1721-1728.

[27] Daillant, J., Gibaud, A. *X-ray and neutron reflectivity : principles and applications*; Springer Verlag: Berlin, Germany , 1999.

[28] Pedersen, J. S., *Adv. Colloid. Interfac.* 1997, *70*, 171-210.

[29] Dreiss, C. A.; Jack, K. S.; Parker, A. P., *J. Appl. Crystallogr.* 2006, *39*, 32-38.

[30] Imperor-Clerc, M.; Davidson, P.; Davidson, A., *J. Am. Chem. Soc.* 2000, *122*, 11925-11933.

[31] Manet, S.; Schmitt, J.; Impéror-Clerc, M.; Zholobenko, V.; Durand, D.; Oliveira, C. L. P.; Pedersen, J. S.; Gervais, C.; Baccile, N.; Babonneau, F.; Grillo, I.; Meneau, F.; Rochas, C., *The Journal of Physical Chemistry B* 2011, null -null.

[32] Schmitt, J.; Imperor-Clerc, M.; Michaux, F.; Blin, J. L.; Stebe, M. J.; Pedersen, J. S.; Meneau, F., *Langmuir* 2013, *29*, 2007-2023.

[33] Albouy, P. A.; Ayral, A., *Chem. Mater.* 2002, *14*, 3391-3397.

[34] Findenegg, G. H.; Jahnert, S.; Muter, D.; Prass, J.; Paris, O., *Phys. Chem. Chem. Phys.* 2010, *12*, 7211-7220.

[35] Mascotto, S.; Wallacher, D.; Brandt, A.; Hauss, T.; Thommes, M.; Zickler, G. A.; Funari, S. S.; Timmann, A.; Smarsly, B. M., *Langmuir* 2009, *25*, 12670-12681.

[36] Rouessac, V.; van der Lee, A.; Bosc, F.; Durand, J.; Ayral, A., *Micropor. Mesopor. Mat.* 2008, *111*, 417-428.

[37] Grosso, D.; Balkenende, A. R.; Albouy, P. A.; Lavergne, M.; Mazerolles, L.; Babonneau, F., *J. Mater. Chem.* 2000, *10*, 2085-2089.

[38] Grosso, D.; Cagnol, F.; Soler-Illia, G. J. D. A.; Crepaldi, E. L.; Amenitsch, H.; Brunet-Bruneau, A.; Bourgeois, A.; Sanchez, C., *Adv. Funct. Mater.* 2004, *14*, 309-322.

[39] Grosso, D.; Babonneau, F.; Sanchez, C.; Soler-Illia, G. J. D. A.; Crepaldi, E. L.; Albouy, P. A.; Amenitsch, H.; Balkenende, A. R.; Brunet-Bruneau, A., *J. Sol-Gel. Sci. Techn.* 2003, *26*, 561-565.

[40] Grosso, D.; Soler-Illia, G. J. D. A.; Crepaldi, E. L.; Cagnol, F.; Sinturel, C.; Bourgeois, A.; Brunet-Bruneau, A.; Amenitsch, H.; Albouy, P. A.; Sanchez, C., *Chem. Mater.* 2003, *15*, 4562-4570.

[41] Flodstrom, K.; Teixeira, C. V.; Amenitsch, H.; Alfredsson, V.; Linden, M., *Langmuir* 2004, *20*, 4885-4891.

[42] Khodakov, A. Y.; Zholobenko, V. L.; Imperor-Clerc, M.; Durand, D., *J. Phys. Chem. B.* 2005, *109*, 22780-22790.

[43] Sundblom, A.; Oliveira, C. L. P.; Palmqvist, A. E. C.; Pedersen, J. S., *J. Phys. Chem. C.* 2009, *113*, 7706-7713.

[44] Sundblom, A.; Oliveira, C. L. P.; Pedersen, J. S.; Palmqvist, A. E. C., *J. Phys. Chem. C.* 2010, *114*, 3483-3492.

[45] Manet, S.; Schmitt, J.; Imperor-Clerc, M.; Zholobenko, V.; Durand, D.; Oliveira, C. L. P.; Pedersen, J. S.; Gervais, C.; Baccile, N.; Babonneau, F.; Grillo, I.; Meneau, F.; Rochas, C., *J. Phys. Chem. B* 2011, *115*, 11330-11344.

[46] Teixeira, C. V.; Amenitsch, H.; Linton, P.; Linden, M.; Alfredsson, V., *Langmuir* 2011, *27*, 7121-7131.

[47] Imperor-Clerc, M.; Grillo, I.; Khodakov, A. Y.; Durand, D.; Zholobenko, V. L., *Chemical Communications* 2007, 834-836.

[48] Linton, P.; Rennie, A. R.; Alfredsson, V., *Solid State Sci* 2011, *13*, 793-799.

[49] Zhao, D.; Hhuo, Q.; Feng, J.; Chmelka, B. F.; Stucky, G. D., *J. Am. Chem. Soc.* 1998, *120*, 6024.

[50] Linton, P.; Rennie, A. R.; Zackrisson, M.; Alfredsson, V., *Langmuir* 2009, *25*, 4685-4691.

[51] Zholobenko, V. L.; Khodakov, A. Y.; Imperor-Clerc, M.; Durand, D.; Grillo, I., *Advances in Colloid and Interface Science* 2008, *142*, 67-74.

[52] Jammaer, J.; van Erp, T. S.; Aerts, A.; Kirschhock, C. E. A.; Martens, J. A., *J. Am. Chem. Soc.* 2011, *133*, 13737-13745.

[53] Lee, H. I.; Kim, J. H.; Stucky, G. D.; Shi, Y. F.; Pak, C.; Kim, J. M., *J. Mater. Chem.* 2010, *20*, 8483-8487.

[54] Michaux, F.; Baccile, N.; Imperor-Clerc, M.; Malfatti, L.; Folliet, N.; Gervais, C.; Manet, S.; Meneau, F.; Pedersen, J. S.; Babonneau, F., *Langmuir* 2012, *28*, 17477-17493.

In: Comprehensive Guide for Mesoporous Materials. Volume 1 ISBN: 978-1-63463-958-3
Editor: Mahmood Aliofkhazraei

Chapter 14

SYNTHESIS OF MESOPOROUS CARBON BY STABILIZATION OF THE NANOPARTICLES FORMING THE RESORCINOL/ FORMALDEHYDE GELS

Cesar A. Barbero, Mariano M. Bruno, Juan Balach, Angelica Baena Moncada, Gabriel A. Planes and Maria C. Miras
Departamento de Química, Universidad Nacional de Río Cuarto,
Córdoba, Argentina

ABSTRACT

The reaction of resorcinol and formaldehyde , catalyzed by base , produce a porous gel formed by chain-like aggregates of resin nanoparticles . Conventional drying caused aggregation of the constituent nanoparticles, with collapse of the pores, producing a resin without open pores (xerogel). The pyrolysis of the resin produces a compact carbon . The usual way to avoid pore collapse is to avoid water evaporation and instead drying using supercritical fluids (aerogels), lyophilization (cryogels) or using low surface tension solvents (ambigels). In all cases, the procedures are complex and expensive, including solvent exchange and/or long curing steps. In this chapter we describe the use of different species (micelles , polyelectrolytes , fibers or nanoparticles) to stabilize the resin nanoparticles against aggregation during drying. In that way, porous dry resin can be produced by air drying without complex experimental steps. Pyrolysis of the dry porous resins produces porous carbons which are suitable for applications in energy storage (supercapacitors), water purification (deionizers), energy conversion (fuel cell s) or catalysis . Moreover, the size and distribution of pores can be controlled by the concentration of soluble stabilizers . On the other hand, hierarchical porous carbons can be produced by hard templating the resin with silica nanoparticles or cellulose fibers. The materials are characterized and properties relevant to technological applications are measured.

INTRODUCTION

Porous carbon materials are widely used as electrode materials of supercapacitors ,[1] deionizers,[2] or fuel cell s.[3] The advantages of carbon for those applications is its low cost , good electrical conductivity and relative chemical inertness . Using different materials, both large BET surface area s (m^2/g) and large specific capacitances (F/g) can be achieved (Table 1)

Table 1. Properties of different carbon materials as electrode materials in electrochemical cells (adapted from ref.[4])

Material	Capacitance surface density ($\mu F/cm^2$)	BET area (m^2/g)	Testing solution
Black carbon	8	80-230	Aqueous (acid)
Black carbon	10	80-230	Aqueous(basic)
Carbon fibre cloth	6.9	1630	Organic
HOPG (basal plane)	3	< 1	Aqueous (neutral)
HOPG (edge plane)	50-70	< 1	Aqueous (neutral)
Graphite powder	35	4	Aqueous (neutral)
Carbon cloth	10.7	630	Aqueous (neutral)
Glassy carbon	13	< 1	Aqueous (neutral)
Carbon aerogel	23	650	Aqueous (basic)

Pyrolysis of cellulosic materials render non graphitizable or "polymeric" carbon which do not contain graphitic domains.[5] Therefore, the material will not suffer intercalation of ions in a electrochemical cell. Therefore, it can be used in electrochemical systems where reversible charging/discharging of the double layer is expected. On the other hand, carbons produced by pyrolysis of pitch or other polymers (e.g., polyethylene) contain graphitic domains and are not suitable to be used in electrochemical capacitors, while are useful in Li-ion batteries . The biggest drawback of cellulosic material as source of porous carbon is its poor processability.

Table 2. Surface areas of activated carbon s produced from different cellulosic materials

Cellullosic material	Surface area BET (m^2/g)	References
Apricot stones	1190	[6]
Cherry stones	875	[6]
Almon shells	998	[6]
Grape seeds	497	[6]
Nut shells	743	[6]
Pistachio nut shells	778	[7]
Macadamia nut shells	1718	[8]
Peanut hulls	1177	[9]
Corn cob	960	[10]
Olive seed	1339	[11]
Corn stover	334	[12]
Rice straw	2410	[13]

On the other hand, in Table 3 are summarized typical values of textural and electrochemical properties of porous carbons produced from petroleum pitch . As it can be seen, the specific capacitance could reach large values (ca. 100 F/g) for the largest BET surface area s (>2000 m^2/g). However, the measured capacitance surface density is low (1-5 μF/cm^2) compared with the values depicted in Table 1. It is likely that a large part of the surface area is not accessible to the solution electrolyte and therefore does not contribute to the material capacitance. Therefore, the BET surface area cannot be used alone to assess the usefulness of porous carbon materials .

Table 3. Properties of activated carbon s produced by activation of pitch carbon (adapted from ref. [14])

Activated carbon	BET area (m^2/g)	Pore size (nm)	Total Pore volume (cm^3/g)	Specific Capacitance (F/g)	Capacitance surface density (μF/cm^2)
M-10	1370	0.91	0.5	55.95	4.08
M-15C	1518	0.98	0.6	63.34	4.17
M-15A	1800	0.92	0.63	78.1	4.34
SACF-20	1839	0.97	0.7	48.8	2.65
M-20	2130	1.47	0.71	100	4.7
SACF-25	2371	1.19	0.98	27.9	1.18
M-30	2571	1.5	1.23	62.9	2.45

An alternative way to produce non graphitizable carbon involves pyrolysis of phenolic resins (PR) which renders the so called "glassy carbon", which is widely used in electrochemistry . [15] Phenol-formaldehyde resins can be used to produce electrodes for electrochemical supercapacitors when a thin layer of carbon is deposited onto mesoporous inorganic matrices [16]. The polymerization reaction of phenols with aldehydes (and related reactions [17, 18]) in solution produces a three dimensional polymer matrix which trap a significant amount of solvent. The growth occurs through a sol-gel process which usually renders a nanoporous solid. [19] If the spontaneous porosity of the gel could be maintained into the carbon, a porous glassy carbon can be easily produced by drying and pirolysis. However, the reaction of phenol with formaldehyde produces a material with occluded pores which has a small exposed surface. [20] The material can be activated by electrochemical oxidation to produce large surface capacitances . [21]

Table 4. Textural properties of carbons produced by pyrolysis of resins made by condensation of different phenol /aldehyde reactants

Phenol (P)	Aldehyde (A)	Surface area BET (m^2/g)	Pore size (nm)	Ref
Resorcinol	Formaldehyde	920	8	[23]
Phenol	Formaldehyde	625	28	[23]
Cresol	Formaldehyde	705	10	[24]
Tannin	Furfural	420	9.6	[25]
Tannin	Formaldehyde	715	8.5	[26]

Polymerization of reactive phenols (resorcinol , [27] cresol[24], [28] tannin,[29] etc.) produces a gel having open pores. Indeed, different phenols and aldehydes can be used to produce porous gels (Table 4) However, if a phenol -aldehyde gel is dried in ambient air , the porous structure collapses and the resulting material (xerogel) show small surface area s. The collapse is mainly due to the effect of large surface tension forces at the water /air interface inside the pore during drying . To avoid collapse, the gels could be dried using supercritical liquids (aerogels , [30]) since no liquid/gas interface exists in this condition. Unfortunately, the procedure is expensive and requires replacement of the water inside the pores by a molecule where supercritical conditions can be easily achieved. The most commonly used supercritical liquid, carbon dioxide , is immiscible with water. Therefore, water has to be exchanged by a solvent miscible with liquid CO_2 (e.g., acetone). Then, this solvent is replaced by liquid CO_2. Then, the pressure and temperature are adjusted to supercritical conditions (T= 304 K, P= 73.8 atm) and the pressure is slowly decreased to extract the CO_2 as a gas without liquid formation . Alternatively, the water miscible solvent (e.g., acetone) can be set to supercritical conditions (T = 508 K, P = 43.8 atm), which are more difficult to attain. [31]

Alternatively, the water in the pores is exchanged by a weakly interacting solvent (e.g., heptane) which is evaporated from an interface with low surface tension forces, producing ambigels . [32] Again, water cannot be directly exchanged by the final solvent because they are immiscible and several intermediate solvent exchanges have to be performed. Another way to avoid pore collapse involves freezing the solvent inside the pores, followed by sublimation (lyophilization) of the solvent, giving cryogels . [33] Again, water cannot be frozen inside the pores because ice crystals destroy the nanoporosity creating macropores . Therefore, water has to be exchanged by less structured solvents (e.g., tert-butanol [34]) before lyophilization. In Table 5 are described the properties of porous carbon produced by pyrolysis of resorcinol -formaldehyde resins , dried using different methods.

Table 5. Properties of porous carbon produced by pyrolysis of resorcinol -formaldehyde resins , dried using different methods to maintain porosity

Drying method	Name of the material	Surface Area BET (m^2/g)	Specific Capacitance (F/g)	Capacitance surface density ($\mu F/cm^2$)	Ref.
Supercritical drying	Aerogel	680	28	4.1	[35]
Freeze drying	Cryogel	739	128	17.3	[36]
Low surface tension solvents	Ambigel	709	81	11.4	[37]

As it can be seen in the preceding brief description, all the procedures involving special drying are experimentally complex and some loss of porosity still occurs during drying or solvent exchange. Therefore, another way to maintain porosity during drying should be found, which should be based on a better understanding of the gel formation mechanism. A schematic view of the gel´s growth mechanism of the gel is described in scheme 1. [38]

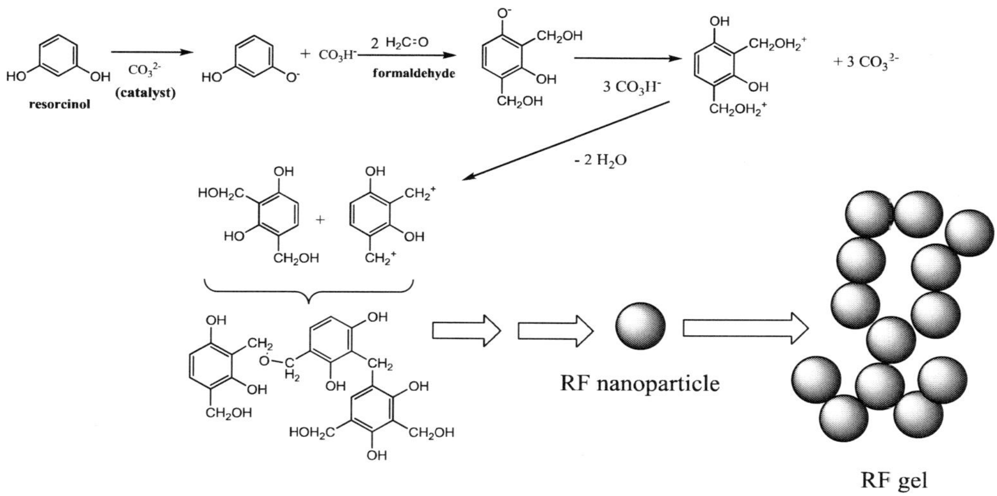

Scheme 1. Polymerization and growth of a RF gel .

The process begins with the deprotonation of resorcinol to give the more reactive fenoxide ion, which is electrophilically attacked by formaldehyde to form the methylol derivative. The alcohol becomes protonated and form the carbocation which can give an electrophylic attack on the ring, giving a methylene bridge, or attack the oxygen , giving an ether bridge. Note that carbonate act as catalyst because the species in equilibrium could remove or add protons . However, in water other bases (e.g., OH^-) could also act as catalysts due to the buffering effect of water. The process continues to produce nanoparticles or clusters of crosslinked polymer (7 to 10 nm in diameter).[39] The nanoparticles which aggregate into chains creating a three dimensional network. [40] (Scheme 1)

The building blocks of the gel are nanoparticles which aggregate loosely with each other. Upon conventional drying , the surface tension forces reorder the particles and they collapse into each other, closing the pores. (Scheme 2) The process is similar to the aggregation of nanoparticle dispersions. [41] Therefore, the same agents which stabilize nanoparticle dispersions could stabilize the porous structure of the gel. [42, 43, 44, 45] In this chapter, we will describe different ways to stabilize the gel nanoparticles against aggregation as a way to maintain the porosity during drying in ambient air .

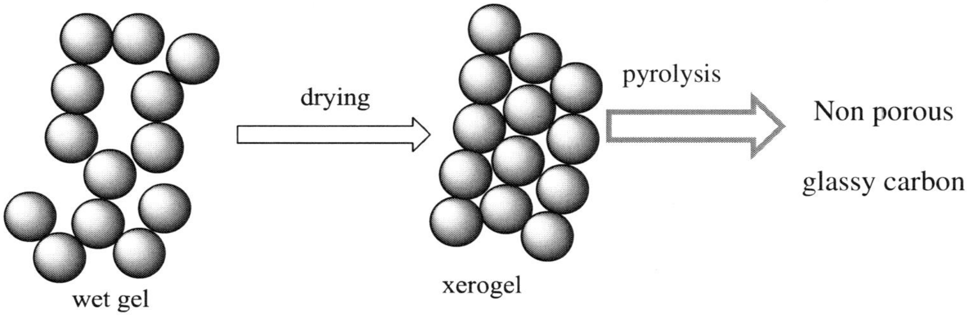

Scheme 2. Formation of non porous dry resin (xerogel) by conventional drying without stabilizers .

The earliest form of the mechanism was found while studying a soft template method to create resin porosity . Bell and Dietz reported the preparation of porous RF resins using cationic surfactant (e.g., cetyl, trimethylammoniuym bromide, CTAB) micelles as soft templates which are eliminated during carbonization , [46] a well known method to induce porosity of inorganic solids. [47, 48, 49] Others also report independently the use of the method, proposing similar mechanisms of pore formation . [50, 51, 52]

Table 6. Properties of porous carbon produced by pyrolysis of resorcinol -formol resins using cetyl, thimethylammoniuym bromide (CTAB) as template forming agent

Proposed soft template	Surface area BET (m^2/g)	Pore size (nm)	Specific Capacitance (F/g)	Capacitance surface density (mF/cm^2)	Ref
Spherical micelles	651	11	135	20.2	[46]
Microemulsion	607	15	101	16.6	[53]
Spherical micelles	501	7.4	--	--	[52]
Vesicular polymer*	2193	10	202	9.2	[54]

* With tert-butanol and mesitylene as co-surfactants .

As it can be seen, the procedure produces porous dry resins by drying in ambient conditions. The resins can be pyrolyzed to form porous carbon materials , which have large BET areas (Table 6) and can be used as electrode materials in electrochemical double layer capacitors (EDLC) having a reasonable capacitance per unit area (> 9 F/cm^2).

While the procedure indeed produces porous materials , there are strong evidences against a soft templating mechanism. The material does not show an ordered array of cylindrical pores as expected by the presence of an aggregate of cylindrical micelles but show an amorphous packing of aggregates, which it is more similar to those carbons produced by supercritical drying . On the other hand, the threshold concentration of surfactant required for porous structure to appear is much lower than the critical micelle concentration (CMC) for the formation of cylindrical micelles by CTAB and it is closer to the CMC for spherical micelles. Therefore, it is more likely that spherical micelles are the stabilizing agent of the nanoparticles forming the porous surface. The gel nanoparticles will bear negative charge due to the ionization of the phenoxide groups in the basic media . Therefore, CTAB micelles could adsorb on the negative nanoparticle surface, stabilizing them against aggregation . Such effect stabilizes the pores. We studied the method in detail and proposed a different mechanism of pore formation. The study allows proposing the porogenic effect of cationic polyelectrolytes .

Colloidal Stabilizer: Nanoparticle Stabilization by Adsorption of Spherical Micelles

The porous resin was prepared by reaction of resorcinol (R) with formaldehyde (F) in water (W) in the presence of a surfactant (S, CTAB). [55] To the solution , Na_2CO_3 (C) was added as catalyst . In a typical experiment, the molar ratio on the solution was fixed to F/R = 2, C/R = 5 x 10^{-3} and W/R = 10. The S/R molar ratio was set to maintain the CTAB concentration above its critical micelle concentration (CMC), to assure the presence of cationic micelles in the polymerization solution. Therefore, the minimal CTAB concentration is 0.077 M. Below that limit, non porous xerogel is produced. The maximum CTAB concentration used is 0.61 M because macroscopic phase separation is observed at higher concentrations. The mixture was stirred and heated was heated at 70 ºC for 24 h at atmospheric pressure . The brown monolithic RF polymer was dried at 70 ºC in air for 6 hours. No solvent exchange or special curing step is required. The dry monolithic resins are carbonized under a Ar or N_2 atmosphere at different temperatures (typically above 800 oC) reached with a heating rate of 40 K h^{-1}. The carbon materials produced in this way will be called micelle stabilized gel (MSgel).

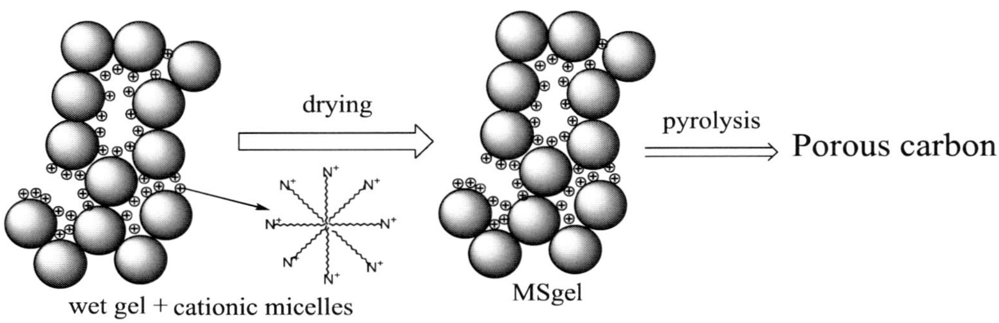

Scheme 3. Formation of a porous dry resin (MSgel), by cationic micelle stabilization , followed by pyrolysis to produce porous carbon .

The textural properties of the porous carbon were determined by N_2 adsorption isotherms . The BET surface area of the samples was measured to be 671 ± 4 m²/g (C = 0.077 M), and 576 ± 3 m²/g (C= 0.61 M). The values are comparable to those reported for carbon aerogels . [56] The BJH method, applied to the adsorption isotherms , renders a monomodal distribution of mesopores with maximum at ca. 30 nm. However, the adsorption at low pressures indicates that a finite contribution of micropores to the surface area could exist. The results suggests that it is possible to produce porous carbon by conventional drying , using cationic micelles as stabilizer. To try to find out the true nanostructure of the carbons, we used scanning electronic microscopy (SEM). Figure 1 shows the SEM micrography of a fractured porous carbon. As it can be seen, there is no defined order, porous channel system or distinct superstructures on the carbon, as it would be expected with templates made of surfactant micelles. On the other hand, spherical nanoparticles are clearly appreciated.

Figure 1. SEM micrograph of the fractured surface of a porous carbon prepared by carbonization of a RF resin synthesized in the presence of cationic micelles .

The nanoparticles have sizes varying from 20 to 50 nm (average 45 nm). The thin slits between the packed nanoparticles seems to account fo the carbon surface area . By TGA - MS, it is shown that CTAB ions are still retained in the RF matrix and decomposes during carbonization . The carbonizations processes of porous and compact RF resin are similar but the porous material carbonizes at lower temperatures. The Raman spectra of the MSgel show two broad bands at ca. 1350 cm^{-1} y 1580 cm^{-1}. The band at 1580 cm^{-1} (band G) is assigned to the in-plane vibration of the double bonds (C=C) between sp^2 carbon atoms , which is characteristic of graphitic layers. The band at 1350 cm^{-1} (band D) is assigned to the dangling bonds present in the borders and defects of the graphite layers. [57] Pure crystalline graphite shows a sharp and intense G band. [58] Therefore, the relative intensity of both peaks (I_D /I_G) has been used as a measure of the degree of carbon disorder .[59] The results observed here, with a ratio I_D/I_G = 1.1 indicate a medium degree of disorder, similar to glassy carbon and already observed in carbon aerogels . [60, 61] The X-ray diffractogram carbon MSgel show two broad bands at $2\theta = 24°$ and 43°, which are related to the reflections on the plane (002) and (100), respectively in the graphite crystals . Instead of sharp peaks, the reflections present themselves as broad band related with the disordered nature of the material. [62] Moreover, the diffractogram of the precursor resin shows a featureless diffractogram.

The material contracts during carbonization due to the densification inherent to the transformation of crosslinked RF polymer to a graphitic mass . In the present case, it was found a 10% contraction from wet gel to dry resin and a ca. 30 % contraction during carbonization. As expected, the higher the carbonization temperature , the larger the contraction. The small contraction during drying suggests that the porous structure has not collapsed. The conductivity of the carbon material depends on the carbonization temperature. It seems that higher temperatures allow formation of more ordered domains, with higher electronic conductivity. The higher conductivity obtained is of ca. 10 S/cm. The value is comparable with those reported before for porous carbon, [63] and high enough to allow the application of the material as electrode in electrochemical devices. The textural properties of the porous carbons depend strongly on the synthetic conditions (Table 7), mainly the relative amount of surfactant (S/R) and of catalyst (R/C).

Table 7. Textural Properties of porous carbon produced by pyrolysis of resorcinol - formol resins using cationic micelles as gel nanoparticle stabilizer

Synthetic conditions	Surface area BET (m^2/g)	Area µpore[b] (m^2/g)	Area mesopores[b] (m^2/g)	Volume pores[c] (cm^3/g)	Volume µpores[c] (cm^3/g)	Pore size[d] (nm)
R/C =200 S/R = 6	536	396	140	0.46	0.27	15
R/C = 600 S/R = 12	505	384	121	0.45	0.26	18
R/C = 600 S/R = 6	119	103	6	0.16	0.11	60

[a] Determined from BET theory, [b]determined from t-plot, [c]determined at p/p_0 = 0,97, [d]determined by Barret-Joyner-Halenda method (BJH).

The electrochemical properties were evaluated using cyclic voltammetry (CV) and Electrochemical Impedance Spectroscopy (EIS). The CV response for a carbon MSgel in 1 M H_2SO_4 (scan rate: 1 mV/s show a broad wave with maxima at ca. 0.43 $V_{Ag/AgCl}$ and minima at ca. 0.31 V. The average specific capacitance is of ca. 173 F/g. The wave observed is likely to be due to the oxidation /reduction of quinone -like functionalities (eqn (1)) present on the carbon surface.

$$QH_2 \rightleftharpoons Q + 2\,H^+ + 2\,e^- \qquad (1)$$

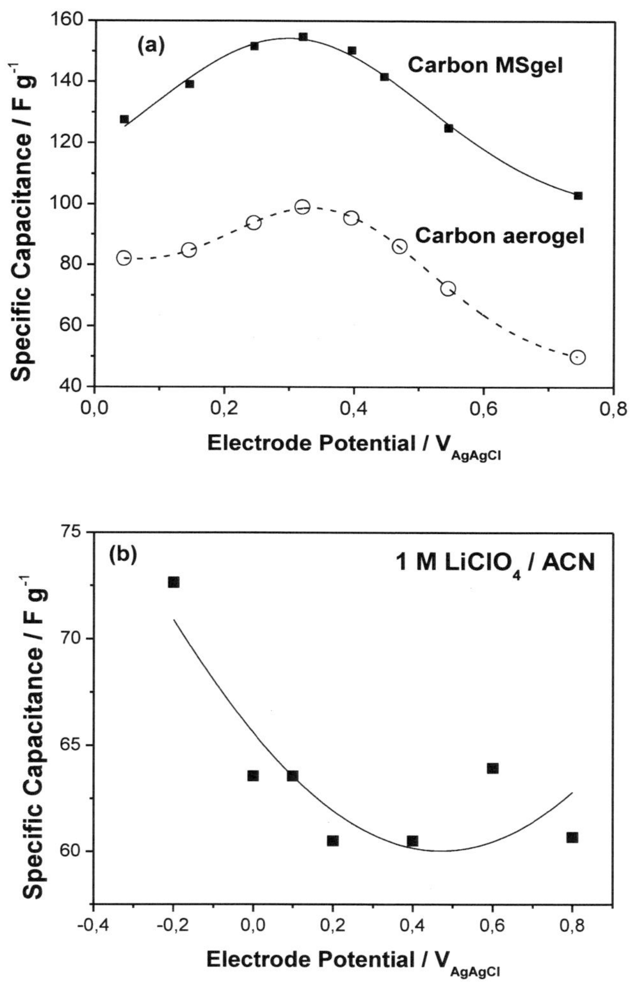

Figure 2. Specific capacitance values of: 1) carbon MSgel (full squares and full line) and carbon aerogel (open circles and dashed line) in: 1 M H_2SO_4/H_2O ; 2) carbon MSgel in 1 M $LiClO_4$/acetonitrile .

A small scan rate used is necessary to avoid the effects of slow response which alter the CV at higher scan rates. The slow response is due to the large time constant of the electrode, related to the slow mass transport inside the pores.

A better way to evaluate the maximum charge which can be stored inside the material involves EIS measurements (Figure 2). At high frequency, only the outermost area is accessed by the measurement . But at low frequency, all the surface area is accessed by the measurement. [64] Therefore, the maximum capacitance was calculated from the impedance values measured at low frequency (3 mHz): The capacitance varies with the electrode potential. Therefore, the measurements were all made at 0.35 V, in 1 M H_2SO_4. Figure 2 shows the values of specific capacitance (at 3 mHz) as a function of the electrode potential in 1 M H_2SO_4. The specific capacitance shows an inverted parabola with a maximum value of ca. 145 F/g at 0.35 $V_{Ag/AgCl}$. For comparison, a commercial aerogel (Maketech), produced using supercritical drying , was measured in acid media . The shape of the curve is similar to that of carbon mesogel but the maximum capacitance is only ca. 100 F/g. Therefore, the MSgel is as good as or better than those produced using supercritical drying. The bell shape of SC vs potential curve is likely due to the presence of pseudocapacitance related surface redox groups , which is apparent in the CV . To measure only the double layer charge, we used NaF as electrolyte . The ions in this electrolyte have not specific adsorption on carbon and the quinones surface groups are inactive at neutral pH in the range of potentials measured. [65] The capacitance vs potential graph has a parabolic shape, with a minimum value at ca. 0.3 V Ag/AgCl which should correspond to the potential of minimum charge of the material. The values of specific capacitance at that potential are smaller than in acid, being ca. 55 F/g in NaF/H_2O. The same measurement carried out in 1 M $LiClO_4$/acetonitrile (Figure 2.b), shows a parabolic shape as predicted by theory. [66] The value of specific capacitance is of ca. 60 F/g. The cyclic voltammogram measured in $LiClO_4$/acetonitrile show a typical ''butterfly'' appearance, with minimum aboslute value of current (both anodic and cathodic) at ca. 0.25 VAg/AgCl. Similar results were observed for carbon electrodes in Et_4NBF_4/acetonitrile. [67, 68] The shape can be attributed to electronic conductance of carbon, [69] or electrostatic interactions of ionic solutions side in the formation of double layer (Gouy Chapman theory). [70] By comparison between the capacitance values obtained in acid sulfuric and sodium fluoride media is possible to calculate the pseudocapacitance contribution which is around 50 % of the total capacitance in acid media. Therefore, the presence of redox ox ygenated groups (e.g., quinones) generates a significant faradaic pseudocapacitance. [71] The ion exchange during the charging/discharging of the double layer was studied using Probe Beam Deflection (PBD) techniques . [72] To avoid the interference of faradaic reactions , which have been shown to involve protons , [35] the measurement is carried out in a binary neutral electrolyte (KNO_3). The cyclic deflectogram shows a bipolar behaviour. In the anodic scan, at potentials more negative than ca. 0.1 $V_{Ag/AgCl}$, a negative going deflection is observed. This indicates the expulsion of ions from the electrode, likely to be cations expelled from the double layer. At potentials more positive than ca. 0.2 V, a positive going deflection indicates ion insertion , likely anions being incorporated to the double layer. The behaviour is the opposite in the cathodic scan. The region of mimimum charge (0-0.2 $V_{Ag/AgCl}$) agrees with the minimum in the capacitance-potential curve in NaF) and the depression in the "buterfly" CV shape indicating that the ion exchange is directly linked to the double layer charging/discharging process.

As it can be seen in Table 8, the synthetic conditions (S/R and R/C ratios) affect strongly the specific capacitance of the porous carbons. Specific capacitrances as high as 111 F/g (CV) and 272 F/g (EIS) can be obtained. The clear difference is related to the fact that cyclic voltammetry (CV) access only those larger pores where ions can be inserted from solution at larger rates. On the other hand, the electrochemical impedance spectroscopy at low frequencies (like 1-10 mHz) access the whole porous surface.

Table 8. Electrochemical properties of porous glassy carbon prepared by pyrolysis of resorcinol -formol resins prepared in the presence of CTAB

Synthetic ratios		**Specific Capacitance (F/g)**	
S/R	R/C	CV*	EIS#
6	100	-	100
6	200	-	144
6	300	-	9
6	400	-	1
6	500	-	2
6	600	-	<1
8	200	68	137
10	200	41	60
12	200	58	174
14	200	20	114
8	600	<1	20
10	600	24	69
12	600	111	272
14	600	30	96

Measured in 3 M H_2SO_4.
* Cyclic voltammetry (CV). Scan rate = 50 mV/s.
# Electrochemical impedance spectroscopy (EIS). Frequency = 2.8 mHz, E= 0.25 Vsce.
S= surfactant (CTAB), C= catalyst (Na_2CO_3), R = resorcinol .

Since different models have been proposed to explain the effect of CTAB, the existence of a minimal concentration of the cationic surfactant could indicate that micellization of the surfactant is necessary to form nanostructurated materials . However, just above the CMC only spherical micelles are formed with this surfactant concentration and those cannot template elongated pores. It seems more likely that stabilization of RF clusters by spherical micelles is the mechanism behind the templating effect of CTAB. It is possible to compare the results with the variation of BET surface area with CTAB concentration. While the surface area decreases ca. 100 m^2/g (20 % reduction) when S/R is varied from 0.015 to 0.12, the specific capacitance diminish only 5 %. It is likely that some compensation effect allows maintaining the specific capacitance in a high value. In any case, the results show that a wide range of CTAB concentration produces carbon mesogels with good specific capacitance.

To obtain the highest specific capacity, the effect of both catalyst and CTAB concentration was studied and the optimum concentrations was found using focused ion beam based nanotomography (FIB-nt).[73] The technique is performed by a serial sectioning procedure with a dual beam FIB (Ga ions)-scanning electron microscopy (SEM) instrument. In that way, a visual 3D morphology representation of disordered porous carbons is obtained. The pore size distribution of the same MSgel carbons was determined both by FIB-nt and nitrogen sorption isotherm methods in order to compare both techniques . The obtained MSgel carbon exhibit pore sizes from 4 to 120 nm. The specific area is quite low because the resolution of the measurement (2-10 nm) only allows measuring the mesopores and macropores (Table 9).

Table 9. Porosity of glassy carbon prepared by pyrolysis of resorcinol -formol resins (R/C=600) prepared in the presence of CTAB, measured by a Focused ion Beam-Scanning Electron Microscope (FIB-SEM)

Synthetic conditions	Total volume (μm^3)[a]	Area (μm^2)	Pore volume (μm^3)	Euler number	Total Porosity (%)[b]	Pore size (nm)	Specific Area (m^2/g)[c]
S/R =08	0.0625	1.446	0.0048	975	7.6	4	25.7
S/R = 10	0.0625	2.564	0.0118	-76	18.9	9	45.6
S/R = 12	0.0625	2.597	0.0214	-436	34.2	13	46.2
S/R = 14	0.0625	0.939	0.0283	-9	45.3	122	16.7

S= surfactant (CTAB), R= Resorcinol, C= Na_2CO_3

[a] analyzed volume, [b] PT = (V/VT)*100, [c] Surface area of 1 g of carbon calculated with density = 0.9 g/cm^3.

Besides monolithic carbon , microparticles can be prepared by controlled milling and sieving or using an inverse microemulsion as microreactor. The microparticles can be step by step self assembled into layers with specific capacitances of 194 F/g and high areal capacitance of 152 mF/cm^2. [74]

The carbonization plays an important role on the formation of the porous structure . The carbonization process was studied using thermogravimetric (TGA) analysis coupled with mass spectroscopy . The TGA curve of CTAB presents a mass loss around 270 °C due to its thermal degradation . The weight loss around 270 °C was observed for the CTAB compound and for the mesogel resin. This result suggests that this peak in the RF-surfactant curve can be attributed to CTAB decomposition . The production of the trimethylammonium ion (m/z=59) during thermal decomposition of the MSgel resin and CTAB alone is detected by coupled mass spectrometry . The profiles show that trimethylammonium ion detection begins at the same temperature in the mesogel resin and CTAB alone. However, the profile of the mesogel shows a tail at higher temperatures indicating that relatively free and more trapped CTAB ions are present in the resin. Fragments of larger masses (90 to 105 m/z) were detected RF mesogel carbonization, probably be due to the destruction of crosslink and oligomers in the resin.

Polymeric Stabilizer: Stabilization by Adsorption of Polyelectrolytes

The stabilization by cationic micelles of the RF gel nanostructure only requires the formation of the micelles and the adsorption on the RF gel nanoparticles , without apparent influence of the shape or size of the micelles. Therefore, other chemical species which could adsorb on negatively charged nanoparticles could produce the desired effect. The most obvious choice is a cationic polyelectrolyte bearing positive charges in any pH . Therefore, we tested poly(diallyl, dimethylammonium) chloride (PADAMAC) as nanoparticle stabilizing agent. The materials are called polyelectrogels. [75] (Scheme 4)

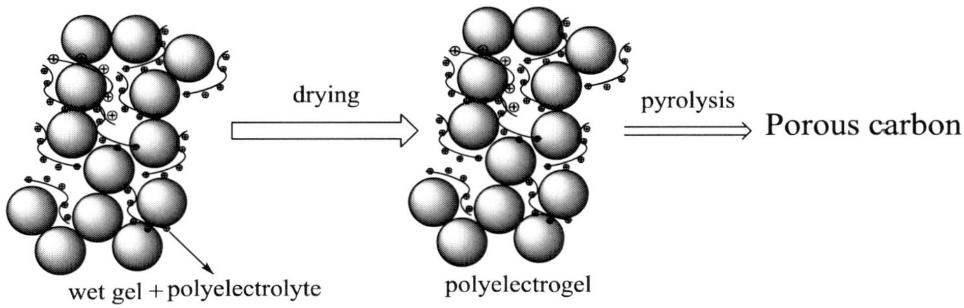

Scheme 4. Formation of a porous dry resin (polyelectrogel), stabilized by a cationic polyelectrolyte, followed by pyrolysis to produce porous carbon .

Resorcinol and formaldehyde were polymerized by heating at 70 °C during 24 hours, using CO_3Na_2 as catalyst in the presence of PDAMAC. In a typical experiment, the molar ratio of the different reactants are R:F:C:PDAMAC = 1:2:5 x 10^{-3}:1.5 x 10^{-2}. (concentrations = [4 M][10 M][0.03 M][0.064 M]). The monolithic gels were then dried in air. Finally, the material was carbonized by heat treatment under N_2 or Ar atmosphere at 800 °C. To test the effect of additive charge on the stabilization , other RF gels were fabricated using a neutral (poly(vinyl alcohol), PVA) and a negatively charged (poly(styrenesulfonate) sodium salt, PSS), hydrophilic polymer as additive.

The N_2 adsorption /desorption isotherms of the carbonized material show a specific surface area, calculated using the BET method, of ca. 675 m^2g^{-1} for the carbon prepared using PDAMAC (Table 10) while those prepared using PVA or PSS show surface area s below 5 m^2/g.

Table 10. Textural properties of porous carbon produced by pyrolysis of resorcinol - formol resins using a cationic polyelectrolyte as gel nanoparticle stabilizer

Synthetic conditions	Surface area BET (m^2/g)	Area μpores[b] (m^2/g)	Area mesopores[b] (m^2/g)	Volume pores[c] (cm^3/g)	Volum μpores[c] (cm^3/g)	Pore size[d] (nm)
R/C = 200 P/R = 7	675	364	311	0.76	0.18	11

[a] Determined from BET theory, [b]determined from t-plot, [c]determined at p/p_0 = 0,97, [d]determined by Barret-Joyner-Halenda (BJH) method.

The same negligible surface area is obtained if the polymerization is carried out without additives . These results suggest that stabilization occurs by interaction of the positive charges in PDAMAC with the negatively charged (in basic media) RF gel nanoparticles . The type-H_1 shape of the isotherm is compatible with the existence of agglomerates of uniform spherical particles having narrow pore size distribution . [76] The BJH pore size distribution plots determined from the adsorption branch of the N_2 isotherms , and present a single maximum at ca. 10 nm of diameter.

The SEM micrograph of a fractured surface of a carbon polyelectrogel shows a packing of globular forms, with an average diameter ca. 70 nm. The topography is quite similar to mesoporous RF carbon obtained suing cationic micelles as stabilizers .

It is likely that the globular forms observed in SEM are in fact agglomeration of small carbon nanoparticles (ca. 5.5 nm of diameter). [77] Because the surface area of a packed array of 70 nm particles, as those observed in SEM, would have an area of only 60 m^2/g. The porosity arises from the interstices between packed nanoparticles which form the monolithic material. [78] The material seems to comprise of a packed array of carbon nanoparticles which are themselves made by carbonization of the RF nanoparticles stabilized by the cationic polyelectrolyte.

The cyclic voltammograms of different carbons, measured at 1 mVs^{-1} in H_2SO_4 1 M, show a maximum specific capacitance of ca. 275 F/g and an average specific capacitance is ca. 210 F/g. On the other hand, by using PVA or PSS in the synthesis of RF resins , carbons with negligible specific capacitance are produced, as expected by the small surface area s. The same negligible capacitance is obtained if the polymerization is made without additives . In addition to the double layer charging, a broad wave around of 0.4 $V_{Ag/AgCl}$ is observed, suggesting that quinone -like surface electroactive groups are also used to store charge. [79] A similar dependence of specific capacitance with potential is observed when the capacitance is measured from the low frequency (2.8 mHz) values of AC impedance. In this case, the maximum specific capacitance (at 0.35 V) is ca. 280 Fg^{-1} while the average specific capacitance is ca. 210 Fg^{-1}. It can be seen that in acid media the capacitance has a strong dependence on potential, due to the effect of surface redox groups . Therefore, the capacitance was measured in neutral media where such groups are inactive

The cyclic voltammogram of the porous carbon in 1 M KNO_3 shows the typical "butterfly" shape . [46] The average specific capacitance is of ca. 110 Fg^{-1}. Therefore, it seems that ca. 50% of the specific capacitance in acid media is due to pseudocapacitive contribution of surface groups. Taking into account the BET surface area , the measured specific capacitance corresponds to capacitance surface densities of 14.5 $\mu F/cm^2$ (neutral) and 28.9 $\mu F/cm^2$ (acid), which are comparable with those measured in other porous carbons. [80, 81]

As it can be seen in Table 11, the synthetic conditions (P/R and R/C ratios) affect strongly the specific capacitance of the porous carbons. Specific capacitrances as high as 104 F/g (CV) and 197 F/g (EIS) can be obtained. The clear difference is related to the fact that cyclic voltammetry (CV) access only those larger pores where ions can be inserted from solution at larger rates. On the other hand, the electrochemical impedance spectroscopy at low frequencies (like 1-10 mHz) access the whole porous surface.

The nature of ions used up to compensate the charge stored in the porous matrix is studied using Probe Beam Deflection techniques . The cyclic deflectogram in 1 M KNO_3 shows first negative and then positive deflection when the potential is scanned from negative to positive potentials, with a reverse response in the backward scan. The data suggest that ions are first expelled from the material and then, when potentials more positive than the potential of minimum charge are reached, ions are inserted. This is a typical behaviour of the ion exchange associated with double layer charging of porous carbon electrodes in neutral media .

Table 11. Electrochemical properties of porous carbon produced by pyrolysis of resorcinol -formol resins using poly(diallyl, dimethylammonium chloride) as gel nanoparticle stabilizer

Synthetic Conditions		**Specific capacitance (F/g)**	
P/R	R/C	CV*	EIS#
3.5	200	27	129
7	200	104	197
14	200	19	84
21	200	11	60
28	200	10	46
7	100	10	42
7	300	70	195
7	400	167	104
7	500	14	55
7	600	11	58

Measured in 3 M H_2SO_4.
* Cyclic voltammetry (CV). Scan rate = 50 mV/s.
# Electrochemical impedance spectroscopy (EIS). Frequency = 2.8 mHz, E= 0.25 Vsce.
S= surfactant (CTAB), C= catalyst (Na_2CO_3), R = resorcinol .

The ability of a cationic polyelectrolyte, which does not form micelles , to maintain the porosity is clear evidence against the need of a soft-template mechanism for pore formation . The method was then used in more stringent conditions. It has been shown that water in oil emulsions , stabilized by emulsifiers (SPAN80) could be used as microreactors to produce aerogel resin microparticles . [82] The porosity is maintained by solvent exchange and lyophilization . Further carbonization renders porous carbon microparticles. Using PDAMAC as soft template agent and SPAN80 as emulsifier we were able to produce spherical dry resin microparticles which are carbonized to a high surface area (550 m^2/g) carbon with specific capacitances of ca. 80 F/g. The results indicate that the method is robust enough to be used in the presence of surface active agents. On the other hand, we tried to use another soft-templating method, which uses a cationic surfactant (CTAB), [8, 9] to the same purpose but we only obtain non spherical microparticles which show negligible specific capacitance.

Table 12. Textural properties of porous carbon microparticles produced by pyrolysis of resorcinol -formol resins nanoparticles , made by inverse emulsion using PDAMAC as gel nanoparticle stabilizer

Mean particle size (μm)	**Surface area BET[a] (m^2/g)**	**Area μpores[b] (m^2/g)**	**Area mpores[b] (m^2/g)**	**Vpores[c] (cm^3/g)**	**Vμpores[c] (cm^3/g)**	**Pore size[d] (nm)**
4	555	411	144	0.38	0.18	20

[a] Determined from BET theory, [b]determined from t-plot, [c]determined at p/p_0 = 0,97, [d]determined by Barret-Joyner-Halenda (BJH) method.

It is likely that interference between both surface active agents (SPAN80 and CTAB) preclude both the formation of spherical microparticles and the ability of the surfactant to maintain porosity . The use of a cationic polyelectrolyte as structruring agent in the polymerization media provides a new and simple method to produce porous dry resins , without complex drying procedures, and porous carbon materials by carbonization of the resins. Since the cationic polyelectrolyte has a covalent backbone, it should be less sensitive to the effect of experimental condition s (solvent, temperature , concentration, presence of other surfactive molecules) than organized systems based on cationic surfactants . The specific capacitance can be further increased by tuning the catalyst and stabilizer (PDAMAC) concentration. [83]

Table 13. Electrochemical properties of porous carbon microparticles produced by pyrolysis of resorcinol -formol resin microparticles made by inverse emulsion (using PDAMAC as gel nanoparticle stabilizer)

Synthetic Conditions		Specific capacitance (F/g)
P/R	R/W	CV*
31	0.62	13
31	1.5	81
48	0.5	5
38	1.35	8

Measured in 3 M H_2SO_4.
* Cyclic voltammetry (CV). Scan rate = 50 mV/s.
# Electrochemical impedance spectroscopy (EIS). Frequency = 2.8 mHz, E= 0.25 Vsce.
S= surfactant (CTAB), C= catalyst (Na_2CO_3), R = resorcinol .

The porosity of carbon was also studied using nanotomography. [84] Using 3D reconstruction and analysis software it was possible to visualize the non ordered pore morphology and quantitatively characterize their physical properties of polylectrogels. Mesopore size and mesopore surface area rendered by the technique correlate well with the nitrogen adsorption isotherm measurements .

The synthetic method has been used extensively to produce porous carbon materials for fuel cell electrodes . [85, 86, 87, 88, 89, 90, 91]

Surface Stabilizer: Stabilization by Fibers

Another way to obtain porous carbon materials with conventional drying involves the carbonization of resorcinol –formaldehyde (RF) resins formed on a cellulose fiber . [92] While the original purpose was to hold the resin with the fiber, to improve the mechanical properties , it was found that a highly porous composite material is formed without the use of templates or special drying procedures. The method seems to produce a carbon having both nanometric sized pores (meso and micropores) due to the resin and micrometric sized pores (macropores) due to the fiber arrangement, making it a hierarchical structure . While the nanopores gives the material the large surface area which could be used to store electrical charge in an electrochemical system, the macropores allows fast access of the electrolyte making the charge storage faster than in monolithic materials . The materials are called fibergels. Fricke and coworkers, [93] suggested that hydrophilic reinforcing surfaces act as equivalent to a locally increased catalyst concentration. The higher catalyst concentration provides more nucleation sites and more stable nanoparticle network. However, we found (see below) that either cellulosic fiber and carbonized fibers act as stabilizer of nanoparticles during drying, suggesting that even hydrophobic surfaces promote porosity .

The resorcinol -formaldehyde resins were produced using the molar ratios of resorcinol to formaldehyde (R:F = 1:2.9) and catalyst (Na_2CO_3) to resorcinol (C:R = 0.005:1). The viscous solution was impregnated onto the support material (e.g., cellulosic fabric), until saturation . It is then possible to create multilayer assemblies by adhesion of individual fabric layers. To cure the resin, the materials are left by 24 hrs at 70 °C inside a closed glass chamber in presence of a water saturated atmosphere . Then, the resins were dried in an open oven at 70 °C during 24 hrs. The dry resins are carbonised under a nitrogen stream at 800 °C. Complex shapes, multilayers and extensive (up to 15 cm dia.) flat pieces could be easily assembled by the method. Visually, it can be seen that the resin decorates the cellulosic fiber . Since RF resins are widely used as adhesives , it is possible to build complex shapes by binding fabric layers with RF resin. The shape is maintained during carbonisation, with an obvious contraction. Moreover, multilayers of fabric can render millimetre thick specimens. The macropores allow release of the decomposition gases without breaking of the material, as it happen with monolithic carbons. The low resolution SEM micrographs of the carbonized fabric looks quite similar to the cellullosic fabric carbonised in the same conditions, suggesting that the fabric fibre morphology is maintained during the sol-gel process, as it has been shown to occur with titania . [94]

The specific surface area of a typical composite (558 +/-3 m^2/g) was calculated from the nitrogen adsorption (– 196 °C) isotherm, using the BET method. The hysteresis loop s are closer to H4 type, These suggest that the material has large mesopores embedded in a matrix with pores of much smaller size. [95] The BJH profile suggests that, while micropores contribute to the surface area , a significant portion of the pores has size of ca. 40 nm (mesopores).

The carbon fibergels were tested as electrode materials of electrochemical super-capacitors. Small pieces of the material were tested in aqueous solution (1 M H_2SO_4), in a conventional three electrode cell. The cyclic voltammetric response shows a high capacitance with a broad wave around ca. 0.35 Vsce.

The wave could be related with the oxidation /reduction of quinone functionalities present on the carbon surface. [96, 97, 98] At low scan rates, maximum specific capacitances of up to 170 F/g are measured. It is well known that cellulosic fibers , properly activated could be used to store charge. [99] Therefore, it is a question if the charge is stored only in the porous resin or on the cabonized fabric, ii) which is the mechanism of templating or stabilization of the RF gel which originates the large surface area of the carbon. While the carbonized fabric is not subjected here to any activation, a reaction with the RF resin could cause the activation. To check that, we tested an electrode formed by the carbonized fabric alone. The profile (Figure 3, line a) show a different voltametric response which seems related to ion intercalation instead than than to double layer charging. Moreover, if a non fibrous cellulosic film (Celofan®) is covered with RF resin and carbonized, a carbon film is obtained which shows a CV profile resembling the one obtained with the carbonized fabric alone. (Figure 3, line b). On the other hand, if we cover this carbonized fabric with RF resin and carbonized the composite, we obtain a material with a voltammetric profile quite similar to the one prepared before by supporting RF resin onto cellulosic fabric and then carbonize the composite.

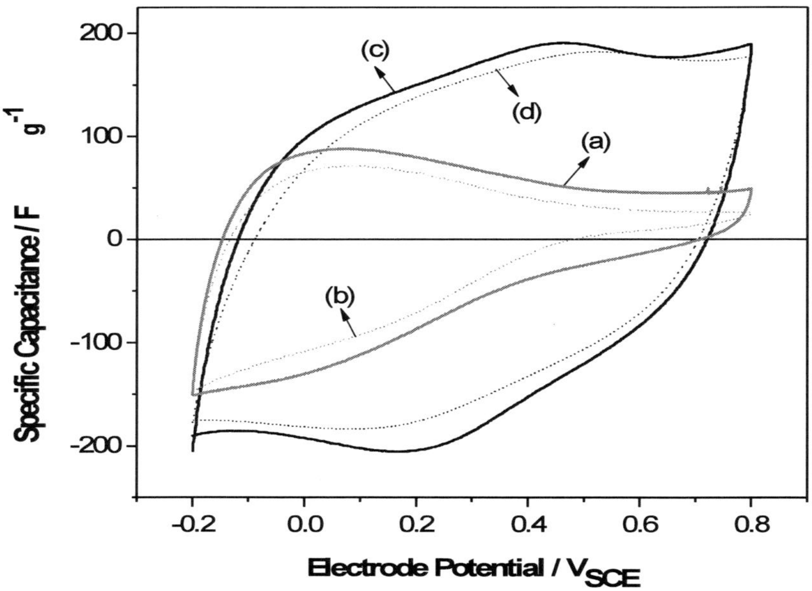

Figure 3. Cyclic voltammograms of carbon materials produced by carbonisation of: (a) cellulose fabric; (b) a RF resin deposited on a thin film of cellulose: (c) RF resin deposited on a carbonized fabric; (d) RF resin deposited on a carbonized fabric. Scan rate = 1 mV/s. Solution = 1 M H_2SO_4.

It seems that cellullose properties (e.g., hydrophilicity) are not necessary to stabilize the porous RF gel . The results are significant because it suggests that other micrometric or subnanometric fibers /particles could produce a templating or stabilizing effect similar to the fabric.

The ESI profiles with frequency and potential are different to those reported before for carbon aerogels . However, the complex plane plot show constant impedance values (line parallel to the imaginary axis), at low frequency. Therefore, the specific capacitance could be calculated from the impedance values measured at low frequency, when most surface area of the pores is accessed. It is noteworthy that covering the fabric with RF resin which contains a soft template (cetyl, trimethylammonium bromide) renders a material with a specific capacitance of ca.130 F/g. The ion exchange during charge storage was studied using Probe Beam Deflection techniques . The cyclic deflectogram shows a negative deflection when the potential is scanned from negative to positive potentials, with a reverse response in the backward scan. The data suggest that ions are expelled from the material when more positive potentials are applied to the electrode. It is likely that the dominant ion exchange observed is related to the oxidation of quinone functionalities on the carbon surface (eq. 1). A different situation exists when neutral electrolytes are used. In KNO_3 the deflectometric signal is positive during the anodic scan and negative during the cathodic scan. This is indicative of ion insertion dring the anodic scan and ion expulsion during the anodic scan. This is to be expected from a double layer charging with a potential of minimum charge (PMC) of ca. 0.0 $V_{Ag/AgCl}$. While cyclic voltadeflectometry gives information on the steady state ion flux, the time response is more difficult to interpret. Therfore, we monitored the PBD signal during time when the potential is stepped between two different potentials. The chronodeflectometric data obtained with the composite electrode could not be fitted by a continuous flux of ions. On the other hand, the data could be fitted well with the chronodeflectometric profile of a discontinuous process.

For comparison, the data corresponding to a monolithic porous electrode shows a good agreement with a continuous process. Such behaviour is similar to that observed before in fibre supported aerogel materials . Hierarchical porous materials containing macropores are faster than monolithic ones when mobile species have to reach the whole pore surface area . Both in the restricted diffusion and transmission line model, the relevant parameter is the pore length. By shortening the pore length, it is possible to decrease the time to charge/discharge the whole porous matrix . If the pores are charged/discharged in a time span negligible to the chronodeflectometric measurement , the ions flux to the electrode in a discontinuous process. Therefore, hierarchically porous materials show faster movement of mobile species. This is seen here in the ion movement of a supercapacitor electrode. However, the effect will also influence the kinetic behaviour in adsorption , deionization electrodes and drug release devices. Additionally the macropores could be used as reservoir of electrolyte in supercapacitor devices.

Hierarchical structures can be also produced using other polymer fibers as sacrificial templates to left macroporous channels inside the monolithic carbon matrix . In such case, the small ratio of fiber surface to carbon volume makes the stabilizing effect non effective and a polyelectrolyte has to be used to produce nanoporous carbon. [100]

Stabilization and Hard Templating by Silica Nanoparticles

As it was shown above, hierarchical porous carbons are very interesting carbon -based porous materials , [101] because the materials have a multimodal porous size distribution , with each pore size contributing to the desired properties. Instead of using a macroporous template, it is possible to polymerize the RF inside a matrix formed by an inorganic template which is removed after polymerization . The whole process result in a reverse copy of the template. [102] We used the method with a silica nanoparticle opal as hard template. The SiO_2 nanoparticles of 400 nm in diameter were synthesized by Stober method, TEOS hydrolysis in basic media , [103] giving sets of nearly monodisperse nanoparticles. Then, the SiO_2 nanoparticles were deposited by sedimentation in the bottom of a vial [104]. After solvent evaporation , the resulting weak opal was taken out of the vial and then treated at 1000 ºC for 4 hours to form a connected matrix of SiO_2 nanospheres. The template was then impregnated by the carbon precursor, a mixture of resorcinol (1 g), formaldehyde (1.6 ml), and sodium carbonate (0.4 ml, 0.1 M) as catalyst . The impregnated SiO_2 opal was dried in an oven and polymerized at (105 ºC) to obtain a resin (RF) containing the silica nanospheres. With the purpose of studying the effect of the template removal on carbon porosity , the SiO_2 in some samples were removed before pyrolysis by treatment with hydrofluoric acid solution (early removal, ER). The porous resin material was then pyrolized into carbon. Another samples (late removal, LR), was pyrolyzed in the presence of the SiO_2 hard template. The pyrolysis of both samples was carried out at 850 ºC for 24 h.

Table 14. Electrochemical properties (measured by cyclic voltammetry) of porous glassy carbon (LR) prepared by pyrolysis of resorcinol -formol resins prepared in the presence of of SiO_2 nanoparticles (400 nm)

	Specific Capacitance (F/g)			
Solution =	**H_2SO_4 1 M**	**KCl 1 M**	**KCl 0.1 M**	**KCl 0.01M**
Scan rate (V/s^{-1})				
0.010	104	149	73	44
0.025	89	60	65	32
0.050	81	51	52	24
0.075	76	43	38	19
0.100	72	41	40	16
0.250	62	31	26	9
0.500	53	24	17	5
0.750	48	21	13	3
1.000	42	19	6	2

As stated before, the volume of a RF sample decreases during carbonization . When a sample contains a rigid template, the contraction observed in the macroscale is low due to the presence of a non compressible array of nanometric SiO_2 nanoparticles in close contact between them. The HRTEM image of ER samples show that the presence of SiO_2 nanoparticles induces, in the carbon matrix , the formation of spherical macropores with size close to that of the original silica nanoparticles. The pore walls of ER samples are smooth and seem almost free of pores. In contrast, when the carbon is pyrolyzed in the presence of the SiO_2 template (sample LR) the HR-TEM image shows a carbon skeleton composed of macropores (average diameter of ~400 nm) delimited by permeable mesoporous walls. A further increase in the magnification of the image shows that those porous walls are really composed of reticular carbon shapes), with thickness of the carbon branches in the scale of ~10 nanometers .

The composition of polymeric precursor and the sample treatment is the same in both experiences until solidification of the RF resin inside the silica template. The difference in morphologies of the pore wall should not be a direct consequence of dissimilar condensation /agglomeration of the RF sol inside the cavities of the SiO_2 templates, since this process is identical in both samples. However, it is reasonable to assume that the existence of a high contact surface, in addition to the small volume of the RF network, can change the way in which RF nanoparticles agglomerates to copy the templates. However, in terms of stabilization , the expected result should be the inverse. High extended contact areas between RF and the SiO_2, present during pyrolysis , should promote the appearance of a smooth surface, mainly due to the additional stabilization of the RF precursor during the first step of carbonization . We interpret this fact as follow. The mechanical strength and symmetry of the silica array prevents, to the RF precursor, of any local displacement or contraction on the macroscale. For this reason, the only functional mechanism for structural relaxation during pyrolysis operates over the pore wall dimension and, the most important fact; relax only short carbon domains. The final result of the process is the formation of an interconnected carbon mesh, quite similar to that observed in aerogels . Nevertheless, a simple comparison between both HPC samples leads to conclude that the function of the templates largely exceed the simplest patterning of the shape on the host structure.

The CV of LR carbon in 1 M H_2SO_4 shows an almost constant current in the range 0.0 to 1.0 Vsce with a specific capacitance of 130 F/g, associated with a large surface area. This capacitance value is quite similar to that reported before for the CMK type porous carbon [105]. However, for the ER carbon, the CV shows a maximum capacitance of only 30 F/g. From these values, a surface area of ca. 867 m^2/g can be estimated for LR and only 200 m^2/g for ER (based on 15 $\mu F/cm^2$) [106] It should be noted that the insignificant pseudocapacitive (faradaic) contribution allows relating directly the specific capacitance with the surface area. Since nanoparticles of the same size were used, the difference should be related to the effect or early or late removal of the hard template. On one side, the surface area of both materials is larger than the geometrical area of a porous material with pores of ca. 400 nm, which should be of ca. 10 m^2/g. Moreover, the area of LR carbon is almost 4 times larger than ER carbon. The larger surface area of LR carbon is related to the creation of microfractures upon carbonization due to the fact that the carbon contract while the silica nanoparticles does not change its size. On the other hand, the large surface area of ER cannot be assigned to different contraction because the hard template was removed before pyrolisis. However, the effect of the silica nanoparticles surface on the aggregation RF gel nanoparticles could maintain enough porosity during drying to sustain a surface area of ca. 200 m^2/g.

CONCLUSION

Using chemical (cationic micelles, polyelectrolytes) or physical (fibers, silica nanoparticles) as stabilizers of the resorcinol -formaldehyde gel nanoparticles it is possible to maintain and control de porosity during conventional air drying . The porous dry resin is then converted into porous carbon by pyrolysis .

The pore distribution analysis of N_2 isotherm data shows that the material presents micro and mesopores . Changing the concentration of CTAB during the synthesis it is possible to form nanostructurated carbon with specific surface areas (Surface Area BET) from 576 to 671 m^2/g. CTAB ions are still retained in the RF dry matrix which only decomposes during carbonization . The carbonizations of porous and compact RF resin are similar but the porous material carbonizes at lower temperatures. Both cyclic voltammetry and EIS allows measuring specific capacitances as high as 150 F/g in acid media . The values in neutral or aprotic media are ca. 50% lower due to the absence of faradaic contributtion of surface groups.

In an analogous way, the presence of a cationic polyelectrolyte (poly (diallydimethyl) ammonium chloride, PDAMAC) in the polymerization solution stabilizes the gel nanoparticles and allows the porosity to be maintained during drying . In that way, high (>700 m^2/g) surface area carbon materials can be obtained by carbonization of the RF gel dried in air. It seems that the cationic polyelectrolyte adsorbs onto the negatively charged polymer nanoparticles which form the gel structure , stabilizing the porous matrix against aggregation during drying. The specific effect of the cationic polyelectrolyte is supported by the fact that anionic or neutral polymer has no significant effect. Since the cationic additive has a covalent backbone, it is less sensitive to the effect of experimental condition s (solvent, temperature , concentration) than organized systems based on cationic surfactants . The fact that organic solvents are not required makes the process friendlier to the environment . The porous carbon material present large electrochemical capacitances (ca. 270 F/g in acid media and 140 F/g in neutral media) making it a useful material in supercapacitor applications. The Probe Beam Deflection study of ion exchange suggests that the electrode behaves as a double layer structure which exchange anions or cations depending on the applied potential with respect to the potential of minimum charge. The determined potential of minimum charge is of ca. −0.17 V.

Additionally, it is possible to produce porous composite materials by polymerizing resorcinol -formaldehyde onto a cellulosic fibre fabric. The porous composite resin , dried in air, is carbonized to give a carbon -carbon composite which has surface area s of more than 550 m^2/g. The TGA -MS mesaurenents reveal that both the cellulosic fabric and the RF resin area converted into carbon materials . Comparison of the composite with RF resin alone, deposited onto cellulose films and already carbonized fibers suggests that the fiber nature is determinant to the stabilization of the RF gel porosity durung drying . The composite could be used as supercapacitor electrode in acid aqueus media having a specificic capacitance of more than 160 F/g. In that media, the ions exchanged during charge/discharge of the electrode seems to be mainly protons . In neutral media, the ions exchanged depend on the potential applied against the potential of minimum charge (PMC). Using cyclic deflectometry, sampled deflection deflectometry and chronocoulometry it was found a value of PMC of ca. 0.0 $V_{Ag/ClAg}$. The ions move faster than in monolithic materials, as shown in the chronodeflectograms and chronoameropmetric measurements . The synthetic procedure is simple and could be used to produce large pieces and complex shapes. The results suggest that, using a simple synthetic procedure, it is possible to produce materials with good electrochemical, electronic and mechanical properties which can be applied as electrodes in EDLC devices.

It is possible to hard template the RF resin using silica nanoparticles . Besides the spherical pores, smaller pores are produced by two phenomena. If the silica particles are removed after carbonization , the contraction of the resin matrix into carbon produces microfractures which make up for additional porosity . However, even if the hard template is removed at the dry resin state porosity larger than expected is obtained, likely due to the gel stabilizing effect of the silica nanoparticle surfaces.

These results show that it is possible to synthesize a useful and complex porous structures starting from a simple host matrix , i.e., double patterning by using a single template. The key factor is the fact that hard templates obstruct contraction, while soft templates (or its previous elimination) allow such contraction. This strategy can be used for reduction of time and efforts in the synthesis of complex hierarchical precursors.

Besides the use in supercapacitors , the porous carbon materials described above has been modified with metal or metal oxide nanoparticles and used in fuel cell electrodes , [107] and electrochemical sensors . [108]

The maximum specific capacitance (F/g), a value quite important for supercapacitor applications, of the different carbon materials described in this chapter are summarized in Table 15.

Table 15. Summary of electrochemical properties of the different porous carbon materials described in the chapter

Stabilizer agent	Name	Solid form	Synthetic conditions	Specific Capacitance (F/g)[a]
Micelles of cationic surfactant	MSgel	Monolithic	S/R = 12 C/R = 600	272
Cationic polyelectrolyte	polyelectrogel	Monolithic	P/R = 7 R/C = 200	197
Cationic polyelectrolyte	Porous carbon microparticles	Microparticles	P/R = 31 R/W = 1.5	85
Cellulosic fiber	Fibergels	Hierarchical porous solid	--	130
Silica nanospheres	Hard template gels	Hierarchical porous solid	--	104

[a] measured by ESI at 2.8 mHz.

As it can be seen, the flexibility of the stabilization with cationic micelles or cationic polyelectrolyte allows tuning the carbon porosity to obtain the largest specific capacitances . On the other hand, the spontaneous stabilization by cellulosic fibers or hard silica spheres allows producing hierarchical porous carbons (having macropores and micro/mesopores) but cannot be so easily tuned to obtain larger specific capacitances.

ACKNOWLEDGMENTS

C.A. Barbero, M.M. Bruno and G.A. Planes are permanent research fellows of CONICET. The work was funded by FONCYT, CONICET, SECYT-UNRC and J.S. Guggenheim Foundation. A. Baena Moncada and J. Balach thank CONICET for graduate studies fellowships. The authors wish to thank the EU (IRSES project 'SUMA2-Network', Pr. No: 318903) for the financial support of the scientist's mobility.

GLOSSARY

CTAB: cetyl, trimethylammonium bromide
PDAMAC: poly(diallyl,dimethylammonium chloride)
EIS : electrochemical impedance spectroscopy
CV : cyclic voltammetry
RF: resorcinol /formaldehyde
C: basic catalyst for phenol -aldehyde condensation (usually Na_2CO_3)
CMC : crytical micelle concentration
P: PDAMAC
R: resorcinol
W: water
S: surfactant

BET: Brunnaer-Emmet-Teller theory of adsorption . Used to calculate surface area s from gas (usually nitrogen) adsorption isotherms .
FIB-SEM: Focused ion Beam-Scanning Electron Microscope.
FIB-nt: FIB nanotomography.
TEOS : tetraethylorthosilicate
ER: carbon pyrolized after removal of the silica nanoparticles used as hard template.
LR: carbon pyrolized in the presence of the silica nanoparticles used as hard template.
CMK: Carbon Mesostructures from Korea .
EDLC: electrochemical double layer capacitor.
PMC: potential of minimum charge (equivalent to the potential of zero charge defined in single crystal surfaces).
PBD: Probe Beam Deflection
BJH: Barret-Joyner-Halenda
SPAN80: sorbitan monoleate
MSgel = gel made in the presence of cationic surfactant micelles .
Vsce = electrode potential (Volts) measured using a saturated calomel electrode as reference
VAg/AgCl = electrode potential (Volts) measured using a silver -silver chloride electrode as reference

REFERENCES

[1] Frackowiak, E; Beguin,F. *Carbon* 2001, 39,937–950.
[2] Porada,S.; Zhao,R.; van der Wal,A.; Presser, V.; Biesheuvel, P.M. *Prog. Mater. Sci.*, 2013, 58, 1388-1442.
[3] Candelaria, S.L.; Shao, Y.; Zhou, W.; Li, X.; Xiao, J.; Zhang,J-G.; Wang, Y.; Liu,Y.; Li,J.; Cao, G.; *Nano Energy*, 2012, 1, 195-220.
[4] A.G. Pandolfo, A.F. Hollenkamp, Journal of Power Sources 157 (2006) 11–27.
[5] Harris, P.J.F. *Philos. Mag.* 2004, 84, 3159–3167.
[6] Savova,D; Apak, E; Ekinci, E; Yardim, F.; Petrova, N; Budinova,T; *Biomass Bioenergy,* 2001, 21, 133–42.
[7] Lua, A.C.; Yang, T.; Guo, J.; *J. Anal. Appl. Pyrol.* 2004,72, 279–87.
[8] Ahmadroup, A.; Do, D.D.; *Carbon* 1997,35, 1723–32.
[9] Girgis, B.S., Yunis,S.S.; Soliman, A.M. *Mater Lett.* 2002,57, 164–72.
[10] Tsai, C.W.T.; Chang,Y.; Lee, S.L. *Carbon* 35(1997,35, 198–200.
[11] Stavropoulos, G.G.; Zabaniotou, A.A. *Micropor. Mesopor. Mater.,* 2005,82, 79–85.
[12] Fan, M.; Marshall,W.; Daugaard, D.; Brown,R.C.; *Bioresource Technol.* 2004,93, 103–107.
[13] Oh, G.H,; Park, C.R. *Fuel* 2002,81, 327–336.
[14] Qu, D.; Shi, H. J. *Power Sources* 1998,74, 99–107.
[15] Kinoshita,K.; Carbon: Electrochemical and Physicochemical Properties Wiley-Interscience: Weinheim, GE, 1988.
[16] Darmstadt, H.; Roy, C.; S. Kaliaguine, T-W. Kim, R. Ryoo, *Chem. Mater.*, 2003, 15, 3300-3307.
[17] Jada, S.S; *J. Appl. Polym. Sci.*1988, 35, 1573–1592.

[18] Pasch,H.; Dairanieh,I.S.; *Macromolecules*, 1991, 24, 671–677.

[19] Rivero,R.E; Molina, M.A.; Rivarola, C.R.; Barbero, C.A. Sensor. *Actuat. B-Chem.* 2014, 190, 270-278.

[20] Jenkins, G.M.; Kawamura,K. Polymeric Carbons. Carbon fibre, glass and char., Cambridge Univ. Press, Cambridge, 1972.

[21] Barbero, C.; Kötz, R.; J. *Electrochem. Soc.* 1993,140, 1-6.

[22] Tamon, H.; Ishizaka,H; Yamamoto, T.; Suzuki, T. *Carbon.* 1999, 37, 2049–2055.

[23] Szczurek, A.; Jurewicz, K.; Amaral-Labat, G.; Fierro, V.; Pizzi,A.; Celzard, A, *Carbon*, 2010,48, 3874 –3883.

[24] Li, W.; Reichenauer, G.; Fricke, J.; *Carbon* 2002, 40, 2955–2959.

[25] Kraiwattanawong, K.; Mukai, S.R.; Tamon, H.; Lothongkum, A.W. *Microporous and Mesoporous Materials* 2007, 98, 258–266.

[26] Szczurek, A.; Amaral-Labat, G.; Fierro, V; Pizzi, A.; Masson, E.; Celzard, A. *Carbon*, 2011,49, 2773 –2784.

[27] Elkhatat A.M.; Al-Muhtaseb S.A.; *Adv. Mater.* 2011, 23, 887-903.

[28] Zeng, S.; Guo,L.; Zhang,L.; Cui,F.; Zhou,J.; Gao,Z.; Chen,Y.; Shi, J.; *Macromol. Chem. Physic.* 2010, 211, 845–853,

[29] Amaral-Labat, G.; Szczurek, A.; Vanessa Fierro, V. Pizzi, A. Celzard, A. *Sci. Technol. Adv. Mater.* 2013, 14, 1-13.

[30] Pekala, R.W.; *J. Mater. Sci.,* 1989, 24, 3221-3225.

[31] Liang, L.; Sha, G.; Guo, S.; *J. Non-Cryst. Solids*, 2000, 271,167-170.

[32] Mayer, S.T.; Kaschmitter, J.L.; Pekala, R.W. US Patent Nr., 5,420,168, 1995.

[33] Yamamoto, T.; Sugimoto, T.; Suzuki, S.R.; Mukai, H. Tamon,H.; *Carbon*, 2002, 40, 1345-1351.

[34] Yamamoto, T.; Endo, A.; Ohmori, T.; Nakaiwa, M.; *Carbon*, 2005,43, 1231-1238.

[35] Schmitt, C.; Probstle,H.; Fricke, J. J. Non-Cryst. *Solids*, 2001, 285, 277-282.

[36] Babi,C.; Kaluderovi,B.; Vracar, L.; Krstaji, N.; *Carbon* 2004,42, 2617–2624.

[37] Lee, Y.J.; J. Jung, C.; Yi, J.; Baeck, S-H.; Yoon,J.R.; Song, I.K. *Current Applied Physics* 2010,10, 682–686.

[38] Lin, C. Ritter, J.A.; *Carbon*, 1997, 35, 1271-1278.

[39] Pekala, R.W.; Farmer, J.C.; Alviso, C.T.; Tran, T.D.; Mayer, S.T.; Miller, J.M.; Dunn, B.; *J. Non-Cryst. Solids*, 1998, 225, 74–80.

[40] Al-Muhtaseb, S.A.; Ritter, J.A.; *Advanced Materials*, 2003, 15, 101-114.

[41] Zhang, W.; Crittenden, J.; Li, K.; Chen, Y.; Environ. *Sci. Technol.* 2012, 46,7054-7062.

[42] Gibson, M.I.; Danial, M.; Klok, H-A.; *ACS Comb. Sci.* 2011, 13, 286-297

[43] Lin, S.; Cheng, Y.; Liu, J.; Wiesner, M.R.; *Langmuir,* 2012, 28, 4178-86.

[44] Rozenberg, B.A.; Tenne, R.; *Prog. Polym. Sci.* 2008, 33, 40–112.

[45] Kleinen, J.; Richtering, W.; *Macromolecules*, 2008, 41, 1785-1790.

[46] Bell, W.; Dietz, S.; *US Patent* 6297293, 2001.

[47] Antonietti, M. J. *Colloid Interface Sci.* 2001,6,244.

[48] Kresge, C.T.; Leonowicz, M.E.; Roth, W.J.; Vartuli, J.C.; Beck, J.S. *Nature*, 1992,359,710-713.

[49] Beck, J.S.; Vartuli, J.C.; Roth, W.J.; Leonowicz, M.E.; Kresge, C.T.; Schmitt, K.D.; Chu, T.W.; Olson, D.H.; Sheppard, E.W.; McCullen, S.B.; Higgins, J.B.; Schlenker, J.L. *J. Am. Chem. Soc.* 1992,114,10834-10838.

[50] Lee, K.T.; Oh, S.M.; *Chem. Commun.* 2002, 2722-2723.

[51] Fujikawa, D.; Uota, M.; Yoshimura, T.; Sakai, G.; Kijima, T.; Carbon, 2007,45,1289.

[52] Wu, D.; Fu, R.; Dresselhaus, M.S.; Dresselhaus, G. *Carbon*, 2006, 44, 675-681.

[53] Wang, J.; Yang,X.; Wu, D.; Fu, R.; Dresselhaus,M.S.; Dresselhaus,G. J. *Power Sources* 2008,185, 589–594.

[54] Tashimaa, D.; Taniguchi, M.; Fujikawa, D.; Kijima,T.; Otsubo, M. *Materials Chemistry and Physics* 2009, 115, 69–73.

[55] Bruno, M.M.; Cotella, N.G; Miras, M.C.; Barbero, C.A. *Colloid. Surface.* A 2010, 362, 28-32.

[56] Worsley, M.A.; Satcher, J.H.; Baumann, T.F. *Langmuir*, 2008, 24, 9763-9766.

[57] Su, F.; Zeng, J.; Bao, X.; Yu, Y.; Lee, J.Y.; Zhao, X. S. *Chem. Mater.,* 2005, 17, 3960-3967.

[58] Beyssac, O.; Goffé, B.; Petitet, J.P.; Froigneux, E.; Moreau, M.; Rouzaud, J.N.; Spectrochim. *Acta Part A*, 2003, 59, 2267-2276.

[59] McCreery, R.L., Carbon Electrodes: Structural Effects on Electron Transfer Kinetics; Electroanalytical Chemistry 17; Bard, A.J. Ed., Dekker, New York, USA , 1991.

[60] Reynolds, G.A.M.; Fung, A.W.P., Wang, Z.H.; Dresselhaus, M.S.; Pekala, R.W. J. Non-Cryst. *Solids*, 1995, 188, 27-33.

[61] Even, W.R.; Crocker, R.W.; Hunter , M.C.; Yang, N.C.; Headley, T.J.; J. Non-Cryst. Solids, 1995,186, 191-199, Li, J.; Wang, X.; Huang, Q.; Gamboa, S.; Sebastian, P.J. *J. Power Sources,* 2006, 158, 784-788.

[62] Li, J.; Wang, X.; Huang, Q.; Gamboa, S.; Sebastian, P.J. *J. Power Sources*, 2006, 158, 784-788.

[63] Li, J.; Wang, X.; Wang, Y.; Huang,Q.; Dai, Ch.; Gamboa, S.; Sebastian, P.J. *J. Non-Cryst. Solids*, 2008, 354, 19-24.

[64] Planes, G.A.; Miras, M.C.; Barbero, C.A.; *Chem. Commun.*, 2005, 2146-2147.

[65] Ying, T.Y.; Yang, K.L.; Yiacoumi, S.; Tsouris, C. *J. Colloid Interface Sci.,* 2002, 250,18-27.

[66] Conway, B.E.; Electrochemical Supercapacitors: Scientific Fundamentals and Technological Applications, *Springer*, Weinheim, GE, 1999.

[67] Barbieri, O.; Hahn, M.; Herzog, A.; Kötz, R.; Carbon, 2005, 43,1303-1310.

[68] Lust, E.; Nurk, G.; Jänes, A.; Arulepp, M.; Permann, J.; Nigu, P.; Möller, P. *Condens. Matter. Phys.,* 2002,5, 307.

[69] Hahn, M.; Bärtsch, M.; Barbieri, O.; Sauter, J-C.; Kötz, R.; Gallay, R. *Electrochem. and Solid-State Lett.*, 2004,7,A33-A36.

[70] Bard, A.J. Faulkner, L.R. Electrochemical Methods: Fundamentals and Applications, 2nd Ed., Wiley, New York, US, 2000.

[71] Lozano-Castelló, D.; Cazorla-Amorós, D.; Linares-Solano, A.; Shiraishi, H.; Kurihara, S.; Oya, A. Carbon, 2003,41,1765-1755.

[72] Barbero, C.; *Phys. Chem. Chem. Phys.*, 2005, 7,1885-1899.

[73] Balach,J.; Soldera, F.; Acevedo, D.F.; Mücklich, F.; Barbero, C.A.; *Microsc. Microanal.* 2013, 19, 745-750.

[74] Balach, J.; Bruno, M.M.; Cotella, N.G.; Acevedo, D.F.; Barbero, C.A. *J. Power Sources*, 2012, 199, 386-394.

[75] Bruno, M.M.; Cotella, N.G.; Miras, M.C.; Barbero, C.A.; *Colloid. Surfaces* A, 2010, 362, 28-32.

[76] Gregg S.J.; Sing, K.S.W. Adsorption, Surface Area and Porosity, 2nd ed., Academic Press: London, UK , 1982.
[77] Gavalda, S.; Gubbins, K.E.; Hanzawa, Y.; Kaneko, K.; Thompson, K.T. *Langmuir,* 2002, 18, 2141-2151.
[78] Yamamoto, T.; Mukai, S.R.; Endo, A.; Nakaiwa, M.; Tamon, H. J. *Colloid Inteface Sci.,* 2003, 26, 532-537.
[79] Bruno, M.M.; Cotella, N.G.; Miras, M. C. and C. A. Barbero, *Chem. Commun.*, 2005, 5896-5898.
[80] Centeno, T.A.; Hahn, M.; Fernández, J.A.; Kötz, R.; Stoeckli, F. *Electrochem. Comm.*, 2007, 9, 1242-1246.
[81] Gryglewicz, G.; Machnikowski, J.; Lorenc-Grabowska, E.; Lota, G. Frackowiak, E.; *Electrochim. Acta*, 2005, 50, 1197-1206.
[82] Yamamoto, T.; Sugimoto, T.; Suzuki, T.; Mukai, S.R.; Tamon, H. *Carbon*, 2002, 40, 1345-1351.
[83] Balach, J.; Tamborini, L.; Sapag, K.; Acevedo, D.F.; Barbero, C.A. *Colloid. Surfaces-*A:, 2012, 415, 343-348.
[84] Balach, J. Miguek, F.; Soldera, F.; Acevedo, D.F.; Mucklich, F.; Barbero, C.A., J. Microsc-Oxford, 2012, 246, 274–278.
[85] Viva, F.A.; Bruno, M.M.; Jobbágy, M.; Corti, H.R. *J. Phys. Chem-* C, 2012. 116, 4097-4104.
[86] Thomas, Y.R.J.; Bruno, M.M.; Corti, H.R. Micropor. *Mesop. Mater.* 2012, 155, 47-55,
[87] Bruno, M.M.; Franceschini, E.A. Viva, F.A.; Thomas, Y.R.J.; Corti, H.R. *Int. J. Hydrogen Ener.* 2012, 37, 14911-14919.
[88] Veizaga, N.; Fernandez, J.; Bruno,M.; Scelza, O.; de Miguel, S. *Int. J. Hydrogen Ener.* 2012, 37, 17910-17920.
[89] Bruno, M.M., Petruccelli, M.A.; Viva, F.A.; Corti, H.R.; *Int. J. Hydrogen Ener.* 2013, 38, 4116-4123.
[90] Ramos-Sánchez, G.; Bruno, M.M.; Thomas, Y.R.J.; Corti, H.R. Solorza-Feria, O. *Int. J. Hydrogen Ener.* 2012, 37, 31-40.
[91] Viva, F.A.; Bruno, M.M.; Corti, H.R. *ECS Trans.* 2011, 41, 1121-1130.
[92] Bruno, M.M. Cotella, N.G.; Miras, M.C.; Barbero, C.A. *Chem. Commun.*, 2005, 5896-5898.
[93] Petricevic, R.; Glora, M.; Fricke, *J. Carbon*. 2001, 39, 857–867.
[94] Huang, J.; Kunitake, T.; *J. Am. Chem. Soc.,* 2003, 125, 11834-11835.
[95] Kruk,M.; Jaroniec, M. *Chem. Mater.* 2001, 13, 3169−3183.
[96] McCreery, R.L, Carbon Electrodes: Structural Effects on Electron Transfer Kinetics, Electroanalytical Chemistry 17, A. J. Bard, ed., Dekker, New York, US, 1991.
[97] Barbero, C.; Silber, J.J.; Sereno, L. *J. Electroanal. Chem.*, 1988, 248, 321-340.
[98] Barbero, C.; Kotz, R.; *J. Electrochem. Soc.,* 1993, 140, 1-6.
[99] Rufford, T.E., Hulicova-Jurcakova, D.; Khosla, K.; Zhu, Z.; Qing Lu, G. J. *Power Sources*, 2010, 195, 912-918.
[100] Bruno, M.M.; Corti, H.R.; Balach, J.; Cotella, N.G.; Barbero, C.A.; *Funct. Mater. Lett.* 2009, 2, 135-139.
[101] Balach, J.; Bruno,M.M.; Cotella,N.G.; Acevedo,D.F.; Barbero,C.A,; *J. Power Sources.* 2012, 199, 386-393.
[102] Sakintuna, B.; Yürüm, Y.; *Ind. Eng. Chem. Res.* 2005, 44, 2893.

[103] Wang, W.; Gu,B.; Liang, L.; Hamilton, W.; *J. Phys. Chem.* B. 2003, 107, 3400-3408.
[104] Velev, O.D.; Lenhoff, A.M. *Curr. Opin. Colloid In.* 2000, 5, 56-63.
[105] Lufrano, F.; Staiti, P., Int. J. Electrochem. Sci. 2010, 5, 903-916.
[106] Zhao, X.; Tian, H.; Zhu, M.; Tian, K.; Wang, J.J.; Kang, F.; Outlaw. R.A. J. *Power Sources*. 2009, 194, 1208-1212.
[107] Baena-Moncada, A.M.; Coneo-Rodríguez, R.; Calderón, Juan C.; Flórez-Montaño, J.; Barbero, C.A.; Planes, G.A.; Rodríguez, J.L.; Pastor, E., *Int. J. Hydrogen Ener.,*2014, 39, 3964-3969
[108] Coneo-Rodriguez, R.; Baena-Moncada, A.M.; Acevedo, D.F.; Planes, G.A.; Miras. M.; Barbero,C.A. *Faraday Discuss.*, 2013,164, 147-173.

In: Comprehensive Guide for Mesoporous Materials. Volume 1 ISBN: 978-1-63463-958-3
Editor: Mahmood Aliofkhazraei

Chapter 15

POROUS PT-BASED NANOSTRUCTURED ELECTROCATALYSTS FOR OXYGEN REDUCTION REACTION

***You Xu*[1], *Haolan Xu*[2] *and Bin Zhang*[1,•]**
[1]Department of Chemistry, School of Science, Tianjin University, and Collaborative Innovation Center of Chemical Science and Engineering (Tianjin), Tianjin, P. R. China
[2]Ian Wark Research Institute, University of South Australia, Mawson Lakes, SA, Australia

ABSTRACT

Oxygen reduction reaction (ORR) is one of the most interesting research issues in the academia and industries due to its importance in proton exchange membrane fuel cell s (PEMFCs). Despite the advances in all kinds of catalysts, Pt-based materials are still the most efficient electrocatalysts for ORR. In particular, Pt-based nanomaterials with a porous or network-shaped structure possess intriguing physical and chemical properties to generate enormously promising potentials for ORR catalysis . This chapter attempts to summarize the recent advances in the design and synthesis of porous Pt-based nanostructured materials and their application as electrocatalysts for ORR in PEMFCs, including improved understanding of the reaction mechanism s and the development of synthetic strategies for producing catalysts with high activity and stability . Key scientific issues and prospective directions of research in this field are also discussed.

• E-mail: bzhang@tju.edu.cn.

1. INTRODUCTION

The combination of worldwide ever-growing energy demand, depletion of fossil fuels , and increasing environmental pollution has stimulated intense research on alternative renewable and sustainable source of energy [1, 2].

One promising alternative is the fuel cell s, a device which can directly convert the chemical energy of a fuel into electrical energy by electrochemical reactions [3-5]. Since their invention in the middle of the nineteenth century, fuel cells have been identified as one of the most promising energy conversion technology and have been successfully applied in many fields ranging from automotive vehicles to stationary and portable devices. Among various fuel cells, proton exchange membrane fuel cell (PEMFCs), also called polymer electrolyte membrane fuel cells (PEFCs), have attracted intense interests owing to their charming advantages including high energy density , low operation temperature , and low environmental impact . In a PEMFC, both the anodic and cathodic reactions require electrocatalysts to reduce the overpotentials and increase reaction rate s. To make fuel cells a vibrant and competitive force , many issues still need to be solved. Investigations have revealed that the oxygen reduction reaction (ORR) in the cathode of PEMFC is rather sluggish and results in a large overpotential loss under typical PEMFCs operating conditions, which substantially slows down the voltage output and hampered the commercialization of PEMFCs [6-8].

In the past few decades, various materials , especially those nanoscale systems, including inorganic, organic, and inorganic-organic hybrid materials, have been developed and applied as electrocatalysts in fuel cell s [9-18]. Among all kinds of alternative catalysts, Pt-based nanomaterials remain as the most efficient catalysts for use in practical PEM-based fuel cells. Since Pt is a precious metal, the sky-rocketing price of Pt has become a major issue for all commercial technologies based on this precious metal including fuel cell technology . Among the components in a fuel-cell stack, Pt-based electrocatalysts and their associated catalyst layers account for over 55% of the total cost [19]. So, developing effective pathways to address the obstacles of high cost of the Pt catalysts is a straightforward way to reduce the total cost of fuel cells. Except for the performance and the cost, another ineluctable barrier challenging PEMFC's commercialization is the durability of the electrocatalysts. Thus, current efforts are focused on the development of ideal electrocatalysts for ORR to address the obstacles of high cost of the Pt-based catalysts and of declining catalyst activity over time. To achieve this goal, the design of novel catalysts requires not only reducing the amount of Pt used but also enhancing catalytic activity and stability .

Porous nanostructured metal materials composed of an interconnected network of pores or voids have recently attracted intense attention due to their superior physical and chemical properties including unique pore structure , low density, and large specific surface area. These attractive qualities have fuelled up great interests to explore their promising potentials for a wide range of important applications, for example electrochemical catalysis , because the porous metallic structure can bring about numerous catalysis active site s, superior connectivity and mass diffusion properties and, sometimes, good chemical and physical stability [20-22]. Also, the porous characteristics in this kind of catalysts endow them with the ability to interact with reactive species not only at their surfaces, but also throughout the bulk of the materials. Furthermore, the partitions between nanoparticles by the pore structures greatly inhibit particle growth as well as reducing particle aggregation , which can enhance the stability and durability of catalysts in the catalysis reaction. Thus, porous metal materials are generally chosen as the supports for catalysts or used directly as catalysts in many chemical reactions .

In this chapter, we attempt to summarize the recent advances in the design and synthesis of porous Pt-based nanostructured materials and their applications as electrocatalysts for ORR in PEMFCs . Rooting from the basic theoretical understanding on ORR and the commonly used synthesis strategies to fabricate Pt-based nanostructures , recent investigations on porous oxygen reduction electrocatalysts including pure Pt, Pt-based bimetallic and multimetallic nanostructured catalysts are reviewed. The relationship between the parameters such as morphology , composition , as well as structure and the catalytic activity and durability are discussed based on some typical and important examples. Some key scientific issues and prospective directions of research in this field are also discussed.

2. Basic Theoretical Understanding on ORR

Investigations on oxygen reduction mechanisms have revealed that ORR is a very complex reaction process which involving a multi-electron , various intermediates, and many possible reaction pathways [23, 24]. The reaction mechanism depends not only on the experimental condition s, solvent, or pH , but also on the nature of the electrode material.

In fuel-cell reaction conditions, ORR occurs on the surfaces of the catalysts and O_2 is converted into H_2O or OH^- depending on the supporting electrolyte used in the measurements . In an acidic solution , the ORR on catalyst surfaces occurs via one of two mechanisms as below [25]:

$$O_2 + 4H^+ + 4e^- \longrightarrow 2H_2O \text{ (four-electron pathway)}$$

$$O_2 + 4H^+ + 2e^- \longrightarrow H_2O_2 + 2H^+ + 2e^- \longrightarrow 2H_2O \text{ (two-electron pathway)}$$

Water is produced directly when O_2 is reduced through four-electron reduction mechanism. The two-electron reduction, on the other hand, happens at low potentials and produces hydrogen peroxide (H_2O_2) as the intermediates. In alkaline solution , O_2 can be reduced via the following pathways [26, 27]:

$$O_2 + 2H_2O + 4e^- \longrightarrow 4OH^- \text{ (four-electron pathway)}$$

$$O_2 + H_2O + 2e^- \longrightarrow OH^- + HO_2^- ;\ HO_2^- + H_2O + 2e^- \longrightarrow 3OH^- \text{ (two-electron pathway)}$$

Here, through the four-electron process, O_2 can be reduced to form hydroxide (i.e., OH^-); while in a two-electron pathway, O_2 is reduced to form HO_2^- and then OH^-. It is worth noting that in both of acid and alkaline solution , the direct four-electron reduction of O_2 to

H_2O or OH^- is preferred due to its enhanced power generation ; the two-electron process is generally not favored in the practice for ORR.

Pt has long been known as the best electrocatalyst for ORR in acidic medium and the ORR mechanism on Pt and Pt-based catalysts have been widely investigated for several decades. Despite a series of both experimental measurements and computational calculations have been applied to study the oxygen reduction mechanism, the true nature of the ORR process on a Pt catalyst surface is still not well understood due to the reaction process is very complicated. Basically, oxygen reduction on the Pt catalysts involves three major steps: (i) the first step involves chemisorptions of oxygen molecule on the electrocatalyst surface [28]; the adsorbate's structure , bonding type, and energy are key elements in understanding the effects of adsorption on the reaction kinetics ; (ii) oxygen absorbed on the catalysts surface may be converted into different intermediates, such as ox ygenated (O*), hydroxyl (OH *) and superhydroxyl (OOH*) species , through the scission of O-O bond [29, 30]; recent density functional theory (DFT) calculations by Nørskov et al. demonstrated that, at high oxygen coverage, the ORR tends to follow an associative mechanism, in which OOH* is first formed and then O-O is cleaved; whereas, at low oxygen coverage, the ORR follows a dissociation mechanism in which the O-O bond is cleaved before OH* is formed [29]; (iii) the absorbed intermediates species further reacts with protons to produce water . It should be pointed out that more experimental and theoretical studies are needed to provide better understanding of ORR and to help design high efficient electrocatalysts.

3. The Electrochemical Evaluation of ORR

Currently, both rotating disk electrode (RDE) and membrane electrode assembly (MEA) are usually used in the measurements of ORR activity when estimating the performance of an electrocatalyst . Generally, RDE is designed for measuring the catalyst performance based on a half-cell configuration which is suitable for the study of the cathodic half-cell reaction (i.e., ORR) in fuel cell , while MEA, which is consisted of both anodic and cathodic electrodes , diffusion media , and a proton-exchange membrane, is particularly valuable for evaluating the catalyst performance in practical applications of PEMFCs for electric vehicles [9, 31, 32].

To use the RDE to measure the ORR activity of the catalyst , the catalyst sample is usually deposited on a glassy carbon RDE to form a thin-film of catalyst layer as the working electrode. In one popular method for preparing the catalyst ink, the catalyst sample is dispersed in a mixture of water , isopropyl alcohol, and Nafion (5 %) and the mixture was sonicated to form a homogeneous catalyst ink [31]. The isopropanol used here usually can help the dispersity of the catalyst particles in aqueous mixtures; while Nafion is benefit for the bond of the catalyst particles on the electrode surface . It is worth noting that the quality of catalyst film deposited on the surface of the working electrode is of importance for obtaining accurate date on the ORR activity[9]. Thus, people should precisely control the proper amount of catalyst loading on the support, and uniformity in both catalyst ink and electrode film, because these parameters can greatly affect the values obtained. More catalyst loading is not propitious to obtain accurate date because the formation of an overly thick catalyst agglomerate may result in excessive mass -transport loss and large uncertainty in qualifying catalytic activities [33, 34]. Empirically, the catalyst loading is controlled in the range of 10-50 μg cm^{-2} depending on the area of the RDE used [31].

Electrochemically active surface area (ECSA) is an important parameter for evaluating the electrochemical performance of the catalysts. Despite the specific surface area of the catalyst is commonly measured by the multipoint Brunauer-Emmett-Teller (BET) method on the basis of gas adsorption -desorption isotherms , the ECSA is more reasonable to reflect the actual surface area for electrochemical reactions [22, 35]. There are specific ways to measure the ECSA of a given metal catalyst, especially those of noble metal-based electrocatalysts, such as Pt, Pd, and Au. Taking the ORR measurement in acidic solution as an example, the electrochemical properties of the catalyst in an N_2- or Ar-saturated 0.1 M $HClO_4$ or 0.5 M H_2SO_4 solution are first studied by cyclic voltammetry to obtain a cyclic voltammogram (CV) curve. The ECSA of the catalyst can be determined according to the obtained CV curve. Cyclic voltammetric determination of the ECSA of the Pt electrode is based on the charges corresponding to its electrochemical desorption of the hydrogen monolayer . The ECSA of Pt catalyst can be calculated on the basis of the following relation[36]:

$$\text{specific ECSA} = \frac{Q_H}{m \cdot q_H}$$

where Q_H is the charge for hydrogen desorption , m is the Pt loading, and q_H is the charge required for desorbing a monolayer of hydrogen on the Pt surface. The ECSA is an important factor not only for the evaluation of the actual surface area for electrochemical reaction but also for the further qualitatively evaluation of catalytic activity .

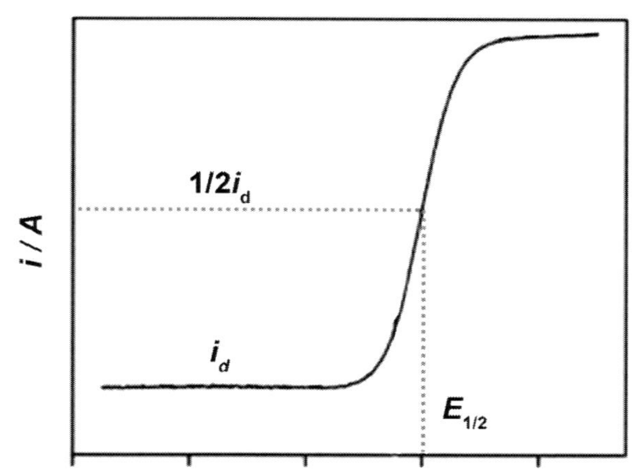

Figure 1. Typical ORR polarization curve of a catalyst and the parameters used to qualitatively evaluate its activity.

ORR activities are reflected on the polarization curve of the electrocatalyst which is recorded in an O_2-saturated supporting electrolyte with the RDE rotating at a certain speed (usually 1600 rpm) and a potential sweep of 5-20 mV s^{-1} to minimize the interference from capacitive currents. The potential scale for sweeping to obtain a polarization curve is usually controlled in the range of 1.05-0 V (vs. RHE). Empirically, for a properly prepared electrode, the polarization curves should be fairly flat in the potential range between 0.2 and 0.5 V (vs. RHE) and the curve density in this range should approach to 5.5 mA cm^{-2} [9]. From the ORR polarization curves of various catalysts under the same catalyst loading and measurement conditions, half-wave potentials ($E_{1/2}$) of various catalysts in the polarization curves can be used to qualitatively determine the catalyst activity (see Figure 1). The higher the $E_{1/2}$ suggests the better the ORR activity.

The kinetic current at a certain potential where ORR takes place (usually 0.9 V vs. RHE) obtained from the ORR polarization curve can be used for a quantitative comparison of the catalytic activities of various catalysts, which can be obtained according to the Levich-Koutecky equation[31]:

$$\frac{1}{i} = \frac{1}{i_k} + \frac{1}{i_d}$$

where i is the current obtained experimentally, i_k is the mass -transport free kinetic current and i_d is the diffusion -limited current. For Pt-based catalyst , the specific and mass activity of the catalyst can be determined by normalizing the i_k with its ECSA or the Pt loading, respectively. The efficiency of a new catalyst can be evaluated by comparing its specific and mass activities with those obtained from the known "standard" Pt/C catalyst under the same measurement conditions. Currently, two standard catalysts are usually used as a reference for comparison of both area specific and mass activities in the ORR measurement [9]. One is the 20% Pt/C from E-TEK which has initial surface area of ~ 100 m^2 $g^{-1}{}_{Pt}$ and specific activity of 0.2 mA $cm^{-2}{}_{Pt}$. Another class of commonly used Pt catalysts is Pt particles in the size range of 3~5 nm supported on TKK porous carbon . The loading of Pt nanoparticles is able to reach 46-50% (*w/w*) with the electrochemical surface area reaching ~65 m^2 $g^{-1}{}_{Pt}$ and specific activity of 0.3-0.4 mA $cm^{-2}{}_{Pt}$.

Except for the activity, the durability is another important indicator should be tested to evaluate the catalytic performance of the catalyst for practical applications. In general, the catalyst durability in acidic solution is tested by accelerated potential cycling over 0.6-1.1 V (vs. RHE) in O_2-saturated supporting electrolyte for more than 4000 cycles . The shift of the ORR polarization curve and the ECSA change of the catalyst before and after accelerated durability test are compared to qualitatively evaluate the durability of the catalyst. Specific and mass activities before and after accelerated durability test calculated according to the kinetic currents at the same potential were commonly used to quantitatively compare the activity change. Moreover, the stability of the catalysts in shape , morphology , structure , and composition is also usually characterized by using scanning electron microscopy (SEM), transmission electron microscopy (TEM), Energy-dispersive X-ray spectroscopic (EDX), and Powder X-ray diffraction (PXRD) when investigating the stability of the catalyst.

4. Strategies for the Synthesis of Porous Pt-Based Nanostructures

The field of fabrication and application of porous Pt-based nanostructured materials is currently experiencing a rapid development with exponentially increasing achievements. Some excellent reviews on the related fields have recently been published. [8, 21, 22, 37-40] In this part, we will give a brief review about two commonly used strategies for the preparation of porous Pt-based nanostructures, namely template synthesis and dealloying. The key points of each synthetic strategy in preparation, morphological and structural features of the products are presented based on some representative and important examples extracted from published investigations.

4.1. Template Synthesis

Template synthesis is a powerful strategy for creating porous nanostructures. In particular, nanoporous materials ranging from microporous to macroporous, and further to hierarchical porous structures can be generated by using template synthesis strategy. In short, template synthesis refers to a method that using a prefabricated material with special architectures as a reference or mould to guide and/or order the placement of building blocks in specific locations during the growth of the target material. [41-43] Generally, according to their nature, the templates can be categorized into three main types, namely, hard, soft, and sacrificial templates.

4.1.1. Hard Template Method

Examples for hard templates used in the synthesis of porous nanostructures are multifarious. Previously, some natural materials with specific pore structures, such as echinoid spins and corals, were recognized and applied as hard templates to generate some interesting porous materials. Recent advances in chemical synthetic methods have led to the formation of many kinds of artificial hard templates such as porous anodic aluminium oxide (AAO) membranes, mesoporous silica, close-packed colloidal crystalline films, which have been widely exploited to create porous structures.

AAO is a kind of widely used hard templates for the fabrication of well-defined 1D nanostructure arrays of desirable materials. The morphology of the resulting replicas from AAO can be tailored through controlling the mother structures of the AAO templates and the synthetic procedures. Generally, partially or completely templating AAO membranes could generate nanotubes and nanowires /nanorods, respectively. [44] Currently, by taking advantage of vertically aligned porous structures of the AAO templates, a variety of Pt-based nanostructures have been created. For example, Liu and co-workers successfully fabricated nanoporous Pt-Co alloy nanowires by electrodeposition of Co-rich Pt_1Co_{99} alloy into AAO membranes followed by a dealloying treatment in 10 wt% H_3PO_4 solution. [45] Figure 2a-c showed the schematic illustration of the preparation process. The resultant Pt-Co alloy nanowires consisted of porous skeletons with tiny pore of 1-5 nm and crystalline ligaments of 2-8 nm (Figure 2d). Based on the study on the evolution process of the nanowires, it was found that both the morphology and composition of the dealloyed porous Pt-Co nanowires could be controlled by simply changing the dealloying time, which was of importance for adjusting the morphology- and composition-depended properties of the target materials. Similar synthetic strategy has been demonstrated to be applicable for the preparation of nanoporous Pt-Ni alloy nanowires. [46]

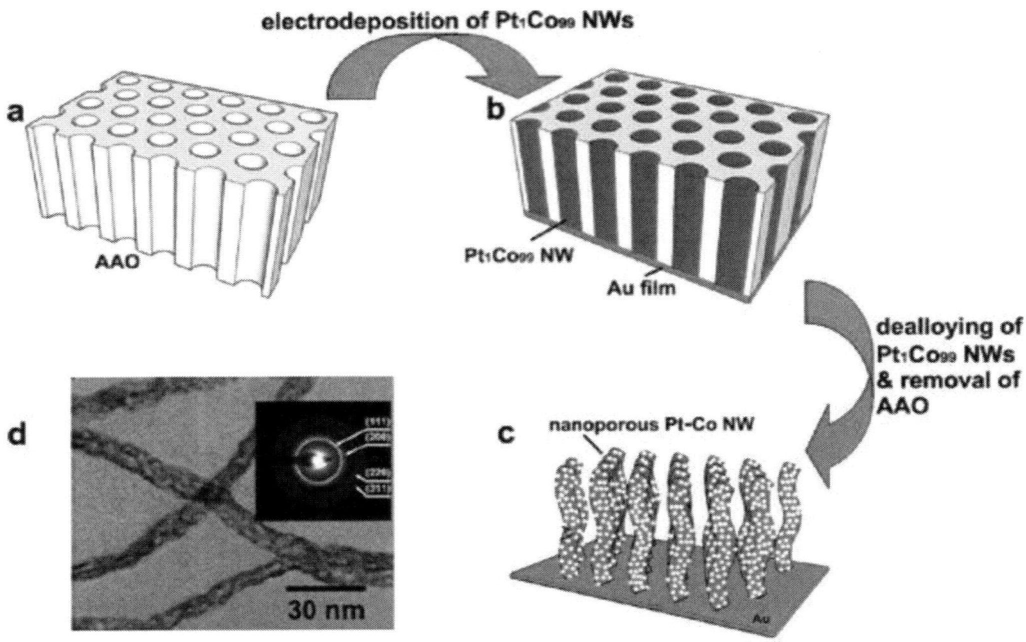

Reprinted with permission from Ref. (45). Copyright 2009 American Chemical Society.

Figure 2. (a-c) Schematic illustration of the fabrication process of nanoporous Pt-Co alloy nanowires . (d) TEM image of the nanoporous Pt-Co alloy nanowires. The inset in Figure 2d is the SAED pattern.

Mesoporous silica is advantageous as a template from the viewpoint of its chemical and mechanical stabilities. With the great advances in synthetic methodology , the pore size , pore geometry , and pore alignment of mesoporous silica can be precisely controlled by changing the synthetic conditions, leading to the formation of a variety of mesoporous silica templates, such as FSM-16, MCM-41, MCM-48, KIT-6 and SBA-15. Many types of porous Pt-based nanostructures have been created by using various mesoporous silica templates. For example, Yamauchi et al. have demonstrated the shape - and size-controlled synthesis of mesoporous monocrystalline Pt nanoparticles by controlled chemical reduction of Pt precursors in special mesoporous silica templates [47].

The detail studies revealed that the strength of the reducing agent played a crucial role on the formation of mesoporous Pt particles with uniform size and the particle size could be controlled by changing the reduction time. In particular, polyhedral- and olive-shaped mesoporous Pt nanoparticles with uniform particle size could be fabricated by selecting 3D bicontinuous mesoporous silica (KIT-6) and 2D mesoporous silica (SBA-15), respectively, as templates (Figure 3).

It is worthy of note that well-defined morphology and uniformity in size is highly desirable and technologically important for the function and utility of these materials in specific applications.

Figure 3. (a, b) SEM images of the mesoporous Pt nanoparticles prepared with different mesoporous silica templates: (a) KIT-6, (b) SBA-15.

Except for AAO and mesoporous silica , close-packed colloidal crystalline film is another ideal starting template to create porous materials , specifically those with pore diameters between 200 and 1000 nm, in a free-standing format with highly ordered void structures [48, 49]. Most commonly, colloidal crystalline templates are formed through self-assembly of silica or polymer spheres such as polystyrene and polymethylethacrylate in the nanometer -to-micrometer size range on solid substrates . As a typical example, Bartlett and co-workers have fabricated highly ordered macroporous metal films (e.g., Pt, Pd and Co) with regular arrays of spherical pores with diameters of 0.40, 0.70 or 1 μm by electrochemical deposition into the interstitial spaces of a colloidal crystal template formed by self-assembly of polystyrene latex spheres on gold substrates. [50] After deposition of the target materials , the polystyrene spheres could be fully removed by washing in toluene to leave a highly periodic, hexagonal close packed, interconnected network of monodisperse spherical pores within the metal film. The authors demonstrated that the size of the pores within the resultant metal film could be manipulated by tailoring the diameter of the polystyrene latex particles used to prepare the template. It is worthy of note that, because of the surface characteristic and the interstitial areas are curved, narrow, and not readily accessible, uniform deposition of the target material into the interstitial regions of a colloidal crystal multilayer is a challenging chemical problem. [49] In this case, surface modification is often an important step to achieve uniformly and completely filling the voids of the template colloidal arrays with the target material of the precursor of it.

4.1.2. Soft Template Method

Soft templates formed by surfactant aggregated can also be used to fabricate porous materials . Lyotropic liquid crystalline (LLC) formed by $C_n(EO)_m$-type surfactant molecules (EO=ethylene oxide) solved in water has been widely applied as a soft template to create Pt-based porous nanostructures . It has been recognized that the structure of the LLCs are largely depended on the concentration of surfactant in the aqueous solution . With increasing surfactant concentration, the structures of LLCs are generally changed from a miceller solution through micellar cubic, hexagonal, bicontinuous cubic, lamellar, to inverse micellar phases. [51] Benefiting from their multifarious morphologies, the LLC-directed template method is a versatile way to fabricate various porous nanostructures. In 1997, Attard and co-workers firstly reported the successful preparation of mesoporous Pt thin films with pore size s over a range of 1.7 to 10 nm using LLC templates. [52] Following this pioneering work, a variety of metal and alloys have be produced by extending the concept to different reaction systems. [53-56] As a typical example, mesoporous Pt-Au binary alloys were electrochemically synthesized from LLC containing corresponding relevant metal species . [53] In this synthesis , the framework composition of the resulting mesoporous Pt-Au alloy could be controlled by changing the compositional ration in the precursor solution. The illustration of the experimental procedure is showed in Figure 4a-d. Importantly, the composition of the alloy largely influenced the morphology of the resultant alloyed nanostructured. Contrast experiments revealed that well-ordered 2D hexagonal mesostructures reflecting the original templates could be found in Pt-Au alloys with low Au content. Figure 4e displays the representative SEM image of the mesoporous $Pt_{90}Au_{10}$ film.

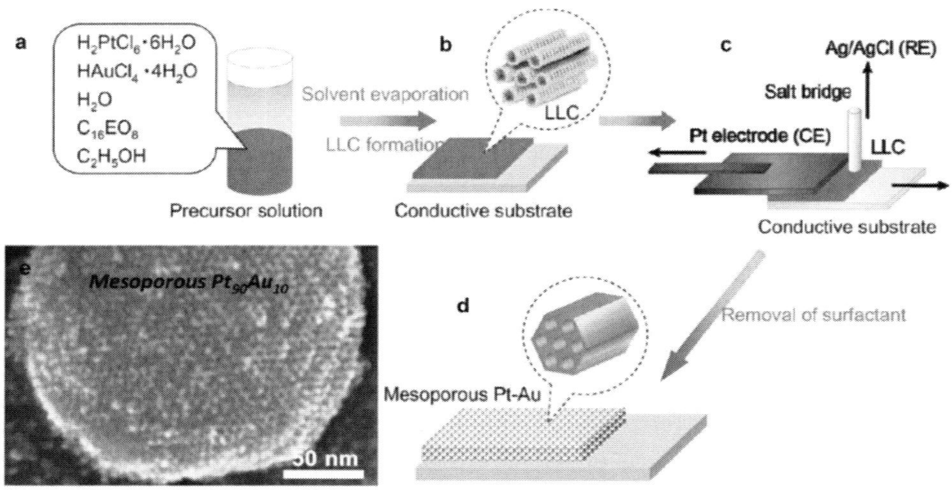

Figure 4. (a-d) Illustration of the experimental procedure for preparation of mesoporous Pt-Au alloy films . (e) SEM image of the mesoporous $Pt_{90}Au_{10}$ film.

With increasing Au content, however, the mesostructural ordering in the resulting alloys gradually decreased and aggregated spherical particles can be generated. It is noted that, for LLC templates-directed synthesis , the rich variability of surfactants in size and composition can form a variety of LLC templates, allowing the rational design of porous structures with respect to pore size , morphology , and distribution in three dimensions. Also, this synthetic methods based on LLCs allows people to realize the microfabrication of porous materials through the solvent-evaporation process, which is not achievable by hard-template approach.

Micelles formed by surfactant molecules solved in water -oil mixture are also another kind of soft templates for the fabrication of porous nanostructures . Song and co-workers evidenced wormlike micelles can be formed in the CTAB/water/chloroform system. [57] By reducing Pt precursors in this two-phase system, Pt network-shaped nanostructures composed of nanowires with an average diameter of 2.2 nm can be produced. The shape selectivity in this synthesis relied on the formation of a network of wormlike micelles within an organic solvent phase that acted as a soft template for metal growth . Since this pioneering work, the concept has been extended to producing a variety of Pt-based alloys analogs [58-61]. For example, Chen *et al.* fabricated Pt-Ru alloy nanowire network via the chemical reduction of Pt and Ru complexes using sodium borohydride ($NaBH_4$) in the presence of CTAB-based soft templates. [59] The authors found that the amount of reducing agent could largely influence the formation of resulting network–like nanostructures. Nanowire networks could be obtained only in the introduction of appropriate amount of reducing agent. Low amount of reducing agent led to the generation of aggregated Pt-Ru nanoparticles due to the metal precursors were not completely reduced. While the products appeared as nanoparticles in the case of excess amount of reducing agent owing to the fact that the metal precursor were reduced too rapidly and CTAB did not act as a soft template. Interestingly, Zhang and co-workers found that the introduction of an appropriate amount of ethylene glycol into CTAB/water/chloroform system could generate Pt-Ni [60] and Pt-Co [61] alloy nanoparticle networks with spongy morphology .

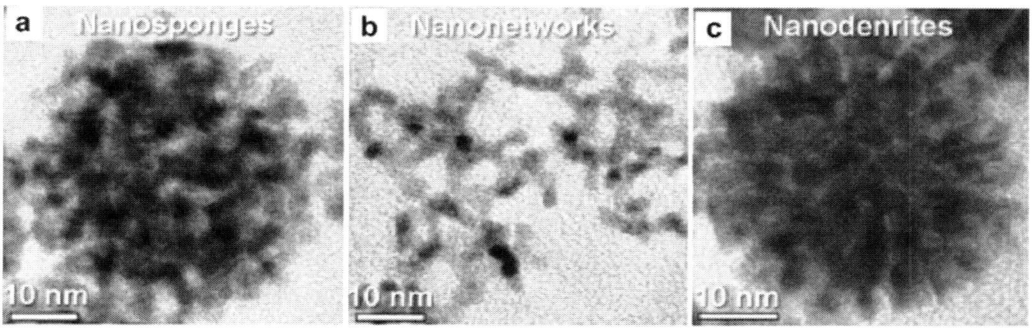

Reprinted with permission from Ref. (63). Copyright 2009 Wiley-VCH.

Figure 5. (a-c) SEM images of various Pt nanostructures obtained through the galvanic replacement reaction of Pt precursor with Te nanowires : (a) nanosponges, (b) nanonetworks, (c) nanodenrites.

They demonstrated the presence of ethylene glycol was crucial for the size control of the nanoparticle building blocks in these spongy networks.

4.1.3. Sacrificial Template Method

Besides hard and soft templates, sacrificial templates are also frequently used in synthesizing porous/hollow nanostructures , where the templates sacrifice themselves in the process of templated target product formation without an additional template removal step. The sacrificial template mechanism is based on galvanic replacement reaction which is driven by the difference of redox potentials between the two different species . In a sacrificial template-directed synthesis , the morphology of templates can act as a reference and guide the nucleation and growth of the depositing metal and final morphology of the resultant products. Most commonly, the shapes of the template materials cannot be preserved due to different valencies and densities among various metals and porous/hollow interiors, frameworks, etc. are usually formed.

Thus, by varying the shapes of the starting template and the amount of the metal precursor added, a variety of morphologies can be obtained. For the fabrication of porous/hollow Pt-based nanostructures , multifarious materials can be used as sacrificial templates including metals (e.g., Cu, Co, Ni, Ag, Pd), semi-metals(e.g., Se, Te), as well as some metal oxides (e.g., Cu_2O, Mn_3O_4).

As an example using metals as sacrificial templates, Ag nanowires have been used as templates for the synthesis of porous Pt nanotubes with an all thickness of 5 nm, an outer diameter of 60 nm, and a length of 5-20 μm. [62] In an example of galvanic replacement of semimetal nanstructures, Chang et al. reported the successful synthesis of various porous Pt nanostructures including nanosponges, nanonetworks, and nanodendrites in aqueous solution from the reduction of Pt precursors via galvanic replacement reactions with Te nanowires, as shown in Figure 5. [63]

In this synthesis , the morphology and architectures of the final porous Pt nanostructures can be controlled by employing different temperatures and concentrations of sodium dodecyl sulfate (SDS) .

Multivalent metal oxide nanostructures can also be used as templates for galvanic replacement. As a typical example, Pt nanoparticles were deposited on Mn_3O_4 via the galvanic replacement process occurring between the Mn_3O_4 surface and $PtCl_4^{2-}$ species . [64]

4.2. Dealloying

Dealloying, also known as selective dissolution has been proved to be another efficient approach to generate porous nanostructures , especially porous metals and alloys. In fact, dealloying refers to a corrosion process in which the active component dissolved selectively in appropriate corrosion conditions from an alloy system that consists of two or multicomponents metals with different chemical activities. According to the dissolution pathway of active component from the alloy precursor, the dealloying methods can be divided into two types: electrochemical dealloying and chemical dealloying.

4.2.1. Electrochemical Dealloying

Electrochemical dealloying refers to an electrochemically driven metallic corrosion process, which allows investigation of the dealloying process in an electrolyte . It is recognized that parting limit and critical potential are two important parameters for dealloying in electrochemical system. The parting limit represents the limiting concentration of the inert component in an alloy to enable dealloying. [38] With an inert component concentration below the parting limit, the outmost layers of the less noble metal get dissolved but no reorganization of the more noble metal takes place. The critical potential represents the electrode potential threshold which has to overcome in order to enable dissolution of the active component from the alloy compound [38].

A number of porous Pt-based nanostructures can be produced by selective dissolution of the active component from proper Pt-based alloy precursors using electrochemical dealloying. [65-70] In a typical example, Jung et al. have reported the successful synthesis of nanoporous Pt thin films . [65] In this synthesis, Pt-Si alloy films were firstly prepared by codeposition of Pt and Si on a highly doped Si wafer. Nanoporous Pt film could be generated by dealloying treatment of these Pt-Si alloy films in HF acid solution under a constant potential. Through simply controlling the Pt-Si alloy composition and the dealloying condition, the morphology , dimension and structure of the finial porous Pt films could be easily adjusted.

4.2.2. Chemical Dealloying

Chemical dealloying is another powerful pathway to access the dissolution of the active component from the alloy system. In a chemical dealloying process, the active component in the alloy precursor can be selectively dissolved naturally in acid or base solution , which is much easier to implement compared to electrochemical dealloying. [71-73] As a typical example, Li et al. have demonstrated that monodisperse nanoporous Pt-Ni alloy nanoparticles could be synthesized by selective dissolution of the Ni component from as-synthesized nanocrystalline Pt-Ni alloy. [71] In this synthesis , Pt-Ni alloyed nanoparticles with an average size of 20 nm and polygonal morphology with clear edges and corner (Figure 6a,b) were firstly synthesized and used as the precursors for converting into nanoporous Pt-Ni nanoparticles by means of dealloying treatment in concentrated nitric acid. After corrosion , the edges and corners of particles disappear and the particle size decreases to about 15 nm (Figure 6c,d). The detail studies reveal that the reaction between nitric acid and Ni was very fast which could be finish with 1 min and about 85% rather than all of Ni in the precursor could be etched away because the surfaces were fully covered by the inert Pt leading to surface passivation .

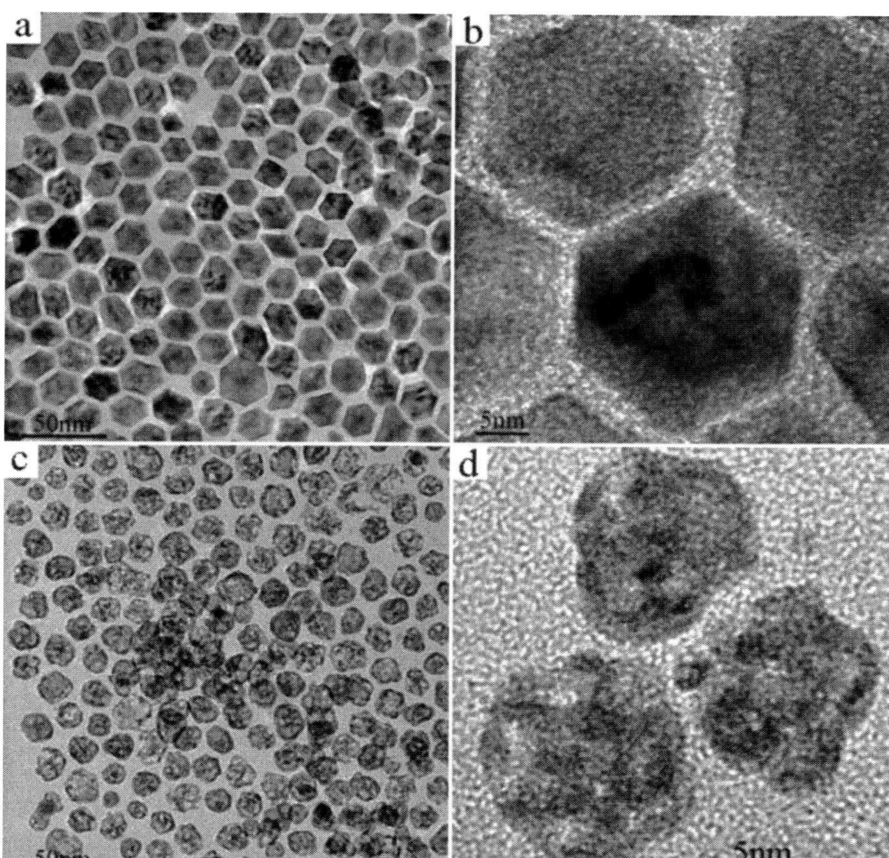

Reprinted with permission from Ref. (71). Copyright 2011Nature Publishing Group.

Figure 6. (a, c) TEM images and HRTEM images (b, d) of the Pt-Ni nanoparticles and nanoporous Pt-Ni alloys: (a, b) Pt-Ni nanoparticles, (b, d) Pt-Ni alloy alloys.

Because the precursor itself possessed small particle size and large surface area (35.3 m^2 g^{-1}), the resulting dealloyed product had not only small pore size below 5 nm, but also much larger surface area (60.1 m^2 g^{-1}).

5. Porous Pt-Based Nanostructured Catalysts for ORR

5.1. Pure Pt Catalysts

At present, the most widely used ORR electrocatalyst system in industry applications is Pt in the form of small nanoparticles (2-5 nm) supported on amorphous carbon materials (Pt/C). Pt/C catalysts are subject to the following issues : [74-77] (1) carbon corrosion which leads to the loss of Pt nanoparticles from the electrical contact, (2) Pt dissolution and redeposition or Ostwald ripening of the Pt nanoparticles, (3) Pt sintering by means of particle migration and agglomeration, and (4) Pt nanoparticle dissolution and subsequent migration of the soluble Pt species with the polymer electrolyte and the precipitation of Pt in the proton-exchange membrance. These issues may result in the decrease of both the active catalyst surface area and the catalytic efficiency, which leads to an undesired increase in overpotentials for ORR. In this regards, current efforts are focused on developing other alternative materials as the catalyst support and synthesizing 3D structures as self-supported electrocatalysts to eliminate the usage of carbon support which is a promising efficient approach to alleviate the above problems and enhance the durability of the cathode electrocatalysts.

Porous structured materials can exhibit high surface area similar to those observed for supported nanoparticles , but offer some attractive advantages for electrochemical applications as compared to those supported nanoparticles. [20-22, 78] On one hand, porous catalysts with stable architectures can effectively presintered, thus minimizing loss of active catalyst surface area due to an agglomeration pathway. On the other hand, omitting the carbon support from the catalyst layer can reduce the thickness of the catalyst layer and reduce the interfacial resistance in a practical device because the catalyst layer would be indirect contact with the gas diffusion layer. What's more, for porous metal structures, metallic frameworks can improve electron mobility in solid ligaments due to the intrinsical good electrical conductivity . These advantages make porous Pt-based materials found enormously promising potentials for ORR catalysis .

Chen et al. reported the one-step synthesis of 3D Pt nanochain network nanostructures and demonstrated their excellent catalytic activity towards ORR. [79] In this synthesis, diethylene triamine penta(methylene phosphonic acid) (DTPMPA) was introduced as ligands to interact with H_2PtCl_6 to generate DTPMPA-Pt^{IV} complex which acted as reaction precursors for creating the Pt nanochain network through a thermal decomposition method. The resulting nanostructures were three-dimensionally interconnected networks with primary Pt nanochains as build blocks and the Pt nanochains grew from the primary spheric Pt nanoparticles via oriented attachment .

Study on their electrocatalytic behaviors revealed that these Pt nanochain network exhibited superior electrocatalytic activity and stability towards ORR compared to commercial Pt black catalyst . They ascribed the enhancement in the electrocatalytic performance to the unique properties of these Pt nanostructures such as the few surface defect site s and the low hydroxyl surface coverage on one-dimensional Pt nanochains, as well as fast O_2 diffusion in 3D structures.

Lou and co-workers compared the electrochemical activities of self-supported interconnected 3D Pt nanoassemblies to 0D Pt nanoparticles towards ORR. [80] The 3D Pt nanoassemblies consisting of 1D single-crystal Pt nanowires were prepared by a one-pot method in which the presence of CTAB was crucial for the formation of a 3D structure . These interconnected 3D Pt nanoassemblies showed a higher resistance to dissolution, migration , Ostwald ripening , and aggregation compared to the 0D Pt nanoparticles. Benefitting from unique structural features, these self-supported interconnected 3D Pt nanoassemblies, when evaluated as electrocatalysts for ORR, manifest high ECSA and significantly enhanced long-term stability compared to the 0D Pt nanoparticles.

Jaramillo et al. found that the activity and stability of Pt for ORR catalysis could be improved by constructing meso-structure (Figure 7). [81]

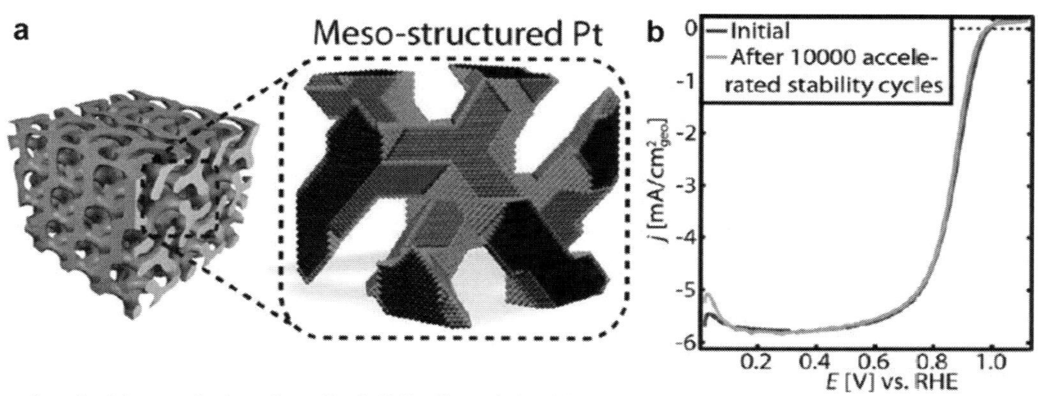

Figure 7. (a) Structural model for the mesoporous double gyroid Pt and (b) electrochemical stability test for mesoporous double gyroid Pt catalyst .

In this study, the high surface area meso-structured Pt thin film was successfully synthesized by electrodeposition of Pt into a mesoporous silica film . The resulting Pt structure was stable after removal of the silica and maintained the interlocked, twisted network structure of the silica template. The unique meso-structure exhibited a surface structure with a lower fraction of undercoordinated Pt atoms and the contiguous nature of the meso-structured Pt could efficiently prevent the pathways for Ostwald repening, migration , and agglomeration of Pt. Benefiting from these advantages, the meso-structured Pt exhibited higher specific activity for ORR than the commercial Pt/C electrocatalyst , as well as a better durability . The unique meso-structure showed excellent specific activity due to a surface structure with a lower fraction of undercoordinated Pt atoms. And the superior stability was ascribed to the contiguous nature of the meso-structured Pt, which could efficiently prevent the pathways for Ostwald ripening , migration, and agglomeration of Pt.

5.2. Pt-Based Bimetallic and Multimetallic Catalysts

Alloying Pt with other transition metal elements is highlighted as a promising approach toward improving ORR electrocatalysis . DFT calculations have demonstrated that alloying an early transition metal with Pt can efficiently downshift the d-band center of Pt, leading to a lower degree of adsorption of ox ygenated spectator (blocking) species and simultaneously increase tremendously the number of catalytic active site s accessible to O_2 molecule. [82-84] The Pt-based alloy catalysts are desirable for ORR, because replacing a fraction of Pt with a cheaper elements can reduce the usage of Pt thus the cost of catalysts and largely enhance the catalytic activity due to electronic, alloying, or strain effects. The last two decades has witnessed an exciting developments for the Pt-based alloy ORR catalysts. This part will focus on several representative and important porous Pt-based bimetallic and multimetallic nanostructures and their applications for ORR catalysis .

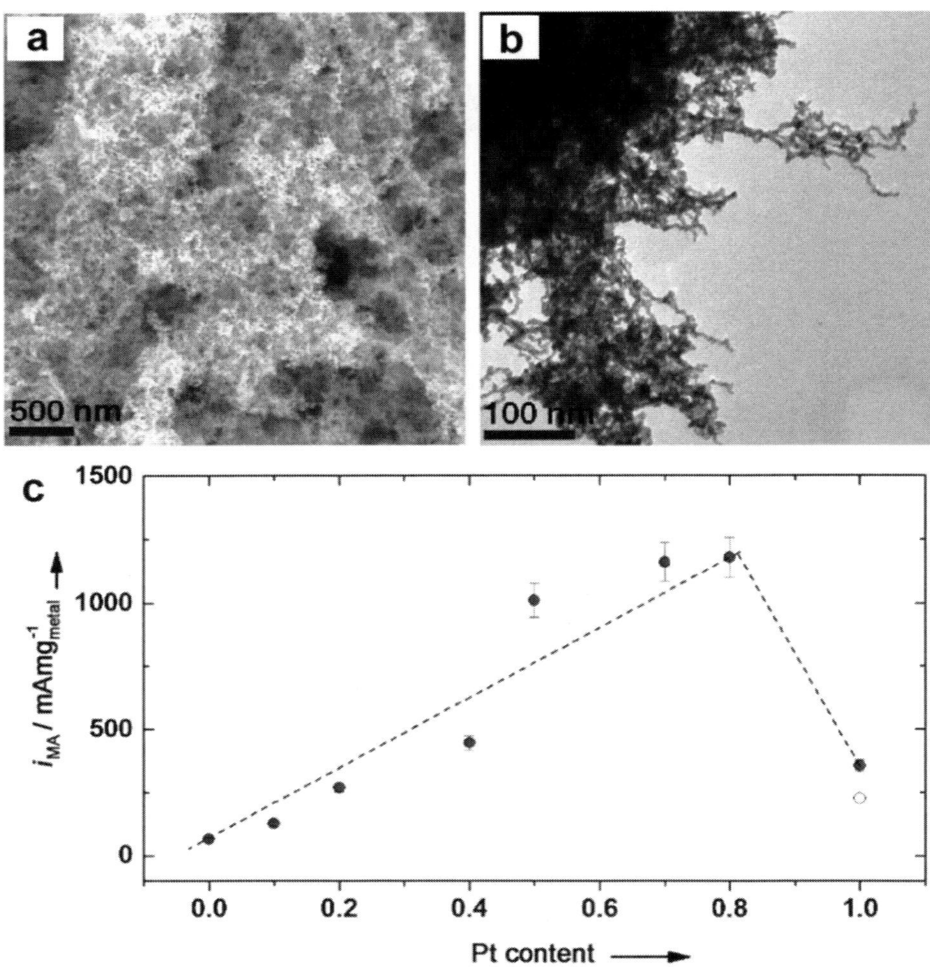

Reprinted with permission from Ref. (86). Copyright 2013 Wiley-VCH.

Figure 8. (a) SEM image and (b) TEM image of the $Pt_{50}Pd_{50}$ aerogel. (c) Volcano plots of the ORR mass activity of the Pt_xPt_y, Pt, and Pd aerogel catalysts as a function of Pt content at 0.9 V.

5.2.1. Pt-Pd

By varying the amount of the additional introduced reducing agent, galvanic replacement of Pd octahedrons and cubes by Pt precursors can produce various Pt-Pd alloy nanostructure including nanocages with porous walls, dendritic hollow nanocages and dendritic core-shell nanoparticles . [85] Study on the ORR activity revealed these hollow Pt-Pd alloy nanocrystals exhibited considerably enhanced oxygen reduction activities compared to those of Pd@Pt core-shell nanocrystals and commercial Pt/C catalysts. Moreover, their electrocatalytic activities were shown to be highly dependent on their morphologies. Among all kinds of counterparts, the Pt-Pd nanocages prepared from octahedral Pd templates exhibited the largest improvement in catalytic performance. In another example, Eychmüller et al. studied the composition -depended electrocatalytic performance of bimetallic Pt_xPd_y aerogels towards ORR. [86] In this case, nanostructured Pt and Pd monometallic and Pt_xPd_y bimetallic aerogels with controlled composition can be prepared through an environmentally benign strategy. The obtained aerogels were composed of 3D nanowire network structures and displayed very high surface area and large porosity (Figure 8a,b).

These self-supported Pt_xPd_y catalysts showed excellent electrocatalytic activity and durability towards ORR, with the $Pt_{80}Pd_{20}$ aerogels performing the best and showing five times higher mass activity than the commercial Pt/C catalyst , as showed in Figure 8c.

5.2.2. Pt-Ni

Huang et al. investigated the electrocatalytic activity of porous Pt_3Ni nanocrystals toward ORR. [87] These porous Pt_3Ni nanocrystals with an average diameter of 28 nm possessed a unique porous structure with highly multibranched subunits can be prepared through simultaneous reduction of $[Pt(acac)_2]$ and $[Ni(acac)_2]$ by using N,N-dimethylformamide (DMF) as both solvent and reducing agent. Figure 9a show the TEM and HAADF-STEM images of these porous Pt_3Ni nanocrystals.

In this synthesis , polyvinylpyrrolidone (PVP) was introduced as surfactant and played a key role in obtaining nanocrystals with good dispersity and well-defined shape and the selective use of phenol as the structure -directing agent was found to be critical for the generation of porous nanocrystals. Benefiting from their intriguing properties including high surface area , rough surface, and porous feature, these porous Pt_3Ni nanocrystals can serve as highly efficient catalysts in ORR with much better performance and durability than commercial Pt black and Pt/C catalysts (Figure 9b-d).

In a recent study, Stamenkovic, Yang and co-workers have demonstrated the excellent ORR activity of Pt_3Ni hollow nanoframes. [88]

In this case, $PtNi_3$ polyhedra with a uniform rhombic dodecahedron morphology and size (20.1 ± 1.9 nm) were firstly prepared and converted into Pt_3Ni hollow nanoframes by interior erosion with unchanged symmetry and size. Both the interior and exterior catalytic surfaces of this open-framwork structure were composed of the nanosegregated Pt-skin structure. Study on their ORR catalytic activity revealed that the Pt_3Ni nanoframe catalyst achieved a factor of 36 enhancements in mass activity and a factor of 22 enhancements in specific activity over state -of-the-art Pt/C catalysts during prolonged exposure to reaction conditions.

5.2.3. Pt-Cu

Chen and co-workers have prepared mesoporous hollow PtCu nanoparticles in an aqueous environment using Cu nanoparticles as sacrificial templates via a modified galvanic replacement approach. [89] The obtained PtCu hollow nanoparticles using this sacrificial templating synthetic method possessed an average diameter of about 20~30 nm and a shell thickness of 2~3 nm. Differ from other common hollow nanostructures , the shells of these PtCu hollow nanoparticles were constructed from discontinuous and seamlessly PtCu alloyed nanoparticles, which generated an incomplete and porous shell of PtCu hollow nanoparticles. Electrochemical tests have shown that these PtCu nanostructures can work as more efficient and durable electrocatalysts for ORR compared with commercially available Pt/C catalysts.

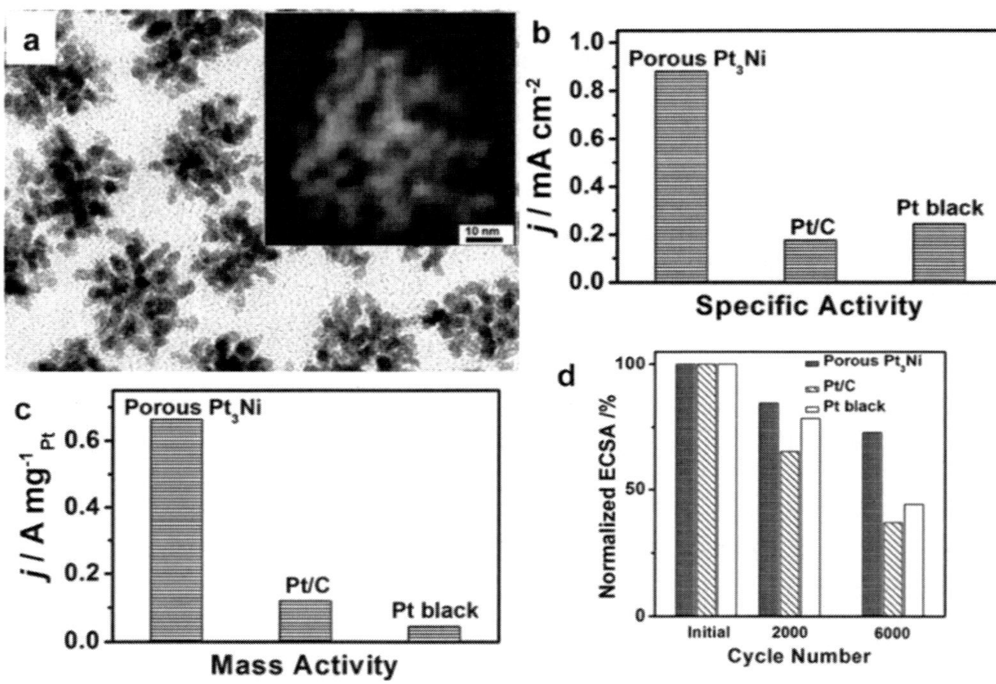

Reprinted with permission from Ref. (87). Copyright 2013 Wiley-VCH.

Figure 9. (a) TEM image of the porous Pt_3Ni nanocrystals . The inset in Figure 9a is the HAADF-STEM image of an individual porous Pt_3Ni nanocrystal. (b) Specific activities and (c) mass activities of porous Pt_3Ni nanocrystals, Pt/C catalyst , and Pt black catalysts at 0.9 V. (d) Loss of ECSA of porous Pt_3Ni nanocrystals, Pt/C catalyst, and Pt black after electrochemical durability test.

In another example, Yu group has demonstrated the design and synthesis of highly active and durable PtCu nanoparticles nanotubes through potential cycling in acid electrolyte [90]. In this study, unsupported porous PtCu nanotubes were firstly prepared through electrodeposition of PtCu into commercial AAO template and used as the initial materials for electrochemical treatments. Their investigation results revealed that when increasing the surface Pt atomic fraction and lattice ordering, fewer Cu atoms were leached out and dissolved into solution but a higher ECSA was obtained after potential cycling due to absorbate-induced morphology restructuring and different surface configuration evolution . The enhanced specific and mass activities for ORR and enhanced long-term stability have been demonstrated by selected those PtCu catalysts with a Pt-rich surface and high lattice ordering as initial materials for potential cycling.

5.2.4. Pt-Fe

Iron (Fe) is also an alternative alloy element for enhancing the electrocatalytic activity of Pt catalyst . As a typical example, Shui and co-workers have demonstrated the ORR activity of nanoporous PtFe alloy nanowires . [91] In this study, for the generation of those nanoporous nanowires, electrospinning was firstly used to make $PtFe_5$ alloy long nanowires with average diameters of 20-30 nm. After dealloying treatment in H_2SO_4 aqueous solution , the initial nanowires were converted into porous nanowires with an overall wire diameter of about 10-20 nm and the ligament diameter of 2-3 nm (Figure 10a,b). Based on observing and studying the evolution of nanoporosity, the authors found that non-uniform composition in the precursor $PtFe_5$ alloy nanowires helped the formation of nanoporous structure and the porosity in the finial nanowires can be adjusted by controlling the acid concentration, treatment duration and temperature . These porous nanowires interweave to form a self-supporting network exhibiting a higher specific activity, over two times higher than that of commercial Pt/C catalysts, and better durability toward ORR, as showed in Figure 10c,d.

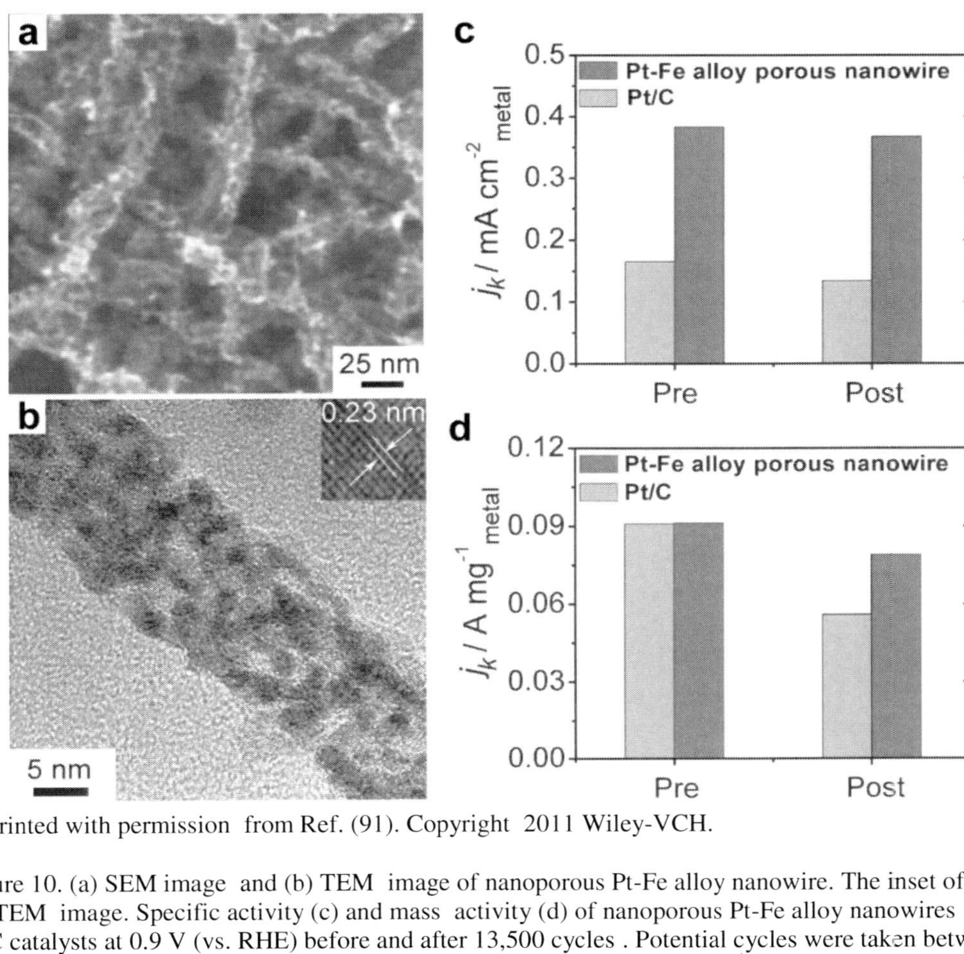

Figure 10. (a) SEM image and (b) TEM image of nanoporous Pt-Fe alloy nanowire. The inset of (b) is HRTEM image. Specific activity (c) and mass activity (d) of nanoporous Pt-Fe alloy nanowires and Pt/C catalysts at 0.9 V (vs. RHE) before and after 13,500 cycles . Potential cycles were taken between 0.6-1 V in Ar-saturated 0.1 M $HClO_4$ solution at a sweep rate of 200 mV s^{-1}.

5.2.5. Pt-Based Multimetalic Catalysts

The catalytic performance of Pt-based alloy catalyst can be further enhanced by modifying Pt with multiplicate metal elements. In contrast to mono- and bi-metallic nanostructures , tri- and multi-metallic nanostructures may deliver even better performance in activity and stability at a much lower Pt content in the catalysts because they offer much more opportunities for tailoring their properties. [92-98]

Yu and co-workers reported the synthesis of PtPdCu alloy nanoparticle nanotubes by a sacrificial template-directed synthetic method and testing of the suitability of these hollow multimetallic nanotubes as effective ORR electrocatalysts. [92] In this synthesis, Cu nanowires with a diameter of 20-40 nm and a length of hundreds of micrometer not only function as templates, but also provide alloying of the remaining Cu atoms with Pt and Pd to form trimetallic alloy nanoparticles . The combination of galvanic replacement reaction and electrochemical dealloying has generated PtPdCu nanoparticle nanotubes with a PtPd-rich shell. It was revealed that the addition of Cu atoms to these alloy catalysts resulted in strain in the surface because of the increased tenability of the lattice parameter between the mixed-Pt shell and the PtPdCu core. Moreover, the introduction of Pd atoms to the topmost layer can modify the surface reactivity . As a result, the resultant PtPdCu nanoparticles showed an enhanced catalytic activity and a superior durability toward ORR compared to PtCu analogue and commercial Pt/C catalysts.

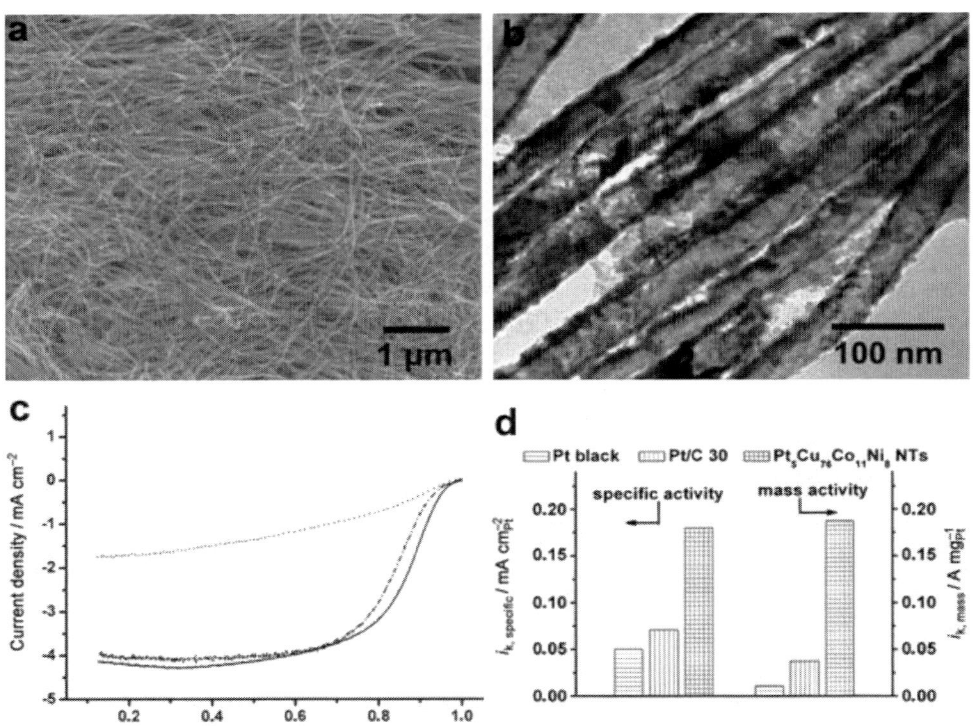

Reprinted with permission from Ref. (93). Copyright 2011 Wiley-VCH.

Figure 11. (a) SEM image and (b) TEM image of hollow PtCuCoNi quaternary nanotubes. (c) Polarization curves recorded in O_2-saturated 0.1 M $HClO_4$ solution at a scan rate of 10 mV s^{-1} and a revolution speed of 1600 rpm. (d) Specific and mass activities of these various at 0.9 V (vs. RHE).

Liu et al. investigated the electrocatalytic activity of quaternary PtCuCoNi hollow nanotubes toward ORR. [93] These low-Pt-content PtCuCoNi hollow nanotubes with an average diameter of 50 nm and a tube wall's thickness ranging from 3 to 10 nm were prepared by means of AAO template-assisted one-step electrodeposition . Figure 11a,d show the SEM and TEM images of the as-generated hollow nanotubes, respectively. Study on the electrochemical behavior s revealed that these quaternary nanotubes exhibit markedly enhanced ORR activity over commercially available Pt black and Pt/C catalysts, with 3.6 and 2.5 times higher of specific activity and 17.5 and 5.0 times higher of mass activity than Pt black and Pt/C catalysts, respectively (Figure 11c,d).

The enhancement in the ORR performance for the quaternary PtCuCoNi hollow nanotubes were attributed to their large ECSA, the lattice strain induced by surface dealloying, the unique combination of micrometer -sized length and hollow configuration , and the highly conductive surface and network-like distribution of these nanotubes on the electrode surface .

CONCLUSION

Driven by the fuel cell popularization and commercialization, great progress and advance in the design, synthesis , and application of electrocatalysts for both anodic and cathodic reactions in fuel cells has been made in the past few decades. Despite the development of various electrocatalysts, the most practical catalysts for fuel cell applications are still Pt-based materials . In particular, porous Pt-based nanostructured materials possess intriguing properties to generate enormously promising potentials for electrochemical catalysis in PEMFCs . In this chapter, recent advancement in the synthesis and ORR catalysis application of porous Pt-based nanostructured materials has been reviewed.

Benefiting from the advances of material science and nanotechnology , efforts in searching for efficient catalysts with high activity and stability for ORR have led to the formation of various with precise control of catalysts morphology , composition and structure . In particular, the use of porous nanostructured catalysts has generated much excitement in ORR catalysis due to their intriguing properties including high specific surface area, superior connectivity and mass diffusion properties and, sometimes, good chemical and physical stability.

Porous Pt-based nanostructured materials have been widely investigated and used as self-supported electrocatalysts in electrochemical oxygen reduction. The Pt-based bimetallic or multimetallic catalyst systems exhibit great synergic effects of each component to further boost the catalytic activities.

Although significant progress has been made in the synthesis and application as electrocatalysts of porous Pt-based nanostructured materials , a number of scientific issues or unsolved challenges still remain.

One of the major challenges is how to scale up the synthesis without compromising control over the morphology , composition , and structure , as well as the uniformity of the catalyst samples. In addition, it has also been a long-standing issue in maintaining the shape or morphology, structure, and composition and thus catalytic activity of the catalysts in fuel cell operating conditions.

Moreover, scientific understanding of the fundamentals, for example the structural effect in the nano-confined space and the nature of the superior catalytic activities towards ORR, are still very challenging.

It appears Pt-based catalysts will continue to play an important role as the choice for the most active ORR catalysts especially under acidic conditions. Further study in this field should focus on the following aspects: (i) optimizing the geometry , composition and structure of porous Pt-based alloy nanostructures to further improve their catalytic activity and stability ; multimetallic alloys are particularly promising; (ii) further exploring effective strategies to address the issue of low retention under fuel cell operation conditions in highly active nanostructures; (iii) developing facile and cost -effective synthesis processes for catalysts preparation to meet the requirement for fuel cell commercialization; and (iv) more theoretical studies and experimental measurements to provide better understanding of ORR mechanism and to help design novel electrocatalysts.

It is worthy believing that such achievements in this area will lead to a future breakthrough in catalyst development, not only for fuel cell , but also for other important chemical reactions .

ACKNOWLEDGMENT

This work was financially supported by the National Natural Science Foundation of China (No. 21373149 and No. 21422104) and the Innovation Foundation of Tianjin University.

REFERENCES

[1] Dresselhaus, M. S.; Thomas, L. L. *Nature,* 2001, 414, 332.
[2] Gasteiger, H. A.; Markovi, N. M. *Science,* 2009, 324, 48.
[3] Jacobson, M. Z.; Colella, W. G.; Golden, D. M. *Science,* 2005, 308, 1901.
[4] Debe, M. K. *Nature,* 2012, 486, 43.
[5] Zhou, X.; Gan, Y.; Du, J.; Tian, D.; Zhang, R.; Yang, C.; Dai, Z. *J. Power Sources,* 2013, 232, 310.
[6] Rabis, A.; Rodriguez, P.; Schmidt, T. J. *ACS Catal.,* 2012, 2, 864.
[7] Liang, Y.; Li, Y.; Wang, H.; Zhou, J.; Wang, J.; Regier, T.; Dai, H. *Nat. Mater.*, 2011, 10, 780.
[8] Guo, S.; Zhang, S.; Sun , S. *Angew. Chem. Int. Ed.*, 2013, 52, 8526.
[9] Wu, J.; Yang, H. *Acc. Chem. Res.*, 2013, 46, 1848.
[10] Fournier, J.; Lalande, G.; Cote, R.; Guay, D.; Dodelet, J. P. *J. Electrochem. Soc.*, 1997, 14, 218.
[11] Brushett, F. R.; Thorum, M. S.; Lioutas, N. S.; Naughton, M. S.; Tornow, C.; Jhong, H.-R. M.; Gewirth, A. A.; Kenis, P. J. A. *J. Am. Chem. Soc.,* 2010, 132, 12185.
[12] Gao, M. R.; Gao, Q.; Jiang, J.; Cui, C. H.; Yao, W. T.; Yu, S. H. *Angew. Chem. Int. Ed.*, 2011, 50, 4905.
[13] Esposito, D. V.; Chen, J. G. *Energy Environ. Sci.*, 2011, 4, 3900.

[14] Ma, X.; Meng, H.; Cai, M.; Shen, P. K. *J. Am. Chem. Soc.,* 2012, 134, 1954.
[15] Xu, Y.; Wang, H.; Zhu, R.; Liu, C.; Wu, X.; Zhang, B. *Chem. Asian J.,* 2013, 6, 1120.
[16] Wu, H.; Chen, W. *J. Am. Chem. Soc.,* 2011, 133, 15236.
[17] Suntivich, J.; Gasteiger, H. A.; Yabuuchi, N.; Nakanishi, H.; Goodenough, J. B.; Shao-Horn, Y. *Nat. Chem.*, 2011, 3, 546.
[18] Suntivich, J.; May, K. J.; Gasteiger, H. A.; Goodenough, J.-B.; Shao-Hom, Y. *Science,* 2011, 334, 1383.
[19] Bing, Y.; Liu, H.; Zhang, L.; Ghosh, D.; Zhang, J. *Chem. Soc. Rev.,* 2010, 39, 2184.
[20] Xu, Y.; Zhang, B. *Chem. Sov. Rev.,* 2014, 43, 2439.
[21] Petkovich, N. D.; Stein, A. *Chem. Soc. Rev.,* 2013, 42, 3721.
[22] Zhang, J.; Li, C. M. *Chem. Sov. Rev.,* 2012, 41, 7016.
[23] Yu, D.; Nagelli, E.; Du, F.; Dai, L. *J. Phys. Chem. Lett.,* 2010, 1, 2165.
[24] Zhang, J.; Sasaki, K.; Sutter, E.; Adzic, R. R. *Science,* 2007, 315, 220.
[25] Markovic, N. M.; Schmidt, T. J.; Stamenkovic, V.; Ross, P. N. *Fuel Cell,* 2001, 1, 105.
[26] Zheng, Y.; Jiao, Y.; Jaroniec, M.; Jin, Y.; Qiao, S. Z. *Small,* 2012, 8, 3550.
[27] Slanac, D. A.; Hardin, W. G.; Johnston, K. P.; Stevenson, K. J. *J. Am. Chem. Soc.,* 2012, 134, 9812.
[28] Anderson, A. B.; Albu, T. V. *J. Am. Chem. Soc.,* 1999, 121, 11855.
[29] Nørskov, J. K.; Rossmeisl, J.; Logadottir, A.; Lindqvist, L.; Kitchin, J. R.; Bligaard, T.; Jónsson, H. *J. Phys. Chem. B,* 2004, 108, 17886.
[30] Nilekar, A. U.; Mavrikakis, M. *Surf. Sci.*, 2008, 602, L89.
[31] Gasteiger, H. A.; Kocha, S. S.; Sompalli, B.; Wagner, F. T. *Appl. Catal. B: Environ.,* 2005, 56, 9.
[32] Thorson, M. R.; Brushett, F. R.; Timberg, C. J.; Kenis, P, J. A. *J. Power Sources,* 2012, 218, 28.
[33] Gloaguen, F.; Convert, P.; Gamburzev, S.; Velev, O. A.; Srinivasan, S. *Electrochim. Acta.,* 1998, 43, 3767.
[34] Perez, J.; Gonzales, E. R.; Ticianelli, E. A. *Electrochim. Acta.,* 1998, 44, 1329.
[35] Trasatti, S.; Petrii, O. A. *Pure Appl. Chem.,* 1991, 63, 711.
[36] Lim, B.; Jiang, M.; Camargo, P. H. C.; Cho, E. C.; Tao, J.; Lu, X.; Zhu, Y.; Xia, Y. *Science,* 2009, 324, 1302.
[37] Chen, A.; Holt-Himdle, P. *Chem. Rev.,* 2010, 110, 3767.
[38] Kloke, A.; Stetten, F. V.; Zengerle R.; Kerzenmacher, S. *Adv. Mater.*, 2011, 23, 4976.
[39] Bing, Y.; Liu, H.; Zhang, L.; Ghosh, D.; Zhang, J. *Chem. Soc. Rve.*, 2010, 39, 2148.
[40] Park, S.; Kim, H. C.; Chung, T. D. *Analyst,* 2012, 137, 3891.
[41] Liu, Y.; Goebl, J.; Yin Y. *Chem. Soc. Rev.*, 2013, 42, 2610.
[42] Jones, M. R.; Osberg K. D.; Macfarlane, R. J.; Langille M. R.; Mirkin, C. A. *Chem. Rev.*, 2011, 111, 3736.
[43] Thomas, A. *Angew. Chem. Int. Ed.*, 2010, 49, 9328.
[44] Platschek, B.; Keilbach, A.; Bein, T. *Adv. Mater.*, 2011, 23, 2395.
[45] Liu, L.; Pippel, E.; Scholz, R.; Gösele, U. *Nano Lett.*, 2009, 9, 4352.
[46] Liu, L.; Scholz, R.; Pippel, E.; Gösele, U. *J. Mater. Chem.*, 2010, 20, 5621.
[47] Wang, H.; Jeong, H. Y.; Imura, M.; Wang, L.; Radhakrishnan, L.; Fujita, N.; Gastle, T.; Terasaki, O.; Yamauchi, Y. *J. Am. Chem. Soc.,* 2011, 133, 14526.
[48] Kulinowski, K. M.; Jiang, P.; Vaswani, H.; Colvin, V. L. *Adv. Mater.*, 2000, 12, 833.
[49] Stein, A.; Li, F.; Denny, N. R. *Chem. Mater.*, 2008, 20, 649.

[50] Bartlett, P. N.; Birkin, P. R.; Ghanem, M. A. *Chem. Commun.*, 2000, 1671.
[51] Yamauchi, Y.; Kuroda, K. *Chem. Asian J.*, 2008, 3, 664.
[52] Attard, G. S.; Bartlett, P. N.; Goleman, N. R. B.; Elliott, J. M.; Owen, J. R.; Wang, H. J. *Science*, 1997, 278, 838.
[53] Yamauchi, Y.; Tonegawa, A.; Komastu, M.; Wang, H.; Wang, L.; Nemoto, Y.; Suzuki, N.; Kuroda, K. *J. Am. Chem. Soc.*, 2012, 134, 5100.
[54] Yamauchi, Y.; Ohsuna, T.; Kuroda, K. *Chem. Mater.*, 2007, 19, 1335.
[55] Liang, Y.-C.; Juan, Y.-W.; Lu, K.-T.; Jeng, U. S.; Chen, S.-A.; Chuang, W.-T.; Su, C.-J.; Liu, C.-L.; Pao, C.-W.; Lee, J.-F.; Sheu, H.-S.; Chen, J.-M. *J. Phys. Chem. C,* 2012, 116, 26649.
[56] Wang, H.; Ishihara, S.; Ariga, K.; Yamauchi, Y. *J. Am. Chem. Soc.,* 2012, 134, 10819.
[57] Song, Y.; Garcia, R. M.; Dorin, R. M.; Wang, H.; Qiu, Y.; Coker, E. N.; Steen, W. A.; Miller, J. E.; Shelnutt, J. A. *Nano Lett.*, 2007, 7, 3650.
[58] Yang, S.; Hong, F.; Wang, L.; Guo, S.; Song, X.; Ding, B.; Yang, Z. *J. Phys. Chem. C,* 2010, 114, 203.
[59] Li, B.; Higgins, D. C.; Zhu, S.; Li, H.; Wang, H.; Ma, J.; Chen. Z. *Catalysis Commun.*, 2012, 18, 51.
[60] Xu, Y.; Hou, S.; Liu, Y.; Zhang, Y.; Wang, H.; Zhang, B. *Chem. Commun.*, 2012, 48, 2665.
[61] Xu, Y.; Yuan, Y.; Ma, A.; Wu, X.; Liu, Y.; Zhang, B. *Chem. Phys. Chem.,* 2012, 13, 2601.
[62] Alia, S. M.; Zhang, G.; Kisailus, D.; Li, D.; Gu, S.; Jensen, K.; Yan, Y. *Adv. Funct. Mater.*, 2010, 20, 3742.
[63] Lin, Z.-H.; Lin, M.-H.; Chang, H.-T. *Chem. Eur. J.*, 2009, 15, 4656.
[64] Kim, K. W.; Kim, S. M.; Choi, S.; Kim, J.; Lee, I. S. *ACS Nano,* 2012, 6, 5122.
[65] Jung, H. Y.; Kim, D. H.; Chun, H. K.; Kim, S. H.; Lim, C. S.; Byun, J. Y.; Jung, Y. J. *Adv. Energy Mater.*, 2011, 1, 1126.
[66] Jin, H. J.; Kramer, D.; Ivanisenko, Y.; Weissmuller. J. *Adv. Eng. Mater.*, 2007, 9, 849.
[67] Huang, J.-F.; Sun , I.-W. *Chem. Mater.*, 2004, 16, 1829.
[68] Snyder, J.; Asanithi, P.; Dalton, A. B.; Erlebacher, J. *Adv. Mater.*, 2008, 20, 4883.
[69] Xu, C. X.; Wang, R. Y.; Chen, M. W.; Zhang, Y.; Ding, Y. *Phys. Chem. Chem. Phys.*, 2010, 12, 239.
[70] Oezaslan, M.; Heggen, M.; Strasser, P. *J. Am. Chem. Soc.,* 2012, 134, 514.
[71] Wang, D.; Zhao, P.; Li, Y. *Sci. Rep.*, 2011, 1, 37.
[72] Zhang, Z.; Wang, Y.; Qi, Z.; Zhang, W.; Qin, J.; Frenzel, J. *J. Phys. Chem. C,* 2009, 113, 12629.
[73] Kerzenmacher, S.; Schroeder, M.; Brämer, R.; Zengerle, R.; Stetten, F. v. *J. Power Sources,* 2010, 195, 6516.
[74] Chen. Z.; Waje, M.; Li, W.; Yan, Y. *Angew. Chem. Int. Ed.,* 2007, 46, 4060.
[75] Wilson, M. S.; Garzon, F. H.; Sickafus, K. E.; Gottesfeld, S. *J. Electrochem. Soc.*, 1993, 140, 2872.
[76] Mayrhofer, K. J. J.; Meier, J. C.; Ashton, S. J.; Wiberg, G. K. H.; Kraus, F.; Hanzlik, M.; Arenz, M. *Electrochem. Commun.*, 2008, 10, 1144.
[77] Perez-Alonso, F. J.; Elkjær, C. F.; Shim, S. S.; Abrams, B. L.; Stephens, I. E. L.; Chorkendorff, I. *J. Power Sources,* 2011, 196, 6085.
[78] Walcarius A. *Chem. Sov. Rev.,* 2013, 42, 4098.

[79] Xu, J.; Fu, G.; Tang, Y.; Zhou, Y.; Chen, Y.; Lu, T. *J. Mater. Chem.*, 2012, 22, 13585.
[80] Xia, B. Y.; Ng, W. T.; Wu, H. B.; Wang, X.; Lou, X. W. *Angew. Chem. Int. Ed.*, 2012, 51, 7213.
[81] Kibsgaard, J.; Gorlin, Y.; Chen, Z.; Jaramillo, T. F. *J. Am. Chem. Soc.*, 2012, 134, 7758.
[82] Stamenkovic, V. R.; Fowler, B.; Mun, B. S.; Wang, G. F.; Ross, P. N.; Lucas, C. A.; Markovic, N. M. *Science*, 2007, 315, 493.
[83] Wang, C.; Markovic, N. M.; Stamenkovic, V. R. *ACS Catal.*, 2012, 2, 891.
[84] Greeley, J.; Stephens, I. E. L.; Bondarenko, A. S.; Johansson, T. P.; Hansen, H. A.; Jaramillo, T. F.; Rossmeisl, J.; Chorkendorff, I.; Nørskov, J. K. *Nat. Mater.*, 2009, 1, 552.
[85] Hong, J. W.; Kang, S. W.; Choi, B.-S.; Kim, D.; Lee, S. B.; Han, S. W. *ACS Nano*, 2012, 6, 2410.
[86] Liu, W.; Rodriguez, P.; Borchardt, L.; Foelske, A.; Yuan, J.; Herrmann, A.-K.; Geiger, D.; Zheng, Z.; Kaskel, S.; Gaponik, N.; Kötz, R.; Schmidt, T. J.; Eychmüller, A. *Angew, Chem. Int. Ed.*, 2013, 52, 9849.
[87] Huang, X.; Zhu, E.; Chen, Y.; Li, Y.; Chiu, C.-Y.; Xu, Y.; Lin, Z.; Duan, X.; Huang, Y. *Adv. Mater.*, 2013, 25, 2974.
[88] Chen, C.; Kang, Y.; Huo Z.; Zhu Z.; Huang, W.; Xin, H. L.; Snyder, J. D.; Li, D.; Herron, J. A.; Mavrikakis, M.; Chi, M.; More, K. L.; Li, Y.; Markovic, N. M.; Somorjai, G. A.; Yang, P.; Stamenkovic, V. R. *Science*, 2014, 343, 1339.
[89] Wang, M.; Zhang, W.; Wang, J.; Minett, A.; Lo, V.; Liu, H.; Chen, J. *J. Mater. Chem. A*, 2013, 1, 2391.
[90] Cui, C.-H.; Li, H.-H.; Liu, X.-J.; Gao, M.-R.; Yu, S.-H. *ACS Catal.*, 2012, 2, 916.
[91] Shui, J.-I.; Chen, C.; Li, J. C. M. *Adv. Funct. Mater.*, 2011, 21, 3357.
[92] Li, H.-H.; Cui, C.-H.; Zhao, S.; Yao, H.-B.; Gao, M.-R.; Fan, F.-J.; Yu, S.-H. *Adv. Energy Mater.*, 2012, 2, 1182.
[93] Liu, L.; Pippel, E. *Angew. Chem. Int. Ed.*, 2011, 50, 2729.
[94] Wanjala, B. N.; Fang, B.; Luo , J.; Chen, Y.; Yin, J.; Engelhard, M. H.; Loukrakpam, R.; Zhong, C.-J. *J. Am. Chem. Soc.*, 2011, 133, 12714.
[95] Wang, C.; Li, D.; Chi. M.; Pearson, J.; Rankin, R. B.; Greeley, J.; Duan, Z.; Wang, G.; Vliet, D. van der; More, K. L.; Markovic, N. M.; Stamenkovic, V. R. *J. Phys. Chem. Lett.*, 2012, 3, 1668.
[96] Srivastava, R.; Mani, P.; Hahn, N.; Strasser, P. *Angew. Chem. Int. Ed.*, 2007, 46, 8988.
[97] Antolini, E. *Appl. Catal. B: Environ.*, 2007, 74, 337.
[98] Zhu, H.; Zhang, S.; Guo, S.; Su D.; Sun , S. *J. Am. Chem. Soc.*, 2013, 135, 7130.

About the Editor

Dr. Mahmood Aliofkhazraei
Assistant Professor
Faculty of Engineering
Department of Materials Engineering
Tarbiat Modares University
P.O. Box 14115-143
maliofkh@yahoo.com

INDEX

#

A

B

C

D

E

F

G

H

I

J

K

L

M

N

O

P

Q

R

S

T

U

V

W

X

Y

Z